In the first hundred years of its history, immunology was mired in the problems of species and specificity both in research and in practice. The old botanical dispute about the nature of species, which has its roots in classical Western thought, reappeared in the late nineteenth century in the disputes of the bacteriologists, and subsequently of their students, the immunologists, immunochemists, and blood group geneticists. The argument centered on the question of unity and diversity. Proponents of unity insisted on the continuity of nature, while those of diversity emphasized the separation and definition of individual species. In the course of this controversy, Pauline Mazumdar argues, five generations of scientific protagonists waged a bitter intellectual war that defined the structure of immunological thought during the first half of the twentieth century. Their science was designed only in part to wrest an answer from nature: it was at least as important to wring an admission of defeat from their opponents.

One of the key figures in the debate was the Austrian immunochemist Karl Landsteiner, whose career provides the central focus for Mazumdar's account. His unitarian views excluded him from promotion within European institutions, where the specificity and pluralism espoused by Robert Koch and Paul Ehrlich were entrenched. Landsteiner himself was forced into a kind of exile at Rockefeller University in New York. Though Landsteiner won a Nobel prize for his work, his inability to gain more widespread acceptance of his views caused him to view his life as a failure.

Species and Specificity

Species and Specificity

An Interpretation of the History of Immunology

PAULINE M.H. MAZUMDAR
University of Toronto

CAMBRIDGE
UNIVERSITY PRESS

PUBLISHED BY THE PRESS SYNDICATE OF THE UNIVERSITY OF CAMBRIDGE
The Pitt Building, Trumpington Street, Cambridge, United Kingdom

CAMBRIDGE UNIVERSITY PRESS
The Edinburgh Building, Cambridge CB2 2RU, UK
40 West 20th Street, New York NY 10011–4211, USA
477 Williamstown Road, Port Melbourne, VIC 3207, Australia
Ruiz de Alarcón 13, 28014 Madrid, Spain
Dock House, The Waterfront, Cape Town 8001, South Africa

http://www.cambridge.org

First published 1995
First paperback edition 2002

A catalogue record for this book is available from the British Library

Library of Congress Cataloguing in Publication data
Mazumdar, Pauline M. H. (Pauline Margaret Hodgson), 1933–
Species and specificity: an interpretation of the history of
immunology / Pauline M. H. Mazumdar
            p.   cm.
Based on the author's thesis (Ph.D. – Johns Hopkins University,
1976)
ISBN 0 521 43172 7 (hardback)
1. Immunology – History.   I. Title.
[DNLM: 1. Allergy and Immunology – history.   2. Species
Specificity.   QW 511.1 M476s 1994]
QR182.M39   1994
574.2'9'09–dc20
DNLM/DLC
for Library of Congress
93-31219   CIP

ISBN 0 521 43172 7 hardback
ISBN 0 521 52523 3 paperback

Contents

Illustrations and Tables

Tables

Preface

This book has evolved from a thesis for the Ph.D. in history of medicine at Johns Hopkins University. The thesis itself, however, began as an idea for a book. I had intended to take two years off my job as assistant pathologist at the North London Blood Transfusion Centre to write it. But my management committee was sure that if I did, I would never come back. I resigned. It did not take long after that to find out that the committee was right. There was more food for the soul in thinking about the history of immunology than in acting in it. And since everyone else at the Johns Hopkins Institute for the History of Medicine was casting their work as a thesis, I did so too.

In the early 1970s, immunology was nearly a hundred years old, yet it had almost no secondary literature: neither immunologists nor historians of medicine nor graduate students nor publishers were interested in it. In the twenty years since then, or perhaps only in the past ten, a readership has begun to collect. In 1986, Bernhard Cinader and I arranged the first Symposium on the History of Immunology when the International Congress for Immunology came to Toronto. Several young historians have recently begun to propose critiques and analyses of this field, and further symposia have been held. I was encouraged to reopen the book on Karl Landsteiner.

Taking the thesis as my starting point, I have reworked a good deal of the old text. New secondary material has meant a reassessment of some areas, and new archival sources have added depth to old judgments. In particular, the materials collected by the New York bacteriologist George M. Mackenzie in the 1940s and 1950s for his biography of Karl Landsteiner have become available at the American Philosophical Society. Soon after Landsteiner's death in 1943, Mackenzie began writing to Landsteiner's old colleagues, people who had known him and collaborated with him in New York and in Vienna. Many of them wrote long and detailed letters in reply. When Mackenzie died in 1952, a good deal of his text was written, but the work was far from finished.

At first, the immunologist Elvin Kabat of Columbia and Gilbert H. Mudge of Johns Hopkins planned to complete it. Sometime later, the collection came into the hands of the Philadelphia anatomist and historian George Corner – now well known for his *History of the Rockefeller Institute* of 1964 – who deposited it with the American Philosophical Society in Philadelphia in 1958. It was followed by the typescript of the biography itself in 1981. The whole collection, however, was under restriction until 1983. I have made use of these important materials, although I do not accept Mackenzie's assessment of the central importance of anti-Semitism in Landsteiner's life, nor, I think, did his informants when Mackenzie suggested it to them. But that assessment was natural enough in the late 1940s: Mackenzie, and Landsteiner too, knew many German-speaking scientists for whom anti-Semitism had been a central fact.

Several of the people who talked to Mackenzie also talked to me, twenty-five years later. One of them was Merrill Chase, who in 1973, when I first met him, still occupied the laboratory at the Rockefeller Institute that he had shared with Landsteiner, and who had cupboards full of files containing Landsteiner's yellow-paper laboratory notes, a bookcase full of Landsteiner's books, and a rack of the test tubes that Landsteiner used for his method of blood grouping. The laboratory notes are now in the Rockefeller Archives in Tarrytown, New York, and the test tubes in the Smithsonian Institution. It was Merrill Chase who prepared a full bibliography of Landsteiner's papers, numbered like Köchl's list of Mozart's works, and who organized the commemoration of the centenary of Landsteiner's birth in 1968. Chase showed me the papers, explained the laboratory shorthand to me, and talked. He gave me a full set of Landsteiner's reprints, still tied up in Landsteiner's string. I have said little about Merrill Chase's work with Landsteiner on cellular hypersensitivity because he himself is planning to write about it.

Many other people who helped me in the early 1970s are dead now. Sadly, Alexander Wiener and Philip Levine, and Robert Race, about whom I write in the last chapter of this work, are among them. But this has allowed me to speak a little more freely about the controversy in which they were involved, and about their relation to one another and to Landsteiner. I think that it was Wiener's passionate commitment to Landsteiner's point of view that first suggested to me the overall form of the century-long controversy that I have written about here.

Without funding, I could have neither written this work in the first place nor rewritten it now. In the 1970s, I was supported by a Macy Fellowship in the History of Medicine and the Biological Sciences, designed to encourage people to take up the history of medicine professionally, which it successfully did. These past three years, my reworking has been funded by the Hannah Institute for the History of Medicine in Toronto. I should acknowledge, too, the staff of the National Library of Medicine in Bethesda, a true healing pool, where I have happily worked every summer for the past twenty years, on this and every other project that I have done; as well as fruitful visits to the Rockefeller Archives Center in Tarrytown, the last in August 1990, when I met and talked with Thomas Rosenbaum.

Finally, I must acknowledge the stimulus of Bernhard (Hardi) Cinader, whose genial harassment shook my memory as a madman shakes a dead geranium. Without that, this book would never have been published.

The logical principle of genera, which postulates identity, is balanced by another principle, namely, that of species, which calls for manifoldness and diversity in things. . . . This two-fold interest manifests itself also among students of nature in the diversity of their ways of thinking. Those who are more especially speculative, are, we may almost say, hostile to heterogeneity, and are always on the watch for the unity of the genus; those on the other hand who are more especially empirical, are constantly endeavouring to differentiate nature in such manifold fashion as to almost extinguish the hope of ever being able to determine its appearances in accordance with universal principles.

Immanuel Kant, *Critique of Pure Reason* (1781), translated 1929 by
Norman Kemp Smith (Toronto: Macmillan, 1965), 540

Introduction

Species and specificity are the concepts that lie at the heart of the modern science of immunology. They make possible the modern immunological understanding of the self and its individuality, of the self's recognition of the other, and of the biological processes that constitute the relationship between them. As concepts, species and specificity go back to the classical roots of Western thought in the work of Aristotle, and they have played a central part in the biological sciences at least since the eighteenth century, with its flowering of classification systems in all the sciences, but especially in the science of botany.

It is the reappearance of the botanical dispute about the nature of species in the work of the bacteriologists, and after them in the work of the immunologists, immunochemists and blood group geneticists, that I shall discuss in this book. In the course of this controversy, five generations of scientific protagonists made themselves aggressively plain. Their science was designed only in part to wrest an answer from nature. It was at least as important to wring an admission of defeat from their opponents – and these were opponents that never admitted defeat.

A controversy is the most useful of all forms of discourse for the historian. Engaged in it, protagonists display themselves with a frankness and enthusiasm that they would never otherwise have needed to make public. Concepts, as Geoffrey Lloyd has pointed out in the case of Greek science, become more explicit when they are part of a debate, a contest with an antagonist.[1] In a world where underlying assumptions are all agreed upon, they may well remain unconscious or unspoken: a single, all-encompassing paradigm is likely to be invisible from within, and undetectable from outside.

There is more: a controversy helps define a style of thought by comparison with its opposite, that which it is not. For the historian, partic-

1 Geoffrey E. R. Lloyd, *Demystifying Mentalities* (Cambridge: Cambridge University Press, 1990), 57–60.

3

ularly the historian of science or medicine, the temptation is always to
compare the thought of the past with that of our own time. How could
they have thought that, when *we* can see it doesn't work? Even if we are
not as naive as all that, our own interpretation of scientific phenomena
may be our only available yardstick: we are left to try to recognize the
specificity of the past in terms of self and not-self, to make use of an
immunological metaphor. A controversy, however, lets us see the past
in terms of its own possibilities: if we know of the alternatives available
to the protagonists, we can understand them in something much closer
to their own terms, so that we may be less inclined intuitively to push
our science forward as the model that they failed to grasp.

The controversy treated in this book was both synchronic and dia-
chronic: it lasted for one hundred years or more, and at any one period, it involved many individuals. Its two teams of thinkers fought over
the fundamentals of their science from the mid-nineteenth century to
the mid-twentieth. Five generations of teachers and students, grandstudents, and great-grandstudents loyally maintained their *diadoche* as their
science metamorphosed from botany through bacteriology to immunology. The group that I have called the pluralists insisted on the separation and definition of species, and the others, the unitarians, on the
continuity of nature. For each science in turn, the problem of species
and their definition came close to being the essential problem of the
science. The two opposing groups were established long before the rise
of immunology, but it is for that science that this controversy has
proved to be so revealing.

In the years before 1957, when the clonal selection theory of Mac-
Farlane Burnett came to replace serology at the centre of the stage, the
applications of serology in medicine and public health were the motor
that drove both practice and the theory that derived from it. The problems of species and specificity were the core problems of both research
and practice in immunology. I present this five-generation controversy
as a key to the structure of immunological thought as it evolved in the
first half of the twentieth century.

Kant was not the first to point out that students of nature fall into two
groups. The first, the more speculative, are always on the watch for the
unity underlying the diversity of nature, and the second are those who,
often more practical or empirical, try to differentiate nature, to accen-
tuate its diversity, to divide into species rather than to unify into genera.
Kant himself, he says, gives each kind of thinking its due. The principle
of homogeneity is balanced by that of specification, and they are joined

by the principle of continuity, which connects the two into a systematic unity. *Datur continuum formarum*, he says, all species border on one another, admitting of no transition *per saltum*. This is the transcendental *lex continui in natura*, a maxim of scholastic logic. But being transcendental, it is a law of the mind only. In experience, species are actually clearly divided, and constitute a *quantum discretum*.

The principles of systematic unity, then, are placed in the order manifoldness, affinity, unity: reason passes beyond experience as it seeks for unity in knowledge.[2] For Kant, the unity is in the human mind, nature itself is manifold. It is to this unifying effort of the human mind that he refers in his famous *Satz*, that each science is scientific only in proportion to the amount of mathematics in it. It is by mathematics that manifoldness in nature becomes continuity in the mind. Whether the unity is to be found in nature as well as in the mind is a different question. Another of Kant's scholastic maxims, *Entia praeter necessitatem non esse multiplicanda*, presumes that it is: chemists, he says, suppose that a step forward was made when the many salts were divided into only two genera, acids and bases; and they are now seeking to show that even these are two varieties of a single fundamental material. They suppose that the unity of reason is justified by the unity of nature itself.

Kant's words suggest that he himself belongs to the group that looks for unity and continuity in thought and in things, but the form of his thought suggests the opposite. It is built up of distinctions and dichotomies and sharp boundaries. His division of the acts of the understanding into categories and those of the reason into regulative ideas performs for the mind what the natural historian of the eighteenth century performed for nature. Linnaeus's division into class, order, genus, species, and variety is a system that divides nature into categories as Kant's system does thought. Kant's categories of the understanding are somewhere near the genera; his regulative ideas correspond to the orders of Linnaeus's system. Both ultimately derive from the five levels of classification of scholastic logic: *genus summum*, *genus intermedium*, *genus proximum*, *species*, and *individuum*.[3]

The problem of species in medieval metaphysics and the essentialist logic with which it was connected are directly related to Linnaeus's classification through the work of the seventeenth-century

2 Immanuel Kant, *Kritik der reinen Vernunft* (1781), translated by Norman Kemp Smith (1929) (Toronto: Macmillan, 1965), 532–549: "The regulative employment of the ideas of pure reason" (pp. 540 ff.).

3 Frans A. Stafleu, *Linnaeus and the Linnaeans: The Spreading of Their Ideas in Systematic Botany* (Utrecht: Oosthoek 1971), 26, 32 (refers only to Linnaeus: the comparison with Kant is mine!).

botanist Andreas Caesalpino.[4] Caesalpino suggested that the essence of a plant lay in the parts serving the functions of its vegetative soul, that is, in growth and reproduction. A truly essentialist classification should therefore take as its *fundamentum divisionis*, the reproductive organs of the plant. The *differentiae* at each of the successive dichotomies of which the classification consists should ideally all refer to this original *fundamentum divisionis*. This, which is Linnaeus's method, and that of the botanists who followed him, is precisely that of the definition *per genus et differentiam* of scholastic logic. The medieval metaphysical problem has become the problem of identifying and classifying plants.

In an essay of 1931, the psychologist Kurt Lewin contrasted the kinds of concept typical of the Aristotelian and the Galilean modes of thinking. Lewin describes as Aristotelian the division of objects into well-defined classes and the use of antithesis and dichotomy; he contrasts this with the unity of the physical world described by Galileo, in which dichotomy and antithesis are replaced by continuity, gradation, and fluid transitions, the class concept by the series concept.[5] Following Lewin, the Ansbachers compared the psychologists Alfred Adler and Sigmund Freud: Freud analyzes, dissects, dichotomizes; he splits mind into death wish and sex drive. Adler rejects such dichotomies, denies the existence of specific categories of mental disease, stresses the unity of the neuroses. He keeps his technical terminology to a minimum, in contrast to Freud, who developed an elaborate vocabulary with many named entities. Adler is a "field theorist," Freud is a "class theorist."[6] They represent the same two groups of students of nature that Kant described in 1781: the field theorist is always on the watch for unity, the class theorist for hidden heterogeneity.

The Linnaean botanists, Aristotelians *sensu stricto*, are Aristotelians in this metaphorical sense too. Their botany is a search for *differentiae* with which to mark off species from each other; existing classes are split up, previously unnoticed dichotomies are brought to light, and species boundaries are rearranged accordingly. When the boundaries are difficult to fix in nature, they are fixed instead by definition. Species differ

4   Philip R. Sloan, "John Locke, John Ray and the natural system," *J. Hist. Biol.* 5(1972): 1–53.
5   Kurt Lewin, "The conflict between Aristotelian and Galilean modes of thought," *J. Gen. Psychol.* 5(1931):141–177; and in Kurt Lewin, *A Dynamic Theory of Personality: Selected Papers*, translated by D. K. Adams and K. E. Zener (New York, N.Y.: McGraw-Hill, 1935).
6   Heinz Ludwig Ansbacher and Rowena R. Ansbacher, *The Individual Psychology of Alfred Adler: A Systematic Presentation in Selections from His Writings* (New York, N.Y.: Basic Books, 1956).

sharply from each other, if they are "good" species – if the botanist, that is, is good.

Linnaean botany had its opponents even in the eighteenth century, especially in France. In 1809 there appeared Jean Baptiste de Lamarck's *Philosophie Zoölogique*, in which classes, orders, families and genera are called *parties de l'art*, or devices introduced by art for human convenience, artificially dividing up nature's continuous series, a series that begins with *Monas* and ascends by infinitesimal differences to man.[7] The lines of demarcation are arbitrary for there are no gaps in nature.

In Germany, Matthias Schleiden seems to have been among the earliest to take up a position opposed to that of systematic botany. His cell theory provided a common lawfulness, a unity underlying the diversity of Linnaean species. In his *Textbook of Botany as an Inductive Science* of 1844, he attacked his Linnaean contemporaries, using Kant's critical philosophy and the heuristic maxims, and with the help of the Kantian philosopher J. F. Fries. Kant's heuristic maxim of unity was one of the regulative ideas of the reason, a category at the highest and most general level of the mind's activity. Although Schleiden found the transcendental *Naturphilosophie* pernicious and absurd, he, like the *Naturphilosophen*, took the search for unity in nature's diversity to be the basic principle of human reason.

It was this desire for unity in scientific thought and the search for it in nature that Schleiden passed on to his otherwise rather disrespectful student Carl von Nägeli. Like Schleiden, Nägeli attacked the Linnaeans, though as a young man he himself had begun by trying to define the species of unicellular algae. Nägeli was a classical field theorist: his unitarianism is a constant feature in everything he wrote from 1853 onwards. In each of the areas in which he worked, it is the *lex continui in natura* that is his leading maxim. He applies it in his phylogeny, which is close to that of Lamarck, in his work on the fine structure of living matter, in his theory of fermentation, in his bacteriology, and in his theory of knowledge. Each of these fields is united to each of the others in a continuous network of thought. *Kontinuität* and *quantitative Abstufung* are the terms he uses to describe the relations of things to each other: there are no sharp boundaries between species. In the case of the "lower fungi," the bacteria, there are not even separate species.

Nägeli's opponent among the Linnaeans was the botanist Ferdinand Cohn of Breslau. Cohn, the class theorist, developed a new classification of simple plants, beginning like Nägeli from an interest in unicellular algae. He then moved on to the bacteria, classifying them along

7 H. Elliott, "Introduction," to *Zoological Philosophy, etc. by Jean-Baptiste de Lamarck*, translated and introduced by H. Elliott (New York, N.Y.: Hafner, 1963), xvii–xcii (p. xxvii).

Linnaean lines by dividing them into four tribes with six genera. His species were mainly defined morphologically, though he recognized that this was probably only a temporary means of classification. The size of the bacteria made classification difficult, and for a Linnaean botanist the absence of special reproductive organs and the difficulty of observing the growth cycle made them especially difficult to deal with. But Cohn was sure that the different species would eventually be properly defined. Where Nägeli saw *Kontinuität*, Cohn saw differentiation: when our microscopes were better, the *differentiae* would come to light. It was because of Cohn's attitude to species that the young Robert Koch wrote to him and not to Nägeli with his offer to demonstrate to him the complete life cycle of the anthrax bacillus.

Koch very quickly became a brilliant bacteriologist whose new morphological technology of stains, solid media, and optical improvements lifted the identification of bacterial species onto a different plane. His adoption by Cohn and his very rapid rise to fame and power within the framework of the new German state set the course of medical bacteriology for generations towards a definition of species of bacteria and their matching species of disease. The growth of Koch's professional power and its institutionalization, together with the active support of the state, first in the Reichsgesundheitsamt in Berlin, and later in the Institut für Infectionskrankheiten, produced a group of enthusiastic students and co-workers, for whom a belief in absolute specificity was an essential mark of group loyalty.

The two kinds of thinking now came into violent conflict. In 1880, in a publication of the Reichsgesundheitsamt, Koch and his students attacked the representatives of *Kontinuität*, who included not only Nägeli himself and his students in Munich but also Louis Pasteur in Paris, and attacked them with extraordinary violence. Nägeli and his group replied, but it was soon clear that they had been defeated. Although this defeat and the enormous growth of Koch's influence ensured that the definition and separation of species of bacteria were generally accepted, Nägeli's thinking was not completely expunged. It continued to live in the minds of the students he had trained. Nägeli's principles of unity, continuity, and *quantitative Abstufung* are found again in the work of Max Gruber.

For Schleiden's generation, and for Nägeli in his earlier days, specificity had been a botanical problem. For Koch and the older Nägeli, it was a problem of bacteriology. For the next generation, it became a problem of immunology: bacterial species, in the absence of visible, morphological differentiae, might be defined by their reaction with specific antisera. Richard Pfeiffer, as Koch's student, maintained that they could be: Gruber, the student of Nägeli, maintained that the specificity of antisera was a matter of *quantitative Abstufung*. The conflict of the earlier generation was repeated by the next: Gruber attacked Koch's student Pfeiffer, and later, as a matter of course, Koch's student Paul Ehrlich.

In the work of Paul Ehrlich, the characteristics of the Linnaean thinker are very well marked. It is all the more interesting that they should be, for Ehrlich had no direct contact with Linnaean botany, apart from his medical training. His thinking can be seen growing along these lines in his juvenile work on dyes, where his use of classes and dichotomy is already well developed. In 1880, in his early work on white blood cells, he treats them exactly as if they were botanical species. It is not surprising that Ehrlich should have been as certain as Koch of the absolute nature of specific differences.

It is this absolute specificity that sets the style of Ehrlich's theory of immunity. Species specificity in immunology is explained by the specificity of affinity chemistry: his receptor theory, which originated as a chemical explanation of dye specificity, was transferred first to immunity and later to chemotherapy. Ehrlich's loyalty to Koch and to the Koch group to which he belonged was no more marked than the loyalty of Max von Gruber to his teacher Carl von Nägeli. Gruber's attack on Koch's specificity passed over to an attack on Ehrlich's specificity and was inherited in turn by Gruber's student Karl Landsteiner, who himself later attacked Ehrlich.

For Landsteiner, the conflict took place in the field of immunochemistry. Very early in his career, Landsteiner began using the diagnostic terms of the field thinker, *Kontinuität* and *quantitative Abstufung*, terms that appeared so often in Nägeli's writing. For this generation the personal struggle, in which loyalties and methodological styles were combined, was reflected in the larger conflict that was taking place between physical chemistry and the affinity chemistry of the structural organic chemists. The new and exciting field of colloid chemistry, the youngest branch of physical chemistry, seemed to suggest that chemical specificity might play no part at all in the reactions that took place in the living organism.

Ehrlich's chemical receptor theory seemed at first to have gone down before the physical or almost-physical conception of antigenic specificity. But Ehrlich's theory was strongly institutionalized: state serum institutes across the world worked along the lines he had laid down, and his hold over the field of practical serology did not die with him. In one area of immunology the receptor theory itself was able to survive intact: this was the field of blood group serology and blood group genetics. Here the receptor theory of immune specificity and the unit-character theory of genetics united to give a picture of absolute specificity that persisted long after both the unit-character theory and the receptor theory had disappeared from their original applications. The receptor theory in blood grouping remained unchallenged until the late 1940s, when there arose a new champion of *Kontinuität*, in Landsteiner's last and youngest student, Alexander Wiener. Wiener's long and bitter controversy with Robert R. Race and Ronald A. Fisher about the terminology for the Rhesus blood group system is the last act in this agon. It has not been resolved.

The conflict has been going on now for five generations. In each generation, new actors have arisen to play the same parts, often in the same words. Each of them has represented not only his own attempt to solve a particular scientific problem, but has acted as spokesman for his group and student of his teacher. This was particularly true of the Koch–Ehrlich group in Germany, where the social and institutional power of their many chairs and directorships was backed by the power of the journals they founded and edited. Landsteiner, excluded from this alliance by his well-known opposition to Ehrlich, was effectively excluded from professional power in the German-speaking world.

In the earlier part of this history, the part played by Kant's heuristic maxim of unity as the model for Schleiden's thought is easy enough to see. Schleiden himself leaves us in no doubt about it. The part played by the physicist-philosopher Ernst Mach later on in the story is not so clear. The statements Mach made about the nature of scientific thought are those of a seeker for unity far more whole-hearted than Kant. There is little lingering Aristotelianism here: in fact, the distinction between Aristotelian statics and Galilean dynamics, one of the sources of Kurt Lewin's essay which was cited earlier, comes from Mach's history of mechanics.[8] The function of science, in Mach's view, is to generalize and simplify, to subsume experience under progressively fewer and simpler laws. This principle of science he calls the principle of the economy of thought.[9] It first appeared in print in 1872, in *Die Geschichte und die Würzel des Satzes von der Erhaltung der Arbeit*. The principle of continuity, he says, which everywhere pervades modern inquiry, simply prescribes a mode of conception that conduces in the highest degree to the economy of thought.[10]

8   Ernst Mach, *Die Mechanik in ihrer Entwicklung: historisch-kritisch dargestellt* (1883), translated by Thomas J. McCormack (Chicago, Ill.: Open Court, 1893).
9   Ernst Mach, *Die Leitgedanken meiner naturwissenschaftlichen Erkenntnislehre und ihre Aufnahme durch die Zeitgenossen*, und *Sinnliche Elemente und naturwissenschaftliche Begriffe, Zwei Aufsätze* (Leipzig: Barth, 1919), 4.
10 Mach, *Mechanik* (1883) (n.8), 490.

As a philosopher, Mach was the mouthpiece of the scientists of his time, particularly of the physical scientists, in whose writing the same emphasis on simplicity and continuity is often to be found. He himself saw this as an essential part of being a philosopher of science, a representative of scientific modes of thought in philosophy. Among physiologists, interest in Mach was widespread, but in many cases those who cited him were more concerned with the biological side of his thought, the sensationalist epistemology, than in the principles of economy and continuity. His influence on the broader cultural life of Vienna in the early decades of the twentieth century was also enormous. Simplicity and economy formed part of the *neue Sachlichkeit*, the outpouring of positivistic thought from science into art and literature that took place in those decades.

In many ways, Mach as the mouthpiece of contemporary physical science codifies and parallels much that is to be found in the writings of Karl Landsteiner, and at first sight, in view of the opportunities for contact in Vienna, it is hard to believe that they were completely independent. But Landsteiner never actually cites Mach, and it is going too far, where there are no such citations or other evidence, to speak of a direct influence, particularly when the direct influence of Gruber is so clear. The case is still an open one, however: the verdict so far is *ignoramus*, "not proven."

It is on the life and work of Karl Landsteiner, extending as it does over so many years of changing thought in immunology, that I have focused this book. Like his predecessors, Carl von Nägeli, Max von Gruber, and Hans Buchner, and his successor, Alexander Wiener, Landsteiner the unitarian was on the losing side. Specificity and pluralism, the legacy of Robert Koch and Paul Ehrlich, were entrenched in the work of the state serum institutes, and it was they, with their practical importance in the world of public health and clinical medicine, who made the rules in the Europe of Landsteiner's day. Only in the protected environment of the Rockefeller Institute in New York was Landsteiner to be free of this powerful opposition. In the course of his twenty years at the Rockefeller Institute, he was both productive and greatly admired, even to the extent of winning a Nobel prize. But he was never to be a happy man: his personal experience was one of failure.

PART I

Specificity and Unitarianism in Nineteenth-Century Botany and Bacteriology

Each of these systems puts the plants in a different order, each one divides them differently, and in each [the author] with justifiable pride, explains that his system is the only truly natural one. I think by now botanists must be like the Roman *Haruspices*, who could not look at each other without laughing.

Matthias Jacob Schleiden, "Methodologische Grundlage: Einleitung," in *Grundzüge der wissenschaftliche Botanik (Leipzig: Engelmann, 2d ed., 1845), 17*

1

The Unitarians: Matthias Schleiden
and Carl von Nägeli

At the beginning of the nineteenth century, botanists accepted Linnaeus's sexual system as one that was scientifically complete in every respect, wrote the historian of botany, Julius von Sachs, in 1875. Any improvements, they thought, could only be in collecting new species and fitting them into Linnaeus's scheme.[1] Linnaeus distinctly declared, wrote Sachs elsewhere, that the highest and only worthy task for a botanist was to know all species of the vegetable kingdom exactly by name, and to this day his school still considers that task to be self-evident. The inevitable result was that botany ceased to be a science: there was an endless accumulation of technical terms, until at length a textbook of botany came to look more like a Latin dictionary than a scientific treatise.[2] Von Sachs's scornful remarks on Linnaean botany are echoes of the anti-Linnaean polemic of a new kind of botany book, Matthias Jacob Schleiden's *Grundzüge der Wissenschaftliche Botanik*, which appeared in 1842–1843.[3]

Schleiden was born in Hamburg in 1804 and first studied law. In 1831, in a fit of hatred for his profession, he tried to commit suicide by shooting himself in the head. When he recovered, he gave up law and at the age of thirty took instead to science in Göttingen and Berlin. In Berlin, through his uncle, Johannes Horkel, professor of comparative physiology, he came in contact with Alexander von Humboldt and his circle, including such men as the physiologist Johannes Müller and Christian Gottfried Ehrenberg. At the time, Ehrenberg was working on the one-celled organisms called *Infusoria*. Here, it seems, Schleiden was already becoming recognized: he was elected to the Leopoldina, the Royal Society of Imperial Prussia. Soon after obtaining his degree in 1839, he was

1 Julius von Sachs, *Geschichte der Botanik von 1530–1860* (1875) translated by H. E. F. Garnsey, revised by I. B. Balfour (Oxford: Clarendon Press, 1906), 108–109.
2 V. Sachs, *Geschichte der Botanik* (1875) (n. 1), 84.
3 Matthias J. Schleiden, *Grundzüge der wissenschaftliche Botanik, nebst einer methodologischen Einleitung, als Einleitung zum Studium der Pflanze* (Leipzig: Engelmann, 2d ed., 1845), 151.

15

appointed *ausserordentlicher* (associate) professor in Jena, later also director of the Jena Botanical Garden, and in 1850 *Ordinarius* (professor) for
botany. In Jena, he came into contact with the philosopher Jacob Friedrich Fries, whose philosophy he had known since his Göttingen days. It
seems that Heinrich Schleiden, Matthias's younger brother, who was
Fries's student, first introduced him to Fries's work. Schleiden stayed in
Jena until 1863, when he quarrelled with the university authorities and
left for the Baltic university of Dorpat. His career was marked by many
moves. At times, as in Jena and Dorpat, he was a professor of botany, but
more often, as in Dresden, Darmstadt, Frankfurt, and Wiesbaden, he
taught privately, or *privatisirt*. In 1881, in the seventy-eighth year of *seines
bewegten unstäten Lebens*, his wandering and unstable life (the phrase is
Wunschmann's), he died.[4] (See Figure 1.1.)

Schleiden lived at a time when many intellectual radicals of the
German-speaking world were political revolutionaries, manning the bar-
ricades for democracy.[5] Schleiden, the radical, attacked his forebears,
the Linnaean botanists of the Aristotelian tradition, just as his older
contemporary Johannes Müller attacked his forebears in physiology,
the *Naturphilosophen*, and as Karl Marx and the "young Hegelians" attacked theirs in philosophy and theology.[6] Emil du Bois-Raymond,
Müller's student, wrote of Johannes Müller's *Handbuch der Physiologie*,
which appeared in its complete form in 1840,

> Müller's *Physiologie* started a new epoch; it ended one period and began another
> . . . a period of sceptical shaking of everything long since believed-in; time hon-
> oured problems were assailed with a boldness of research unheard-of till then,
> before which they fell like mediaeval burghs before the new machines of war.[7]

4 Ilse Jahn, "Matthias Jacob Schleiden an der Universitt Jena," *Naturwissenschaft, Tradition, Fortschritt*, supplement to *N.T.M.: Z. f. Geschichte der Naturwissenschaften* (1963),
   63–72; Ulrich Charpa, "Introduction," *Matthias Jacob Schleiden, Wissenschaftsphilosophische Schriften mit kommentierenden Texten von Jakob Friedrich Fries, Christian G. Nees von
   Eesenbeck und Gerd Buchdahl* (Cologne: Dinter, 1989), 9–43; E. Wunschmann, "Schleiden, Matthias Jacob," in *Allgemeine deutsche Biographie* (Leipzig: Dunker and Hum-
   boldt/k. Akad. der Wissenschaften, 1930), v. 13, 417–421. For Julius von Sachs's
   discussion of the *Grundzüge*, see v. Sachs, *Geschichte der Botanik* (1875) (n. 1), 188 ff.
5 William L. Langer, *Political and Social Upheaval 1832–1852*, in series, "Rise of Modern
   Europe" (New York: Harper, 1969), 107.
6 The dates of some of these revolutionary works are: David Strauss, *Das Leben Jesu*
   (1835); Bruno Bauer, *Kritik der Synoptiker* (1841); Ludwig Feuerbach, *Das Wesen des
   Christentums* (1841); Karl Marx and Friedrich Engels, *Manifest der kommunistischen Partei*
   (1848); Johannes Müller, *Handbuch der Physiologie des Menschen fur Vorlesungen* (1840),
   which appeared in parts from 1833.
7 Emil du Bois-Reymond, "Gedächtnissrede auf Johannes Müller, gehalten in der Aka-
   deme der Wissenschaften am 8 Juli 1858," in Estelle du Bois-Reymond, ed., *Reden von
   Emil du Bois-Reymond* (Leipzig: Veit, 1912), v. 2, 143–334 (p. 215); and quoted by

Figure 1.1. Matthias Jacob Schleiden, portrait engraved from a photograph, published by Wilhelm Engelmann of Leipzig. It shows Schleiden with his microscope and a trousse for carrying slides. (Photograph by Instructional Media Services, University of Toronto)

Müller's new machine of war was experimental physiology, with which he attacked the speculations of the *Naturphilosophen*. Schleiden's was the principle of unity, which he used against the "mediaeval burgh" of Linnaean botany: "The basic principle of human reason, its inescapable striving for unity in its knowledge, is as valid for the study of the organism as it is elsewhere in science."[8] Botany must be an inductive science, whose aim should be to find unifying principles under which to subsume the manifold diversity of nature.

Schleiden's *Grundzüge der wissenschaftliche Botanik* begins along the same lines with the famous *methodologische Einleitung*, the philosophical introduction in which he attacks the prevailing Linnaean systematic botany of scholastic diaeresis and sets up in its place a new science, the leading principle of which is not subdivision, but the search for underlying unity, built upon the common basis of the single cell. This new botany text is not to be a manual for identifying plant species as successfully as possible. Schleiden sarcastically quotes the eighteenth-century physician Hermann Boerhaave, who was Linnaeus's friend and contemporary, as saying that the aim of botany as a science is to "recognize and remember plants as successfully as possible, and with the least possible bother."[9] And he says that from now on, this kind of systematic botany will be reduced to its proper *rôle* as that of an unskilled labourer to the science.

Schleiden and his colleague, Theodor Schwann, Müller's assistant, had been the first to consider the cell to be the common building block of both plants and animals: beneath the complicated particularities of the plant and animal kingdoms lay the ubiquitous simple cell. According to Schwann, the cell was not only the starting point of the organic series but a bridge to the organic world. It forms in its mother-liquor in the same way as an inorganic crystal grows in a solution. This analogy between cell formation and crystallization had far-reaching significance: it was the hallmark of the materialism that replaced the speculative *Naturphilosophie* of the early part of the century with concepts grounded in a materialistic play of physical forces. Materialism did not, however, imply the replacement of unity and analogy as heuristic devices, which had been typical of the *Naturphilosophic* style.[10]

Schleiden's work, instead of naming and classifying, starts with a general section on the organic and inorganic materials of plants and goes

8   Matthias J. Schleiden, "Beiträge zur Phytogenese," *Archiv für Anatomie und Physiologie* (Müllers) 2(1838):137–176 (p. 137).
9   Schleiden, *Grundzüge der wissenschaftliche Botanik* (1845) (n. 2), 2.
10 Ilse Jahn, *Grundzüge der Biologiegeschichte* (Jena: Fischer, 1990), 334–335, points out the significance of this analogy between cell form and crystallization in the writing of the period. The cell-as-microcosm continued to serve as the model for scientific unity into twentieth-century biology: see Natasha X. Jacobs, "From unit to unity: protozoölogy, cell theory and the new concept of life," *J. Hist. Biol.* 22(1989):215–242. For introducing me to the work of Ilse Jahn, I thank Margaret Ghattas of the University of Toronto Library, who is preparing a translation and commentary on Matthias Schleiden's writings.

on to discuss the plant cell, the tissues that are formed from it, and the life of the cell in growth and assimilation and in death. Only in the second volume is there any kind of systematic species-by-species discussion, and even here the purpose is not to provide a means of recognizing species, but to discuss their development and comparative morphology, rather as his contemporary Johannes Müller was doing for the animal kingdom – with a mixture of anatomy, physiology, and embryology.[11] Incidentally, Schleiden's drawings are beautiful: realistic and at the same time just diagrammatic enough (Figure 1.2).

In the *methodologische Einleitung*, Schleiden the revolutionary takes a slashing sweep at Aristotelian scholasticism, at the philosophy of Friedrich W. J. Schelling (although he had by no means shaken himself quite free of Schelling's thinking), and at that of Hegel. His aggressively adversarial style can perhaps be traced to his earlier career as a lawyer. This is one of the earliest, as well as one of the most complete and articulate, statements of the philosophy of science, that strain of critical positivism originating in a lopped-off version of Kant's critical philosophy: in Kant's phenomenalism separated from his noumenal world. Schleiden's criticism of Hegel and Schelling has a decidedly positivist tinge. Hegel, he says, we can forget about, thank God, because as far as science goes, he has disappeared without a trace. Unfortunately, we must admit that Schelling's influence has been as important as it was harmful: just look at the confusion of psychology and ignorance with which all the Schellingites mix up imagery, dreams, and poetry and call it thinking![12]

For a botanist however, Schleiden's criticism of what he calls dogmatism – that is, systematic Linnaean botany – is more important. The botanists *say* that their science is inductive; but they carry on with the same division and subdivision of definitions and systematic analysis of names in the scholastic manner that we have inherited from the middle ages. Sometimes they do not even bother to give examples of the forms they are supposed to be describing.[13] The morphological view of the

11 Johannes Müller, letter to Minister von Altenstein, in which he proposed himself for a chair in Berlin, lists this combination as his (by 1832) contribution to the new science. Quoted by du Bois-Reymond, "Gedächtnissrede auf Johannes Müller" (1858) (n. 7), 186–187.

12 Schleiden, *Grundzüge der wissenschaftliche Botanik* (1845) (n. 3), pp. 18, 20, 22; Schleiden, "Schellings und Hegels Verhltnis zur Naturwissenschaft" (1843), in Charpa, *Schleidens wissenschaftliche Schriften* (1989) (n. 4), 199–264. Here he attacks the *Naturphilosoph*'s imagery of parallels and polarities.

13 Schleiden, *Grundzüge der wissenschaftliche Botanik* (1845) (n. 3), 22–23. Schleiden's favourite example of scholastic Linnaean systematics is the (then new) textbook of Endlicher and Unger: see Stephan [Istvan Laszlo] Endlicher and Franz Unger, *Grundzüge der Botanik, entworfen von Stephan Endlicher u. Franz Unger* (Vienna: Gerold, 1843). Endlicher, an Austro-Hungarian botanist (1804–1849), had also recently published his *Genera plantarum secundum ordines naturales disposita, auctore Stephano Endlicher* (Vindo-

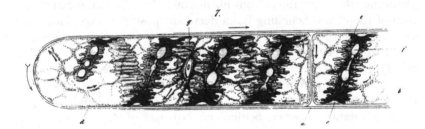

8.

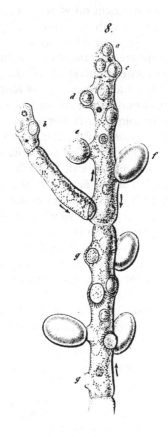

Figure 1.2 (*above and opposite*). Delicately beautiful drawings by Matthias Schleiden.
Opposite: *Spirogyra quinina*, a simple thread-like alga. Schleiden points out three layers of
cell wall: *a* is a gelatinous coat that covers the true cell membrane, *b*. Both are as trans-
parent as water and are only separated by a fine darker line between them. The cell
membrane is lined with a thin layer of pale yellow, half-fluid slimy substance *d*, made up
of protein that is just clearly visible. On the inner surface of this layer, there is a zig-zag
band of chlorophyll, sticking to the edges of the cells, *e*, and probably made of wax. These
bands have a groove on the outside, *c*, which contains some solid substance that stains
with iodine. In continuity with this water-clear substance there are some larger and
smaller grains, *f*, that seem to be starch grains. In the middle, there is a long cytoblast,
g, with clear nuclei (*Kernkörperchen*) surrounded by an open space filled with slime, from
which fine strands of slime reach to the cell walls. Here, in the nitrogen-containing layer,
there is a network of circulating streams, the directions of one of which is shown by the
arrow. Above: A thread of a mould, growing on a stalk of *Passiflora alata*. Upper part of
the plant, with a side branch. Here, too, the nitrogen-containing layer circulates in tiny
streams. The plantlet shows an apparently complete formation-series of fungal spores, if
the steps *a, b, c, d, e, f* are compared, while *g-g* represent the scars of fallen spores. From
Schleiden's *Grundzüge der wissenschaftliche Botanik* (3d illustr. ed., 1850) (n.3), plate I, figs.
7 and 8. (Photograph by National Library of Medicine, Bethesda, Md.)

world, in which forms are essences on which classifications are built, is purely subjective and is valid only for "me," the subjective observer. The things-in-themselves, for each other, as it were, are neither green nor red; their relationships are only those of mass and movement, space and time. Natural knowledge is made up of the laws of nature that express these relationships mathematically. As Kant says, every science is scientific only in proportion to the amount of mathematics in it.[14]

This rejection of morphology in favour of mathematics, Schleiden himself tells us, is Kant's philosophy of science. It stems originally from Kant's *Metaphysische Anfangsgrunde der Naturwissenschaft* of 1794.[15] It came to Schleiden through the work of Jacob Friedrich Fries: "We . . . call ourselves followers of Kant and Fries," he says, and elsewhere refers to "our Fries."[16] Schleiden's reliance on Fries dated not, as might have been supposed, from the personal contact he had with Fries at Jena, which lasted only from late 1839 until Fries's death in 1843. According to Ilse Jahn, Schleiden was already making Fries's *Mathematisches Natur-philosophie* of 1822 the basis of his philosophy of science, from his Göttingen days onward, even before his studies in Berlin brought him into the field of botany and physiology.[17] It is in the use of "regulative principles," particularly the principle of unity, which Schleiden called "the basic principle of human reason," that Kant and Fries really make their presence felt in Schleiden's methodology.[18]

bonae: apud Fr. Beck, 1836–1840). In 1881, he edited a letter from Linnaeus to N. J. Jacquin. He was the author of "Introduction to Chinese Grammar." Many of his publications on the botany of various areas (Czechoslovakia, Brazil) are in Latin.

14   Immanuel Kant, *Metaphysische Anfangsgrunde der Naturwissenschaft* (Leipzig, 1794), "Vorrede," x; Schleiden, *Grundzüge der wissenschaftliche Botanik* (1845) (n. 3), 39; Kant's famous *Satz* is not directly quoted by Schleiden, but it is part of the Kantian argument that he is rehearsing here. On the interpretation of the relationships between the "things-in-themselves" and between the "things" and their observer, see Gerd Buchdahl, "A key to the problem of affection," *Proceedings of the VIIth International Kant Congress* (in preparation).

15   Kant, *Metaphysische Anfangsgrunde der Naturwissenschaft* (1794) (n. 14), iv–x.

16   Schleiden, *Grundzüge der wissenschaftliche Botanik* (1845) (n. 3), 20, 30. For literature on Jacob Friedrich Fries, Ernst Friedrich Apelt, and their relations with Schleiden, see Thomas Glasmacher, *Fries-Apelt-Schleiden: Verzeichnis der Primär und Sekundärliteratur 1798–1988* (Cologne: Dinter, 1989). For easily accessible accounts of Fries, see H. M. Nobis, "Fries, Jacob Friedrich," in Charles C. Gillispie, ed., *Dictionary of Scientific Biography* (New York: Scribner's, 1972), v. 5; Alexander P. D. Mourelatos, "Fries, Jacob Friedrich," in Paul Edwards, ed., *Encyclopedia of Philosophy* (London: Macmillan, 1969), v. 3.

17   On the relationship between Schleiden and Fries, see Ilse Jahn, "The influence of Jacob Friedrich Fries on Matthias Schleiden," in William Ray Woodward and Robert S. Cohen, eds., *World Views and Scientific Discipline Formation* (Dordrecht: Kluwer, 1991), 357–365; Charpa, "Introduction," *Schleidens wissenschaftliche Schriften* (1989) (n. 4), 13–16.

18   Gerd Buchdahl, "Leading principles and induction: the methodology of Matthias Schleiden," in Ronald N. Giere and Richard S. Westfall, eds., *Foundations of Scientific Method* (Bloomington, Ind.: Indiana University Press, 1973), 23–52.

The idea of unity as the goal of all human mental activity plays an important part in Kant's critical philosophy. Kant conceived the stages in the mind's work on nature as successive efforts to unify the data more and more completely: just as the understanding unifies the manifold in the object by means of concepts, positing a certain collective unity as the goal of the activities of the understanding, so reason unifies the manifold of concepts by means of ideas.[19] But the drive towards unity, which is a product of the mind, is balanced by recognition of the diversity in nature: "The logical principle of genera which postulates identity, is balanced by another principle, that of species, which calls for manifoldness and diversity."[20] According to Kant, three principles can be used to guide the empirical employment of reason, the three heuristic maxims:

Reason thus prepares the field for the understanding:
1. Through a principle of the *homogeneity of the manifold* under higher genera.
2. Through a principle of the *variety of the homogeneous* under the lower species, and
3. In order to complete the systematic unity, a further law, that of the affinity of all concepts – a law which prescribes that we proceed from each species to every other by a gradual increase in activity.[21]

Fries restates these three heuristic maxims in his *System der Philosophie als evidente Wissenschaft* of 1804.[22] They appear in Schleiden's *methodologishe Einleitung* in scarcely altered form. But it is the maxim of unity on which he, and Fries, lay most weight.

It was an aspect of Kantian thought that appealed to a generation of post-Kantian philosophers of science, among them the *Naturphilosoph* Friedrich Wilhelm Schelling, whose fascinating lectures at the university of Jena started a new wave of unitarian enthusiasm. At the turn of the century, Fries was trying unsuccessfully to compete with Schelling as a lecturer in philosophy at Jena. Fries was critical of Schelling, but he was also enthralled by him. For Fries, as for Schelling, unity was the basis of scientific thought. But, Fries said, it should not be used as Schelling used it, purely as speculation. The heuristic maxims should enable us in practice to bring the manifoldness of nature under unifying general laws of science.[23]

19 Immanuel Kant, *Kritik der reinen Vernunft* (1781), translated by Norman Kemp Smith as *Critique of Pure Reason* (New York: St. Martin's, 1965), 532–549 (pp. 533–534, 644).
20 Kant, *Critique of Pure Reason* (1781) (n. 19), 540.
21 Kant, *Critique of Pure Reason* (1781) (n. 19), 541–546.
22 Jacob F. Fries, *System der Philosophie als evidente Wissenschaft aufgestellt* (Leipzig: Hinricks, 1804), 257–258.
23 Frederick Gregory, "Die Kritik von J. F. Fries an Schellings Naturphilosophie," *Sudhoffs Archiv* 67(1983):145–157.

Schleiden's version of this adds both practical and polemic emphasis. The botanist who hopes, *un*like the Linnaeans, to do more than just collect facts must, Schleiden says, be educated in the critical philosophy. Unity and lawlikeness are the goals of an inductive science. Philosophy, the mathematical natural philosophy, reminds us that there is only One Nature and One Science.[24] Schleiden's feeling is that traditional Linnaean botany, with its emphasis on the recognition of species and on division and subdivision, is the very antithesis of the science that he admires, the science of leading maxims. Schleiden's inductive botany is a science in which the particular, the manifold given of the plant world, is unified by reason under general laws. And for Schleiden, the unity underlying the diversity of the Linnaean species is a single cell crystallizing out in a slimy mother-liquor.

The question is, to what extent did these methodological ideas exert a sustained influence on subsequent philosophers of science?[25] One who clearly was influenced was the philosopher Ernst Friedrich Apelt, Fries's student and his successor at Jena, who, like Fries, was a close friend of Schleiden's. After Fries's death, Schleiden and Apelt started a journal called *Abhandlungen der Fries'schen Schule*, which both of them, but mainly Apelt, wrote. In its original form, it lasted only two years, from 1847 to 1849, though it was revived in 1904. Apelt's Friesian textbook, *Theorie der Induktion* (1854), continues to make use of the leading maxims.[26] An extant copy of Apelt's book is heavily annotated by Ernst Mach, a philosopher who was also to associate himself with the principle of unity in science.[27] The biologists of the nineteenth century, however, had other ways of reading Kant. This late unitarian strain, with its echo of *Naturphilosophie*, was but one of them.[28]

The insistence on unity as the basis of scientific thought reappears in the work of Carl von Nägeli, passed on, it would seem, through Schleiden. Nägeli (Figure 1.3) was born in 1817 in Switzerland, near

24 Schleiden, *Grundzüge der wissenschaftliche Botanik* (1845) (n. 2), 131, 135.
25 Buchdahl, "Leading principles" (n. 18), 31.
26 Ernst Friedrich Apelt, *Theorie der Induktion* (Leipzig: Engelmann, 1854), 50–53, "Die leitende Maximen." For bibliographic information and secondary literature on Apelt, see Glasmacher, *Fries-Apelt-Schleiden* (1989) (n. 16), 88–100; L. L. (Larry) Laudan, "Apelt, Ernst Friedrich," in Charles C. Gillispie, ed., *Dictionary of Scientific Biography*, v. 1.
27 Buchdahl, "Leading principles" (1973) (n. 18), 48, n. 44.
28 Timothy Lenoir, *The Strategy of Life: Vital Materialism in XIXth Century German Physiology* (Dordrecht: Reidel, 1983), 17–53, 240–245. The "teleological mechanists," led by the Göttingen zoologist Johann Friedrich Blumenbach, who was in contact with Kant himself, drew their conception of biological causality from the *Kritik der Urteilskraft* (Lenoir, p. 24). The reductionists, who included Schleiden and the physiologist Emil du Bois-Reymond, make no use of this book and see no need for teleology (Lenoir, p. 242).

Figure 1.3. Carl von Nägeli, portrait photograph, n.d. (Courtesy of the Institut für Geschichte der Medizin, Munich)

Zurich. He studied in Zurich for some time and heard the *Naturphilos-opher* Lorenz Oken lecture there, then went to Geneva to read botany under Pyrame de Candolle, and later moved to Berlin. He spent the years 1840–1842 with Schleiden in Jena. Most of Nägeli's working life, however, was spent in Munich, as director of the newly founded Botanical Institute. Nägeli's colleague Simon Schwendener says that the invitation to Munich, and the building of this luxurious institute, with plant houses, laboratories, and lecture rooms, was part of the plan of Maximilian II, king of Bavaria, to equal the importance in scientific and artistic patronage of his predecessor, Ludwig I. Maximilian II died in 1864 and was succeeded by Ludwig II, who continued the tradition by inviting Richard Wagner to Munich and by building the Bayreuth opera house for him.[29]

It is impossible to say with certainty that Nägeli was indebted to Schleiden for his unitarianism, which was the most striking feature of both men's thought. Of all the "students" mentioned in this book, Nägeli alone had only a few words of rather grudging praise for Schleiden, his teacher. He felt himself generally to be "going beyond" Schleiden, negating him, as he might have put it, but his work shows many parallels with Schleiden's thought.

The Swiss zoologist Albert Kölliker, the friend of Nägeli's youth, tells in his autobiography of the years from 1836 on, when he and Nägeli were together at the University of Zurich, enthusiastically collecting plants around the countryside, attending the lectures of Lorenz Oken on his *Naturphilosophie* and on zoology: *die so sehr anregenden Vorträge Okens*, "Oken's lectures, which were so exciting," says Kölliker.[30] Forty years later, Nägeli himself remembered Oken's lectures as speaking to something already present within him. In 1877 he wrote:

When I was a student, just beginning to be interested in natural science, I had the desire to try to relate accepted notions to each other and to see things from the general point of view. This natural leaning was encouraged by Oken's lectures on natural history and I was led to look for the universal in everything. Luckily, this was balanced by another, equally strong, natural leaning, to be critical. . . . So, though I was full of enthusiasm for Oken's ideal intentions, I

29 No full-length study of this important figure has yet been done. The most detailed of the short biographies is that of his colleague Simon Schwendener, "Carl Wilhelm von Nägeli," *Ber. der deutsche botanische Gesellschaft* 9(1891):(26)–(42), which contains a list of his students and a bibliography. See also C. Cramer, "Prof. C. v. Nägeli," *Actes de la Soc. Helv. des Sciences Naturelles* (Fribourg), C. R. 74ième Séance (1890–1891), 184–188; Robert Olby, "Nägeli, Carl Wilhelm von," in Charles C. Gillispie, ed., *Dictionary of Scientific Biography* (New York: Scribner's, 1974), v. 9, 600–602. John S. Wilkie's work on Nägeli is referred to in n. 53 of this chapter.

30 R. Albert Kölliker, *Erinnerungen aus meinem Leben* (Leipzig: Engelmann, 1899), 6–7. On Kölliker, see Erich Hintzsche, "Kölliker, Rudolf Albert von," in Charles C. Gillispie, ed., *Dictionary of Scientific Biography* (New York: Scribner's, 1973), v. 7, 437–440.

could not accept the arbitrarily schematic way he worked them out, any more than I could listen to his *Naturphilosophie!*[31]

After finishing his degree in 1840 in Geneva, Nägeli joined Kölliker in Berlin, where they lived next door to each other in the Dorotheen-strasse, and where, says Kölliker, they discussed and criticized all that they heard. Johannes Müller was lecturing on comparative anatomy and pathological anatomy, Jacob Henle on histology. They also went to the lectures of some of the Hegelians.[32] Nägeli says that although he tried to grasp the ideas in Hegel's writing, it was a completely fruitless effort.[33]

From the point of view of 1877, it may well have seemed like that, but in 1844 Nägeli had actually written:

Every empirical science advances by negation, in that the earlier is replaced by the newer and better; but the negation never goes so far that the earlier formulation is completely cancelled out [*nie . . . aufgehoben wird*]: there is always something left unchanged.[34]

It has a very Hegelian sound to it, though the Hegelianism may not have gone very deeply. Nägeli's philosophizing usually follows different lines. In the spring of 1841, Kölliker and Nägeli spent a fortnight in Jena on their way home from Berlin, getting to know Matthias Schleiden.[35] Thereafter Nägeli came back alone, and spent a year and a half in Jena working with him. Schleiden had only been there since the autumn of 1839, and Jacob Fries was still alive. In the summer semester of 1841, Schleiden gave a course entitled "Philosophical Botany," as well as one on the use of the microscope. He was then working

---

31 Carl von Nägeli, "Die Schranken der naturwissenschaftlichen Erkenntniss," lecture to the 50 Versammlung deutscher Naturforscher und Aerzte, Munich, 1877, in his *Mechanisch-physiologische Theorie der Abstammungslehre, mit einem Anhang.* 1. *Die Schranken der naturwissenschaftlicher Erkenntniss.* 2. *Kräfte und Gestaltungen im Molecularen Gebiet* (Munich: Oldenbourg, 1884), 553–680 (pp. 555–556).

32 Kölliker, *Erinnerungen* (1899) (n. 30), 80.

33 Nägeli, "Schranken der naturwissenschaftliche Erkenntniss" (1877) (n. 31), 566.

34 Carl von Nägeli, "Ueber die gegenwartige Aufgabe der Naturgeschichte, insbesondere der Botanik," *Z. f. wiss. Bot. 1*(i)(1844):1-33, *1*(ii)(1844):1–45; citation in *1*(i):5.

35 Kölliker's later work was concerned with interpreting Schwann's version of the cell theory, the derivation of complex animal forms from the cell unit, rather than Schleiden's, which saw the cell as the symbol of the whole. Kölliker followed—in Duchesneau's neat phrase—an "epistemological model of development," according to which the properties of the cell units directed the form and the function and the embryological development of the complex organism. See François Duchesneau, *Genèse de la Théorie Cellulaire* (Montreal: Bellarmin, 1987), 233–253. Like Nägeli, Kölliker conceived of nature as being unified by the cell.

on the *Grundzüge*, which came out in 1842.[36] Nägeli was exposed to a concentrated dose of Schleiden's philosophy.

According to Nägeli's biographer and co-worker, Schwendener, this stay had no effect on Nägeli: he thought Schleiden's work on the microscope was not rigorous enough. But it was the occasion for the founding of the *Zeitschrift für wissenschaftliche Botanik*, which was edited by Schleiden and Nägeli together – though it was mostly written by the latter. Schleiden says that he only put his name on it to please Nägeli.[37] But just because of the material in this journal, it is hardly possible to say that this stay in Jena had no effect, or that Nägeli was not deeply involved in Schleiden's views.[38]

The article that leads off the new journal is Nägeli's version of a *methodologische Einleitung*: it reflects and comments on, and argues against, the first (the general) section of Schleiden's *Grundzüge*. It is a dialogue with Schleiden, which, in Nägeli's own Hegelian terms, both negates and goes beyond him. Schleiden is indeed *aufgehoben*: *Mein Freund Nägeli ist Hegelianer*, he said.[39] Nägeli writes first of "the gulf between Schleiden's *Grundzüge* and all that went before it, such that it should really be acclaimed as the beginning of a new era."[40] Then he critically examines each of Schleiden's basic botanical ideas. He does not go into the philosophy of the *methodologische Einleitung*, much less its Kantian background: this journal is intended to advance scientific botany by empirical means, he says.[41] But Nägeli is building directly upon those foundations.

Everything empirical, he writes, begins with observations of individuals and advances by uniting them under universals. The plan of this study is to begin with the individual and its developmental history and to advance step by step through the concept of species to that of the plant and animal kingdoms:

Individuals are related to each other in the same way as successive states of the same individual. They are continuous with each other, every boundary is arbitrary, the whole movement is *unendlich Theilbar*, infinitely divisible. In the same

36 Jahn, "Schleiden in Jena" (1963) (n. 4), 68.
37 Schwendener, "Nägeli" (1891) (n. 29), 27; Jahn, "Schleiden in Jena" (1963) (n. 4), 71.
38 Olby, "Nägeli" (1974) (n. 29), 600, says, "Like Schleiden, Nägeli began with a philosophical essay . . . in which he eschewed compilations of empirical data since science is concerned not with the changing characteristics of individuals but with unchanging law," and elsewhere, "These studies of cell formation illustrate N.'s striving for general laws [and] the strong influence of Schleiden upon him." I think so, too, in spite of Schwendener's opinion, which may very well represent Nägeli's own.
39 Nägeli, "Schranken der naturwissenschaftlichen Erkenntniss" (1877) (n. 31), 566.
40 Nägeli, "Gegenwartige Aufgabe der Naturgeschichte" (1844) (n. 34), 27.
41 Nägeli, "Gegenwartige Aufgabe der Naturgeschichte" (1844) (n. 34), 2.

way infinitely numerous individuals are possible and the differences between them blend continuously into each other. . . . Individual differences cannot therefore be the object of science, as they are variable and infinitely divisible. Science seeks the constant among these related individuals. This is the concept of species.[42]

Upon this foundation of a seamless nature, continuous in every dimension, science – that is, the human mind – builds a series of more absolute concepts. Species are concepts, which, unlike nature itself, can have absolute boundaries. The facts must lose their real existence in order to attain their ideal existence in our consciousness, as *Vorstellungen* or representations. Knowledge of them is indirect only, conditioned by the intellectual viewpoint of the observer.[43] Individuals of the same species differ from each other by small stepwise differences. The point is, whether the concept of species is absolute, or whether the concepts, too, only differ quantitatively.[44]

It is a Kantian theory of knowledge in which the conclusions of science are a mental structure only and are imposed on nature through our *Vorstellungen*, or concepts. Nägeli does not relate it to Kant himself, but produces it in conversation, as it were, with Schleiden. In one respect, however, it is the reverse of the Kantian picture: for Nägeli, nature is continuous, and science – that is, the work of the mind – produces separate species. Since they are only concepts, they can be as sharply defined as the thinker pleases. Science is one, Schleiden said, but for Nägeli at this time science could be as many as necessary.

This position did not lead Nägeli to take sides with the Linnaean botanists. He chose the same anti-heroes among the Linnaeans as Schleiden had done: the contemporary Viennese botanists Endlicher and Unger and their *Grundzüge der Botanik*. He adduces Schleiden's "first special leading maxim for botanists," the developmental history of the plant.[45] Endlicher's criteria of definition seem at first sight to be sharp enough, but if the plants are examined from Schleiden's developmental point of view, the differences lose their sharpness and shade into one another. Instead of well-defined species, one finds only gradual differences, *graduelle Abstufungen*.

But Schleiden's criteria fail for Nägeli as much as Endlicher's. By using his developmental maxim, Schleiden focuses on the one-celled stage of the life cycle of plants. He looks at a plant at a point where specific differences are simply not yet present, and since there is no

42 Nägeli, "Gegenwartige Aufgabe der Naturgeschichte" (1844) (n. 34), 9.
43 Nägeli, "Gegenwartige Aufgabe der Naturgeschichte" (1844) (n. 34), 3.
44 Nägeli, "Gegenwartige Aufgabe der Naturgeschichte" (1844) (n. 34) 9.
45 Schleiden, *Grundzüge der wissenschaftliche Botanik* (1845) (n. 2), 135–136.

mathematical law describing the rest of its development, Schleiden's attempt to use it to define species fails: at this one-celled stage all plants are one. Species are impossible to define by the methods of developmental history, just as they are by those of Linnaean botany. Nägeli, at this early stage in his own development, does not seem to be aware that that is why Schleiden was interested in a one-celled stage, and that he was *not* trying to define species. Although the definition of species has eluded him at each point, Nägeli still feels that with more research he will be able to overcome the intractable continuity of nature and make absolute divisions that take the whole life history of the plant into consideration.

A fine example of this early, absolute Nägeli is his monograph of 1848, *Gattungen einzelliger Algen*. One-celled algae represent plants with a very simple life history, in which the "whole concept of the species" – the essence, that is – is realized within the single cell.[46] For these, Schleiden's single-celled stage of development should be enough. But even here the continuity of nature cannot be gainsaid. Some forms are one-celled and live alone, some live in colonies or conglomerates, some are a multicellular individual. If those forms are defined as one-celled in which the "whole concept of the species" is realized in a single cell, the definition excludes those one-celled forms in which the whole concept is completed only after several generations.[47]

Nägeli's solution here was to choose as an example an organism that was certainly one-celled and by "analogy and natural relationships" to make decisions in the difficult cases using all possible knowledge, especially of cell formation. It is an empirical solution, one might say, to a dogmatic problem, but it did not satisfy him very long. The monograph on the algae and their systematic relationships came out in 1848, but by 1853, as he put it in his autobiographical fragment of 1877, he was persuaded that species as well as individuals shaded gradually into each other.[48] As Schleiden had said, not only nature but science too was one. For Kant, it was the unity of human reason that led us to suppose a unity in nature:

[A]ll the manifold genera are divisions of one single highest universal genus ... there are no species or subspecies which are the nearest possible to each other. ... This logical law of the *continuum specierum (formum logicarum)* presupposes however a transcendental law (*lex continui in natura*).[49]

46   Carl von Nägeli, *Gattungen einzelliger Algen, physiologisch und systematisch bearbeitet* (Zurich: Schulthess, 1848), 9.
47   Nägeli, *Gattungen einzelliger Algen* (1848) (n. 46), 3.
48   Nägeli, "Die Schranken der narurwissenschaftlichen Erkenntniss" (1877) (n. 31), 557.
49   Kant, *Critique of Pure Reason* (1877) (n. 31), 557.

But it is a transcendental law only. Continuity of forms is a work of the mind, and nature itself is divided. For Nägeli, unlike Kant, it was *nature's* continuity that forced him to accept the continuity of forms in the mind.

From this time onward, the *lex continui in natura* became Nägeli's leading maxim. He applied it in each one of the fields in which he worked. In each one, it is no longer the *differentiae* that he tries to seek out, but the transitions between forms, the *quantitative Abstufungen* that connect them. In his work on the fine structure of living matter, for example, he examines the transition between living and nonliving; in his phylogeny he sees a continuous flow of forms passing into each other from an origin at this transition point. His theory of fermentation and his bacteriology are built upon *specifische quantitative Abstufungen*: species differ from each other only quantitatively, by gradual transitions. Everything is connected to everything else: each of these fields is linked in nature to the others. Nägeli covers nature in a continuous network of thought.

Nägeli's systematic examination of nature may be said to have originated in his criticism of 1844 of the cell theories of Schleiden and of Theodor Schwann.[50] The cell theories are speculative attempts to provide a mechanistic explanation of the transition between inorganic and organic matter. In both, cell formation is a crystallization of matter around a solid nucleus: for Schwann, the organic crystal differs from an inorganic one only in that it is permeable to water, and it can therefore grow by intussusception, whereas the inorganic one grows by apposition.

Nägeli broadly approves of this notion, although he says it fails to account for the difference in chemical composition between the cell nucleus and its wall.[51] In Schleiden's theory, this difference is accounted for: only the nucleus is formed by crystallization, while the

50 John R. Baker, "The cell theory: a restatement, history and critique," *Q. J. Microsc. Sci.* 89(1948):103–125 (p. 103). Baker discusses Schleiden's influence on Theodor Schwann, which he thinks was a bad one, leading Schwann to a wrong view of the origin of cells. Schwann's theory of 1839 is an "exposition of the idea that the general principle of construction of organic products is that of cell-formation." Nature first joins the inorganic molecules together to form a cell, which then differentiates into a fibre, a tube, etc. This reductionist account of cell formation as an intermediary stage between inorganic and organic was taken over from Schleiden and attacked by Virchow in his *Satz, Omnis cellula e cellula*. It persists as the basis of Nägeli's phylogeny. A more recent discussion of the two men, which takes into account Buchdahl's work, is Duchesneau, *Genèse de la Théorie Cellulaire* (1987) (n. 35), 137–151.
51 Carl von Nägeli, "Die Stärkekörner," in Carl von Nägeli and C. Cramer, *Pflanzenphysiologische Untersuchungen* (Zurich: Schulthess, 1855–1858), pt. 2 (1858), 1–623 (pp. 1–13).

membrane is a secondary formation, a result of the catalytic effect of the nucleus on its surrounding mother-liquor. The cell nucleus itself is a simple organic crystal, like the other solid bodies inside a cell, such as the starch grains and spiral fibres.

At this stage of Nägeli's development – this was written in 1844 – he was not ready to accept the implication of Schleiden's theory that there is really no boundary between the living and the nonliving. If the difference is only one of permeability, the properties of the living cell should be those that depend on permeability and nothing more. But they are not: they include not only growth, which could be explained thus, but also metamorphosis and reproduction. It is wrong to conclude, he writes, that the inorganic and the organic differ from each other only stepwise.[52] A few years later, he changed his mind.

It is against this background of contemporary theory, with its emphasis on the transition between living and nonliving, that Nägeli worked on the fine structure of living matter. The problem of the analogy between the mineral crystal and the living one, and how far it might extend, how far the similarities and contrasts in their properties might explain each other, is the starting point for his work on the structure and growth of starch grains, which was so carefully dealt with by John Wilkie in 1960. In Wilkie's view, Nägeli arrived at the (mistaken) theory that starch grains grow by intussusception and not by apposition purely by observation and because of imperfections in his microscope and his technique. But it seems much more probable that he did so as a contemporary and critic of Schwann's and particularly of Schleiden's cell theories, for which the growth of starch grains modelled that of the growth of cells themselves, at the transition between the organic and the inorganic worlds (Figure 1.4).[53]

This work on starch grains was published in 1858 but was mostly done in the three years between 1852 and 1855, while Nägeli was *Ordinarius*

---

52 Nägeli, "Die Stärkekörner" (1858) (n. 51), 37.
53 John S. Wilkie, "Nägeli's work on the fine structure of living matter," *Ann. Sci.* 16(1960):11–14, 171–207, 209–239 (1960); 17(1961):27–62. In the introduction to this paper, Wilkie writes: "The observations on the growth of starch grains are most elaborate and appear highly reliable. In fact, however they are almost certainly systematically erroneous. This can be attributed to some extent perhaps to the imperfection of N's microscope but he was certainly too easily satisfied with simple observation, interpreting various inferences by experiment. . . . He asserts that small grains are dense throughout and unstratified whereas larger grains are stratified and their central regions are occupied by a soft kernel. Hence he argues that growth cannot be by apposition. He thus arrived at his theory of growth by intussusception. . . . It is noteworthy that this theory though mistakenly argued for the starch grain certainly does hold in the case of the cell wall which N. mentions only in passing." On the significance of this for contemporaries, which is not discussed by Wilkie, see Ilse Jahn, *Biologiegeschichte* (1990) (n. 10), 334–335.

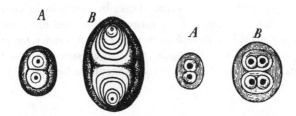

Figure 1.4. The growth of starch-grains. Carl von Nägeli's comparison of the growth of starch grains (left) and analogous developmental stages of a *Gloeacapsa* colony (right). In both, the peripheral hull does not change much, while thickening takes place in the interior layers. Both grow by intussception, not by apposition from outside. From Nägeli and Schwendener, *Das Mikroskop* (1877) (n.54), figs. 242, 243, p. 543. (Photograph by National Library of Medicine, Bethesda, Md.)

at Freiburg-im-Breisgau.[54] According to Schwendener, it was a happy and fruitful time for him, surrounded by grateful and gifted students and by many local friends.[55] It was also the date that Nägeli himself gives (1853) for his abandonment of the idea of absolute differences in nature. He refers to a paper entitled *"Systematische Uebersicht der Erscheinung im Pflanzenreich"* of 1853, which is listed by Schwendener but which I have not been able to trace.[56] In this paper, by his own account, he says, "the development of a starch grain or a cell seem to me to be no different, although they are more complicated than the development of a crystal" – an idea that became the starting point for his work on the fine structure of matter.

The same idea, of a gradual transition between inorganic and organic, in which the inorganic molecule becomes ever more elaborate and complicated, also gave Nägeli the model for his genetic theory of descent, a theory that appeared, fairly fully worked out, in 1856. According to this theory, there is a continuous production of simple organic beings from inorganic materials; the beings that come into existence in this way are the simplest unicellular plants – in our time,

54 Carl von Nägeli, *Die Stärkekörner: eine morphologisch, physiologisch chemisch-physikalisch und systematisch-botanische Monographie* (Zurich: Schulthess, 1858), 332–377; Carl von Nägeli and Simon Schwendener, *Das Mikroskop: Theorie und Anwendung desselben* (Leipzig: Engelmann, 2d ed., 1877), 532–548 (the figures of the starch grains and the cells are from p. 543).
55 Schwendener, "Nägeli" (1891) (n. 29), 28.
56 Carl von Nägeli, "Systematische Uebersicht der Erscheinung im Pflanzenreich" (1853), referred to by Nägeli in the work cited in n. 57, p. 174; "Diess habe ich in meiner Schrift, "Systematische Uebersicht, etc." weiter ausgeführt." Full reference not found.

that is, the fungi and moulds. From these, by continuous elaboration, are developed stage by stage all the more complicated forms of the two kingdoms. He calls this a *Vervollkommnungs-princip*, a drive towards ever-increasing perfection, which implies that organisms always develop in the direction of increasing complexity. It accounts for the relationships between species:

Each species does not develop by itself, and with no causal relationship to the rest of nature, and disappear, fruitless and without trace. On the contrary, the organisms, as they are related anatomically and physiologically, are also genetically related.[57]

Species develop, increase in complexity, and flow continuously upward. The species itself is an individual composed of other individuals, as a tree is of cells. The species and the genus are as real as the individual, and like individuals, they shade into each other with no sharp differences. An earlier distinction between the continuity of individuals and the absolute differences between species, genera, and orders has disappeared. All is flowing and changing, species and genera as much as the individuals, which are composed of smaller units, like the higher classes. *Unendlich Theilbarkeit* is true of the whole of nature; and so is continuous change.

The driving force of this theory of descent, the *Vervollkomnungsprincip* or drive to perfection, has a respectable ancestry in the dynamic natural philosophy of the late eighteenth- and early nineteenth-century German physiologists.[58] This German transformism is far from being dependent on the ideas of Charles Darwin; it appears that Darwin himself may have drawn upon it.[59] It is rooted in Nägeli's own experience and antedates Darwin's theory by some years.

Nägeli analyzed the differences between Darwin's later theory and his own in a long paper he gave in Munich upon being admitted to the königliche Akademie der Wissenschaften in 1865, after the appearance of Darwin's book. Both theories, he says, explain the relation between higher and lower species in the same way: the higher have devel-

57 Carl von Nägeli, "Die Individualität in der Natur, mit vorzüglicher Berücksichtigung des Pflanzenreiches," *Monatschrift des wissenschaftlichen Vereins in Zürich* 2(1856):171–212.

58 Timothy F. DeJager, "G. R. Treviranus and the Biology of a World in Transition" (Ph.D. diss., University of Toronto, 1990), 1–23; Philip R. Sloan, "Buffon, German biology and the historical interpretation of biological species," *Brit. J. Hist. Sci.* 12(1979):109–153.

59 Philip R. Sloan, "Darwin, vital matter and the transformation of species," *J. Hist. Biol.* 19(1986):369–445.

oped out of the lower.[60] Darwin accounts for this exclusively by the mechanism of natural selection. Nägeli maintains that, in addition to competition and selection, there is a second mechanism, that of the drive to increasing complexity, which he had postulated as his own theory in 1856. Only this drive, he felt, could account for the fact that evolution took place always in a forwards direction, and never backwards, although conditions in an environment might well regress and repeat those of an earlier time. He takes as his most distinguished forbear in this the French zoologist, Jean-Baptiste de Lamarck, and refers to his *Zoölogie Philosophique* of 1805.[61] Nägeli's attitude to species was fully formed before Darwin's book appeared. Its criticism of Darwin is not that of a man struggling to show that he can "go beyond" the popular theory. In fact, Darwin himself seems to have had some difficulty in dealing with Nägeli's argument that there were many nonadaptive characters, which could not have been produced by natural selection.[62]

The result of Nägeli's change of heart on the species problem and of his work on the transition between the organic and inorganic is his enormous monograph *Mechanisch-physiologisch Theorie der Abstammungslehre* of 1884. It is, as he says in the "Foreword," a monument to the principle of *unendlich Theilbarkeit*, or infinite divisibility.[63] It originated in a paper that Nägeli gave at the fiftieth anniversary meeting of the Naturforscher-versammlung in Munich in 1877.[64]

It was originally written from pure inspiration, while Nägeli was on holiday in the alps. He was asked at a moment's notice to step into a gap in the festival programme. There were other famous names on it: those of the biologist Ernst Haeckel, and the pathologists Rudolf Virchow and Edwin Klebs. Haeckel's paper, like Nägeli's, was on the theory of evolution. He gave an outline of a monistic *Weltanschauung* that included in its sweep the development of the life of the spirit, from its beginnings in the attraction and repulsion between molecules to the formation of consciousness in man, a combination of natural science

---

60 Carl von Nägeli, *Entstehung und Begriff der naturhistorischen Art* (Munich: Verlag der K. Akad., 1865).

61 Nägeli, *Begriff der naturhistorischen Art* (1865) (n. 60), 9.

62 Peter J. Bowler, *The Eclipse of Darwinism: Anti-Darwinian Evolution Theories in the Decades around 1900* (Baltimore, Md.: Johns Hopkins University Press, 1983), 149–150.

63 Nägeli, *Mechanisch-physiologisch Theorie* (1884) (n. 31), "Foreword."

64 Anon., "The German association at Munich," report of the proceedings of the fiftieth anniversary meeting in *Nature*, 16(1877):491–492; translation of Nägeli's paper, "On the limits of natural knowledge," *Nature* 16(1877):531–535, 559–563; translation of Ernst Haeckel, "The present position of the evolution theory," *Nature* 16(1877): 492–296; Rudolf Virchow, in *Amtlicher Bericht der 50 Versammlung Deutscher Naturforscher und Aerzt*, Munich, 17–22 September 1877 (Munich: Akad. Buchdruckerei, 1877), 65–77; 41–55 (contribution of Edwin Klebs).

and *Geisteswissenschaft* in a single whole. Haeckel started his speech, appropriately enough, by figuratively laying a wreath on the tomb of the *Naturphilosoph* Oken, whose enthusiasm for unity in science had led to the founding of the Naturforscherversammlung.[65] Nägeli's contribution took the form of an answer to the famous paper, "On the Limits of Natural Knowledge," read to the Naturforscherversammlung by the mechanistic Berlin physiologist Emil du Bois-Reymond at its meeting of 1872.[66]

Natural knowledge, says Nägeli, is relative not absolute: it consists in the measurement of natural phenomena, according to a measure deduced from themselves. The saying of Kant, that in each special natural science there is only as much real science as there is mathematics in it, is after all still true. We cannot conceive of absolutely different properties: only relative or quantitative differences can be understood by mathematic methods. The consequence is, as far as our measuring advances without gaps, that we will find no absolute differences in nature. There will be no chasms that cannot be filled. Like Haeckel, Nägeli traces a continuity from chemical molecules to crystals, to the parts of the cell and the cell itself, through the plant and the animal kingdom to human consciousness.

At the lowest and simplest stage of material organization which we know, we therefore find on the whole the same phenomenon as we do at the highest stage, where it appears as conscious sensation. The difference is only one of gradation.[67]

Like Haeckel, he presents a monistic, an *einheitlich* or unitarian picture of science and of nature.

Just in the same way as divisibility is endless, we must suppose by analogy, what we find confirmed in the whole domain of our experience, that combination too of individual separate particles continues endlessly downwards. In like manner we are forced to suppose an endless combination upwards in always larger groups. The heavenly bodies are bodies which unite into groups of lower and higher order, and the whole of our system of fixed stars is only a molecule group of an infinitely larger whole, which we must suppose to be a single [*einheitlich*] organism, and at the same time, part of a still larger whole.[68]

65 Haeckel, "The present position of the evolution theory" (1877) (n. 64).
66 Emil du Bois-Reymond, "*Ueber die Grenzen des Naturerkennens,*" and *"Die sieben Welträthsel,"* *zwei Vorträge* (Leipzig: Veit, 1898); the first of these *Vorträge* was given to the 45 Versammlung Deutscher Naturforscher und Aerzte, Leipzig, 14 August 1872.
67 Nägeli, "Limits of natural knowledge" (1877) (n. 64), 562.
68 Nägeli, "Schranken der naturwissenschaftlichen Erkenntniss" (1877) (n. 31), 572; translation altered from (n. 64), 534.

For the naturalist, says Nägeli, it is a logical necessity to admit only differences of degree in finite nature. But natural science must be exact, and it must rigidly avoid everything that oversteps the limit of the finite and the intelligible, and the transcendental: "The power of education and habit was, up to the most recent period, an obstacle in the way of the separation of these two domains, and yet it is certain that every metaphysical addition turns natural science and natural investigation into a muddy alloy."[69]

In this speech one can see *in parvo* the transformation of the Kantian philosophy of science – with its emphasis on mathematics, its leading maxim of unity, and its separation of the phenomenal from the transcendental – into the positivism of the later part of the century, whose tenets were very much the same, with the addition of the rejection of metaphysics. The time has passed, says Nägeli, when it was possible to demand that the scientist be educated in the "critical philosophy."[70] We are coming close to the *Satz* of Ernst Mach, positivist philosopher and scientist, written in 1905: "The land of the transcendental is closed to me. . . . I am *definitely not a philosopher, only a scientist.*"[71]

Among the many examples of Nägeli's unitarianism to be found in his mature work, at least four indicate his speculations depend of necessity on his view of the *unendlich Theilbarkeit*, the *lex continui in natura*. One of these forms the starting point of his phylogeny of 1884. As he himself put it,

If in the material world everything is causally related to everything else, if all phenomena have a natural origin, if all organisms are made up of the very same materials as inorganic nature, and disintegrate into them again at the end, then they must originally arise out of inorganic compounds. To deny spontaneous generation is to call for a miracle.[72]

The transition from the inorganic to the organic takes place in two stages; the first stage is the synthesis of albumin compounds and their organization into micellae to form the primordial plasma mass. The

69 Nägeli, "Limits of natural knowledge" (1877) (n. 64), 563.
70 The "demand" probably represents Schleiden's remark in the *methologische Einleitung:* "The first thing that we demand from anyone who hopes to be successful in getting botany over the stage of simple facts, is that he be educated in the critical [that is, Kantian] philosophy, and that he be able to appreciate the need for general laws." *Grundzüge der wissenschaftliche Botanik* (1845) (n. 2), 132. Nägeli needed no encouragement to look for general laws, but he may well have learned that from Schleiden.
71 Ernst Mach, *Erkenntnis und Irrtum: Skizzen zur Psychologie der Forschung* (Leipzig: Barth, 1905), vii.
72 Nägeli, *Mechanisch-physiologisch Theorie* (1884) (n. 31), 83.

second stage, which leads up to the simplest known organisms, is the stage of the *Probien*, beings that precede life.[73]

Nägeli's investigation of the stored proteins of the Brazil nut gave him an example of the first stage – a living model, as it were, of the nonliving. The proteins form crystalloids, rather than true crystals. Because they have interstices into which fluid can penetrate, they grow by intussusception, not by apposition. They swell irregularly and crack when fluid is imbibed, and they change shape. The protein-crystalloids are like crystals in shape, but they differ from them in every other way, whereas they closely resemble starch grains and cell walls. In other words, the protein-crystalloids are the "elementary organs" of life.[74]

It is remarkable that this is the very claim denied by Nägeli in 1844 in his criticism of the cell theories of Schleiden and Schwann: if the cell and its organs are to be regarded as no more than a permeable crystal, then all its properties must be referable to permeability alone.[75] The Nägeli of the *Absolutheit der Begriffe* of 1844 denied it.[76] The unitarian Nägeli of 1863 asserts it.[77] By 1884, it had become the cornerstone of his theory of spontaneous generation: "Other organised substances behave exactly like starch grains: I maintain that what I have said about the formation of starch grains is true point for point about the spontaneous generation of plasma masses."[78] Under the influence of molecular forces, new modifications of the proteins appear, together with unorganized ferments. By its own inner laws, the makeup of the plasma mass becomes steadily more complex, in the same way as do inorganic materials.[79]

---

73 Nägeli, *Mechanisch-physiologisch Theorie* (1884) (n. 31), 90.
74 Carl von Nägeli, "Ueber die aus Proteinsubstanzen bestehenden Crystalloide in der Paranuss," *Bot. Mitth.* (Munich) *1*(1863):217; cited by Wilkie, "Nägeli's work on the fine structure of living matter" (1960) (n. 53), 33; Wilkie comments in his n. 68 (p. 33) that N. used the word *Elementarorgane*, which corresponds approximately with "cell organs," though it seems unusual to refer to starch grains and to cell walls as organs of the cell, since they have not the same kind of autonomy as that possessed by plastids or nuclei. My feeling is that Nägeli implies that they are closely related and do have that same kind of autonomy. Wilkie is not aware of this current in Nägeli's thought, which originates in Schleiden's cell theory.
75 Nägeli, "Gegenwartige Aufgabe der Naturgeschichte" (1844) (n. 34), pt. 2, 1–13.
76 Nägeli, "Gegenwartige Aufgabe der Naturgeschichte" (1844) (n. 34), pt. 2, 9.
77 Nägeli, "Ueber die aus Proteinsubstanzen bestehenden Crystalloide in der Paranuss" (1863) (n. 74).
78 Nägeli, *Mechanische-physiologische Theorie* (1884) (n. 31), 97.
79 Nägeli, *Mechanische-physiologische Theorie* (1884) (n. 31), 342. This upward flow of continually increasing complexity, generated by the *Vervollkomnungstrieb*, was, Nägeli thought, responsible for phylogenesis. His *Abstammungsungslehre* makes this an alternative to Darwin's theory, in which competition and natural selection have that effect. In Nägeli's opinion, Darwinian *Concurrenz* and *Zuchtwahl* (competition and natural selection) only weed out the unfittest; new phylogenetic advances, in contrast, depend on the *Vervollkommnungstrieb* (the drive to perfection).

It seemed impossible to Nägeli that this could have happened only once in the earth's history. Governed as it was by general laws, by the properties of the inorganic world, the appearance of organic compounds and simple organisms at the lowest stage of development *must* occur wherever the right conditions come together:

The vegetable kingdom in its historic totality is therefore not a single many-branched trunk, nor many trunks which originated at the same time from identical beginnings, and so could be looked on as branches of the same tree. It is, on the contrary . . . composed of countless phylogenetic stems, which have originated at all times and in the most scattered places on the earth's surface.[80]

Nägeli's evolutionary theory, as he expresses it here, supposes a whole series of different lines of descent, all originating under the same conditions, and subject to the same laws. It is these conditions and laws that have determined the similarity of organisms at the same stages of development. Contrasted with the Darwinian "umbelliferous" family tree, Nägeli's is more like a group of water weeds, independent of each other, floating on the surface of the present, and hanging down to various depths into the past. Branching does take place as well, but it is difficult – it should be impossible – to distinguish true blood relationship, *Blutverwandschaft*, of this kind, from cases of parallel development under the same laws. The likelihood of similarities being due to *Blutverwandschaft* is greater in the higher species, where a greater number of coincidences over a longer period would have been necessary to bring them to the same point, but among simple, young species it becomes more likely that they are of multiple origin and are passing through the same stage of development.

The overall *Blutverwandschaft* of the living organisms of the present day and of the whole phylogenetic system is actually nothing but a beautiful dream. But because of the unity of the developmental laws which cover the whole of the organic kingdom, it can be accepted as a general norm. Even if the organisms are not genetically connected, they behave by and large as if their relationship to each other was a true one.[81]

At the bottom of the phylogenetic system, in succession to the stage of *Probien*, are the schizophytes, the lower fungi. It is these that are produced everywhere and at all times, by spontaneous generation; these are the youngest, the most recent graduates, as it were, from the inorganic world. Their characteristics, Nägeli felt, were just what might be expected from a youthful group: their multiple forms, so similar to

80 Nägeli, *Mechanisch-physiologische Theorie* (1884) (n. 31), 468.
81 Nägeli, *Mechanische-physiologische Theorie* (1884) (n. 31), 470.

each other, and so easily transforming into each other, make varieties, species, and genera impossible to fix. Whether they actually do develop any further, he could not say.[82]

That the fungi originated by spontaneous generation was not a new idea for Nägeli. In his earliest systematic work, the product of his absolute period, he distinguished the fungi from the unicellular algae by, among other things, the manner of their origin. In contrast to the algae, the fungal cells contain no chlorophyll, starch, nor colouring matter, and they arise not from germs but by spontaneous generation "from fermenting, putrefying or disintegrating organic substances."[83] He did not, however, continue to maintain this last notion: it was replaced later by the more sophisticated version of spontaneous generation that followed his work on the fine structure of living matter and that was integrated into the unitarian system.

The connections between Nägeli's unitarianism and the monism of Ernst Haeckel are not easily untangled. At first sight, they look rather alike. They are both based on evolutionary transformism, and are both mechanistic in style; they both presuppose the unity of nature and that of science. The German biologists had clearly linked evolution and spontaneous generation: together they unified the organic and the inorganic into a single fundamental conception of nature.[84] Haeckel's roots lie clearly in Darwinism itself, even though his Darwinism is of a rather loose kind, and includes several un-Darwinian features, such as the so-called bio-genetic law.[85] Haeckel states that his *Generelle Morphologie der Organismen*, which was written in 1866 when he was thirty-three, was "the first attempt to carry the newly founded developmental theory through the whole field of the forms of living organisms."[86] His natural history of creation, the *Natürliche Schopfungsgeschichte*, was a shorter, more accessible exposition of the same unified view of nature. Haeckel recognized two Kants, one critical, the other transcendental.[87] The first Kant, the critical author of the *Pure Reason*, held a "monistic" cosmogony derived from Newtonian physics and knew that the dogmas of metaphysics could not be reached by pure reason. The second Kant, of whom Haeckel did not approve, was the dualistic author of the *Critique of Judgement* and the dogmatic *Critique of Practical Reason*, the

82 Nägeli, *Mechanische-physiologische Theorie* (1884) (n. 31), 467.
83 Nägeli, *Gattungen einzelliger Algen* (1848) (n. 47), 1.
84 John Farley, *The Spontaneous Generation Controversy from Descartes to Oparin* (Baltimore, Md.: Johns Hopkins University Press, 1974), 142.
85 Bowler, *Eclipse of Darwinism* (1983) (n. 62), 35–37.
86 Ernst Haeckel, *Die Welträthsel: Gemeinverständliche Studien über monistische Philosophie* (1899) (Born: Emil Strauss, Volksausgabe, 1903), 4.
87 Haeckel, *Die Welträthsel* (1899) (n. 86), 156–157.

discoverer of the two different worlds, the phenomenal and the noumenal. But for both Kants, the *Anfangsgrunde der Naturwissenschaft*, the
basic premises of mathematics as an *a priori* science, are metaphysical.
Haeckel does not seem ever to mention the Kant – perhaps one might
call him a third Kant – of the heuristic maxims and the unitarianism
derived from this aspect of his writing.

Nägeli's scientific unitarianism originates in the Schleiden-Schwann
cell theory; his catchwords *unendlich Theilbarkeit* and *specifische quantitative Abstufung* are not part of Haeckel's vocabulary, though they are
not inconsistent with his thought. Haeckel approved of Nägeli's theory
of spontaneous generation: "I absolutely agree with his *Satz*, 'To deny
spontaneous generation is to call for a miracle.' "[88] But Nägeli's critical
attitude to the Darwinian evolutionary tree is different from Haeckel's
more orthodox Darwinian single-trunk version. The difference can be
traced to his work on the fine structure of living matter and the micellar
theory, and it leads to his opinion on the specificity of fermentations
and of bacterial forms.

In his epistemology, too, Nägeli comes quite close to Haeckel. He
denies Kant's *a priori* as applied to the law of quantition and to mathematics, and he denies the existence of inborn ideas, such as those of
space, time, and causality, just as Haeckel does. But he *does* accept that
these are the necessary forms of thought, that our mental activity can
take no other form. We are a part of nature, and in nature space, time,
and causality rule. Our nervous system is subject to the same laws as
operate outside it. Not the ideas themselves, but the activity that necessarily, eventually, by *allmähliche Abstufung*, leads to them, is actually
inborn: these are indeed laws of thought, as they are of nature. As such
they are *a priori* in the philosophical sense, and not arrived at by induction. But they do not precede experience either ontogenetically or
phylogenetically: experience takes place in the animal world long before it reaches the complexity of conscious thought. So Nägeli manages
to retain the *a priori* nature of causality, while escaping from the metaphysical: thinking causally has a natural history.

In spite of all these concurrences, Nägeli is not a Haeckelian monist.
The wilder shores of monism, Haeckel's monistic religion, and such
remarks of Haeckel's as "World (= Nature = substance = cosmos-
universe = God)" are, like the world of the transcendental for Mach,
closed to him.[89] There is for Nägeli no monistic religion or monistic
ethics. Haeckel's monism covers both the phenomenal world of science

88 Haeckel, *Die Welträthsel* (1899) (n. 86), 104.
89 Haeckel, *Die Welträthsel* (1899) (n. 86), 93, cited from his own *Monismus* (1892), 18,
    42.

and nature, and also the noumenal world of God and morality, in its attempt to get away from Kantian dualism. Nägeli keeps to the phenomenal world: every metaphysical addition only muddies the clarity of science.[90] His answer to du Bois-Reymond's *Ignorabimus*, itself an attempt to divide the physical from the metaphysical, is *Wir wissen und wir werden wissen,* "We *do* know, and we *will* know," within the bounds of the finite world.[91] The scientist may interest himself in the transcendental, but when he does, he is not acting as a scientist. The world of knowledge, of *Wissenschaft*, is the finite world, and that world must be monistic. The other, the infinite world, may be either monistic or dualistic as one pleases, but that has nothing to do with science.[92]

Nägeli's mature position appears in a short and rather simply-written book for "laymen" in the public-health field, that is to say, physicians rather than botanists. He admits three true divisions of what he calls the "lower fungi:" moulds, yeasts, and "fission fungi." The last group contains what are usually called "bacteria," but Nägeli found that too restrictive in its suggestion of rod-like shapes. He includes organisms that are rod-shaped, round, comma-shaped, and spiral in the same group. His drawing shows them as he saw them, loosely and intimately twined together as they grew in nature (Figure 1.5).[93] Across the three groups – moulds, yeasts, and fission fungi – transformation did not take place, at least, not within the time span of experiments; but within the groups, especially within the fission-fungi, he felt that it was very likely that it did.

In this view of the relationships between the "lower fungi," Nägeli's opponent was the Linnaean botanist Ferdinand Cohn of Breslau, whose classification of these same organisms had appeared in 1875. Cohn postulated four separate types within the group *Bacteria*, which in turn he divided into six separate genera.[94] Cohn's divisions rested on morphology. His four main groups were named *Sphaerobacteria, Microbac-*

---

90 Nägeli, "Limits of natural knowledge" (1877) (n. 64), 563.
91 Nägeli, "Schranken der naturwissenschaftlichen Erkenntniss" (1877) (n. 31), 602; translation altered from "Limits of natural knowledge" (1877) (n. 64), 563.
92 Nägeli, "Schranken der naturwissenschaftlichen Erkenntniss" (1877) (n. 31), 600. This does not occur in the original speech: the references to monism and dualism, like all his references to Haeckel, date from the 1884 edition, thus post-dating his contact with Haeckel at the meeting of 1877.
93 Carl von Nägeli, *Die niederen Pilze in ihren Beziehungen zu den Infektionskrankheiten und der Gesundheitspflege* (Munich: Oldenbourg, 1877), 1. The three terms are *Schimmelpilze* or moulds; *Sprosspilze,* "sprouting fungi" or yeasts; and *Spaltpilze,* "splitting fungi," which include the forms known as bacterium, coccus, vibrio and spirillum.
94 Ferdinand Cohn, "Untersuchung über Bacterien," *Beiträge zur Biologie der Pflanzen* *1*(ii) (1872, published in 1875):127–222.

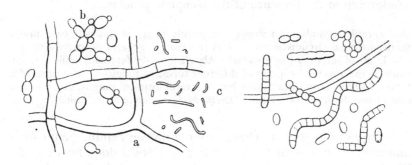

Figure 1.5. Relationships between the "lower fungi," according to Carl von Nägeli. The drawing on the left shows: a. *Schimmelpilze*, the threads of a mould; b. *Sprosspilze*, "sprouting fungi" or yeasts; c. *Spaltpilze*, "splitting fungi" or bacteria.

The drawing on the right shows the *Spaltpilze* further enlarged: the rod shapes and spirals are made up of small, round forms, which can also exist as separate units. From Nägeli, *Die niederen Pilze* (1877) (n.93), 4. (Photograph by National Library of Medicine, Bethesda, Md.)

teria, Desmobacteria, and *Spirobacteria,* with the implication of round, rod-shaped, thread and spiral forms. Nägeli's feeling was that the differences lay only in size, and that after division the cells either broke loose or remained attached to each other as rods or threads, and these sometimes became more or less screw-shaped.[95] In the case of the yeasts, he thought that an individual cell was not transformed from one kind of yeast into another, but that in the course of a few generations, the yeast that produced alcohol from sugar could transform into what turned the alcohol into acetic acid. Between the two pure forms lay a number of transformation stages, which gave the appearance of a number of different species.

Nägeli elaborated this theory in his longer work on the theory of fermentation, which appeared two years later. He distinguishes between *Fermentwirkung* or enzyme activity, the activity of diastase and invertin, for example, and *Gärwirkung* or fermentation proper, the activity of the yeast cells, which takes place only in or near the yeast cell itself, and which is a variety of the slow combustion common to all the lower fungi.[96] In some cases, they also have the ability to produce a *partial*

95 Nägeli, *Die niederen Pilze* (1877) (n. 93), 21.
96 Carl von Nägeli, *Theorie der Gärung: ein Beitrag zur Molecularphysiologie* (Munich: Oldenbourg, 1879), 29.

combustion such as acetic fermentation, an ability that has an inverse relationship to the presence of the complete process:

Slow complete combustion shows a very wide range of intensity even under similar external circumstances. . . . It is therefore a general property with specific levels of intensity [*mit spezifischer Abstufung in der Intensität*] while fermentation is the specific property of different forms of fungus. . . . If a particular fermentation is found not in one but in several forms of fungus, this, too, is a case of specific levels of intensity [*specifische Abstufung in der Intensität*].[97]

Some types of fungus can change from a non-fermenting mould-form into a fermenting yeast form, as the genus *Mucor* does, for example. Without doubt, says Nägeli, the yeasts have arisen phylogenetically from the moulds, with the development of the faculty of fermentation. It is not yet clear how the fission fungi are related to the others: their forms and properties are labile and easily altered by changes in culture conditions; their fermentations are easily lost, or pass over into each other.

Nägeli's unitarianism is the consistent thread to be found in all this work. The *unendlich Theilbarkeit* which he found in Nature in his earliest days, though not, at first, in science, he finds again in the lower fungi. The description of the mass of individuals in nature that he gave in the first paper in the journal that he and Schleiden founded in 1844 could equally well apply to the bacteria. He wrote in 1844:

Individuals are related to each other in the same way as successive states of the same individual. They are continuous with each other, every boundary is arbitrary, the whole movement is infinitely divisible [*unendlich Theilbar*]. In the same way infinitely numerous individuals are possible and the differences between them blend continuously into each other [*findet ein allmäliger Uebergang der Unterschiede statt*].[98]

It could apply to the *allmäliger Uebergang* between the organic crystals, the micellar bodies, and the starch grains, and to the development of consciousness in the animal kingdom. The change may be over time, and be developmental, as in his *Abstammungslehre*, or across species, as in the *specifische Abstufung in der Intensität* between organisms which carry on apparently different kinds of fermentation. For the naturalist, for Nägeli, that is, it is a logical necessity to admit only differences of degree, *nur gradweise Unterschiede*. It is no accident that Nägeli, the field theorist, did *not* see different species of bacteria when he looked at

97 Nägeli, *Theorie der Gärung* (1879) (n. 96), 116.
98 Farley, *Spontaneous Generation* (1974) (n. 84), 144–146.

them under a microscope. It is equally no accident that Cohn, who could not see them either, knew they were there. Neither spontaneous generation nor unrestrained pleomorphism would be acceptable to the new and powerful science of bacteriology, which took its stand on Cohn's point of view.[99]

99 Nägeli, "Gegenwartige Aufgabe der Naturgeschichte" (1844) (n. 33), 9.

2

The Linnaeans: Ferdinand Cohn and Robert Koch

The new botany of the second half of the nineteenth century grew out of the Schleiden cell theory and easily adopted Darwinism or some transformist variant of it. But if one were inclined to think that the Linnaean tradition of definition and diaeresis was thereby made obsolete – how could it have survived Schleiden's scorn?[1] – one has only to read the work of Ferdinand Cohn (Figure 2.1) to know that it was still fully alive. Cohn himself was quite conscious of the difference between his own traditional thinking and the popular transformism of such famous modern biologists as Ernst Haeckel, Thomas H. Huxley, and Carl von Nägeli. He was aware of it, but he never polemicized. There is none of Schleiden's exhilarating knockabout in Cohn's writing.

Cohn was born in Breslau, where his father was the first of his family to move out from the closed Jewish community of the Breslau ghetto. He was educated partly in Breslau and partly in Berlin, where, like Matthias Schleiden, he was in contact with Johannes Müller and Christian Ehrenberg, the biologist whose work on microscopical organisms interested him from this time onwards.[2] He was in Berlin during the 1848 revolution, but in 1849 he came back to Breslau, where he spent the rest of his working life. He got his appointment as *ausserorden-*

---

1 Matthias J. Schleiden, "Methodologische Grundlage: Einleitung," in *Grundzüge der wissenschaftliche Botanik, etc.* (Leipzig: Engelmann, 2d ed., 1845) 17; he describes the contradictory systems of the different systematic botanists, each of whom claims that his system is the only truly natural one. "I think that by now the botanists must be like those Roman soothsayers who foretold the future by examining the entrails, who could not look at each other without laughing." He is very aptly quoting Cicero, who says, "mirari, Cato se aiebat, quod non rideret haruspex haruspicem cum vidisset" (Cic. Div. 2 24, 51).
2 Ilse Jahn, "Ehrenberg, Christian Gottfried," in Charles C. Gillespie, ed., *Dictionary of Scientific Biography* (New York: Scribner's, 1971), v. 4, 288–292; Mary P. Winsor, *Starfish, Jellyfish and the Order of Life: Issues in XIX Century Science* (New Haven, Conn.: Yale University Press, 1976), 28–43; Frederick B. Churchill, "Infusoria from Ehrenberg to Bütschli," *J. Hist. Biol.* 22(189):189–213.

Figure 2.1. Ferdinand J. Cohn, portrait, n.d. (Photograph courtesy of the National Library of Medicine, Bethesda, Md.)

tlicher professor in 1859, and as *Ordinarius* for botany in 1872. In 1866, after many years of effort, he persuaded the university to set up an institute of plant physiology.[3]

3 Pauline Cohn, *Ferdinand Cohn: Blätter der Erinnerung zusammengestellt von seiner Gattin Pauline Cohn mit Beiträgen von Professor F. Rosen* (Breslau: Kerns, 1901). This is a beau-

Cohn's thinking, like Nägeli's, began with the classification of the lower plants. Simple unicellular forms had begun to arouse scientific interest owing to the availability of the microscope and the speculations of Schleiden and Schwann on the origin of plant cells. Schleiden himself had suggested analyzing plant forms at the unicellular stage of their life-history, before they diverged into different species, an idea that Nägeli, in his early days of absolutism, had found to present needless difficulties for the systematist.[4] In 1847 Nägeli produced his first systematic analysis of single-celled algae, which he developed further in his monograph of 1849.[5]

Cohn's first papers on the lower organisms were on the unicellular *Protococcus*, and later on the life history of the microscopic algae and fungi. A long paper that appeared in 1854 began, like Nägeli's monograph on the same subject, with the statement that the study of the cell is the basis for the understanding of plant life and that unicellular organisms afford a good opportunity for it. For this reason, he said, there had been a great deal of recent work on these organisms, especially on the problem of the systematic definition of species.[6] His concern in this monograph was to disentangle the systematic relationships and arrive at sharply defined "good" species.

Among these unicellular forms was the family that Christian Ehrenberg called *Vibrio*.[7] Ehrenberg recognized four genera, *Bacterium, Spirillum, Spirochaet,* and *Spirodiscus,* which, on account of their vigorous movements, he placed in the animal kingdom as "infusion animalcules." Cohn, however, felt that he had conclusive proof that they should be classed as plants, and he demonstrated this by describing a full life cycle of one organism, called *Vibrio lineola* by Ehrenberg and *Bacterium termo* by Félix Dujardin.[8] He found this dot or comma-shaped

tifully produced book with plant decoration at the chapter heads by Max Wislicenus of Breslau. It contains memoires on Cohn's life and work and a bibliography; Gerald J. Geison, "Cohn, Ferdinand Julius," in Charles C. Gillispie, ed., *Dictionary of Scientific Biography* (New York: Schribner's, 1971), v. 3, 336–341. Geison assesses Cohn's botany from a point of view slightly different from mine.

4 Carl von Nägeli, "Ueber die gegenwartige Aufgabe der Naturgeschichte, insbesondere der Botanik," *Z. f. wiss. Bot. 1*(i):1–33; *1*(ii):1–45 (1844).

5 Carl von Nägeli, "Die neueren Algensysteme und Versuch zur Begründung eines eigenen Systems der Algen und Florideen," *Schweizer Gesellsch. N. Denkschr., 9*(1847). I have not been able to locate a copy of this.

6 Ferdinand J. Cohn, "Untersuchung über die Entwicklungsgeschichte der microscopischen Algen und Pilze," *Acad. Caes. Leop. Nova Acta 24*(1854):101–256, p. 102.

7 Christian Gottfried Ehrenberg, *Die Infusionsthierchen als vollkommene Organismen: ein Blick in das tiefere organische Leben der Natur* (Leipzig: Voss, 1838); see also Frederick B. Churchill, "The guts of the matter: infusoria from Ehrenburg to Bütschli, 1838–1876," *J. Hist. Biol. 22*(1989):189–213.

8 Cohn, "Entwicklungsgeschichte der microscopischen Algen u. Pilze" (1854) (n. 6), 116–117.

body in putrefying material, where it formed masses of jelly, resembling those of the alga *Palmella*. The jelly contained the organisms, bound together; as they matured, they escaped from the mass and became independent, as they did in *Palmella*.[9]

I believe, therefore . . . that they should be recognised as belonging to a new genus, for which I suggest the name of *Zoogloea*. The diagnosis is as follows:
Zoogloea cellulae minimae, bacilliformes, hyalinae, gelatina hyalina in massas mucosas globosas, uvaeformes, mox membranaceas consociatae, dein singulae elapsae per aquam vacillantes. *Zoogloea termo*. cellulis liberis mobilibus, rectis, 1/2000–1/700' " aequantibus . . . *Bacterium termo* (Duj.) *Vibrio lineola* (Ehr.).[10]

Because of the similarity to *Palmella*, which is obviously a plant, he places the vibrios in the vegetable kingdom. As "colourless microscopical plants, living in infusions," they should be placed in the group of water fungi or *Mycophycae*. But this order, the *Mycophycae*, did not really satisfy him; he saw it not as a natural order, but an artificial union of plants belonging to different families and genera. He felt it should be scrapped and the members distributed among the *Algae*, which they more nearly resemble, in spite of their colourlessness. The group of *Spirulina*, and the straight rods, which he called *Vibrio bacillus*, are close to the straight filamentous algae, particularly to the colourless *Beggiatoa*, a member of the group of *Oscillariae*.

Beginning in this way, Cohn's thinking grew out of his interest in the cell-theory, and was moulded by his Linnaean methodology. Both these elements of his intellectual ancestry remained with him throughout his life, modified very little by the advent of Darwinism in 1860.

Unlike Ehrenberg, Cohn was by no means opposed to Darwinism. His talk to the Schlesischen Gesellschaft für vaterländische Cultur on the origin of the Silesian flora, which he gave in 1860, only a month or two after the publication of the *Origin of Species*, shows that he recognized Darwin's importance immediately. The struggle for existence seemed to be a useful idea:

The Darwinian view of nature is undoubtedly of importance for plant geography. It explains why one and the same species spreads to different degrees in different localities, according to whether its rivals are more or less successful. . . . The old rule, "Each plant spreads as far as it is able to find suitable climate

9　Cohn, "Entwicklungsgeschichte der microscopischen Algen u. Pilze" (1854) (n. 6), 123.
10　Cohn, "Entwicklungsgeschichte der microscopischen Algen u. Pilze" (1854) (n. 6), 123.

and soil," must be changed to "Each species can spread only insofar as through climate and soil it has an advantage over its competitors."[11]

He enthusiastically compared the importance of Darwin's hypothesis for biology with that of the atomic hypothesis for chemistry, the hypothesis of the ether for physics, or that of cosmogony for geology. These ideas were the bases of their sciences.

When Cohn and his wife Pauline travelled to London in 1876, they paid a most respectful visit to Charles Darwin. They had lunch with the Darwins at Down House.[12] But an appreciation of Darwin's explanation of the relationships between species did not deter Cohn from trying to define them, and for this his Linnaean essentialism was a perfectly adequate methodology.

In 1862 Cohn found microscopical algae – filamentous *Oscillarinae*, some of which were blue-green – living in highly saline water in the Carlsbad spring at temperatures of more than 50° C. He also found these algae in the hot efflux of some Breslau factories.[13] He suggested that these might be the earliest inhabitants of the earth, adapted as they were to conditions perhaps like those of the *Urmeer*, the waters that covered the cooling earth.

In 1867, he was still interested in the classification of these lower organisms:

If like Nägeli one tries to derive all animals and plants by gradual improvement (*Vervollkommnung*) splitting and natural selection (*natürliche Züchtung*) from a single original form, one would look for that common starting point for the series of developing and diverging animal and plant forms in the *Phycochrom-algae*.[14]

Although Cohn quotes Nägeli's theory of development, it is not the later Nägeli of seamlessly flowing nature on whom he builds his ideas of the definition and separation of species, but the earlier absolute Nägeli. Nägeli divided the algae possessing either chlorophyll or some other pigment, from the fungi, which were colourless. Within the algae, he divided the green, chlorophyll-containing forms such as *Protococcus* from the *Chroöcoccaceae* and *Nostochaceae*, which contained blue-green or orange pigments.

11 Ferdinand J. Cohn, "Ueber den Ursprung der Schlesischen Flora," *Jahres Ber. d. Schlesisch. Ges. f. vaterl. Cult.* *38* 101–126(1860):114.
12 Pauline Cohn, *Blätter der Erinnerung* (1901) (n. 3), 204–209.
13 Ferdinand J. Cohn, "Ueber die Algen des Carlsbader Sprudels, und deren Antheil an der Bildung des Sprudelsinters," *Jahres Ber. d. Schlesisch. Ges. f. vaterl. Cult.* *40*(1862):65–67.
14 Ferdinand J. Cohn, "Beiträge zur Physiologie der Phycochromaceen und Florideen," *Arch f. microsc. Anat. 3* 1–60(1867):3–5.

Cohn had set up a seawater aquarium in the Breslau Institute. Early in the summer of 1865 it filled with vivid purple-red filamentous algae, which later in the year produced blue-green threads. They grew enormously, covering everything else in the aquarium: the gravel on the bottom disappeared, the rocks and even the other inhabitants, madrepores and live shells, were smothered in them. It was a "new" species of *Spirulina*, which had not been described before. Cohn called it *Spirulina versicolor* and wrote the traditional Linnaean "diagnosis:"

Spirulina versicolor, nov. spec. *S. vivide* mobilis filamentis praelongis violaceis in stratum gelantinosum vel mucilaginosum atropurpureum vel nigro chalybeum intricatis anfractibus densissime . . . Conchas calcareas nec non lapides in Aquario marino induit ibique immense multiplicatur. Hamburgae et Vratislaviae; etiam sponte crescens in lapidibus submarinis prope Helgoland; color inter purpureum et aerugineum vagatur.[15]

Because of this abundant growth, Cohn was able to investigate the nature of the pigment the alga produced. By means of a kind of paper chromatography, he found that the pigment was not a single "phycochrome," as Nägeli had thought. Rather, the algae contained chlorophyll, as well as a bluey-green or red pigment, which unlike chlorophyll was water soluble, but insoluble in alcohol. When the algae were placed on filter paper, it leaked out as a bluey-green or red halo around the dying plant, while the chlorophyll remained within the cells. A drop of the bluey-green pigment on a piece of paper separated further into a spreading circle of dark blue, separated by a colourless circle from an inner ring of pale red, which disappeared on exposure.

Cohn's analysis of the pigments led him to discover new relationships between the various types of algae. Having found that the blue-green and the red filaments of Spirulina both contained chlorophyll, along with the other pigment (which he called phycocyan), he was able to divide the algae into two groups, the *Chlorosporaceae*, which contained only chlorophyll, or a modification of it, and the *Phycochromaceae*, which contained chlorophyll together with another pigment, blue or red phycocyan. The blue-green *Oscillariae* – which included *Spirulina*, as well as some colourless forms, *Spirochaet, Spirillum, Vibrio*, and *Beggiatoa* – were also linked up with the *Florideae*, the red-brown algae, which contained chlorophyll together with the phycoerythrin, a third red or purple pigment. The *Oscillariae* were also brought into relationship with the *Chroöcoccaceae*, a group that included the organism that produced the "miracle of the blood," whereby bread seemed to become flesh: Cohn

15 Cohn, "Algen des Carlsbader Sprudels" (1862) (n. 13), 9.

describes this organism growing in the damp windowless larder of a house in Breslau, and on potato dumplings in the pastor's house at Bennstadt-bei-Halle. It was hard to say whether the red phycochrome of this so-called *Monas prodigiosa* was a red phycocyan or a phycoerythrin; it formed a link with the *Bacteriae* and both *Oscillariae* and *Florideae.*

It is very difficult, on reading Cohn's paper, to disentangle from the web of relationships anything resembling either a dichotomous or a common-ancestor classification. The schema is approximately as outlined in Figure 2.2. The *Fundamentum divisionis* would appear to be the pigmentation, with the mode of reproduction a secondary feature introducing a developmental aspect into the relationships. The *Chroöcocci* and the lower forms of *Oscillariae* (which include *Spirulina* and *Spirochaet, Spirillum,* and *Vibrio*) reproduce by simple fission. The higher forms (*Nostoc, Rivularia,* and *Scytonema*) show alternation of generations, which links them to the *Florideae,* the red-brown algae, to which they are also linked by the pigment that they have in common. "From all this" Cohn concluded "that the *Oscillarinae* in a natural system should be separated from the *Chlorosporae* near which they are usually placed, and should be seen as the first stage of a different line of organization leading directly to the *Florideae.*"[16]

It has been said that Cohn's idea about the primitive conditions in which the *Oscillarinae* could live led him to base a classification system on the new Darwinian evolutionary principles.[17] But Cohn did not expand this discussion of physiological relationships into a complete system. The basic dichotomy between the green and brown algae compared to the blue-green or red led him to see two distinct groups of algae, each with a separate line of development. But in 1872, when he came to produce a fully worked-out systematization for the whole family

16 Cohn, "Algen des Carlsbader Sprudels" (1862) (n. 13), 36.
17 Geison, "Cohn" (1971) (n. 321), 338; *pace* Geison, the use of the term "natural system" is not a pointer to Darwinism. The "essence" of scholastic essentialism is the essence in nature, as much as the essence in the mind. The question was discussed by St. Thomas Aquinas, *De Ente et Essentia* (1250), and by Linnaeus, *Philosophia Botanica* (1751). See Philip R. Sloan, "John Locke, John Ray and the Natural System," *J. Hist. Biol.* 5(1971):1–53.

Figure 2.2 (*opposite*). Ferdinand Cohn's "classification" of the *Algae.* The relationships are based mainly on pigmentation, with a glance at reproduction. From Cohn, "Beiträge zur Physiologie der Phycochromaceen und Florideen" (1867) (n. 16).

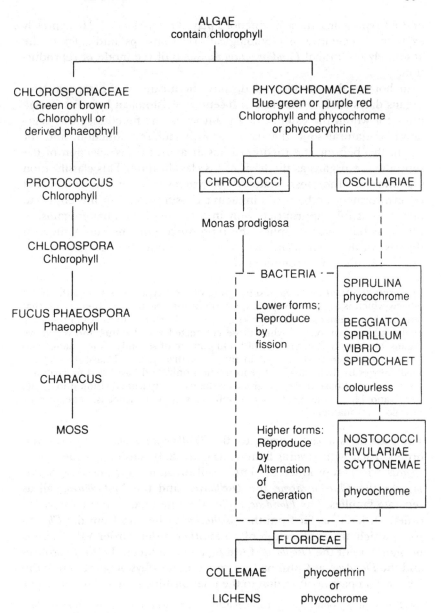

of the Cryptogams, these distinctions played no part in it.[18] He expressly excluded "vegetative or physiological" relationships, and stuck to the absolutely traditional *Fundamentum divisionis* of the mode of reproduction.

In botany, the tradition of dividing by means of the reproductive organs dates back to the sixteenth-century Aristotelian botanist Caesalpino, for whom the essence of a plant lay in the functions of its vegetative or natural soul, that is, in growth and reproduction, and for whom this became the theoretical justification for his selection of the reproductive organs as the basis of his classification. This classification was therefore the most "natural" of systems.[19] It was this system that in turn formed the basis of Linnaeus's classification. But the Linnaean method was fully appropriate only in the case of the Phanerogams, in which, as their name implies, the reproductive organs were visible. In the case of the Cryptogams, where they were not, the classification was more difficult. Cohn commented:

> The present attempt at a natural ordering of the Cryptogams adopts the point of view which is accepted without question in the system of Phanerogams, that only markers connected with reproduction and development are of importance in the higher divisions . . . while those connected with habitus, the vegetative organs, anatomy and the means of livelihood are of secondary importance and can only be taken into account in the lower divisions. . . . I have divided the Thallophytes by their means of reproduction only, and have tried to carry this right through; I have therefore given up the usual tripartite division into Algae, Fungi and Lichens as being based only on secondary (that is, vegetative or phytological) markers.[20]

The class he is here referring to, the *Thallophyta*, is divided into seven orders, each with several families (Figure 2.3). Order I – the *Schizosporeae*, which reproduce by simple cell division – includes the *Schizomycetae*, the *Chroöcoccaceae*, the *Oscillariae*, and the *Nostochaceae*, all as separate families. The *Florideae*, under their new name, form a separate order, Order V, as far from the *Oscillariae* as they are from the *Chlorosporae*, which are a subsection of yet another order, Order VII. There is no sign here of the *Oscillariae* forming a link between the *Chroöcoccaceae* and the *Florideae*, as Cohn had suggested in his physiological paper; the later system pays no attention to the relationships sketched in the paper

18 Ferdinand J. Cohn, "Conspectus familiarum cryptogamarum secundum methodum naturalem dispositarum auctore Ferdinand Cohn," *Hedwigia 11*(1872):17–20.
19 Andreas Caesalpino, *De plantis libri XVI* (1583); the significance of Caesalpino's choice of the reproductive organs, its effect on later botany, and its relation to Aristotelianism are discussed by Sloan, "John Locke, John Ray" (1972) (n. 17).
20 Cohn, "Conspectus familiarum cryptogamarum" (1872) (n. 18), 18; cf. Caesalpino, *De Plantis* XIII.

CLASS *THALLOPHYTA*

ORDERS	FAMILIES

I. *Schizosporeae:*	Six families: *Schizomycetae,* *Chroococcacea* *Oscillariaceae,* *Nostococcaceae* *Rivulariaceae,* *Scytonemaceae*

II. *Zygosporeae:*	Four families, including *Diatomaceae*

III. *Basidiosporeae:*	Two subdivisions, seven families in all, including *Agaricaceae*

IV. *Ascosporeae:*	Six families including Lichens

V. *Tetrasporeae:*	Eight families including *Florideae*

VI. *Zoosporeae:*	Six families including *Laminariaceae*

VII. *Oosporeae:*	Three subdivisions: *Levcosporeae,* *Chlorosporeae* *Phaeosporeae*

Figure 2.3. Ferdinand Cohn's arrangement of the class *Thallophyta*. Modified from Cohn, "Conspectus familiarum Cryptogamarum" (1872) (n. 18).

on the pigments. Cohn clearly did not intend that as a complete system, Darwinian or otherwise.

Cohn is a botanist *der alten Schrot und Korn*: a man of definitions, diaeresis, and beautiful Latinity, and of a loving enthusiasm for the species of the plant kingdom, like Linnaeus himself. His "new" species are appropriately named and elegantly defined: see the vivid description of *Spirulina versicolor*, quoted above, or the fungus that he found in 1854 attacking house-flies in autumn.

Finally, I give the diagnosis of the new species:
Empusa: Entophyta, e tribus constans cellulis quarum infima in insecti eijusdam

alvo evoluta, mycelii instar fortuosa, pare ramificata superne prolongatur in mediam . . . the characteristics of the single species of *Empusa* known at present I give as:

> *Empusa muscae* n.s. cellula mycelliformi 1/200' " lata, sursum in califormen 1/100' " latum excurrente, spora campanuliformi 1/200' " In muscae domesticae morbo quodam letali abdominis inflati cavitatem explet, apicibus cellularum claviformium post muscae mortem segmentorum membranes perforantibus, demum sporiferis, annulos semi-circulares molles albos componentibus.[21]

He has chosen the name *Empusa*, he says in a footnote, as it is the name of a Greek plague spirit, which, like the modern vampire, was supposed to suck the blood and sometimes took the form of a bluebottle fly.

In 1886 Cohn spoke to the fifty-ninth Versammlung deutscher Naturforscher und Aerzte, the body that had heard Emil du Bois-Reymond's "On the boundaries of natural knowledge" in 1872, and Nägeli and Haeckel in 1877. It is perhaps no accident that Cohn, the traditionalist, the believer in species and differentiation, should have talked about the Aristotelian soul as the principle of life: "Aristotle was right then, when he called the soul the principle of all life, and ascribed to plants only those powers of the soul which underlie the activities of nutrition and reproduction, while they lack the power of sensation and of thought."[22] All Darwin's well-known mechanisms that earlier showed promise of explaining everything about life, Cohn said, turned out to be dependent upon forces present only in the living. The cleft between life and death, organic and inorganic, had been effectively bridged by none of the new hypotheses.

Perhaps it was the intense Darwinian enthusiasm of the transformist-unitarian school that lessened Cohn's enthusiasm for Darwinism. Cohn's thinking was as internally consistent as that of Nägeli. The cleft between life and death, animate and inanimate, would have been inconsistent with the acceptance of spontaneous generation; the transformation of one species into another would have been equally inconsistent with his belief in "good" natural species. Nowhere, even in his discussion of the multilateral relationships of the *Oscillariae* and the *Chroöcoccaceae*, does he suggest that there was ever actual transformation of one into the other. And against the trend of his time, he believed that the *Bacteria* could be divided into species as good and distinct, and as natural, and as distinct as those of the other lower plants

---

21 Ferdinand J. Cohn, "Ueber Pilze als Ursache von Thierkrankheiten," *Jahres Ber. der Schlesisch. Ges. f. Vaterl. Cult 32* 43–48(1854):48.

22 Ferdinand J. Cohn, *Lebensfragen: Rede gehalten am 22 sept. 1886 in der 2 allgemeinen Sitzung der 59 Versammlung deutscher Naturforscher und Aerzte zu Berlin* (Berlin: Hirschwald, 1887), 19.

and animals. But Cohn was not blind to the difficulties, nor to the arguments against dividing the bacteria into species:

> The question may be asked, whether there really are species of bacteria, in the same sense as there are of the higher organisms. Even if one pays no attention to the transformism of those mycologists who think that everything can arise out of anything, and develop into anything else, one can still feel a sense of doubt at the sight of a mass of bacteria, that these innumerable minute bodies of every conceivable form could be sorted out into natural species. . . .
> Even so, I am convinced that the bacteria can be divided into species as good and as distinct as those of the other lower plants and animals; it is only their exceptionally small size, their habit of living with many different types in association with each other, and the variability of the types, that in many cases makes differentiation impossible for us with our present means.[23]

For a proper Linnaean botanist like Cohn, it was difficult to create a satisfying classification of the *Bacteria* into "good" species and genera, because they had apparently no special means of reproduction, nor could a complete life cycle be followed out by watching a single individual. The only means of classification available were those of morphology and physiology, secondary markers as Cohn called them in his *Hedwigia* paper of the same year.[24] Even morphology in many cases was wanting; one was left with a classification based on pigment or ferment production, a basis that he had already abandoned for the *Oscillariae* and *Florideae*. In spite of these doubts, Cohn produced a classification with four tribes and six genera, based mainly on morphological criteria (Figure 2.4). His definition of the group *Bacteria* was not this time in Latin, perhaps an indication of his uncertainty about it: "The *Bacteria* are chlorophyll-less cells, of round, oblong, cylindrical, pinched-in or crooked shape, which reproduce exclusively by cell division, and live either isolated or in families of cells."[25] But he stated his belief that though it might be too soon yet, separation of *Bacteria* into species would eventually be possible.

> I believe that it is not yet time to give the final decision on this question. But in any case, it is *not* true that the same germ (*Bacterien-Keim*) according to whether it grows in urine or in wine, turns the one alkaline and the other "ropy," or that the same bacteria form butyric acid here and transmit anthrax there, or produce a red fleck on a potato and a case of diphtheria in a human windpipe. It is far more likely that when we have better microscopes, many

23 Ferdinand J. Cohn, "Untersuchung über Bacterien," *Beiträge z. Biol. d. Pflanzen 1* 127–222(1875):133.
24 Cohn, "Conspectus familiarum cryptogamarum" (1872) (n. 18), 18.
25 Cohn, "Untersuchung über Bacterien" (1875) (n. 23), 136.

THE GROUP *BACTERIA*

TRIBE	GENUS
I. *Sphaerobacteria*	i. *Micrococcus* (Char. emend.)
II. *Microbacteria*	ii. *Bacterium* (Char. emend.)
III. *Desmobacteria*	iii. *Bacillus* (n.g.)
	iv. *Vibrio* (Char. emend.)
IV. *Spirobacteria*	v. *Spirillum* (Ehr.)
	vi. *Spirochaet* (Ehr.)

Figure 2.4. Ferdinand Cohn, classification of the *Bacteria*. From Cohn, "Untersuchung über Bakterien" (1875) (n. 24).

organisms that now look alike will be recognised to have morphological differences upon which primary species differences can be based.[26]

It was at this time that Robert Koch (Figure 2.5), *Kreis-physikus* from Kreis Wollstein, wrote Cohn a letter, dated 22 April 1876:

Hoch Geehrter Herr Professor!
Stimulated by your work on bacteria in the *Beiträge zur Biologie der Pflanzen*, and since I am able to get plenty of the necessary material, I have been working for some time on the contagium of anthrax. After many attempts, I have finally managed to follow the complete life-history of the anthrax bacillus. I think that my results are reliable, as the experiments have been repeated many times. Before I publish them, I would be most grateful if you, *hoch geehrten Herr Professor*, as the best expert on bacteria would give me your opinion on the findings.[27]

26 Cohn, "Untersuchung über Bacterien" (1875) (n. 23), 135.
27 Robert Koch, letter to Ferdinand Cohn, 22 April 1876; quoted in Bruno Heymann, *Robert Koch I Teil 1843–1882* (Leipzig: Akademische Verlag, 1932), 148. This and many other letters are quoted by Heymann in his well researched biography, which unfortunately was never finished. Heymann was born in 1871 in Breslau and died in 1943 in Berlin. He studied in Breslau and in Freiburg-im-Breisgau and became assistant to Carl Flügge at the Institute of Hygiene in Breslau. He followed Flügge to Berlin and took over from him the leadership of the research department. After the

Figure 2.5. The youthful Robert Koch: portrait drawing. (Photograph courtesy of the National Library of Medicine, Bethesda, Md.)

Although Cohn, as he said, did not expect very much of this completely unknown country doctor, he replied that it would give him great

Nazis came to power, he was no longer allowed to teach and in 1935 was stripped of his medical post. He worked and published in all fields of hygiene and microbiology. In 1931 the first part of his brilliant biography of Koch came out, but the

pleasure if the doctor would visit the Institute at Breslau and demonstrate his findings.[28]

The astonishing effect of this demonstration is reported in several published and unpublished accounts by eyewitnesses. Both Cohn, the director of the Institute of Plant Physiology, and Julius Cohnheim, director of the neighbouring Institute of Pathology, were deeply impressed.[29] Cohnheim, coming back from the demonstration, said to his assistants, "Nun lassen Sie alles stehen und liegen, und gehen Sie zu Koch – dieser Mann hat eine grossartige Entdeckung gemacht" (Drop everything and go and see Koch – this man has made a great discovery).[30] Like the two Institute directors, everyone else who could boast of having been present was equally admiring:

> As though they were dazzled by the beams of this blaze of genius, and completely under his spell, they readily lent him their aid. Cohnheim himself, an experimentor of the first rank, was not embarrassed to act as his assistant. . . . The high point of Koch's proceeding was when, watched by the bystanders, the anthrax bacilli sprouted from the spores, and Koch and Cohn . . . independently drew the event from the same preparation, and the drawings agreed perfectly with each other.[31]

It is not difficult to see why Cohn was so deeply impressed. Koch's demonstration of the life cycle of the anthrax bacillus, its growth from

manuscript of the second part disappeared. He was not arrested as a Jew until January 1943; but before he could be deported he died of tuberculosis of the knee joint in the Jewish Hospital in Berlin. Biographical material is from *Die neue Deutsche Biographie.*

28 Cohn, cited by Heymann, *Robert Koch I* (1932) (n. 27) 150–151.

29 Heymann, *Robert Koch I* (1932) (n. 27) 149–153. Heymann's account of the relationship between Koch and Cohn is detailed and well documented. He quotes Koch's letters to Cohn, for the most part whole letters with no lacunae (in the possession in 1932 of Cohn's nephew, G. Landberg of Berlin, p. 148), also from Koch's personal diary (in the possession of the Institut für Infektionskrankheiten, "Robert Koch" in Berlin, p. 129); from Koch's Wollstein notebooks (also in the Institute, p. 137), and from Cohn's Daybook for the Institute of Plant Physiology, Breslau (in the possession of that Institute, p. 150). A more recent English-language biography of Koch—Thomas D. Brock, *Robert Koch: A Life in Medicine and Bacteriology* (Madison, Wisc.: Science Tech, 1988)—still uses Heymann as a source for Koch's letters and diary. Where the points I have made are by Heymann, I have noted this. Many of them support from alternative evidence opinions that I had already formed on other grounds, particularly, as I shall point out below, one point that is pivotal for my argument: Koch's interest in species and specific differences.

30 Willy Kühne, "Zur Erinnerung an Julius Cohnheim," in E. Wagner, ed., *Gesammelte Abhandlung von Julius Cohnheim* (Berlin: Hirschwald, 1885), xxxv, quoted by Heymann, *Robert Koch I* (1932) (n. 28), 151. Heymann noted that Kühne had the date of the event wrong, but he himself misprinted 1855 for 1885, the date of Kühne's publication.

31 Heymann, *Robert Koch I* (1932) (n. 27), 151.

spores, and the conditions under which spore formation took place, as well as the demonstration of the relationship of the bacilli and the spores to the disease – all this answered the questions that Cohn had asked, as he would have liked them answered: the large bacilli did not vary outside prescribed boundaries, they had a well-defined life cycle, they produced a well-defined disease – they belonged to a single species.

Cohn's own earlier attempts to demonstrate an organism in cholera stool, which he reported in 1867, had not been successful, but he had been able to pinpoint the problems involved. The most important of them from his point of view, which was that of the Linnaean botanist, was the specific identity of a given organism, and the limits of the variability at different stages of its life cycle. According to the botanist Hallier, whose opinion had been the occasion for the discussion, the yeast of beer, the ergot of rye, the rust of wheat, the *Urocystis* of rice and other swamp plants, and all parasites of human mucous membrane and hair – in short, all possible forms of fungus – were identical with the cholera organism. Its life history included cystic, yeast, and *Penicillium* stages, all of which Cohn had placed in different families, not to mention different species. Cohn called this "pure fiction unworthy of the respect of botanists." But, he said, physicians had allowed themselves to be imposed on by Hallier's confident assertions.[32] Although he himself had not succeeded, he knew what was needed, and he immediately knew that Koch had done it: "Koch came to my Institute on the 30th April, and I can pride myself that I recognised in the first hour that he was an unsurpassed master of scientific research."[33]

Koch in turn chose Cohn for his demonstration because "almost the only botanist who was seriously interested, and who tried to order the bacteria systematically, was Ferdinand Cohn, while the others would have nothing to do with the separation of Bacteria into well-defined species."[34] Cohn's support encouraged Koch to go on in the way he had begun, and also put him in touch with high-level university medicine: he made contact with Julius Cohnheim, director of the Institute of Pathology at Breslau, and with Carl Weigert, Paul Ehrlich's cousin, then Cohnheim's assistant. Weigert was developing staining methods

32 Ferdinand J. Cohn, in discussion following a paper by Schneider, "Ueber Hallier's Cholerapilze und dessen Entwicklung," *Jahresbericht d. Schlesisch. Ges. f. vaterl. Cult.* 45 114–125 (1867):119–125, esp. 123.
33 Cohn, cited in Heyman, *Robert Koch I* (1932) (n. 27), 150–151.
34 Robert Koch, "Antrittsrede in der Akademie der Wissenschaften am 1 Juli 1909," in Georg Gaffky, W. Pfuhl, and J. Schwalbe, eds., *Gesammelte Werke von Robert Koch* (Leipzig: Thieme, 1912), v. 1, 1–4 (p. 1).

for bacteria in animal tissues using the new aniline dyes, a point that Koch mentions in his diary.[35] Koch returned home after his four days in Breslau glowing with happiness and full of new self-confidence, to prepare his paper for publication in Cohn's journal.[36]

The importance of this support may be gauged by the result of Gregor Mendel's very similar approach to that of Nägeli.[37] Mendel, whose work implied discontinuity and sharp-edged differences, tried to interest Nägeli, the believer in seamlessness and unity, and was discouraged forever by the problems Nägeli raised. Koch's choice was luckier, or more insightful: Koch himself was probably aware of the differences between these two schools of botanical thought.

The next stage in Koch's work makes even more explicit his consciousness of his position on the problem of specific differences. Koch's enormous improvements in microscopical technique, his new dry, stained preparations, and his microphotography date from this period, 1876–1878. By 5 March 1877, he was able to send Cohn some negatives of bacterial preparations that had been taken using an improved Seibert und Krafft optical system. In the accompanying letter, after describing and identifying each of the photographs in turn, he writes:

I hope that you, *sehr geehrter Professor*, will be convinced by these pictures, that I have managed by these methods to fix and preserve Bacteria of various different species [*Arten*] so that they can be duplicated by photography at any time. . . . My plans for future work are as follows. Last summer and autumn I made a collection of preparations, which contains a mass of different characteristic Schizophytes . . . and first I shall photograph everything that seems to me a proof of the richness of the Schizophyte flora in different species [*Arten*]. Unfortunately some of these species [*Arten*] were put up in Canada balsam, as I found out too late that this would not do for photography.[38]

And on 17 March 1877 he writes,

Do you think, *geehrter Herr Professor*, that the photographs will do as proof of the existence of different species [*Artverschiedenheit*] of bacteria? The longer I work on the Schizophytes, the more I feel my ignorance of the related lower plants. To improve my knowledge of botany at least a little, may I respectfully request you to help me by telling me of suitable books. Perhaps through your kindness

35 Heymann, *Robert Koch I* (1932) (n. 27), 154.
36 Heymann, *Robert Koch I* (1932) (n. 27), 154.
37 Peter Bowler, *The Mendelian Revolution: The Emergence of Hereditarian Concepts in Modern Science and Society* (Baltimore, Md.: Johns Hopkins, 1989) 84–85. Some of the Nägeli-Mendel correspondence has been published in Curt Stern and E. R. Sherwood, *The Origin of Genetics: a Mendel Sourcebook* (San Francisco, Calif.: Freeman, 1966).
38 Koch, letter to Cohn, 5 March 1877, quoted in Heymann *Robert Koch I* (1932) (n. 27), 184.

I could get some of the larger books on the *Algae*, ones with pictures in, out of the University library? [39]

As Koch's biographer Heymann remarks, through all the discussion of optical systems and other technical matters in these letters, Koch never loses sight of the goal, which is the solution of the problem of unity or specificity of the bacteria.[40]

Each one of the technical problems attacked by Koch in his paper of 1877, "Procedures for the investigation, preservation and photography of bacteria," was part of this greater problem. The bacteria were in continual motion, they grew mixed swarms, and the same object could never be seen by two different observers: drawings were outlines of forms in the mind's eye, whose relation to the object was not clear. Koch's procedure was to smear the bacteria out on a microscope slide, to separate them, to dry the smear to keep the organisms still for observation, to stain them with the aniline dyes to make their appearance clearer, and finally to photograph them instead of making drawings (Figure 2.6).[41]

His comments on the published photographs make the position even clearer than do his letters to Cohn:

In the same putrid blood, then, we can distinguish large, medium, small and very small *Micrococci*. Each form is found in a more or less circumscribed group, with a few *Micrococci* of other types around the edges, as is only to be expected with this method of preparation. However, these can easily be distinguished from those that form part of the group. Undoubted intermediate forms between these different groups are not present.[42]

Koch took the opportunity to attack Nägeli, whose book *Die niederen Pilze* of 1877 had just appeared. One of Nägeli's diagrams shows, in simplified form for laymen, the shapes of different kinds of *Spaltpilze*; he drew a wavy line made up of linked cells, implying that the spiral forms consisted of cords of *Cocci* (see Figure 1.5).[43] Koch had been looking for evidence of this but had never found it:

39 Koch, letter to Cohn d. March 17, 1877, quoted in Heymann, *Robert Koch I* (1932) (n. 27), 186.
40 Heymann, *Robert Koch I* (1932) (n. 27), 186.
41 Robert Koch, "Verfahren zur Untersuchung, zum Konservieren und Photographieren der Bakterien," *Beitr. z. Biol. d. Pflanzen*, 2, 399–434 (1877); and in *Ges. Werke* (n. 34) *1*, 27–60, and Plates II and III; for a more detailed description of the techniques, see Brock, *Koch* (1988) (n. 29), 70–113.
42 Koch, "Untersuchung zum Konservieren und Photographieren der Bakterien" (1877) (n. 41), 43–44.
43 Carl von Nägeli, *Die niederen Pilze in ihren Beziehungen zu den Infektionskrankheiten und der Gesundheits-pflege* (Munich, Oldenbourg 1877), 4, fig. 2.

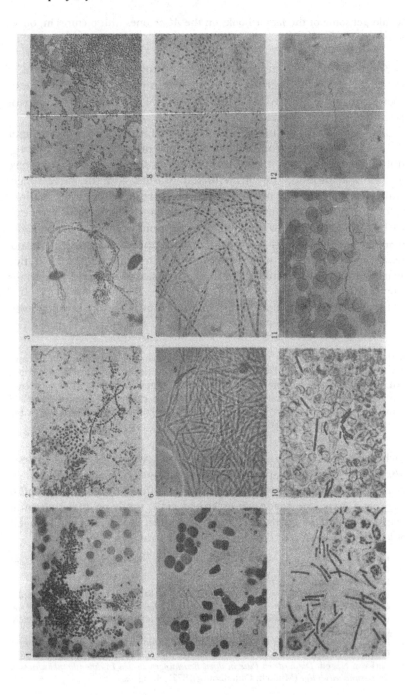

I have kept a particular watch for this from the beginning of my investigations,
. . . [for] the disintegration of *Bacilli* into *Micrococci* and vice-versa, the formation
of rods from *Micrococci*, according to whether this is confirmed or shown to be
an error, our view of the bacteria must be completely different. This, then, is
the most important question in bacteriology. It must be solved if any understanding is to be reached among bacteriologists, and every one of them should
do all he can to solve it. My experience is based on thousands of dried preparations . . . and contradicts Nägeli's observations.[44]

It is a measure of the respect that Koch's new methodology had so
suddenly earned that the editor of the *Deutsche medizinishe Wochenschrift*

44 Koch, "Untersuchung zum Konservieren und Photographieren der Bakterien"
    (1877) (n. 41), 47

Figure 2.6 (*opposite*). Robert Koch's natural history and classification of micro-
organisms in photographs. We must fix things, at least in a preliminary way,
says Koch, and photographs are an excellent way to do it. They will get rid of
numerous *wilde Schössling*, shots into the air. *1*. Sheep's blood, kept at 8–10° C
for four days, aniline brown stain. Koch points out that this shows four types
of micrococcus, with no intermediate forms. *2*. Same after four weeks: the
groups of cocci have disappeared, but there are now cocci in chains. *3*. A
scraping from a tongue: shows chains of micrococci, with spaces between each
two or three. At the end of the chain, a group of very small cocci are forming
a thick *Zoögloea* united by slime. *4*. This organism was found only once, in the
spring of 1877. It was forming a fine pellicule on the surface of water, which
also contained a *Gomphonema* embedded in slime. Koch comments that if the
bacteria are left in their natural arrangement and not mixed up, no intermediate forms appear. Each species forms a separate colony, growing from a centre outwards, finally meeting and pressing against other forms. They will finally
mix, since they are mobile. *5*. *Bacillus anthracis* in mouse blood. *6*. Same after
standing twenty-four hours at 18–20° C: now a thick felted mass of anthrax
bacilli. *7*. and *8*. Five-year-old dried spores of anthrax that successfully infected
experimental animals. Grown in aqueous humour. Three groups of organisms
visible: bacilli with spherical spores, bacilli with bubble-like spores, and a separate group of mixed *B. anthracis* and *B. subtilis*, which form long threads. *9*.
Spleen substance with *B. anthracis* stained aniline brown. *B. anthracis* show spot-
ting with light spots; Koch notes that this distinguishes it from *B. subtilis*, which
otherwise looks very like it. They differ not only physiologically, but also in
appearance – they are quite different organisms. *10*. Blood from a two-day old
cadaver. Organisms are *Coccobacteria septica* (Billroth) or *Streptobacteria gigas*.
Later one finds thinner, smaller bacilli and later still micrococci, *B. termo* (Duj.)
&c. Failed injections with anthrax are often due to mistaking these for *B. anthracis*. Koch says these two photographs allow him to remark on Nägeli's claim
that staff shapes and threads are made by joining up bacilli and spirilli, and
that bacilli will break up into micrococci. He emphasizes that he has been
keeping his eyes open for this from the beginning and has never seen it. This
is a matter of principle that must be solved if bacteriology is to get onto a firm
footing: bacteriology stands or falls on its truth or falsity. His experience, says
Koch, contradicts Nägeli's. *11* and *12*. Spirochaetes from cases of recurrent
fever. From Koch, "Verfahrung zur Untersuchung zum konservieren und photographieren der Bakterien," *Gesammelte Werke*, v. 1, Plate III (n. 34) (1877).

asked him to review Nägeli's book.[45] His review is judicious and careful, but his opinion was given more shortly in a letter to Cohn: "I have read Nägeli's book. . . . I have seldom come across any book which contained so much error and nonsense, and nothing at all which contributes to our knowledge." [46]

Koch's next advance was the use of the new illumination system developed by the physicist Ernst Abbé for the Jena firm of Carl Zeiss. The improvement in Koch's technology that this made possible was the visualization of bacteria in animal tissues, free from the interfering shadows of the tissues themselves. The result was the papers on the aetiology of wound infections of 1878, in which Koch again contrasted his own and Cohn's findings with Nägeli's. This time, he added a new dimension to his definition of the bacterial species, that of the species of disease with which it is associated: the disease and the bacteria define each other:

One fact was so prominent that I must regard it as constant, and as it helps to remove most of the obstacles to the admission of the existence of a *Contagium vivum* for traumatic infectious disease, I look on it as the most important result of my work. I refer to the differences which exist between the pathogenic bacteria, and the constancy of their characters. A distinct bacteric form corresponds, as we have seen, to each disease, and this form always remains the same, however often the disease is transmitted from one animal to another. Further, when we succeed in reproducing the same disease *de novo* by the injection of putrid substances, only the same bacterial form occurs which was before found to be specific for that disease. In addition, the differences between these bacteria are as great as could be expected between particles which border on the invisible.[47]

In Koch's experiments, the body of the animal was itself an apparatus for producing a pure culture of the organism, and at the same time an indicator of the properties of the species cultured: the disease was one of the *differentia* of the species, as constant as its length and breadth.

Finally, Koch demolishes another adversary. The French bacteriologist Davaine had reported that when bacteria were serially transferred from one experimental animal to the next, the injections became increasingly lethal. Davaine's explanation was that the organisms became

45 Heymann, *Robert Koch I* (1932) (n. 27), 240. Koch says this in a letter to Cohn.
46 Koch, letter to Cohn, 11 November in Heymann, Robert Koch I (1932) (n. 27), 218.
47 Robert Koch, *Untersuchungen über die Aetiologie der Wundinfektionskrankheiten* (Leipzig: Vogel, 1878); and in *Ges. Werke* (n. 34), v. 1, 60–108, and plates IV and V, 101; translated by W. Watson Cheyne, *Investigations into the Etiology of Traumatic Infective Disease* (London: New Sydenham Society, 1880), 65. This passage is a statement of the principles known as Koch's postulates. They are discussed further in Chapter 3.

increasingly virulent.[48] Koch found it impossible to accept this: it did not accord with his own conception of the strict constancy of the properties of a bacterial species, and he attributed it instead to the overgrowth of a single more virulent species in the animals' blood. "We do not need the magic wand of adaptation and inheritance," he says.[49] It obviously, and distastefully, reminded him of Nägeli.

48 Koch, letter to Cohn, 11 November 1877 in Heymann, *Robert Koch I* (1932) (n. 27), 218; Koch quotes Davaine's work from the German summary that appeared in *Med. Jahrb.*, 166, p. 174 (Koch's n. 1, 69); the first part of Koch's article is a thorough survey of the literature, including the French, where it was available in German summary.
49 Koch, *Aetiologie der Wundinfectionskrankheiten* (1878) (n. 47), 105. Watson Cheyne translates *Anpassung und Vererbung* as "natural selection," pp. 71, 72.

3

The Dominance of Specificity: Koch and His Adversaries

Robert Koch's views on specificity and constancy in micro-organisms and his belief that there was a constant and invariable relationship between a bacterial species and the disease it caused were born into a world in which the established hygienists were profoundly opposed to any such idea. Nonetheless, with the backing of the state, Koch's views changed the nature of hygiene in Germany and forced all opposing thinkers into a defensive position.

Max von Pettenkofer (Figure 3.1) was a leading figure in the German sanitary establishment.[1] It was he who had used his influence at the Bavarian court to have hygiene installed as a recognized subject in the three Bavarian universities and who had pressured the Bavarian government to build the Institute of Hygiene in Munich, which opened in 1878. His was a powerful school of students and admirers, who, like those of Robert Koch at a later date, stood at the head of about thirty similar institutes in Germany and elsewhere. The group edited two journals, the *Zeitschrift für Biologie*, founded in 1865, and the *Archiv für Hygiene*, split off from it in 1883. The flow of Munich thinking from biological unitarianism through to the sanitary movement, with its emphasis on statistical data and economics, can be traced in the two journals. The editors of the *Zeitschrift für Biologie* introduced their reborn journal in 1883 with a vow to take a unified view of the plant world, especially since evolution had made transformation a more important aspect of biology than constancy of form. And in the foreword to the *Archiv für Hygiene*, the editors wrote that health is the highest of human rights, and in practice it is an economic good. One of the goals of their journal was to present research on the economic aspect of hygiene –

1 On the sanitarian movement at its origin in Britain, see Michael W. Flinn, Introduction to *Report on the Sanitary Condition of the Labouring Population of Great Britain by Edwin Chadwick* (Edinburgh: University Press, 1965), 1–73; Michael J. Cullen, *The Statistical Movement in Early Victorian Britain: The Foundations of Empirical Social Research* (Hassochs, Sussex: Harvester, 1975), 53–64; Anthony S. Wohl, *Endangered Lives: Public Health in Victorian Britain* (London: Methuen, 1983), 1–9, 166–204.

Figure 3.1. Max von Pettenkofer, portrait, n.d. (Photograph courtesy of National Library of Medicine, Bethesda, Md.)

thus emphasizing the importance of statistical data for decision-making.[2]

Pettenkofer's work as a hygienist was akin to that of the British sanitary movement. He emphasized water supplies, sewerage, ventilation, and the removal of fermenting and putrefying nuisances from the city streets, and the collection of data relating to health and economic fac-

2 The Munich professors L. Buhl, Max von Pettenkofer, Ludwig Radlkofer, and Carl von Voit were the founding editors of the *Zeitschrift für Biologie* (1865). In 1883 the *Zeitschrift* split into two: a new *Zeitschrift für Biologie*, edited by the physiologists Willy Kühne and Carl von Voit, and the *Archiv für Hygiene*, edited by J. Forster, Friedrich Hofmann, and Max von Pettenkofer.

tors.[3] Pettenkofer's system of thought was consonant with liberal bour-
geois theories of the state. Health was a form of property or capital: by
investing in prevention, the hygienically educated individual would cut
the costs of sickness. Pettenkofer's calculations were based on local statistics, and his programme pressed for local urban improvement in
cleansing, water, and sewerage; he argued against any attempt by the
newly formed central government of the Reich to enforce a general code
of health legislation. Persuaded by Pettenkofer's localist views, the Reich
and the Prussian health authorities relaxed quarantines by the early
1870s, ceased to control cross-border movements of traffic on roads and
rivers, and concentrated more on the need for street-cleaning and data
collecting, and "keeping the soil free from impurities," an important
feature of the Pettenkofer localist system. The effect was, however, more
pronounced in the liberal mercantile centres such as Hamburg, with its
history of independence as a Hanseatic port, and much less so in the mil-
itaristic state of Prussia, famous for its efficient bureaucracy.[4]

Pettenkofer's theory of the transmission of epidemic cholera came
from the work of the British statistician Robert Farr. It was based on
the statistical finding of seasonal changes in cases reported, which Pettenkofer related to local changes in temperature, the level of ground-
water, and height above sea level. Of these factors, groundwater was
the most important; it had to be kept clear of sewage contamination.
Paradoxically, Pettenkofer did not believe that a piped water supply
could transmit the disease.[5]

The Imperial Bureau of Health had been set up in 1876 with the
mandate to collect statistics and disseminate information on public
health, and to prepare statistical material for public health legislation.
Its small staff consisted of a director, a senior physician and a senior
veterinarian, a secretary, and a laboratory technician. The medical ap-

3 Henry E. Sigerist, ed. and trans., " 'The value of health to a city,' Two lectures deliv-
   ered in 1873 by Max von Pettenkofer," *Bull. Hist. Med.* 10(1941):473–503, 593–613.
4 Richard J. Evans, *Death in Hamburg: Society and Politics in the Cholera Years 1830–1910*
   (Oxford: Clarendon Press, 1987), 263–264. Evans, a political historian, deals brilliantly
   with the relationship of Pettenkofer, Koch, society, and state in the newly united
   Second Reich.
5 Edgar E. Hume, *Max von Pettenkofer: His Theory of the Etiology of Cholera, Typhoid Fever
   and Intestinal Disease: A Review of his Arguments and Evidence* (New York: Hoeber, 1927).
   For the sources of the Pettenkofer theory, see Margaret Pelling, *Cholera, Fever and
   English Medicine* (Oxford: Oxford University Press, 1978), chap. 4, on Justus von Liebig
   and fermentation; John M. Eyler, *Victorian Social Medicine: the Ideas and Methods of
   William Farr* (Baltimore, Md.: Johns Hopkins University Press, 1979), 97–122, on Farr's
   methods and the elevation law. For a carefully argued comparison and critique of the
   Koch and Pettenkofer–Farr theories of cholera transmission by a Koch supporter, see
   Carl Flügge, *Micro-organisms with Special Reference to the Etiology of Infectious Disease,* trans.
   William Watson Cheyne, from the second edition of Flügge's *Fermente und Mikropar-
   asiten* (London: The Sydenham Society, 1890), 415–470 (especially pp. 467–470); on
   Flügge, see n. 30.

pointees were all men with positions in the Prussian civil and military health bureaucracy: *Oberstabs-und-Regimentsarzt Sanitätsrat* Dr. Heinrich Struck, the first director; the Prussian *Medizinalrat* Professor Dr. Finkelnburg of Berlin University, a statistician; and the Prussian *Departmentsthierarzt* Professor Friedrich Heinrich Roloff of Halle University for veterinary matters.[6] Pettenkofer was first invited to head the bureau, but he did not want to leave Munich and agreed only to act as adviser. The bureau's initial function, however, matched his views on the subject: it was to collect sanitary statistics and to prepare for health legislation on the basis of the data collected.[7]

The bureau was soon expanding its work. By 1879, there was also a laboratory for food inspection, directed by Professor Eugen Sell of Berlin University, and a public health laboratory directed by Professor Gustav Wolfhügel, who had been Pettenkofer's first assistant at the Institute of Hygiene in Munich. In addition, there was a body of outside advisers from various parts of Germany, one of whom was Max von Pettenkofer himself.

In 1880, both Robert Koch and Ferdinand Cohn were appointed advisers. The same year, the senior physician retired, and Dr. Koch (Figures 3.2 and 3.3), *Kreisphysikus* from *Kreis* Wollstein replaced him. A bacteriology laboratory was then added, and Koch was placed at its head, with the "entry-level" civil rank of *Regierungsrat*.[8]

Friedrich Loeffler (Figure 3.4), Koch's young assistant (he calls Koch *unser aller Lehrer und Meister,* teacher and master of all of us) notes that the tiny room in which they were set up was much smaller than the one used by the other senior men. Koch started work and immediately produced some cooked potatoes on which the fabulous blood-red *Micrococcus prodigiosus* had performed its miracle of the body and blood. Spellbound, Loeffler, *Assistentarzt I Klasse* of the Prussian army, asked to be *kommandirt,* or seconded, to Koch, as did *Assistentarzt I Klasse* Georg Gaffky. There they sat, Loeffler and Gaffky, one on each side of Koch, and every day new wonders of bacteriology arose before their astonished eyes. Since there were now three of them, they got a bigger room. As Loeffler wrote, bacteriology burst its constricting sheath and conquered the space that was its due.[9]

The problem of bacterial specificity was also under discussion in

6 *Das Reichsgesundheitsamt 1876–1929: Festschrift herausgeben vom Reichsgesundheitsamte aus Anlass seines fünfzig-jährigen Bestehens* (Berlin: Springer, 1926), 2–6.
7 Evans, *Death in Hamburg* (1987) (n. 4), 239–240.
8 Bruno Heyman, *Robert Koch I Teil 1843–1882* (Leipzig: Akad. Verlag, 1932), 288.
9 Friedrich Loeffler, "Robert Koch zum 60 Geburtstage," *Deutsche med. Wschr.* 29(1903): 937–943. In 1904, Gaffky was to succeed Koch as director of the Institut für Infektionskrankheiten Robert-Koch in Berlin; when Gaffky retired in 1913, Loeffler succeeded him, only to die two years later. See Heyman, *Koch* (1932) (n. 8), 295.

Figure 3.2. Friedrich Loeffler, portrait photograph. (Courtesy of National Library of Medicine, Bethesda, Md.)

France during the 1880s, but the discussion there bore almost no relationship to that in Germany. The positions marked out as opposite poles in Germany were irrelevant in France. But for the Berlin group, the work of Louis Pasteur was very close to the transformism of their enemies at home: they classed Pasteur with Nägeli.

Figure 3.3. Robert Koch, portrait, 1887. (Photograph courtesy of National Library of Medicine, Bethesda, Md.)

It did not seem like that in France. Emile Duclaux, who succeeded Louis Pasteur as director of the Institut Pasteur, felt that the whole idea of bacterial specificity was Pasteur's invention.[10] Pasteur was not part of the "naturalist tradition," and morphology and nomenclature there-

10 Emile Duclaux, *Traité de Microbiologie*, 4 v. (Paris: Masson, 1898), v. 1, 18–20.

Figure 3.4. Robert Koch, portrait with his second wife, on a visit to Tokyo in 1903. (Photograph courtesy of National Library of Medicine, Bethesda, Md.)

fore were of little concern to him.[11] Indeed, that was one of the things that the properly trained botanist Ferdinand Cohn found particularly irritating about Pasteur:

> [He] speaks sometimes of *végétaux cryptogames microscopiques*, sometimes of *animalcules*, or of *champignons* or *infusoires*, and without any proper differentiation calls the same things *torulacées, bactéries, vibrioniens, monades*, which elsewhere are called *Mycoderma, Mucor, mucédinées* or yeasts (*lévures*). . . . With sovereign arbitrariness, Pasteur sets himself above the rules of botanical nomenclature.[12]

How could he think specifically if he used names so loosely? Pasteur classified his organisms broadly in terms of the diseases they produced, although many of the diseases in which he was interested – purulent infections, septicaemia, bacteraemia and so forth – could be caused by a number of different *contagia*. The transformation of species, and spontaneous generation, he rejected on experimental grounds. Pasteur's experimental disproof of spontaneous generation took place against a background in which Darwinian transformism was a politico-theological doctrine allied with republican and anti-clerical radicalism.[13]

In 1888, however, Pasteur read to the Académie des Sciences a paper in which he claimed to have found that by a change in culture conditions, the virulence of an organism might be diminished. Inoculation of an animal with this attenuated microbe would produce a mild illness, which would protect it against infection with a fully virulent form of the same microbe. He had found it difficult at first to accept the change in behaviour of the organism, since it flew in the face of everything he believed about the constancy of life forms. This was not true transformation of species, but it was transformation of a kind. He argued that the fundamental nature of the organism remained the same. The disease caused by the transformed organism might be mild or severe, but it was not really a different disease.

How, then, could it be argued that Pasteur did not believe in specificity? The answer to that question lies with the German workers. In Germany, arguments about specificity had polarized around the posi-

11 Gerald J. Geison, "Pasteur, Louis," in Charles C. Gillespie, ed., *Dictionary of Scientific Biography* (New York: Scribner's, 1974), v. 10, 350–416; Geison, personal communication.

12 Ferdinand J. Cohn, "Ueber Pilze als Ursache von Thierkrankheiten," *Jahres Ber. der Schles. Ges. f. Vaterl. Cult. 32*(1854):43–48 (p. 48).

13 John Farley and Gerald L. Geison, "Science, politics and the spontaneous generation debate in XIXth century France," *Bull. Hist. Med. 48*(1974)161–198; Farley, *The Spontaneous Generation Controversy from Descartes to Oparin* (Baltimore, Md.: Johns Hopkins, 1977), 92–120.

tions taken by Koch, on the one hand, who stood for the strictest one-to-one monomorphic relationship of the disease and its organism, and Nägeli, who stood for a broader unity and transformism. Either of them could in theory have had grounds for criticizing Pasteur, but it was Koch, backed by the German state, who attacked Pasteur as the embodiment of all that he despised.[14] In 1881, shortly after the appearance of Pasteur's work on the attenuation of virulence and immunization, the first volume of the *Mittheilungen aus dem Kaiserlichen Gesundheitsamte*, the Communications from the Imperial Bureau of Health, was published.

The first article was Koch's, a *résumé* of all the powerful new technology that was to be his legacy to bacteriology. It included the staining methods that he had derived from the work of Paul Ehrlich on dyes, and the equally important solid-medium technique for the growth of pure cultures of bacteria. On the solid nutrient medium, originally a slice of cooked potato, the organisms formed separate colonies, each of one type only, rather than the mixed swarms in which they grew in broth. Much of this article was devoted to an attack on Louis Pasteur, on the basis of Koch's confidence in the new pure cultures. In Koch's eyes, Pasteur's claims for a pure culture were worthless: he was still using wet cultures, in which the bacteria to be studied were only very roughly separated out by the methods of serial culture or of doubling dilution with a Pasteur pipette. For the same reason, the mixed cultures of Nägeli and his student Hans Buchner allowed them to imagine that one species could be transformed into another by changing conditions in the culture. Koch repeated, even more confidently, his statement about the division of bacteria into good species.[15]

Koch's two young lieutenants, Gaffky and Loeffler, seem to have been delegated to examine and refute different parts of Pasteur's work. The enthusiasm of their attack suggests a Prussian cavalry charge, sabres drawn, under Koch's command. Appropriately enough, since the Franco-Prussian war was still fresh in memory, it was against a French enemy. Theirs was a well-disciplined troop:

The investigations underlying this [Gaffky's own] are carried out from the point of view of *Herr Regierungsrat* Dr Koch, under his leadership and control, in the

14 On the relations between Koch and Pasteur, see H. H. Mollaret, "Contribution à la connaissance des relations entre Koch et Pasteur," *NTM: Schriftenreihe f. Naturwissensch. Tech. u. Med.* 20(1983):57–65; K. Codell Carter, "The Koch–Pasteur dispute on establishing the cause of Anthrax," *Bull. Hist. Med.* 62(1988):42–57.
15 Robert Koch, "Zur Untersuchung von pathogenen Organismen," *Mitth. aus dem K. Gesundheitsamte* 1 Art 1 (1881); and in Georg Gaffky, E. Pfuhl, and J. Schwalbe, eds., *Gesammelte Werke von Robert Koch*, 2 v. (Leipzig: Thieme, 1912), v. 1, 112–163.

laboratory of the Imperial Bureau of Health. The more the reader sees between the lines the influence of my esteemed teacher, the more I can hope that my work is justified.[16]

The position that Gaffky attacks is shared by Pasteur, Nägeli, Nägeli's student Hans Buchner, and a certain Albrecht Wernich, a Berlin physician active in public health, who had just published a book. Gaffky sets his stage by quoting Nägeli's statement that the fungi of the miasmas develop under the influence of their surrounding conditions, which range from those of simple putrefaction, outside the body, to those of disease in the body. They remain in a pathogenic state if they are continuously transmitted from the sick to the well, but they revert to ordinary fungi if they are forced to live outside the body.[17] Gaffky then illustrates the implications of this view by quoting from Wernich, who had suggested that the organisms of cholera, typhoid, and the dysenteries can arise from the normal putrefactive gut organisms; hence his reintroduction of the old name "putrid fever."[18] The third proponent of these views is Hans Buchner (Figure 3.5), who recently, says Gaffky, has tried to support Nägeli's position experimentally by transforming the harmless hay-bacillus into *B. anthracis* (Figure 3.6a-c) by serial cultures.[19] *Es liegt auf der Hand*, he says, how perfectly Pasteur's law of the progressive increase of virulence by the passage of septicaemic blood fits into Nägeli's frame. Gaffky goes on to quote Nägeli's most famous passage:

If my opinion on the nature of the fission-fungi is correct the whole species in the course of generations takes on different and changeable morphological and physiological forms, which in the course of decades produces sometimes souring of milk, sometimes butyric fermentation in sauerkraut, sometimes the aging of wines, sometimes the putrefaction of proteins . . . and sometimes diphtheria, sometimes typhoid, sometimes recurrent fever, sometimes cholera. . . . When

16 Georg Gaffky, "Experimentelle erzeugte Septicämie, mit Rücksicht auf progressive Virulenz und accommodative Züchtung," *Mitth. aus dem K. Gesundheitsamte*, 1 Article no. 3 (1881), 1–54 (p. 4).

17 Carl von Nägeli, *Die niederen Pilze in ihren Beziehung zu den Infektionskrankheiten und der Gesundheitspflege* (Munich: Oldenburg, 1877), 92.

18 L. Albrecht Wernich, *Die Entwicklung der organisirten Krankheitsgifte, nebst einem offenem Briefe an Herrn Professor Klebs in Prag* (Berlin: Reimer, 1880), 89.

19 Hans Buchner, *Ueber die experimentelle Erzeugung des Milzbrandcontagiums aus den Heupilzen, nebst Versuchen über die Entstehung des Milzbrands durch Einathmung: Der Medizinischen Facultät der kgl. Ludwig-Maximilians-Universität München. Pro venia legendi, vorgelegt von Dr. Hans Buchner, k.b. Assistenzarzt* (Munich: Straub, 1880); Buchner, "Ueber die experimentelle Erzeugung des Milzbrandcontagiums aus den Heupilzen I. u. II. Mittheilung," and Buchner, "Beiträge zur Morphologie der Spaltpilze," in Carl von Nägeli, *Untersuchungen über niedere Pilze aus dem Pflanzenphysiologische Institut in München* (Munich: Oldenbourg, 1882), 140–185, 186–204, 205–224, and figs. 3–9, reproduced here in Figure 6a, b, and c.

Figure 3.5. Hans Buchner, portrait photograph. (Courtesy of the Institut der Geschichte der Medizin, Munich)

one form of this fungus-species gets into a new medium, it adapts itself progressively to the new conditions, and becomes more characteristic the longer it is in the same medium . . . it becomes more or less weakened, and may even lose all its characteristic properties, if it does not pass directly from one patient to another.[20]

20 Nägeli, *Die niederen Pilze* (1877) (n. 17), xii; on the Pettenkofer–Farr "ground-water theory," see Nägeli's Chapter 8, "Hygienische Eigenschaften des Bodens."

Only a few years later, the passage could be used to raise a laugh from certain audiences, and Nägeli's name had become a taunt. But at this time it was a striking statement of a powerful opposing theory. It was a statement in perfect accord with the reigning localist sanitary theory in Munich, according to which the normal soil organisms under certain conditions of temperature and dryness might become pathogenic and start a local epidemic. In fact, Nägeli expressly backs the "groundwater theory" in his book, with a detailed exposition aimed at public health workers. All this seemed very close to Pasteur's claim that organisms weakened by a culture medium might be the cause not of disease, but of immunity to disease, and might be restored to virulence by serial passage in guinea-pigs. There were other parallels: neither Pasteur nor Nägeli used Koch's new techniques. Both used the old wet cultures and live, unstained preparations. Neither was interested in the details of morphology, or the definition of species. It was easy for the Berlin group to interpret Pasteur's changes in virulence, and Buchner's preposterous transformed hay-bacillus, as simply the result of the over-growth of another organism.[21] (For Koch's argument against the transformation of *B. subtilis* into *B. anthracis*, see his photographs in Figure 2.6, 1–12).

Gaffky concludes his attack triumphantly:

The basic conditions for unassailable [*unanfechtbar*] results are work with pure cultures, and constant microscopic control.... On these grounds, and supported by numerous personal observations, we make the following *Satz*: The pathogenic fungi are specific beings, which arise only from their own kind, and in turn produce only their own kind.[22]

Pasteur was both surprised and insulted by this criticism.[23] It is possible that he had never heard of either Nägeli or Buchner until then. But his acceptance of changes in the virulence of an organism in relation to its environment brought him very near to the point of view of the Munich school. In 1881 Pasteur had written:

There are other virulent diseases that appear spontaneously in all lands; camp typhus is one of them. There is no doubt that the germs of the microbes that cause these diseases are found everywhere. Man carries them on himself or in

21 Friedrich Loeffler, *Vorlesungen über die geschichtliche Entwicklung der Lehre von den Bacterien für Aerzte und Studirende. Erster Theil: bis zum Jahre 1878* (1887); introduced by Hanspeter Mochmann and Werner Köhler (Leipzig: Zentralantiquariat der Deutschen Demokratischen Republik, 1983), 241–244.
22 Gaffky, "Experimentelle erzeugte Septicämie" (1881) (n. 16), 54.
23 Louis Pasteur, "De l'attenuation des virus," P. L. Dunant, ed., *C. R. et Mémoires du IVième Congrès International d'Hygiène et de Démographie* (1882) (Geneva: Georg, 1883),

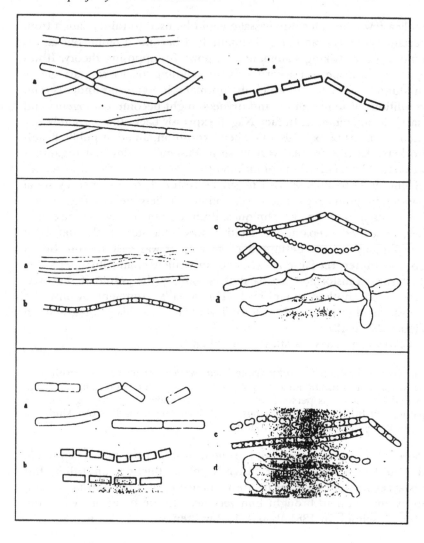

Figure 3.6. Hans Buchner's transformation of the hay bacillus, *B. subtilis,* into the anthrax bacilus, *B. anthracis.* Top: The hay bacillus, *B. subtilis,* from a hay infusion, allowed to stand 4 hours at 36°, showing (a) the long rods of the normal bacillus, and (b) the same after iodine add (his Figure 4). Middle: Involution forms of the hay bacillus: (a) and (b) grown in meat extract with 10 percent sugar; (c) and (d) in asparagine with 10 percent sugar (his Figure 7). Bottom: The anthrax bacillus, *B. anthracis,* showing (a) the long rods grown in meat extract; (b) the effect of adding iodine; and (c) and (d) involution forms of the anthrax bacillus (his Fig. 9). The hay bacillus grown under some conditions starts to look very much like *B. anthracis.* Buchner's drawings

his intestinal canal where they do no harm, but are always ready to become dangerous . . . when their virulence can become progressively reinforced.[24]

For the German reader at least, this was the theory of disease transmission advocated by Nägeli, Wernich, and Max von Pettenkofer, director of the Institute of Hygiene in Munich.

For Pasteur, the relationship of virus and disease entered a period of instability during the early 1880s. The invisible "virus" of rabies, in particular, was difficult to control. In his early attempts to identify an organism, Pasteur was prepared to believe that though the disease in the experimental animal did not resemble that in the patient, the difference might possibly be due to changes in the state of the organism.[25] It was not until 1884 that the flux of virulence and disease-picture in rabies was brought under experimental control and the range of variation in each determined. He was sure at least that there was no practical question of spontaneous generation:

Rabies is never spontaneous . . . to argue that there must have been a first case is to say nothing . . . it is to bring in unnecessarily the problem of the origin of life. That science which knows its own limitations knows that nothing is to be gained by talking about the origins of things.[26]

It has recently been argued that the immunization programme of the *pastoriens*, with its viruses attenuated by environmental manipulation, did indeed imply a certain Lamarkian transformism. It fills that role more particularly by contrast with the position taken by the Berlin group, namely, that specificity is a rigid and complete one-to-one relationship, according to which a given organism, constant as to mor-

v. 1, 127–145; and in Pasteur Vallery-Radot, ed., *Oeuvres de Pasteur Réunies par Vallery-Radot*, 7 v. (Paris: Masson, 1933), v. 6, 391–411 (p. 403).

24 Louis Pasteur, "De l'attenuation des virus et de leur retour à la virulence," *C. R. de l'Acad. des Sciences* 92(1881):429–435; and in Vallery-Radot, ed., *Oeuvres de Pasteur* (1933) (n. 23), v. 6, 332–338 (p. 337).

25 Louis Pasteur, "Expériences faites avec la salive d'un enfant mort de la rage," *Réc. de Méd. Vet. (Bull. de la Soc. de Méd. Vet.)* 58(1881):150–155; and in Vallery-Radot, ed., *Oeuvres de Pasteur* (1933) (n. 23), v. 6, 553–558 (p. 556).

26 Louis Pasteur, "Microbes pathogènes et vaccins," *Congrès Périodique International des Sciences Médicales* (1884) (Copenhagen: 1886), v. 1, 19–28; and in Vallery-Radot, ed., *Oeuvres de Pasteur* (1933) (n. 23), v. 6, 593.

show his feeling that there was very little difference between the two forms; the hay bacillus had only to begin producing a toxin for the change to be complete. The style of his drawings is very close to that of Nägeli (Figure 1.5). Drawings of a series of cultures, from Buchner, "Beiträge zur Morphologie der Spaltpilze," in Nägeli, ed., *Untersuchungen* (1883) (n. 19), 205–224.

phology and physiology, causes a given disease. The Lamarkism of the Institut Pasteur is not, however, morphological and physiological transformism of the Nägeli type, in which an organism can arise anew from inorganic materials, assume a wide range of forms, and produce sometimes cholera and sometimes sauerkraut. But the *pastoriens* were less interested in the theoretical problems, the problems of immunity, or of the morphology of the organisms, than in finding practical means of making protective vaccines.[27]

Koch's move to the Imperial Bureau of Health was the beginning of his triumph. It was from this safe position that he attacked his enemies in Paris and Munich, protected by official Berlin. It was here that he discovered the tubercle bacillus, and it was from the bureau, accompanied by Gaffky, that he went to Egypt, and then to Calcutta, to compete with a French team in the attempt to discover a specific organism for cholera. Koch's success in capturing and naming *Vibrio cholerae* made him a national figure.[28] He was recognized by the state with a steady promotion through both civil and military ranks: from *Regierungsrat* to *Geheimer Regierungsrat* in 1882, to *Geheimer Medizinalrat* and *Ordinarius* for Hygiene in the new Institute for Hygiene in 1885, and in 1891 given his own institute, the Institute for Infectious Disease. In 1884 he was appointed to the Prussian Privy Council and to the Cholera Commission, which had previously been controlled by Pettenkofer but was now quickly dominated by Koch. At the same time, he was promoted through the military ranks of the Sanitary Corps: *Oberstabsarzt* in 1883, *Oberstabsarzt I Klasse* in 1884, *Generalarzt II Klasse* in 1887, *Generalarzt* with the rank of major-general in 1901. His final civil rank was *Wirklicher Geheimer Rat*, with the title of *Excellenz* conferred in 1908.

Koch's biographer Bernhard Möllers, himself one of the many students who owed their success to Koch's influence, gives information on about fifty-three students and co-workers. He points out their connection with the Sanitary Corps: many were trained at the Kaiser-Wilhelms-Akademie, the military medical school, and many were officers in the Sanitary Corps when *kommandiert* to Koch. Koch's tremendous power was due in part to his close connection to the Prussian state and its military arm and in part to the fact that he and his students and associates (Figure 3.7) occupied a large number of newly founded chairs and institutes of hygiene controlled by the group (Figure 3.7). Of the

27 Anne-Marie Moulin, *Le Dernier Langage de la Médecine: Histoire de l'Immunologie de Pasteur au SIDA* (Paris: Presses Universitaires de France, 1991), 27–48 (p. 47).
28 Bernhard Möllers, *Robert Koch, Persönlichkeit und Lebenswerk 1843–1910* (Hannover: Schmorl, 1950), 151; William B. Coleman, "Koch's comma bacillus: the first year," *Bull. Hist. Med. 61*(1987):315–342.

Figure 3.7. Robert Koch's first post-graduate bacteriology course, Berlin, May 1891. The students came from all over the world. Front row (left to right): Rosencrantz, San Francisco, Calif.; Feigel, Lemberg; Pfeiffer, Berlin; Koch, Berlin; Frosch, Berlin; Stüler, Berlin; Costa-Pruneda, Chile; Stanitta, Messina. Row 2 (left to right): Wolff, Berlin, Heflebower, Cincinnati, Ohio; Frank, Meiningen; Mannheim, Berlin; Sobernheim, Berlin; Scholl, Trepow a.R.; Connell, Fall River, Mass.; Mlady, Meran. Row 3 (left to right): Scott, Edinburgh; v. Szydlowski, Stanislau; Sanderson, San Francisco, Calif.; Bond-Stowe, Chicago, Ill.; Metternich, Mainz; Tobold, Berlin; Rügenberg, Spandau. Row 4 (left to right): Ransom, Cambridge; Weber, New York; Crooker, Providence, R.I.; Dahmen, Cologne; Straub, Brooklyn, N.Y.; Hargrove, Jacksonville, Ill.; Küstenmacher, Stettin; Fricke, Münster i.W. Row 5 (left to right): Adamson, London; Burnett, Buffalo, N.Y.; Heine, Berlin; Mackenzie, Fall River, Mass.; Bliesenick, Berlin.

fifty-three first-generation students, only fifteen held no office. Between them, the other thirty-three held fifty-five chairs and directorships.[29]

Another expression of Koch's power was a journal, the *Zeitschrift für Hygiene und Infektionskrankheiten*, which began in 1885 with Koch and Carl Flügge as editors.[30] The policy statement of the new journal demonstrates the change in the approach to public health that had been

29 Möllers, *Robert Koch* (1950) (n. 28), 367–402.
30 Carl Flügge (1847–1923) was a close associate of Koch, though never a student. Like Koch, he studied at Göttingen and was appointed to the chair of hygiene there in 1881. In 1887, he moved to Breslau as Professor of Hygiene. Koch wanted to have him as successor in the Berlin chair, but the faculty chose Max Rubner, a Pettenkofer

brought about through Koch's influence: "Teaching and research in hygiene has undergone an important change in the past decade, in that along with empirical observation, upon which it has been almost exclusively based until now, scientific observation and experiment is coming more and more into use."[31] The British-style hygiene of "empirical observation," of vital statistics, life tables, and death rates, which had come into being in the middle of the century and was represented by Pettenkofer, was now being replaced by that of bacteriology, represented by Koch. The new journal was intended for the publication of research material in the form of detailed protocols, with control series, tables, and above all photographs, which were capable of giving the fine morphological detail needed for absolutely true-to-nature illustrations. The importance of photography for pinpointing morphological specificity in the Koch manner can be seen by comparing the drawings of Nägeli and Buchner with the photographs that Koch used to refute their claims (Figures 1.5, 2.6, and 3.6). The journal was not meant to be restricted to bacteriology, however, but was to cover the whole of hygiene. Of course, say the editors, bacteriology occupies the foreground in hygiene today. Perhaps this new Berlin journal of 1885, with its fighting manifesto, was a response to the Munich journal, the *Archiv für Hygiene*, started in 1883 under Pettenkofer's editorship, which published *very* little bacteriology.

The new hygiene of the bacteriologists trained in Berlin in Koch's methods quickly came to overshadow the old localist hygiene of Pettenkofer and Munich. The argument centred on the water supply as a source of infection. In 1892, in what might have been the group *harakiri* of a discredited *ancien régime*, the seventy-four-year-old Pettenkofer and his students drank water containing live cholera cultures obtained from Georg Gaffky, to demonstrate their disbelief in bacteriology. Luckily, Munich being famous for its bad water – it had a reputation for typhoid, as Pettenkofer's own lectures point out – they were all at least partly immune.[32] Perhaps, too, Gaffky sent them a rather dilute suspension. In any case, no one died. Rudolf Emmerich, who had the worst illness following this episode, was a frequent contributor to the *Archiv der Hygiene*.[33]

student instead. Koch got his way in 1906, when Rubner moved to the chair of physiology, and Flügge finally got the Berlin chair of hygiene. See Isidor Fischer, "Flügge, Karl," *Biographisches Lexikon*, 2 v. (Munich: Urban, 1961), v. 1. For Flügge's critique of Pettenkofer's theory of the transmission of cholera, see Flügge, *Microorganisms* (1890) (n. 5).

31   Robert Koch and Carl Flügge, "Zur Einführung," *Z. f. Hygiene u. Infektionskrankheiten* 1(1885):1–2.

32   Pettenkofer, "Ueber den Werth der Gesundheit" (1873) (n. 3).

33   Rudolf Emmerich, *Max Pettenkofers Bodenlehre der Cholera indica, experimentell begründet und weiter ausgebaut von Rudolf Emmerich mit Beiträgen von Ernst Angerer [et al.] Jubi-*

Koch's cholera programme was the opposite of Pettenkofer's in every respect. He pushed for border controls, quarantines, boiled water, strong state intervention, and health legislation on a national scale. The civil authorities supported Koch to the hilt in these moves. Berlin imposed Koch's methods on the quasi-independent Hanseatic port of Hamburg in 1892–1893, and subsequently upon the whole country through a National Epidemic Law. Koch's sudden and complete eclipse of Pettenkofer at the Bureau of Health may have been due less to his "correct" theory of disease transmission, than to its consonance with the desire of the central government to impose itself upon its constituent states. At every turn, Koch was confronted by Pettenkofer, and Pettenkofer was worsted.[34] In 1901, the eighty-three-year-old Pettenkofer shot himself.

The overwhelming victory of the Koch group was to colour the history of bacteriology, and of immunology, for the next half-century. Until recently, it also greatly influenced the historian's approach to the history of bacteriology. History, beginning with Loeffler's account, was written by the winners in this conflict.

It was not only the old hygiene but also the bacteriology of Nägeli and his students, the other group of Koch foes, that lost to Koch. Transformism, once at the spearhead of modern biology, became the losing side in the medicine of the eighties.[35] Specificity was in, unitarianism was out. But a few supporters remained loyal to Nägeli, among them Max von Gruber, professor of hygiene at the Institute for Hygiene in Vienna, who along with Hans Buchner, had been a student of Nägeli in Munich. His *Gedächtnissrede* for Buchner tells us what it was like to be on this losing side in the eighties. Even though "Buchner, supported by few others, had unshakeable proof of the inconstancy of form of microbes," it seemed that "the masses, more papist than the Pope, swore to absolute constancy of form as to a dogma, so that every discussion of a change in form as a result of change in living conditions was taken without any further ado as a proof of the unreliability of the author."[36]

Buchner had continued to defend his thesis of 1880 on the disappearance of virulence from the anthrax bacillus on serial culture – he called the new form an intermediate form of hay bacillus, not *Bacillus subtilis*, the true hay bacillus, and he adduced the work of others in his

läumsschrift zum 50-jährigen Gedenken der Begründung der lokalistischen Lehre Max Pettenkofers (Munich: Lehmann, 1910).

34 Evans, *Death in Hamburg* (1987) (n. 4), 311–314, 490–507.

35 Compare the foreword, "An unsere Leser," of the editors of the *Zeitschrift für Biologie* of 1883: transformism had not lost ground outside bacteriology.

36 Max von Gruber, "Hans Buchners Anteil an der Entwicklung der Bakteriologie," *Münch. med. Wsch.* 50(1903):564–568 (p. 566).

support.[37] His writing has a rather pathetic air of special pleading, of defensiveness, of the embarrassment of being defeated. The support he quotes in his papers of 1885 is that of Prazumowsky, an extremely obscure author writing in Polish.[38]

Some of the difficulty, he claimed, had been due to the use of the same name for species and for growth forms of the organisms, as if each species had only one form. This led to the assumption that where a bacillus takes on the form of a chain of cocci, this represents the transformation of a *Bacillus* into a *Micrococcus*. Two questions arise here: they concern the constancy of species and the constancy of growth forms. Species differences, he says, are not effaced within the human time span. There are only two known examples of qualitative change, that of the loss of the fermenting ability by fermenting organisms and that of the transformation of the anthrax bacillus into a non-pathogenic form. Both changes take place without any change in morphology. In contrast, the anthrax bacillus is equally infectious whether it is in the form of short rods or long ones, or spores. These examples, he says, are not compatible with a monomorphic theory like Cohn's or Koch's. Buchner calls this monomorphic theory the result of a childish fear that if it is given up, bacteriology will sink helplessly into a chaos of indeterminate forms – a needless fear, for the forms are constant under constant cultural conditions, and diagnosis is in no danger.[39]

Ten years later, Max von Gruber in his memorial to Pasteur, who died in 1895, is still trying to defend Nägeli and Buchner, by making use of Pasteur's undoubted fame and success. He takes up the assimilation of Pasteur's thinking with theirs, as had Koch and Gaffky in the attack on all three of them by the Kaiserliche Gesundheitsamt in 1881. Gruber comments on Koch's claims:

The discovery of large numbers of pathogenic species of bacteria, to which were ascribed absolute constancy of form and of physiological properties, led in the late seventies and early eighties to an unscientific conception of the nature of infection. It can hardly be denied that . . . the old *Entia morborum* seemed to have come to life again in medicine. The microbe was the essence of the disease. . . . Pasteur . . . recognised that diseases themselves are not species in the natural history sense, and that the cause of the disease is not its essence.[40]

37 Hans Buchner, *Ueber die experimentelle Erzeugung des Milzbrandcontagiums* (1880) (n. 19).
38 Hans Buchner, *Sitzungsber. d. Ges. f. Morph. u. Physiol.* (Munich), *1*(1885):27–30; Buchner, *Sitzungsber. d. Ges. f. Morph. u. Physiol.* (Munich), *1*(1885):121–129.
39 Buchner, *Erzeugung des Milzbrandcontagiums aus den Heupilzen* (1880) (n. 19), 122–124.
40 Max von Gruber, "Pasteurs Lebenswerk im Zusammenhang mit der gesammten Entwicklung der Microbiologie," *Wiener klin. Wschr,* 8(1895):823–828, 844–848, 836–866 (p. 865).

Edwin Klebs, professor of pathological anatomy at Prague, had said exactly that at the fiftieth meeting of the Naturforscherversammlung in Munich, where he spoke along with Nägeli and Ernst Haeckel. Like Ferdinand Cohn, Klebs was convinced that bacteria would eventually be separated into several well-defined independent species, and, like Koch, he saw them as causing specific diseases. It would soon be possible, he said, to classify diseases by the genera and species of the infecting organism. The same *Fundamentum divisionis* would do for both: "If we seek a principle upon which to classify infectious disease, it must necessarily depend upon our knowledge of these processes [of infection]. I am persuaded that a time will come when the Genus and Species of the organisms which are their cause can be used as such a principle."[41]

Historians have seen in this programmatic statement a prefiguration of Koch's so-called postulates.[42] There is one essence, but two existences: the *Ens bacteriorum* has the same essence as the *Ens morborum*, precisely as Gruber points out. The form of the argument is that of Saint Thomas Aquinas, in which he says that essence has two existences, one in nature and one in the soul, and genus and species are accidents of the one in the soul.[43]

Gruber traces Pasteur's change of attitude, from his early belief in specificity through his acceptance of the variability of virulence in 1880. Variability had been demonstrated a year or two before by Hans Buchner in 1878, as an extension of Nägeli's theory, showing how fruitful the theory had been: "It was Pasteur, of all the bacteriologists, who showed us the way. He had however, forerunners in his correct conception of the process of infection, for example Nägeli, and especially Buchner, in Germany."[44]

But only an authority like Pasteur, head of a great school, in command of any amount of material, a great experimenter, bold and brilliant, could have broken through to the truth. We need not fear that we are doing anyone an injustice when we say that Pasteur showed us

41 Edwin Klebs, "Ueber die Umgestaltung der medizinischen Anschauungen in den letzten drei Jahrzehnten," *Gesellschaft deutscher Naturforscher und Aerzte, Amtlicher Bericht der 50 Versammlung* (1877) (Munich: 1877), 41–55 (p. 47).

42 Heyman, *Koch* (1932) (n. 8), 118, points out that this statement is very close to the form that Koch was to give his famous "postulates," which first appeared the year after Klebs's talk, in Koch, *Untersuchungen über die tiologie der Wundinfektionskrankheiten* (Leipzig: Vogel, 1878), and in his *Gesammelte Werke*, v. 1, 61–108 (p. 72). On "Koch's postulates," see K. Codell Carter, "Koch's postulates in relation to the work of Jacob Henle and Edwin Klebs," *Med. Hist.* 29(1985):353–374.

43 Saint Thomas Aquinas, "De Ente et Essentia," Robert P. Goodwin, ed., *Selected Writings of St. Thomas Aquinas* (New York: Bobbs-Merrill, 1965), 33–67 (pp. 47–48).

44 Gruber, "Pasteurs Lebenswerk" (1895) (n. 40), 866.

the way. Gruber suggests that all bacteriology, with its emphasis on the infecting organism, may just have been a detour around which pathology had to pass before getting back on the main road again. There is more to public health than a few laws for sewage disposal and disinfection, important though they may be. We must remember the infected organism and its powers of resistance, as well as the infecting one.

This *Gedächtnisrede* for Pasteur, given as it was by a Nägeli student, was instantly recognizable as an attack on Koch. Accordingly, Richard Pfeiffer (Figure 3.8) of the Institute for Infectious Disease rose as his champion, with Koch's gage in his cap.[45] Pfeiffer angrily rebutted the charge against Koch, which he recognized in Gruber's words about the *Entia morborum*: Koch, that sober researcher, who *never* made fantastic speculations of that *Natur*-philosophic kind, had *never* mistaken the cause of disease for its essence. Gruber does his friend Buchner no service, he says, in reminding us of his absurd experiment with the hay bacillus. It was Pasteur who discovered attenuation.[46]

Gruber offered a response in the next issue, of course. Loyal to his own side, he appeals to history for justice:

I do not want to argue with Pfeiffer about whether Koch had the right idea from the beginning about the essence of the process of infection, and the host-parasite relationship. These things are historical facts, and every effort of the Koch school to deny their earlier errors is bound to be fruitless.[47]

The same feeling is evident in Pfeiffer's final words:

I am far from keen to get into this kind of polemic with Gruber, as long as it is only on my own account. But I cannot be silent when in giving the French scientist the honour that is his due, Gruber does an actual injustice, to the father of German bacteriology [*dem deutschen Altmeister der Bacteriologie*].[48]

45   Richard Pfeiffer (b. 1858) was a military physician *kommandiert* to Koch in 1887 at the Institute for Hygiene of the University of Berlin. In 1891 he went with Koch to the Institute for Infectious Disease. In 1899 he became professor and director of the Institute for Hygiene in Königsberg, and in 1909 the same in Breslau. Most of his work was on cholera immunity. He was best known as the discoverer of "Pfeiffer's phenomenon": 92 n.57.
46   Richard Pfeiffer, "Kritische Bemerkungen zu dem Aufsatze Max Grubers 'Pasteurs Lebenswerk in Zusammenhang mit der gesammten Entwicklung der Mikrobiologie,'" *Deutsche med. Wschr.* 22(1896):15–16.
47   Max von Gruber, "Erwiderung auf R. Pfeiffers Kritik meines Vortrag, 'Pasteurs Lebenswerk, &c.,'" *Deutsche med. Wschr.* 22(1896):94–95.
48   Richard Pfeiffer, "Bemerkung zu vorstehender Erwiderung," *Deutsche med. Wschr.* 22(1896):95.

Figure 3.8. Richard Pfeiffer, portrait photograph. (Courtesy of National Library of Medicine, Bethesda, Md.)

There were two final rejoinders, although the editor had already closed the correspondence. There was a mild one from Buchner, who by this time was director of the Institute of Hygiene in Munich, after Pettenkofer.[49] There was also a surprisingly cool one from Gruber, pointing

49 Replies to Pfeiffer from Hans Buchner and Max von Gruber, *Deutsche med. Wschr.* *22*(1896):128.

out only that he had not failed completely to mention Koch's contributions in his memorial to Pasteur.[50] Gruber's pacific reply is rather surprising, because he was already engaged in an attack on the Koch school on other grounds.

Gruber's work at this time and Gruber's thinking, as much as that of his friend Buchner, show the influence of Nägeli's style. In his *Gedächnisrede* for Buchner, he speaks of Buchner's relationship with Nägeli, and could equally well be describing his own love and admiration for him:

So lived Buchner within the spell of this powerful mind. With joy he recognised in the great scientist an intellectual nature akin to his own: the same desire for encompassing order and harmony within a natural whole, the same bold pushing of thought to the very limits of natural knowledge, the same love of speculation, and of drawing the last possible logical consequences from observations, the same tendency to use a general theory as a source of experimental questioning.[51]

It is the "desire for an encompassing order and harmony within a natural whole" that is so strong in Nägeli's thought. It is also present both in Buchner's modulating anthrax bacillus and in Gruber's work on cholera, in which Nägeli's influence appears so clearly. Neither Buchner nor Gruber were willing to admit to sharply defined and impassable boundaries: they hoped *not* to be able to divide species from each other, and this hope influenced their experiments and their results.

Gruber's paper of 1894, to the Eighth International Congress for Hygiene and Demography (the title of the Congress is an indication of the earlier close association of these two fields), takes up an easily recognizable position, a criticism of Koch's claim that he can distinguish which among the many common *Vibrio* types is the real *Vibrio cholerae*. All the available methods of discrimination (by this time they include some that would have worried Caesalpino) – microscopical, cultural, chemical, the smell of the cultures, the so-called cholera red reaction – do not allow a certain diagnosis to be made. All Gruber has managed is to distinguish one group of organisms containing among them the "true" one, from another group; all the distinctions do not suffice to call them two species. Not only is it impossible to

50   Max von Gruber (1853–1927) had been a student of Pettenkofer and of Nägeli, and of Carl von Voit also, in Munich. In 1884 he became *a. o. Professor*, and in 1887, *Ordinarius* for Hygiene in Graz, and in 1891 accepted the new chair of hygiene in Vienna; and in 1902, following Buchner's death, he succeeded him in Munich.
51   Max von Gruber, "Hans Buchners Anteil an der Entwicklung der Bakteriologie," *Münch. med. Wschr. 50*(1903):564–568.

distinguish one *Vibrio* species from another, but the disease pictures they produce when injected intra-peritoneally are indistinguishable, too: neither the organism nor the disease is truly specific.[52] Even immunity, which Pfeiffer claims to be specific, in Gruber's hands is not. He has immunized guinea-pigs with *Vibrio danubicus, V. massauah, V. seine-versailles, Bacillus coli,* and *B. typhi* and tested them by injecting *V. st. goarshausen.* The typhoid-immune one died; the others became sick but recovered. There was no specific difference to be detected. He concludes that we cannot say whether all these vibrios belong to a single species or to many, whether the real cholera vibrio is different from the others or not.

By this time, the argument had descended to the third generation. Pfeiffer maintained exact specificity: immunization protects *only* against the vibrio species used to immunize; immunity is specific. Gruber maintained that immunity is equally present to a whole group of similar organisms: immunity is no more specific than morphology. Neither the organism itself nor the response of the host is specific. Nägeli's frame had been narrowed; as Gruber said, Nägeli went too far.[53] It is no longer the whole of nature that is seamlessly one, only the vibrios. In spite of their Linnaean names, they are indistinguishable from each other. "The foundations of Koch's teachings on cholera," says Gruber with pleasure, "are shaking."[54]

It was a phrase that Pfeiffer picked out for an angry critical remark in some *kritische Bemerkungen zu Grubers Theorie* of the following year.[55] This, however, formed part of a different controversy. The combatants were, again, Richard Pfeiffer, with the backing of the Koch group and the assistance of the young Wilhelm Kolle (Figure 3.9) in Berlin, and Gruber and his two English students, Herbert Durham and Arthur S. F. Grünbaum, in Vienna.[56]

In 1896, Pfeiffer, who had been working on cholera at the Berlin Institute for Infectious Disease since it started in 1891, came to prominence with a paper on what he himself recognized as "*ein neues Grundgesetz der Immunität*," a new basic law of immunity, and which later

---

52 Max von Gruber, "Ueber den augenblicklichen Stand der Bakteriologie der Cholera," Sigismond de Gerloczy, ed., *Congrès International VIII d'Hygiene et Démographie, Comptes Rendus et Mémoires* (Budapest: Pesti Konyvnyomda-Részventárasag 1896), v. 2, pt. 1 (Hygiène), 266–278 (p. 271).
53 Gruber, "Hans Buchners Anteil" (1903) (n. 36), 566.
54 Gruber, "Bakteriologie der Cholera" (1896) (n. 52), 276.
55 Richard Pfeiffer, "Kritische Bemerkungen zu Grubers Theorie der aktiven und passiven Immunität gegen Cholera, Typhus und verwandte Krankheitsprocesse," *Deutsche med. Wschr.* 22(1896):232–234.
56 Wilhelm Kolle's relationship to Koch was a curiously personal one. Kolle's son, Kurt Kolle (Kurt Kolle, ed., *Robert Koch: Briefe an Wilhelm Kolle* (Stuttgart: Thieme, 1959),

Figure 3.9. Wilhelm Kolle, portrait photograph, n.d. (Courtesy of National Library of Medicine, Bethesda, Md.)

came to be known as "Pfeiffer's phenomenon."[57] His paper starts with a rehearsal of milestones in the history of the study of immunity, which leads up to his own important discovery. It begins with "Pasteur standing on the shoulders of Jenner," and the immunizing effect of the injection of attenuated organisms, which was later extended to dead organisms and to toxins. Emil von Behring had demonstrated that the bacterial toxin was inactivated by antitoxin, and Paul Ehrlich that there were quantitative grades of immunity. In Behring's view, immunity consisted of inactivation of toxins only.

Pfeiffer's *neues Grundgezetz* was that, in the case of immunity to cholera and typhoid, it was not the free toxin, as in diphtheria, but the organisms themselves that were affected by the immune serum: the organisms were actually dissolved by the serum. The cholera "toxin" must be part of the bacterial body, because the symptoms followed the injection of dead bacteria as well as live. The action of the immune serum could be watched by injecting the bacteria into the peritoneal cavity and sampling the fluid at intervals. Pfeiffer wrote:

The vibrios, which in the control animal show a lively motility and continue to increase until the death of the animal, die astonishingly quickly under the influence of the cholera serum. Their motility is lost almost immediately, they begin to swell, they change into small round micro-coccus-like forms, then become paler and paler until they dissolve completely in the peritoneal fluid. . . . The effect of the cholera serum is so eminently specific withal, that it can be used for the differential diagnosis of Koch's *Vibrio* from all the other kinds. Even in a mixture of two species of *Vibrio*, only that one with which the serum donor animal was immunized is destroyed. These facts are a powerful weapon against the open or secret opponents of Koch's teaching on the specificity of the pathogenic organisms.[58]

The work undertaken in Vienna by Gruber and his student Durham was, as Durham wrote, to test the limits of Professor Pfeiffer's method of diagnosing cholera vibrios from other closely allied vibrios – reading

wrote that Koch in his early youth was in love with a very young girl, whose guardian, afraid of the passionate intensity of the young man, forbade their relationship. Many years later, after the girl had married someone else, she again got in touch with Koch, and asked him to advise her son, Wilhelm Kolle, on his career. Koch advised him to read medicine and said that he would look after him. In 1893 Kolle entered the Berlin Institute for Infectious Disease, where he worked with Pfeiffer on cholera, and in 1897 he took Koch's place in Kimberly, South Africa, in his work on Rinderpest. He later became a section leader at the institute and in 1906 was called to the new chair of hygiene at Bern. In 1917 he succeeded Paul Ehrlich at the head of the State Institute for Experimental Therapy (Paul-Ehrlich Institute), Frankfurt-am-Main.

57 Richard Pfeiffer, "Ein neues Grundgesetz der Immunität," *Deutsche med. Wschr.* 22(1896):97–99, 119–122.
58 Pfeiffer, "Ein neues Grundgesetz" (1896) (n. 57), 99.

between the lines here, one can see "show the error of Professor Pfeiffer's method." Durham wrote,

In its essence, Pfeiffer's reaction depends upon the specific action of the serum obtained from immunized animals ... [but] it appears from the experiments ... that the action of these sera is not sufficiently limited to merit the term 'specific;' the term 'special' will be used here, as it appears more fitting to the case. The serum of an animal highly immunized against the cholera vibrio can be regarded as truly 'specific' when tested with such remote groups as those of enteric fever, or the less remote *Vibrio nordhafen* but ... this is not the case when some of the members of the group of vibrios are in question.[59]

In Gruber's later summary of the experiments (he was a kind man and let Durham publish first), the same point is made:

The effect of them [that is, the antibodies] is not sharply specific, but shows stepwise differences [is *graduell abgestufte*] so that each one acts most strongly against its own species. Its effect on other species of bacteria is stronger the more nearly related they are to the species in question.

    The opposite conclusion, drawn by Pfeiffer, that the effect of immune sera is strongly specific, goes beyond the facts of the case.[60]

This short citation reveals both of Gruber's driving motivations: first, his love for Nägeli, in the use of Nägeli's phrases and his thought, "*graduelle Abstufung*," and the boundlessness of species; and second, his dislike of Pfeiffer and the sharp specificity represented by the Koch supporters, his opponents.

    Gruber and Durham had obtained their cultures from Pfeiffer in Berlin and set out to repeat his experiment, working through his experiments of 1894 onwards.[61] When they mixed the immune serum with the organisms in preparation for their first "Pfeiffer's experiment," Gruber noticed that the bacteria agglomerated into clumps, forming flocculi recognizable to the naked eye, which after half an hour or so had sunk to the bottom of the tube. Under the microscope, they could see the clumps forming: if a vibrio hit a clump or another

59 Herbert E. Durham, "On a special action of the serum of highly immunized animals," *J. Path. Bact. 4*(1897):13–43 (p. 13).
60 Max von Gruber, "Theorie der aktiven und passiven Immunität gegen Cholera, Typhus und verwandte Krankheitsprocesse," *Münch. med. Wschr. 43*(1896):206–207 (p. 206).
61 Richard Pfeiffer and Isaeeff, "Ueber die spezifische Bedeutung der Choleraimmunität," *Z. f. Hyg. u. Infektionskr. 17*(1894):355–400; Pfeiffer and Wilhelm Kolle, "Ueber die spezifische Immunitätsreactionen der Typhusbacillen," *Z. f. Hyg. u. Infectionskr. 21*(1896):203–246; Pfeiffer and Kolle, "Zur Differentialdiagnose der Typhusbacillen vermittels Serums der gegen Typhus immunisirten Thiere," *Deutsche med. Wschr. 21*(1896):203–246.

individual, it seemed to become sticky and to be unable to free itself.[62] Gruber therefore named the antibody in the serum *Glabrificin* or *Klebrigmacher*.[63] In his next paper, he had to withdraw this name, calling it a *lapsus calami*; he had doubtless found out that *glaber* means "smooth" not "sticky," though it does sound rather like *klebrig*, and he changed the name to *Verkleber*, or "agglutinin."[64] This, indeed, is a mistake that no Linnaean botanist would ever have made; and there were many of them about to tell him so. Pfeiffer, of course, did not miss it.[65]

In the same paper, Gruber and Durham extended their work to cover anti-typhoid sera. At first, anti-typhoid seemed more specific, but here, too, they were able to find another species, *Bacillus enteritidis* (Gärtner), that was agglutinated by the anti-typhoid serum. Furthermore, there was only a quantitative difference between the reaction of the serum with the real *B. typhi* and with *B. enteritidis*, even though they belonged to different species and had different biochemical properties.

This appeared to invalidate the reaction as a diagnostic test. But Gruber hoped that by manipulating the test conditions he could make use of the *quantitative Abstufung*. The conditions had to be carefully controlled: the anti-serum had to be very high grade, the cultures to be tested had to be 10–20 hours old; one loopful of organisms had to be suspended in 1 cubic centimeter of sterile bouillon. One loopful of serum and one of suspension were to be set side by side on a slide. Once they were mixed, the result needed to be watched under a microscope. For a true positive result, motility would have to cease instantly or within the first minute, and agglutination had to be complete. Conscious that this first attempt to define the conditions and make the test quantitative was still rather coarse, Gruber thought it might be possible to improve it, to make it more precise; apparently, Herr Dr. K. Landsteiner was working on this problem.[66] Karl Landsteiner, then aged 28, who had just finished his training and was in his first job, had joined Gruber at the Institute of Hygiene in Vienna on 1 January 1896. Gruber and Durham's paper appeared in print on March 13. Landsteiner, in his first contact with immunology, immediately focused on Gruber's core problem, the problem of *quantitative Abstufung*.

62 Durham, "On a special action of the serum" (1897) (n. 59), 19.
63 Gruber, "Theorie der aktiven und passiven Immunität" (1896) (n. 60), 206.
64 Max von Gruber and Herbert E. Durham, "Eine neue Method zur raschen Erkennung des Choleravibrios und des Typhusbacillus," *Münch. med. Wschr.* 43(1896):285–286 (p. 285).
65 Pfeiffer, "Kritische Bemerkungen zu Grubers Theorie" (1896) (n. 55), 234.
66 Gruber and Durham, "Eine neue Method zur raschen Erkennung" (1896) (n. 64), 286.

Pfeiffer's critical reply curiously admits, or rather claims, all the same facts, but sees them in a kind of mirror image or inverted interpretation. The antisera *are* specific, he says: assuming that the antibody reacts with one *Vibrio* species to an intensity of 100, and to another with an intensity of 1, if the titres are kept low, these minimal effects will be lost, and tests with the "other" species will be negative.

These facts speak, it is obvious, for the strongly specific nature of the cholera antibodies. The existence of minimal side-reactions with *Vibrio massauah* and *Vibrio weichselbaum* is explicable, as Loeffler first suggested, as an expression of the species relationship [*Artverwandtschaft*] of these vibrios with the cholera organisms. One could even perhaps manage to work out a natural system of the bacteria in this way.[67]

Pfeiffer keeps his serum titres low so that the non-specific reactions disappear, while Gruber insists on "high-grade" sera. The technical difference is a reflection of the interpretation: Pfeiffer wants specificity and arranges the conditions to get it; Gruber does not and keeps his titres high enough to get some agglutination even with the non-specific. An argument about priority in the use of this reaction as a diagnostic test for cholera and typhoid organisms in *dejecta*, as well as the argument about specificity, was mixed into Pfeiffer's paper. As he said, it seemed ironic that his priority should be challenged on the question of the specificity of antibodies by a man who did not believe in specificity.[68]

The arguments both on priority and on specificity were bitter and personal, the positions perfectly predictable. They culminated in a face-to-face confrontation at the 1896 meeting of the Congress for Internal Medicine in Wiesbaden, where both parties appealed to the audience for support in a dramatic dialogue in which each in turn spoke in character, loyally playing the roles that their training had assigned them.[69]

Gruber, and Durham his student, began by demonstrating their agglutinations in test tubes, which were handed round the audience with some photographs by Durham. They emphasized the wide range of the agglutinins and their non-specific nature, with samples of clumped *Vibrio cholerae*, *V. berolinensis*, *V. seine-versailles*, and *V. ivanoff*.[70]

67   Pfeiffer, "Kritische Bemerkungen zu Grubers Theorie" (1896) (n. 55), 232.
68   Pfeiffer, "Kritische Bemerkungen zu Grubers Theorie" (1896) (n. 55), 234.
69   Max von Gruber, "Ueber active und passive Immunität gegen Cholera und Typhus," *Deutscher Congress für Innere Medicin: Verhandlung des XIII Congresses* (Wiesbaden: Bergman, 1896), 207–217.
70   Herbert E. Durham, "Immunitas gegen Cholera und Typhus: Demonstration zu dem Vortrag Grubers," *Deutscher Congress für innere Medizin* (1896) (n. 69), 228–230.

Pfeiffer replied to them with the old suggestion, the same that Koch and Loeffler had used to attack Pasteur, that the immunizing cultures may not have been pure, or that the animals may have been immunized to materials from typhoid and cholera bacilli acquired elsewhere. And he reminded the audience that Gruber came of Nägeli's school, and that therefore the thought must exist somewhere in his mind that specificity in Koch's sense is not true, that bacteria are variable in immeasurable ways. But Gruber must know that all this has been disproved: all cholera cultures behave identically to a cholera-immune anti-serum.[71]

Gruber reminded the listeners in turn that a short time ago *V. ivanoff* was supposed to be distinguishable from *V. cholerae* by any half-baked bacteriologist. Then along came Pfeiffer's reaction, and now all bacteriologists worth anything could see they were no different. When he, Gruber, had written to Professor Pfeiffer in Berlin requesting a genuine cholera culture, Pfeiffer sent him one he had worked with for years. Gruber could only suppose, he said, that Pfeiffer thought it was genuine. But it turned out to be *V. massauah* and did not give the reaction.[72] Gruber did not know how Pfeiffer could be so sure:

I take the position that *Vibrio cholerae* is of one kind only. I only say that there is a relationship between the different kinds; specificity in Koch's sense simply does not exist. The diagnosis of the vibrio is exceptionally difficult. If I handed Koch himself a *Vibrio ivanoff* or *Vibrio sanarelli*, or any other kind, he would not be able to tell with certainty whether it was the cholera vibrio or not. There is no completely specific reaction.[73]

Pfeiffer's final words were that he could not answer these personal attacks in this place, and that he could only protest Gruber's rudeness.

71 Pfeiffer, in Discussion to Gruber, "Immunität gegen Cholera und Typhus" (1896) (n. 69), 218–220.
72 Gruber, in Discussion to Gruber, "Immunität gegen Cholera und Typhus" (1896) (n. 69), 224–227 (p. 224).
73 Gruber, in Discussion to Gruber, "Immunität gegen Cholera und Typhus" (1896) (n. 69), 225.

4

The History of Nineteenth-century Bacteriology from This Point of View

When a war is over, the combatants are judged by the winning side. Until recently, historians of bacteriology tended to see the arguments about the origin of life as being somewhere between science and old-fashioned superstition, symbolized by the biblical story of bees generated from the ear of a dead lion. The debate on the history of spontaneous generation was discussed by Vandervliet in 1971. His chapter headings, such as "The Turnip–Cheese Episode," tend to make the outcome a foregone conclusion, as well as to mock the combatants and their problems.[1] It was not until the 1970s that the problem of spontaneous generation was given proper respect.[2]

The same was true of the few references to the unitarian theory of bacterial species. The history of bacteriology in the nineteenth and early twentieth century is fairly well known. It was treated by William Bulloch in 1938, and its relation to immunology was described by William D. Foster in a short but sympathetic work that has not been as widely read as it deserves. A bibliographic guide to the history of the various topics within the field of bacteriology up to 1958. Accounts of the development of the field, beginning with that of Friedrich Loeffler, Koch's lieutenant, in 1887, generally tended to see the history of bacteriology as a search for ancestors for the Koch group, and for the medical bacteriology that followed the path that they marked out.[3]

1   William G. Vandervliet, *Microbiology and the Spontaneous Generation Debate during the 1870s* (Lawrence, Kans.: Coronado Press, 1971).
2   John Farley, *The Spontaneous Generation Controversy from Descartes to Oparin* (Baltimore, Md.: Johns Hopkins, 1977), 92–120; Farley, "The spontaneous generation controversy (1859–88): British and German reactions to the problem of abiogenesis," *J. Hist. Biol.* 5(1972):285–319; John Farley and Gerald L. Geison, "Science politics and spontaneous generation in XIX century France," *Bull. Hist. Med.* 48(1974):161–198.
3   William B. Bulloch, *The History of Bacteriology* (London: University Press, 1938); William D. Foster, *The History of Medical Bacteriology and Immunology* (London: Heinemann, 1970); Thomas H. Grainger, Jr., *A Guide to the History of Bacteriology* (New York: Ronald Press, 1958). Friedrich Loeffler, *Vorlesungen über die Entwicklung der Lehre von den Bac-*

Loeffler traced his thread from Christian Ehrenberg to Ferdinand Cohn to Robert Koch, as did William Bulloch fifty years later and Koch's most recent biographer, Thomas Brock, writing about Koch's techniques fifty years later still.[4] Lechavalier and Solotorovsky writing in 1965 do the same.[5] There has been a steady growth of "good" information, leading to the description and definition of "good" species of bacteria, with Linnaean names, which could be related to corresponding species of disease. In many cases the bacteria have been given the name of the disease – for example, *Vibrio cholerae* and *Salmonella typhi*, as Klebs suggested – and in others, diseases have come to be called by the names of the bacteria, as for example, streptococcal sore throat and brucellosis. It became difficult to see bacteriology in any other way than as steady progress in this direction: present-day nosology and the terminology to which we are accustomed make it hard to do otherwise.

However, the emphasis changes if the picture includes the powerfully polarizing figure of Carl von Nägeli. His presence converts the interior monologue into a public dialogue: the goal of the bacteriologists of the Koch group was not only to extort an answer from nature, but also an admission of defeat from Nägeli.

The strength of the tradition of Linnaean botany in the work of the Berlin bacteriologists is brought into sharp relief by the comparison with Nägeli and his thinking, along with the connection between the earlier Aristotelian problem of species and that of the later problem of immune specificity. It is the unitarian school of Nägeli that represents the new biology of the post-Darwinian period, that accepts not only transformation of one species into another, but also some kind of spontaneous generation. Both these ideas, which seemed likely to find their empirical support from research on micro-organisms, were accepted by Darwin's most influential supporters, Thomas H. Huxley and Ernst Haeckel. Nägeli's continuous *Urzeugung* and transformism went to the limit along this path, though at the same time he rejected Darwin's proposed mechanism of change and preferred the more German *Vervollkomnungsprincip*. Nägeli, in fact, was a leader in non-Darwinian evolutionary theory. His position, although extreme, was not the eccentric one it seemed to later generations of bacteriologists and to historians of bacteriology accus-

terien: für Aerzte und Studirende I Theil bis zum Jahre 1878 (Leipzig: Vogel, 1887); no further parts published.

4   Thomas D. Brock, *Robert Koch: A Life in Medicine and Bacteriology* (Madison, Wis.: Science Tech, 1988), 70–113.

5   Hubert A. Lechevalier and Morris Solotorovsky, *Three Centuries of Microbiology* (New York: McGraw-Hill, 1965).

tomed to Koch's firm specificity. Rather, it was consistent with the best modern thinking in biology. Nägeli's closeness to Haeckel has already been discussed. His closeness to Huxley is demonstrated in the following passage:

> If it were given to me to look beyond the abyss of geologically recorded time to the still more remote period when the earth was passing through physical and chemical conditions which it can no more see again than a man can recall his infancy, I should expect to be a witness of the evolution of living protoplasm from not-living matter. I should expect to see it appear under forms of great simplicity, endowed like existing fungi, with the power of determining new protoplasm from such matters as ammonium carbonates, oxalates and tartrates, alkaline and earthy phosphates, and water, without the aid of light.[6]

It was this expectation that Huxley thought to have confirmed in his discovery of the formless blob of protoplasm without a nucleus or cell wall that he found living at great depths in the Atlantic and that he named *Bathybius haeckelii*.[7] Both Huxley and Nägeli feel that organisms at this simple beginning stage are not fixed in their nature. Like Nägeli, Huxley was prepared to believe that micro-organisms that looked as if they belonged to different species might simply be at different stages of the same life history, called into existence by changing conditions. He was even prepared to suggest, as Nägeli was not, that a mould, a yeast, and a bacterium might be stages of one and the same life cycle.[8]

More recently, a new generation of historians has begun to try to work out a genealogy for the research in bacterial genetics that developed after the Second World War, and a new body of historical writing has appeared. One writer suggests that the dominance of the Cohn–Koch tradition discouraged the investigation of both morphological variation and inheritance in bacteriology until after the War.[9] Others argue that this was not strictly true: morphological variation, under the name of "dissociation," was the subject of research and scientific dis-

6 Thomas H. Huxley, "Biogenesis and abiogenesis," Presidential address to the British Association for the Advancement of Science for 1870 and in *Discourses Biological and Geological Essays* (New York: Appleton, 1887), 229–271 (p. 256).

7 Thomas H. Huxley, "On some organisms living at great depth in the North Atlantic Ocean," *Q.J. Microsc. Sci.* 8(1868):203–212; this tale and its place in contemporary biology has been discussed by Philip F. Rehbock in "Huxley, Haeckel and the oceanographers: the tragicomedy of *Bathybius haeckelii*," *Isis* 66(1975):504–533.

8 Thomas H. Huxley, "On the relations of *Penicillium, Torula* and *Bacterium*," special report of an address to the Biological Section of the British Association for the Advancement of Science, 13 Sept. 1870, *Q.J.Microsc.Sci.* 10(1870):335–362.

9 Harriet Zuckerman, "Theory choice and problem choice in science;" in Jerry Gaston, ed., *The Sociology of Science* (San Francisco, Calif.: Jossey-Bass, 1978), 65–95; Zuckerman and Joshua Lederberg, "Postmature scientific discovery?" *Nature* 324(1986):629–631.

cussion in the twenties and thirties.[10] As far as clinical laboratories were concerned, however, strict specificity and monomorphism were the conventions of the practical bacteriologist. Interestingly, these can also be seen as conventions that supported themselves in practice by the strictly controlled bacteriological technique that was the Koch tradition of training, and by the use of highly standardized culture media, which limited the appearance of variants. These conventions were supported too by the dominance of German institutions and the German scientific journals.[11] The newer historians place Nägeli at the beginning of this series of research on bacterial variation and bacterial genetics. Pleomorphism, however, like spontaneous generation, was never mainline enough for its advocates to be treated sympathetically, or even seriously, in the histories written by their contemporaries. Discussions of methodology tended to be dismissed as metaphysics: Bulloch wrote that "the perusal of Nägeli's books gives the impression that he was a dialectician and a verbose philosopher rather than an accurate observer in his bacteriological works, whatever he may have been in his other botanical research."[12]

Another interesting feature that emerges when Nägeli and his "other" style of thought are included in the picture is extent to which training and loyalty to a given programme determined the goals of scientific thinking. The collective, as Ludwik Fleck suggested in 1935, developed its own style of thought, and it is this that was learned as the apprentice thinker was inducted into the *Denkkollektiv*.[13] The tightly organized and stylistically uniform Koch school of thought is a good example of the Fleck phenomenon; Fleck himself may even have been a party to the controversy as it continued into succeeding generations. Perhaps it was his experience of the power of the Koch group that suggested the idea of a *Denkkollektiv* to him.[14]

The extent to which a given thinker is influenced by a teacher is difficult to establish, however. Traces of the influence of British phys-

10 William C. Summers, "From culture as organism to organism as cell: historical origins of bacterial genetics," *J. Hist. Biol.* 26(1991):171–190; Olga Amsterdamska, "Stabilizing instability (1991) (n. 11).

11 Olga Amsterdamska, "Medical and biological restraints: early research on variation in bacteriology," *Social Studies of Science 17*(1987):657–687; Amsterdamska, "Stabilizing instability: the controversy over cyclogenic theories of bacterial variation during the interwar period," *J.Hist. Biol.* 24(1991):171–190.

12 Bulloch, *History of Bacteriology* (1938) (n. 3), 200–201.

13 Ludwik Fleck, *Entstehung und Entwicklung einer wissenschaftlichen Tatsache* (1935), translated by Thaddeus J. Trenn as *Genesis and Development of a Scientific Fact*, edited by Trenn and Robert K. Merton (Chicago, Ill.: University of Chicago Press, 1979), 81.

14 Henk van den Belt and Bart Gremmen, "Specificity in the era of Koch and Ehrlich: a generalised interpretation of Ludwik Fleck's 'serological' thought style," *Studies in the Hist. and Phil. of Sci.* 21(1990):463–479.

iologist Michael Foster, for example, are said to be evident in the choice of problem of his students.[15] But the traces are subtle, and indeed, arguable. Such unequivocal transmission can seldom be demonstrated and rarely extends beyond a single generation. In this case, the revolt against dividing began with the work of Matthias Schleiden the unitarian, who attacked the Linnaeans and made fun of their botany of definition and diaeresis, division and subdivision, with the weapons of the critical philosophy, the heuristic maxims of Kant and Fries. The emphasis on the unity of science, and of nature, is found again in the work of his student Nägeli, whose phrase *quantitative Abstufung* expresses his vision of a nature seamless even at the transition between organic and inorganic. Max von Gruber uses the same phrase to describe the specificity of anti-typhoid and anti-cholera antisera. Its more exact investigation was the first task he set for the young Karl Landsteiner.

The same transmission of a style of thought can be found in the opposition, with the emphasis on diversity instead of unity. The definition of species and specific differences is as marked in the work of Richard Pfeiffer, Robert Koch's student, as it is in Koch's own work and in that of the botanist Ferdinand Cohn. Koch's approach to Cohn led to a marriage of exquisitely well suited minds. Koch's visual technology and his emphasis on morphology linked up perfectly with Cohn's traditional botany to form a style that was taken up by the whole Koch school and its later descendants and the goal of which was to arrive at an accurate subdivision of species.

The difference in intellectual outlook and style of thought was mirrored in institutions as well as personal loyalties. Specificity was entrenched in Berlin, at the Kaiserliche Gesundheitsamt and in the *Zeitschrift für Hygiene*, just as unity was centred in Munich. The support of the central government of the Reich contributed to the overwhelming power of the Koch group.[16] Nationalism and Franco-Prussian enmity, as much as opposition to Nägeli's unitarianism or Louis Pasteur's claims about attenuation, underlay the charge of the Kaiserliche Gesundheitsamt.

At each stage, these two groups were conscious of the tension between them. The dialogue – one could almost call it a vendetta – continued from generation to generation with new pairs of antagonists

15 Gerald L. Geison, *Michael Foster and the Cambridge School of Physiology: the Scientific Enterprise in Late Victorian Society* (Princeton, N.J.: Princeton University Press, 1978), 338–355.
16 Richard J. Evans, *Death in Hamburg: Society and Politics in the Cholera Years, 1830–1910* (Oxford: Clarendon Press, 1987), 264–277, 502–509.

playing the same roles: Nägeli versus Cohn, Koch and Gaffky versus Nägeli and Buchner, Gruber versus Pfeiffer, sometimes in print, and sometimes face to face at meetings. Later generations, as I shall show, carried on the tradition. The argument about bacterial species was transformed into an argument about immune specificity.

PART II

The Inherited Controversy: Specificity and Unitarianism in Immunology

The word "specific" can be understood in two ways. Firstly, as the effect of immune substances which work exclusively and very strongly on one particular substrate; and secondly, as the disproportionate effect of a series of similar immune substances on a series of similar substrates.

Karl Landsteiner, "Theories of antibody formation,"
Wiener klin Wschr. 22 (1909):1623–1631

5

Dichotomy and Classification in the Thought of Paul Ehrlich

Paul Ehrlich's personal contact with Robert Koch was long and close. Ehrlich (Figure 5.1) took up the Koch manner of thinking and inherited along with it Koch's opponents, as immunology began to emerge as a branch of bacteriology.

At the very beginning of his career, when he was a student at Strasbourg, Ehrlich was introduced by his older cousin Carl Weigert and the Strasbourg anatomist Heinrich Wilhelm G. Waldeyer, to the technique of staining tissues differentially for microscopic examination by using the then-new aniline dyes coming from the coal-tar industry. Many personal reminiscences of Ehrlich describe him at all stages of his life surrounded by his bottles of dye, experimenting with them, drawing with them, or, when times were bad for him, taking comfort from just looking at them. Throughout his career, he was to maintain close relations with the chemical industry.[1]

Ehrlich's thesis was on the theory and practice of staining. He presented it at Leipzig in 1878, but he spent several semesters at Breslau, where his cousin had become first assistant to Julius Cohnheim, director of the Institute of Pathology. The young pathologist Carl Salomonsen, who came from Copenhagen to spend the summer semester in Breslau in 1877, doing pathological anatomy and bacteriology under Cohnheim and Ferdinand Cohn, has left an affectionate account of Weigert and Ehrlich in Cohnheim's laboratory:

Weigert, Cohnheim's admired and much loved first assistant, came back from his vacation. For his arrival, Lassar [Oscar Lassar, the second assistant] put a garland round both his microscopes, and decorated his chair with flowers; on the window with picric acid, fuchsin and many other aniline dyes he painted a big "Welcome." Under Weigert's leadership, in fact, microscopical staining was pushed particularly fanatically at the Breslau Institute, and the older workers

---

1 Ernst Bäumler, *Paul Ehrlich: Scientist for Life* (New York: Holmes, 1984), 40–44; 148–172; Jonathan Liebenau, "Paul Ehrlich as a commercial scientist and research administrator," *Medical History* 34(1990):65–78.

Figure 5.1. Paul Ehrlich, portrait drawing in pencil. (Courtesy of the National Library of Medicine, Bethesda, Md.)

in the laboratory spoke with a certain pride of the beautiful "Breslau prepa-rations." Only Cohnheim himself was rather cool to the great technical ad-vances that were made possible by the gigantic development of the aniline dye industry. "Painting again, eh?" or "You're the Master Dyer," he would say ironically from time to time.

The real Master Dyer in the laboratory was a blond, twenty-two year old lad,

Paul Ehrlich, who was always working away with nervous zeal. . . . An old Breslau comrade once said to me in Berlin, "Do you remember Ehrlich as a student in Breslau? We used to laugh at him for always running about with blue, yellow, red and green fingers."[2]

This was also the time of Koch's visits to the Institute for Plant Physiology at Breslau. Koch's diary describes the first one of these visits, when he was taken to see Cohnheim and Weigert and the staining procedures.[3] Cohnheim and Weigert too were present at the famous demonstration on the life cycle of the anthrax bacillus.[4] Salomonsen tells the story of a trip to Friedland by train, in a specially booked first-class carriage. It was an expedition of the botanical section of the Schlesische Gesellschaft für vaterländische Kultur to which he and his best friend William Henry Welch from America, both studying at the Institute of Pathology, were invited by Cohn. Cohn, who was the society's secretary, gave a short lecture and demonstration of Koch's photographs of bacteria.[5] (See Figure 2.6.)

It is a picture of close and friendly working relations between the Institutes of Pathology and the Institute of Plant Physiology. As mentioned earlier, Cohn sent across to Cohnheim to come to see Koch's demonstration, and Cohnheim assisted Koch with it later. Cohn the director, Weigert the assistant, Ehrlich the student, and Koch the visitor all had ample opportunity to exchange ideas and did so with full acknowledgements.

Paul Ehrlich's characteristic style of thought is recognizable in his earliest work. In his dissertation of 1878, Ehrlich presents himself with a choice between two possible explanations of the chemistry of dyeing: he chooses the one that implies diversity and specificity, and rejects the one that implies unity and continuity in nature.[6]

2 Carl J. Salomonsen, "Lebenserinnerungen aus dem Breslauer Sommersemester 1877," *Berl. klin. Wschr.* *51*(1914):485–490 (p. 486). On Julius Cohnheim and his work on the relation between white blood cells and pus, see Carl Weigert, "Nekrolog von Julius Cohnheim," *Berl. klin. Wschr.* *21*(1884):564–565; Russell C. Maulitz, "Rudolf Virchow, Julius Cohnheim and the program of pathology," *Bull. Hist. Med.* *52*(1978): 162–182.

3 Robert Koch, cited in Bruno Heymann, *Robert Koch I Teil 1843–1882* (Leipzig: Akad. Verlag, 1932), 154.

4 Heymann, *Robert Koch* (1932) (n. 3), 150–151.

5 Salomonsen, "Aus dem Breslauer Sommersemester 1877" (1914) (n. 2), 488.

6 Paul Ehrlich, "Beiträge zur Theorie und Praxis der histologischen Färbung: Inaugural Dissertation, Universität Leipzig 1878,", in Felix Himmelweit, Martha Marquardt and Sir Henry Dale, eds., *The Collected Papers of Paul Ehrlich, in four Volumes, including a Complete Bibliography* (London: Pergamon, 1956), v.4. The bibliography has never been published; v. 1, 29–64 (German), 65–94 (English). This will be referred to as *Coll. Papers*; citations and quotations are of the English translations in this collection, where they exist. The best bibliography, going to February 1914, within two years of Ehrlich's

He begins with a dichotomy: theories of dye action are of two kinds. First, there are those that suggest that dye uptake is a chemical reaction: dye and tissue unite chemically, to form a lake, which differs from the normal lake of chemistry only in that a metal oxide is replaced by an organic substance. The second type of theory sees dyeing as a physical, surface phenomenon, like the uptake of substances by charcoal. In support of this explanation, it is said that the dye and the tissue do not unite stoichiometrically, that the law of definite proportions, which characterizes a true chemical reaction, does not apply here. Ehrlich's preference is for the first, the chemical type of explanation, and he presents a new one of his own. A chemical compound is formed by dye and tissue: it is a double salt, Ehrlich suggests, rather than the usual lake. He takes for his example the double salt that is formed when silver chloride is mixed with corrosive sublimate (mercuric chloride). This is not a physical effect due to the finely divided state of the silver chloride: other mercuric salts are not taken up by silver chloride, and physical surface phenomena are never confined like this to a single member of a group. The compound is a double salt, but incompletely formed, so that fixed equivalent relations are not found. The effect, Ehrlich is arguing, must be chemically specific because it occurs *only* between *these* two salts. It is analogous to, if not exactly the same as, the specific staining of histological preparations.

Ehrlich classifies the dyes themselves by a series of dichotomies. In order to be active at all as dyes, they must have two kinds of groups in the molecule: a chromophore group to give the substance its colour, and a salt-forming group to bind to the tissue. The binding groups in turn are of two kinds, acidic and basic.

The explanation Ehrlich chooses here is typical of all his later work. He rejects the explanation that suggests unity and continuity, in this case the physical adsorption one, which would have required the differential effects to be quantitative. He prefers the explanation that suggests discontinuity and specificity. We do not need to bring in *quantitative Abstufung* with this theory, because the effect works *only* with *one* of the salts of mercury, the chloride, and not with the others: it either works or it does not.

Ehrlich first applied this insight into the specific nature of the chemistry of dyeing to the classification of white blood cells. The problem of the relationship between white cells and pus, using dye-tagged cells, was an important part of the Breslau experimental work under Julius

death, is by Hans Sachs, in Hugo Apolant et al., *Paul Ehrlich eine Darstellung seines wissenschaftlichen Wirkens: Festschrift zum 60 Geburtstage des Forschers* (Jena: Fischer, 1914) 625–657. This also includes work done under Ehrlich's direction.

Cohnheim. But Ehrlich's methodology made as much use of Ferdinand Cohn's traditional Linnaean or Aristotelian botany – with its diaeresis and systematic classification, its goal of the definition and diagnosis of species – as it did of the programme of the pathologists. It is all the more striking in that he applies the method not to botanical species but to chemical substances and blood cells. His objects of research are not species at all in the usual natural-history sense, but he treats them exactly as if they were. These characteristics of his thought appear in all his work on blood and are particularly clear in the following passages:

> More than ten years ago, Max Schultze indicated that the circulating leucocytes did not form a morphological unity, but could be divided into several anatomically distinct groups. In spite of this stimulating suggestion, there has been no further progress in this direction. . . . In an earlier communication, I showed that it was possible to divide the leucocytes into several sharply characterised separate classes by the use of dyes. Owing to the presence of certain granules, some fine, some coarse, in the protoplasm of the white cells, which can be distinguished by their reaction to certain dyes or groups of dyes, I have been able to divide the morphologically similar elements from each other, on the one hand, and on the other to show that some of different habitus actually belong together. . . . Because of these defining properties, I have called the granules – so far I have found five types of them – specific granulations. My research has convinced me that the granulations I have demonstrated products of specific secretory activity of the cell. . . . [7] I have found that each type of granulation bound only dye stuffs with certain properties. . . . One group is that of the basic aniline dyes. . . . The other is that of the acid dyes. . . .
>
> I found that the granulations which I described earlier as eosinophil . . . were intensively coloured by all acid dye stuffs – and I tried over 30 of them – and not by any of the basic dyes. Exactly the opposite was true of the [other] granulations.[8]

Ehrlich divides the dyes into three groups: basic, neutral, and acidic (Figure 5.2). He then subdivides the acid stains into two subgroups. Group I stains the granules strongly in glycerine solution. It has three sub-subgroups: (a) strongly acid dyes including eosin; (b) strongly acid nitrous substances, such as aurantia; and (c) the sulphonic acid derivatives, which in turn can be divided into two further groups, the first being slowly diffusible water-soluble dyes such as nigrosin, the second newer azo-dyes such as Ponceau red. Group II, the less acidic stains,

7 Max von Gruber, *Münch. med. Wschr.* 50(1903):564–568 (p. 566).
8 Paul Ehrlich, "Methologische Beiträge zur Physiologie und Pathologie der verschiedenen Formen der Leukocyten," *Z. f. klin. Med. 1* (1880):533–560; and in Ehrlich, ed., *Farbenanalytische Untersuchungen zur Histologie und Klinik des Blutes: Gesammelte Mittheilungen* (Berlin: Hirschwald, 1891) 42–50; and in *Coll. Papers* (1956) (n. 6), v. 1, 124–129 (p. 124).

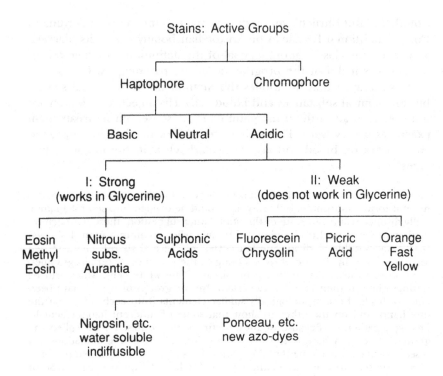

Figure 5.2. Ehrlich's classification of stains. The di- and tri-chotomies of the stains correspond to those of the white cell granulations, from his "Verschiedene Formen der Leukocyten" (1880) (n. 8). Visualized from his dissertation, "Theorie und Praxis der histologische Färbung" (1878) (n. 6).

has three subgroups: (a) fluorescein and chrysolin, (b) ammonium picrate and naphthyl yellow, and (c) orange and fast yellow.

Since a large number of leucocytes were still not stained, the next step was to try these with neutral dyes, which Ehrlich produced by mixing a dye base and an acid. The granules in the remaining leucocytes were then successfully coloured, and he named these cells neutrophils.[9]

The classification of stains matches the corresponding classification of white cells: the differentia of the blood cell species are those of the stains, and they are named accordingly, although the classification of the stains is carried into further dichotomies. Morphological consid-

9   Paul Ehrlich, "Ueber die specifische Granulationen des Blutes," lecture of 15 May, in Verhandlungen der Berliner physiologische Gesellschaft, *Arch f. Anat.u Physiol.* (Physiol. Abt.) (1878–1879), 571–579, and in *Farbenanalytische Unters.* (1891) (n. 8), 5–16; and in *Coll. Papers* (1956) (n. 6), v. 1, 117–123 (p. 118).

erations are subordinated to staining properties: for classification pur-
poses, the essence of the cells is the essence of the stain. We are
reminded of Koch's postulates, in which the disease is defined by the
bacterium, and vice versa, and of Klebs's suggestion that the classification of bacteria can be transferred to diseases.[10]

The similarities between this work of Ehrlich's and that of Koch are
quite striking. Ehrlich repeatedly refers to Koch's new method of making slides of bacteria by smearing them on the slide and drying it in
air.[11] Ehrlich makes his preparations by putting a drop of blood on the
slide and using another slide to spread it thinly. The cells then do not
overlap. Like Koch, he found that the combination of thin film and air
drying gave the best preservation of fine morphological detail:

The following research was set up using Koch's procedure, in which the fluid
(blood) or the parenchyma of the organs (bone marrow, spleen, etc.) were
spread in the thinnest possible film on a slide, dried at room temperature and
then stained for as long as necessary. . . . I soon found that from a purely descriptive standpoint the method gave outstanding results; not only the gross
form but also certain fine and extra-fine structural elements, such as the nuclear
reticulum were strikingly well preserved.[12]

The style of thought, the insistence on differentiation and specificity,
and the techniques are very like those of Koch, who in turn referred
to Ehrlich for his use of stains.[13] The same visual technology is used
for the same Linnaean ends: in the one case specific differences between cells; in the other, between bacteria.

Ehrlich apparently maintained his contacts with Koch after his appointment to the Second Medical Clinic at the Charité Hospital in
Berlin in 1878. He was present, according to Heymann, who was there
himself, at Koch's demonstration of the tubercle bacillus in 1882.[14] A

10 For discussion of this, see Chapter 2, n. 47.
11 Ehrlich, "Verschiedene Formen der Leukocyten" (1880) (n. 8), p. 124.
12 Ehrlich, "Spezifische Granulationen" (1879) (n. 9), p. 117.
13 Robert Koch, "Zur Untersuchung von pathogenen Organismen," *Mitth. aus dem
    Kaiserl. Gesundheitsamte 1*, Art. 1 (1881); and in Georg Gaffky, E. Pfuhl, and J.
    Schwalbe, eds., *Gesammelte Werke von Robert Koch* (Leipzig: Thieme, 1912) v. 1, 112–
    163 (pp. 116–120). Koch also refers to Ehrlich's dried preparations of the granulated
    white blood cells (n. 8) and to his aniline-dye staining. Ehrlich, "Ueber das Methy-
    lenblau und seine klinisch-bakterioskopische Verwertung," *Z. f. klin. Med.* 2(1881):
    710–713; Koch, "Ueber die Atiologie der Tuberkulose," in *Verh. des Kong. f. inn. Med.*
    (Wiesbaden: Bergmann, 1882), 56–68; 68–78, discussion following presentation; and
    in *Ges. Werke*, 446–453 (p. 447) for the earliest staining method for tubercle bacilli.
14 Heymann, *Robert Koch* (1932) (n. 3), p. 324. Ehrlich himself wrote of the deep im-
    pression it made on him, in his "Nachruf auf Koch," in the *Frankfurter Zeitung*, 2
    June 1910: quoted by Max Neisser, "Bakteriologie," in Apolant, *Ehrlich Festschrift*
    (1914) (n. 6), 83–106 (p. 92).

few weeks later, he produced a modification of Koch's staining method, which he had watched at the Kaiserliche Gesundheitsamt.[15] Ehrlich's staining is "specific" for the tubercle bacillus, which, unlike other common organisms of the sputum, is enclosed in a waxy hull. Because of the wax, once stained, the organism resists decolourization by strong mineral acids, which bleach out everything else on the slide. This staining technique, with a minor modification, has come to be called the Ziehl–Nielsen method for the demonstration of tubercle bacilli.

In 1890, after a year and a half in which Ehrlich's work was interrupted by a tuberculous infection – probably picked up in the laboratory, as with so many other pathologists – Koch invited Ehrlich to join him at the new Institute for Infectious Disease. Here he came into contact with Emil von Behring, who in 1890 with the Japanese bacteriologist Kitasato, published his work on passive immunization with immune serum. Upon its discovery, immune serum was quickly adopted as a treatment for diphtheria and was immediately successful in Germany.[16]

The immune serum was also seen as a means of defeating the phagocyte theory of immune defence, proposed by Ilia Ilich (Elie) Mechnikov and supported by the group at the Institut Pasteur in Paris. The debate was dramatic: it was played out in all the international congresses between 1890 and 1910. The phagocyte theory, according to some historians of immunology, was the first intimation that immunity was an active response on the part of an infected host, and the "humoral" or serum-based theory, though it eventually prevailed, was itself dependent on Mechnikov's proposal of an active defence mechanism.[17] It has even been suggested that the debate was, if not one of the causes of the Great War, at least a prodromal sign of it.[18]

In clinical practice, the variable potency of the diphtheria antiserum made it unreliable and difficult to use. These practical problems led to the establishment of the state-supported Institute for Serum Testing and Research in Steglitz near Berlin in 1896. With the help of Robert Koch, Ehrlich was appointed its director. His mandate was to maintain

15   Paul Ehrlich, "Ueber die Färbung der Tuberkelbacillen," Aus den Verein für innere Medizin, 1 May 1882; in *Deutsch med. Wschr.* 8(1882):269–270; and as Ehrlich, "Modification der von Koch angegebenen Methode der Färbung von Tuberkelbacillen," *Coll. Papers* (1956) (n. 6), v. 2, 311–313.
16   William D. Foster, *A History of Medical Bacteriology and Immunology* (London: Heinemann, 1970), 103–104.
17   Alfred I. Tauber and Leon Chernyak, *Metchnikoff and the Origins of Immunology: From Metaphor to Theory* (Oxford: Oxford University Press, 1991), 153–159.
18   Pauline M. H. Mazumdar, "Immunity in 1890," *J. Hist. Med.* 27(1972):312–324; Anne-Marie Moulin, *Le dernier Langage de la Médecine: Histoire de l'Immunologie de Pasteur au SIDA* (Paris: Presses Universitaires de France, 1991), 67–73.

the quality control of the diphtheria serum, which was to be produced by the erstwhile dye-chemistry firm of Hoechst, now entering the field of pharmaceuticals.[19] One year later, Ehrlich brought out the paper on the assay of the diphtheria antiserum that contained the first statement of the *Seitenketten* or side-chain theory. Argument over this theory raged for the next twenty years. In some ways, the side-chain theory even now has not quite disappeared from immunology.[20]

The diphtheria serum initiated an age of international serology. Its importance was guaranteed by its striking clinical effectiveness, when properly standardized. Production, standardization, and distribution of sera became the responsibility of states, through state serum institutes, and later, after the end of the First World War, through the League of Nations Standardisation Commission. Theory followed practice: immunological thought in the first half of the twentieth century had its source in the practical problems of the raising and standardization of antisera, and the nature of the reaction between antigen and antibody.[21]

Ehrlich developed his theory of immunity during the second of the three phases of his life.[22] It has often been pointed out that the ideas used in his discussion of the staining process in the thesis of 1878 are repeated in his later speculations on immunity (the famous side-chain theory), and again in his work on chemotherapy.[23] In each case, an active molecule is visualized as consisting of a central nucleus with two side chains. One of these, the haptophore group, links the molecule to a target; the other, a chromophore group (or toxophore, as the case may be) exerts some effect upon it. The haptophore group provides a

19 Bäumler, *Paul Ehrlich* (1984) (n. 1), 53–63.
20 Paul Ehrlich, "Die Wertbemessung des Diphtherie-heilserums, und deren theoretische Grundlagen," *Klin. Jahrb.* 6(1897–1898):299–326; and in *Coll. Papers* (1956) (n. 6), v. 2, 86–106 (Ger.), 107–125 (Eng.); Pauline M. H. Mazumdar, "Paul Ehrlich and the history of immunology," *Scientific American* (March 1995).
21 Pauline M. H. Mazumdar, "Immunology, a history," in Kenneth F. Kiple, ed., *Cambridge History and Geography of Human Disease* (Cambridge: Cambridge University Press, 1992).
22 Sir Henry Dale, "Introduction" to Coll. Papers (1956) (n. 6), v. 2, 1–6 (p. 1). Dale divides Ehrlich's life and work into three phases, which he represents by his three volumes of collected papers: v. 1 (1877–1891) contains the work on dyes and staining; v. 2 (1891–1905) the work on immunology; and v. 3 (1905–1915) the work on chemotherapy. Dale follows Ehrlich himself (in a letter to Christian A. Herter, published in translation in *Coll. Papers*, v. 1, 9–11) in tracing the enchainment of his ideas, from the "visual, three-dimensional chemistry" of the aniline dyes, to his visualization of the reducing action of cells on different dye stuffs, to the side-chain theory, and thence to the visualization of the chemotherapeutic effect of drugs on organisms.
23 Martha Marquardt, *Paul Ehrlich* (New York: Schumann, 1951), 17–18; Leonor Michaelis, "Zur Erinnerung an Paul Ehrlich: seine wiedergefundene Doktordissertation," *Naturwissenschaften* 7(1919):165–168.

chemical binding to the target in every case. Ehrlich himself expressed this in his later chemotherapy period, by the motto *Corpora non agunt nisi fixata*.[24]

An important feature of each of these theories is the specificity of this fixation. The chemical binding of the haptophore group in each case implies a strictly specific relationship between haptophore and target groups. For Ehrlich, the "purely chemical" is meant to exclude the gradual: there are no smooth, quantitative transitions. Like his theory of stains and white cells, his theory of toxins and antitoxins explained what he saw as the strictly specific nature of antibodies.

Ehrlich's paper on the assay of the diphtheria serum brought into prominence the peculiar relationship of the toxin to its antitoxin. The basic premises of Ehrlich's interpretation of the relationship appear in this article:

The reaction of toxin and antitoxin takes place in accordance with the proportions of simple equivalence, especially since it is possible to titrate the antibodies with great accuracy – in favourable circumstances, the error with the present method is 1%. *A molecule of toxin combines with a definite and unalterable quantity of antibody* [Ehrlich's emphasis]. It must be assumed that this ability to combine with antitoxin is attributable to the presence in the toxin complex of a specific group of atoms with a maximum specific affinity to another group of atoms in the antitoxin complex, the first fitting the second easily, as a key does a lock, to quote Emil Fischer's well known simile.[25]

Ehrlich's assay procedure began by defining a lethal dose of the toxin as the minimum dose needed to kill a 250-gram guinea-pig within four days. One unit of antitoxin consisted of an amount that would neutralize 100 lethal doses (L.D.) of toxin. These values were defined using a standard toxin and a standard antiserum, and new batches of either were assayed by comparison with these values. Thus, the Lo dose of a new toxin was the number of lethal doses neutralized by one unit of the original antitoxin, and the L+ dose of the new toxin was the number of lethal doses just *not* neutralized by one unit of the original an-

24   Paul Ehrlich, "Schlussbetrachtungen," in Ehrlich and Sahashiro Hata, *Die experimentelle Chemotherapie der Spirillosen* (Berlin: Springer, 1910), translated by A. Newbold and revised R. W. Felkin (London: Rebman, 1911), 117–156 (p. 117): "If for chemistry the law applies: *corpora non agunt nisi liquida*, then: *corpora non agunt nisi fixata* is appropriate for chemotherapy." Martha Marquart, Ehrlich (1951) (n. 23), tells of Ehrlich writing this in an autograph book at the request of its owner. It is also quoted by K. Bierbaum, "Salvarsan bei Tierkrankheiten," in Apolant, *Ehrlich Festschrift* (1914) (n. 6), 617–624 (p. 624). There seems to be no record of him saying this any earlier, for example, in his immunology period.

25   Ehrlich, "Wertbemessung des Diphtherieheilserums" (1897) (n. 20), p. 116 (Ger.), p. 93 (Eng.).

titoxin. Theoretically, $L+ - Lo = 1$ L.D. But, in practice, the difference was always greater than one, and as the toxin solutions aged, the difference increased. The Lo dose, a measure of the power of the toxin to neutralize antitoxin, remained constant, but the L+ dose, a measure of its toxicity to guinea-pigs, increased as the toxicity declined. A toxin might be more or less lethal but still neutralize the same amount of antiserum. Ehrlich interpreted this in terms of his side-chain theory to mean that the toxophore groups alone were labile, while the haptophore groups remained unchanged.

Ehrlich titrated the batches of toxin with immune serum by adding a series of aliquots of antiserum to a standard quantity of toxin, and then measuring the toxicity still remaining in terms of its effect on guinea-pigs. In any given toxic broth, successive aliquots of antitoxin had unequal effects on the toxicity. One-fifth of a unit of antiserum, binding 20 percent of the haptophore groups, might not diminish the toxicity at all; two-fifths, while binding 40 percent of the haptophore groups, diminished the toxicity by some other percentage. Ehrlich interprets this as showing that highly avid components of the toxic mixture take up the antitoxin first, and successively less avid later. He illustrates it by a stepped diagram that he calls the *Giftspectrum*. Its profile represents the relation between toxicity and avidity for antiserum at each stage of neutralization (Figure 5.3). For each step, Ehrlich postulates a separate component of the toxin complex.

One of the most convincing pieces of evidence that the toxins contained different components lay in the finding that a mixture of toxin with a large amount of antitoxin was no longer lethal to guinea-pigs, but produced a new and different effect, that of paralysis and oedema. As Thorvald Madsen, who collaborated with Ehrlich in this work at Steglitz, said later, "It was an obvious conclusion that these two actions which differed so much were not the consequence of only a quantitative difference, but that the diphtheria toxin really contained two different components."[26]

The separate components of the mix were each assigned a name according to their position in the *Giftspectrum* and their relative affinity and toxicity. Their complicated relationships were sympathetically explained by the pathologist Ludwig Aschoff in a detailed exposition of the side-chain theory of 1902.[27] Those with the least affinity, Ehrlich at

26 Svante Arrhenius and Thorvald Madsen, "Physical Chemistry applied to toxins and antitoxins," in Carl Julius Salomonsen, ed., *Festskrift ved indvielsen af Statens Seruminstitut 1902: Contributions from the University Laboratory for Medical Bacteriology to Celebrate the Inauguration of the State Serum Institute* (Copenhagen: Olsen, 1902), 1–111, chap. 3 (p. 71).

27 Ludwig Aschoff, *Ehrlich's Seitenketten-theorie und ihre Anwendung auf die künstlichen Im-*

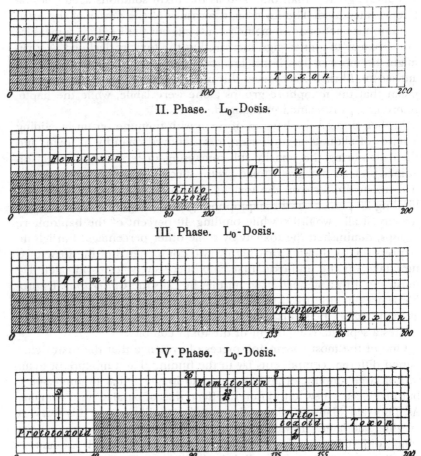

Figure 5.3. Ehrlich's stepped *Giftspektrum* (1897–1898). The neutralization of a deteriorating diphtheria toxin by its antitoxin: four stages in the history of toxin no. V. At each stage, it contains different breakdown products, which react with the antitoxin in order of affinity. Some have lost their toxophore groups, and are no longer toxic, but they still neutralize antiserum. From Ehrlich, "Wertbemessung des Diphtherieheilserums" (1897) (n. 20).

first named epitoxoid, but later toxone. These components had no true lethal toxic effect but produced only paralysis. The active toxins were

munisierungsprozesse: zusammenfassende Darstellung (Jena: Fischer, 1902), p. 79. Reprinted from *Z. für allgem. Physiol.* 1 "Sammelreferate" (1902):69–249 .

divided into three subtypes, according to their affinity for the antitoxin: proto-, deutero-, and trito-toxin, reacting in that order. Each had a sub-subtype, named *alpha* and *beta*. The *alpha*-prototoxin, for example, was more labile than the *beta*-prototoxin and quickly changed into a nontoxic, or toxoid, form; the *beta*-subtypes broke down more slowly. In an old and deteriorated sample, the only remaining toxic component would be the *beta*-deuterotoxin.

This proliferation of names and substances was a consequence of Ehrlich's basic assumptions. If, as he assumed, a molecule of toxin combined with a definite and unalterable quantity of antibody according to a law of simple proportions, the different relationships of toxicity lost to antibody bound had to represent different component substances. The whole construct, as Madsen was to realize later, was an attempt to explain a continuous curve in terms of multiple specific discontinuities.[28] The schema may also be envisaged as a di- or trichotomy, similar to that which Ehrlich had suggested for stains and for white blood cells. The first *fundamentum divisionis* is that of the disease symptoms produced in guinea-pigs, where toxin is lethal, and toxone produces only paralysis. The next one is degrees of affinity for antitoxin, and the third rates of breakdown of toxin to toxoid (Figure 5.4).

Soon after the Steglitz Institute began testing and issuing clinically reliable diphtheria serum, Ehrlich started negotiating with the city of Frankfurt-am-Main to move the institute there. This location had advantages: Emil von Behring was nearby at Marburg, Ehrlich's cousin Carl Weigert was pathologist to the Senckenberg Institute in Frankfurt, and, most persuasive, the chemical firm of Hoechst, the manufacturer of the serum under Ehrlich's direction, was in Frankfurt.[29] The Royal Institute for Experimental Therapy was opened in 1899 in Sachsenhausen, across the river from Frankfurt.

Ehrlich's conception of the antibodies that act on red cells and on bacteria was developed between 1899 and 1901, in a set of six "Communications on haemolysis," produced in collaboration with Julius Morgenroth, who had been at Steglitz, and who in 1898 came with Ehrlich to Frankfurt.[30] In these communications, he applied the side-

28 Arrhenius and Madsen, "Physical chemistry applied to toxins and antitoxins" (1902)
     (n. 26), 58. Madsen had already begun to feel this when, from working with the
     diphtheria toxin, which was assayed by guinea pig toxicity, he went on to the tetanus
     toxin, whose toxicity could be indicated by red cell haemolysis. In his "Ueber tetan-
     olysin," *Z. f.Hyg.u. Infektionskr. 32*(1899):214–237, written while he was working with
     Ehrlich at Steglitz, Madsen says (p. 226) that the effect of the tetanolysin is "probably
     best expressed as a curve falling smoothly from left to right." Nonetheless, he still
     expressed his results in one of Ehrlich's stepped diagrams (p. 225).
29 Bäumler, *Ehrlich* (1984) (n. 1), 68–78.
30 Paul Ehrlich and Julius Morgenroth, "Zur Theorie der Lysinwirkung," *Berl. klin.*

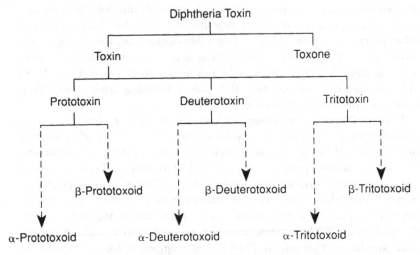

Figure 5.4. Ehrlich's classification of the components of a diphtheria toxin. The scheme is a di- and tri-chotomous tree based on three successive *Fundamenta divisionis.* Visualized from his argument in "Wertbemessung des Diphtherie-heilserums" (1897) (n. 20).

chain theory to antibodies that were produced in the laboratory by cross-immunizing various species of animal with the red cells of other species. In one case, a group of goats was immunized with each others' blood. These experiments are discussed further in Chapter 7.

As usual in Ehrlich's work, his experiments with the antibodies produced a large new vocabulary, much of which is still in use, beginning with the words "haemolysis" and "haemolysin." Antibodies or lysins were divided into auto-, iso-, and heterolysins, according to whether they were to act on the animal's own cells, cells of another animal of the same species, or those of a different species.[31] Ehrlich felt autoan-

Wschr. 36(1899):6–9, and in *Coll. Papers* (n. 6), v. 2, 143–149 (Ger.), 150–155 (Engl.); Ehrlich and Morgenroth, "Ueber Haemolysine: zweite Mittheilung," *Berl. klin. Wschr. 36*(1899):481–486, and in *Coll. Papers* (n. 6), v. 2, 156–164 (Ger.), 165–172 (Engl.); Ehrlich and Morgenroth, "Ueber Haemolysine: dritte Mittheilung," *Berl. klin. Wschr. 37*(1900):453–458, and in *Coll. Papers* (n. 6), v. 2, 196–204 (Ger.), 205–212 (Engl.); Ehrlich and Morgenroth, "Ueber Haemolysine: Vierte Mittheilung," *Berl. klin. Wschr. 37*(1900):681–687, and *Coll. Papers* (n. 1), v. 2, 213–223 (Ger.), 224–233 (Engl.); Ehrlich and Morgenroth, "Ueber Haemolysine: fünfte Mittheilung," *Berl. klin. Wschr. 38*(1901):251–257, and in *Coll. Papers* (n. 6), v. 2, 234–254 (Ger.), 246–255 (Engl.); Ehrlich and Morgenroth, "Ueber Haemolysine: sechste Mittheilung," *Berl. klin. Wschr. 38*(1901):569-574, 598–604, and in *Coll. Papers* (n. 6), v. 2, 256–277 (Ger.), 278–297 (Engl.). Page references to these papers are to the collected edition.
31   Ehrlich and Morgenroth, "Dritte Mittheilung" (1900) (n. 30), 199 (Ger.), 207 (Engl.).

tibodies to be inherently unlikely, dysteleological in the highest degree, a feeling that he ascribed to the body with the phrase *horror autotoxicus*.[32]

In terms of the side-chain theory, an antibody was conceived of as having two haptophore groups, one of which united with a receptor on the cell or bacterium, the other with alexin, or, as Ehrlich preferred to call it, complement.[33] The whole then comprised a lysin, by analogy with a toxin, or with a dye, the group corresponding to chromophore or toxophore, which in this case was detachable from the rest of the molecule. He called the antibody *Zwischenkörper* or *Amboceptor*, as it formed a link between cell and alexin.[34] In this he opposed the theory of Jules Bordet, of the Institut Pasteur, who regarded the antibody as a sensitizer that made the cell itself sensitive to the destructive action of the alexin.[35] According to Ehrlich, both the haptophore groups of the amboceptor were specific, implying necessarily that each antibody had a matching specific alexin or complement. In contrast, Bordet's and Hans Buchner's unitarian theory postulated only a single alexin common to all antibodies.[36] Ehrlich writes:

When Buchner criticises our position by saying that the conception of different substances runs counter to the economy of thought, we reply that our inferences are not speculative but follow inescapably from observations that are incompatible with the assumption of any unitary alexin. It will also be obvious why we prefer to drop the term alexin chosen by Buchner. We have found in our experiments in every case we have investigated carefully, not Buchner's unitary alexin, but the complex haemolysin consisting of amboceptor and complement.[37]

The passage that Ehrlich is paraphrasing here comes from a paper of 1900, in which Buchner criticizes the side-chain theory together with Ehrlich's conception of alexin action. "It obviously contradicts all economy of thought, when two substances are needed for a single effect on

32 Ehrlich and Morgenroth, "Dritte Mittheilung" (1900) (n. 30), 198 (Ger.), 206 (Engl.); and Ehrlich and Morgenroth, "Fünfte Mittheilung" (1901) (n. 30), 242 (Ger.), 253 (Engl.). On Ehrlich's horror autotoxicus, see Arthur M. Silverstein, *A History of Immunology* (New York: Academic Press, 1989), 160–189.

33 Ehrlich and Morgenroth, "Theorie der Lysinwirkung" (1899) (n. 30), 148 (Ger.), 154 (Engl.).

34 Ehrlich and Morgenroth, "Zweite Mittheilung" (1899) (n. 30), 159 (Ger.), 168 (Engl.).

35 Jules Bordet, "Agglutination et dissolution des globules rouges par le sérum: deuxième mémoire," *Ann. de l'Inst. Pasteur* 13(1899):273–297.

36 Hans Buchner, "Zur Kenntniss der Alexine, sowie deren specifisch-bactericiden und specifisch-haemolytischen Wirkung," *Münch. med. Wschr.* 47(1900):277–283.

37 Ehrlich and Morgenroth, "Vierte Mittheilung" (1900) (n. 30), 217 (Ger.), my translation.

a third (here the specific bacteria), to suppose that these two substances can not act directly on the *objectum reactionis*, but before acting on it must first act on and change each other is some dark and secret manner."[38] It is also the sheer number of the different substances involved that Buchner finds unbelievable in the side-chain theory of the origin of antibodies as part of the cell's normal nutrition apparatus:

> The wider our experience of specific immunization becomes, the more difficult I find it to accept that the innumerable specific antibodies which we are finding can be accounted for by particular side-chains of the organism. This goes especially for the specific-haemolytic antibodies. If we consider that apparently – according to the positive findings to date – all or most of the species of warm blooded animals can be immunised against all other species, there seems to be an almost infinitely large number of interrelationships, so that to postulate a preformed chemically different side-chain, which would serve as a basis for each specific antibody, simply will not do.[39]

Two aspects of Buchner's thought appear in these passages. First, he is opposed to Ehrlich. Second, he finds the side-chain theory offensive in its lack of economy: there are too many different alexins, and too many different side chains. Buchner's criticisms, irrespective of their validity or invalidity, point up the most characteristic feature of Ehrlich's speculative thinking: his tendency to postulate a new specific entity to account for every phenomenon. It is no accident that he was unimpressed by Buchner's appeal to economy of thought. He himself called his theory a "pluralistic" one: he postulates a multiplicity of different receptors on cells, a multiplicity of preformed side chains, a multiplicity of different complements.[40] Wanting another Latin motto, he might have chosen *divide et impera*.

---

38 Hans Buchner, "Alexine" (1900) (n. 37), 29: "Denn es widerspricht ersichtlich jeder ökonomie des Denkens," my translation. The *Coll. Papers* version fails to convey what I see as the unitarian aspect of Buchner's thought.
39 Buchner, "Alexine" (1900) (n. 39), 283.
40 Ehrlich and Morgenroth, "Sechste Mittheilung" (1901) (n. 30), 257 (Ger.), 279 (Engl.): "Dieser von uns eingenommene *plurimistische* Standpunkt schafft ja für ein eingehenderes analytisches Arbeiten auf diesem Gebiet zahlreiche Unbequemlichkeite führt aber gleichzeitig zu einem tieferen Eindringen in die verwickelten Probleme."

6

Max von Gruber and Paul Ehrlich

Though Ehrlich was never, strictly speaking, a student of Robert Koch – he was more like a younger colleague – his theories, his manner of thinking, and his career were all closely related to Koch's, and he admired Koch deeply. As he wrote on the occasion of Koch's death in 1910:

The whole stately edifice that today's knowledge of infectious diseases and their related scientific disciplines can call her own rests upon foundations laid by Robert Koch. If one were to remove the pillars and keystones set in place by his own powers, or under the influence of his spirit, the whole proud structure would collapse, as miserable *Stückwerk*.[1]

Stückwerk, which means "patched together out of scraps," also carries another connotation. In Luther's Bible, "We know in part and we prophesy in part," is *Unser Wissen ist Stückwerk*.[2] Ehrlich's feeling would seem to be that in Koch that which is perfect is come, and that which is in part is done away.

It is not surprising that both Hans Buchner and Max von Gruber, who opposed all that Koch stood for, opposed Ehrlich, too. Gruber (Figure 6.1) attacked Ehrlich as energetically and as publicly as he had once attacked Richard Pfeiffer. In this controversy, however, Gruber's criticism had a more distinct sound of animosity and bitterness, and less intellectual content.

While he was still in Vienna (he succeeded Buchner in Munich in 1902), Gruber gave a lecture at the k. k. Gesellschaft der Aerzte, the Royal Society of Medicine of Vienna, the purpose of which was to examine the foundations upon which the Ehrlich theory rested. Even if a theory were false, it might still be useful, but Ehrlich's he felt had

1 Paul Ehrlich, "Robert Koch," *Z. f. Immunitätsf.*, I Teil (orig.) *6* (unpaginated, preceding title page) (1910).
2 Saint Paul, "First Epistle of Saint Paul the Apostle to the Corinthians," *13*:9–10, translated by Martin Luther.

been harmful. It filled in the gaps in knowledge with words, so that people forgot the gaps were there. He examined the theory step by step, beginning with its inception in the toxin–antitoxin relationship. He concluded that it was completely speculative, that it all might be so but Ehrlich's theory could not be proved:

> His method is not capable of giving decisive results. . . . It is quite likely that the toxicity of the toxin and the affinity of toxin for antitoxin rest on two quite different but dependent atom complexes in the toxin molecule. It is a harmless amusement to call one the haptophore and one the toxophore.[3]

His criticism is mainly along these lines – not very damaging, one might think. Only in the second-last paragraph of this section does he touch on a more important problem, but – probably because Gruber himself, as he rather naively admits, had no personal experience of handling toxins – he passes over it without much comment:

> If I understand rightly the material on the decrease in lethality of the toxin solutions – I myself have no experience of it – it seems that of necessity this decrease can not be due simply to a gradual conversion of toxin into toxoid. It appears, from the material already dealt with, as if the lethality measured by the number of fatal doses per unit volume changes not steadily but by leaps, and also in simple proportions. A toxin solution which today has a toxicity of 1 shows after some time toxicity of 2/3, 1/2 or 1/4. There is no indication that this change takes place gradually or steadily. Now if this is so, there is *something missing*, which can be looked for only in chemical processes in which the toxin itself plays no part, which go on continually and lead to sudden alterations in the toxin or the toxicity. It is impossible that a steady influence . . . alone could account for a change that takes place in sudden jumps.[4]

Gruber is prepared to accept both the different components of the toxins and the independent active groups, and also the sudden jumps by which Ehrlich sees the change in toxicity taking place, but he feels that there is an element missing. Though hostile to Ehrlich, Gruber has not really put his finger on the problem. It had not occurred to him yet that the "something missing" was an artefact of Ehrlich's interpretation.

The next section of Gruber's lecture deals with Ehrlich's theory of lysis. Here Gruber, like Buchner, feels that there are too many different

---

3 Max von Gruber, "Zur Theorie der Antikörper, I. Ueber die Antitoxin-Immunität. II. Ueber Bakteriolyse und Haemolyse," *Münch. med. Wschr.* 48(1901):1827–1830, 1294–1297, 1965–1968; Lecture to the k. k. Gesellschaft der Aerzte in Wien, 25 October 1901. Also reported in *Wiener klin. Wschr.* 14(1901):1093–1094, 1142-1143, discussion by Rudolf Kraus (1190–1192), Friedrich Wechsberg (1192–1195), Richard Kretz (1195–1196), Max von Gruber (1214–1215), and Richard Paltauf (1215–1218).

4 Gruber, "Theorie der Antikörper" (1901) (n. 3), 1830.

Figure 6.1. Max von Gruber, portrait photograph. (Courtesy of Institut für Geschichte der Medizin, Munich)

substances postulated: "If, as Ehrlich would have it, for every different lysin an absolutely specific amboceptor must unite with an absolutely specific alexin, it is even more of a riddle where all these substances come from."[5] And, he says, he agrees with Bordet and Buchner that though each species may have its own lysin, he is sure that each has only one. Differences in practice must be due to quantitative or concentration effects.

Where Ehrlich has postulated a large number of specific substances, Gruber suggests an explanation based on quantitative differences. As in his criticism of Pfeiffer's claim that different *Vibrio* types could be sharply and specifically defined by their reaction with antibodies, here again Gruber, like Buchner, rejects specificity and adopts a theory based on *quantitative Abstufung*.

In Gruber's written paper as it appeared in the *Münchener medizinische Wochenschrift*, his opposition to the Ehrlich theory is put plainly enough. In the report of the original lecture that appeared in the *Wiener medizinische Wochenschrift*, his words are even more hard-hitting: "The Ehrlich theory is an aberration that must disappear as soon as may be from the scientific scene."[6]

Gruber's lecture aroused tremendous opposition. At the end of it, Professor Richard Paltauf, director of the Serotherapeutic Institute and of the Institute for Experimental Pathology, and founder of the Royal Society of Medicine in Vienna (Erna Lesky calls him one of the great school-builders of medical history) asked the audience whether they would not like a discussion of Gruber's lecture.[7] He felt it his duty to put his point of view to them, as the scientific work of the Serotherapeutic Institute was based on Ehrlich's concepts. At the following meeting of the society, Paltauf's group, represented by Rudolf Kraus, Friedrich Wechsberg, and Richard Kretz, and by Paltauf himself, presented their reply.[8]

Gruber's attack on the Ehrlich theory polarized thinking on immunology in Vienna into pro- and anti-Ehrlich camps, of which the pro-Ehrlich camp, containing the powerful Paltauf group at the Serotherapeutic Institute, was much the strongest. The supporters whom Gruber quotes, Buchner and Bordet, were not present in Vienna, nor were they likely to be highly respected by his audience. That was par-

5 Gruber, "Theorie der Antikörper" (1901) (n. 3), 1965.
6 Gruber, "Theorie der Antikörper," Reported in *Wiener klin. Wschr.* (1901) (n. 3).
7 Erna Lesky, *Die Wiener medizinische Schule im XIX Jahrhundert* (Graz: Böhlaus, 1965), 578; and see Chapter 12, nn. 2–6. For a discussion of Paltauf's school, see Chapter 12, nn. 2–6; and for his influence on Karl Landsteiner's career in Vienna, Chapter 13, n. 58.
8 Discussion reported in *Wiener klin. Wschr.* (1901) (n. 3), 1190–1196; 1214–1218.

ticularly true of Buchner, notorious for his promotion of the hay bacillus. During the discussion, Gruber was able to add one more
supporter to his side. The physiologist Sigmund Exner objected to the
Ehrlich theory from the point of view of natural selection. How could
separate specific antibodies have evolved to deal with each separate
antigen, he argued, when many of these antigens could never have
been met with by an animal in its normal environment? [9]

Before this pivotal meeting, Gruber had a good reputation as a serologist. The clinician Bernhard Naunyn had just invited him to answer
Ehrlich as co-referent at the seventy-third meeting of the Gesellschaft
Deutscher Naturforscher und Aerzte, in Hamburg in 1901. At this
meeting, Ehrlich gave a long and important paper on the defensive
properties of blood.[10] Gruber, however, was "unable to appear."[11]
What prevented him from appearing was made public two months later,
after the publication of the paper just discussed, which was probably
the "answer to Ehrlich" actually written for presentation at this meeting.

With Naunyn's permission, Gruber sent the editor of the *Münchener
medizinische Wochenschrift* two letters from Naunyn, who had been organizing the meeting. The first invited him to appear with Ehrlich:

As *Referent* I have been able to get Ehrlich . . . and I wonder whether I might
request that you would take over as *Co-referent* because of your work – I am
thinking primarily of your work on the agglutinins – you seem to me to be, like
Ehrlich, the most appropriate man to call upon.[12]

The second letter must have been difficult for Naunyn to write. Ehrlich and Gruber, it seems, could not agree on how to arrange the discussion. Naunyn says that Ehrlich had written to him saying that he
had been persuaded under pressure to appear, and that he was not
going to risk inconvenience (*ungelegenheit*) to do so. It seems likely
Gruber had sent him that very critical paper, and that he was refusing
to accept such criticism in public.

Naunyn's second letter to Gruber reads in part:

9 Sigmund Exner, cited by Gruber, discussion (1901) (n. 8), 1214. I have not been
   able to identify Exner's original.
10 Paul Ehrlich, "Die Schutzstoffe des Blutes," *Verh. d. Ges. Deutsch. Naturf. u. Aerzte,* 73
   Versammlung, Hamburg (Leipzig: Vogel, 1901), pt. 1, 250–275, and Discussion, 275.
   Also in Felix Himmelweit, Martha Marquardt, and Sir Henry Dale, eds., *Collected
   Papers of Paul Ehrlich, etc.* (London, Pergamon, 1957), v. 2, 298–315 (not translated).
11 R. Stintzing, Chairman of the Medical Group, *Verhandlungen* 73 (1901) (n. 10), 249.
12 Bernhard Naunyn to Max von Gruber, first letter published in *Münch. med. Wschr.*
   48(1901):1933.

My conception of the purpose of this discussion is primarily TO MAKE EHR-
LICH'S BRILLIANT CONCEPTS ACCESSIBLE TO A WIDER PUBLIC [Nau-
nyn himself underlined this, says Gruber].... If Ehrlich should withdraw, the
whole thing will be a failure.... I think he is quite right to feel that a thorough-
going criticism of the concepts he brings out would embarrass him in public.
He wants to be sure that an attack like this to which he cannot reply, does not
occur.... I hope, respected colleague, that his trust [in you] will be restored,
without our whole plan being spoiled by one of you withdrawing.[13]

After receiving this deeply insulting letter, Gruber withdrew.

Bernhard Naunyn, the writer of these letters, was born in 1839, and
had lived through a long period of change in clinical medicine, beginning from the time when he was *famulus* to the anatomist Carl Reichert,
Johannes Müller's student and successor in Berlin.[14] The emphasis at
that time was on physiological and chemical explanation, a natural-
scientific medicine built on Johannes Müller's legacy. Over the second
half of the century, the emphasis slowly came to rest instead on what
Naunyn calls *Kasuistic*, the ontological approach to disease as it was
practiced in England and France.[15] The definition of specific disease
pictures came to replace the view of disease as a disturbance of normal
physiology.

In 1900 the *Naturforscherversammlung* was devoted to overviews of the
changes in the various fields of medicine and the sciences over the
preceding century, and Naunyn gave a paper on the developments in
internal medicine, hygiene, and bacteriology. In it, he looks back on
the changes he has experienced – he was now sixty-one – and he pictures the physiological medicine of Johannes Müller's school as basically opposed to the discussion of causes of disease and nosography.
Only from 1870 onwards in Germany did the influence of the "modern
clinicians" become stronger, and the physiological era give way to an
era of the naming of disease and disease syndromes.

With the causes of disease, the study of ontology, nobody was much concerned
in this first flowering period of modern medicine, perhaps from a kind of
embarrassment. It was after all the eternal quest for causes of disease that had
driven the medical world from Paracelsus to Brown into theorizing. It was Pettenkofer in Munich who brought the study of causes back to life, and made it
the central problem of hygiene ... [but] the most powerful advances in aetiol-

13   Bernhard Naunyn to Max von Gruber, second letter, published in *Münch. med. Wschr.*
     48(1901):1993–1994.
14   Bernhard Naunyn, *Erinnerungen Gedanken und Meinungen* (Munich: Bergmann, 1925).
     Naunyn's autobiography covers a very interesting period of change in clinical med-
     icine.
15   Naunyn, *Erinnerungen* (1925) (n. 14), 127.

ogical research were made not by hygiene, but by bacteriology, which developed at about the same time.[16]

Naunyn's historical picture, which was first presented in this lecture and later elaborated in his autobiography, is one of physiological medicine as a reaction to Paracelsian ontology, with its *Entia morborum*, and of Pettenkofer's hygiene as a turning point, a swing back that was completed by the bacteriologists and matched by the clinical describers of disease pictures or syndromes. Naunyn's view of medical history has been accepted as the true picture of nineteenth-century progress in this field. In his enormously influential account of the history of medicine as a history of the progress of nosology, Knud Faber has written Naunyn into the record by quoting him at length.[17]

Naunyn, in his letters to Gruber, instinctively picked Ehrlich, the heir of the bacteriologists, as his hero, rather than Gruber, who did not believe in specificity. To Naunyn, history was on Ehrlich's side. How pleased Gruber must have been to receive the following letter from the young Clemens Freiherr von Pirquet, which read in part, "I have been informed . . . that you will be kind enough to listen to me. . . . Since they [that is, von Pirquet's observations] are incompatible with Ehrlich's hypothesis and rather agree with your interpretation, I am getting in touch with you . . . and trust you will be satisfied with my results."[18]

Von Pirquet, who later became well known in clinical immunology and paediatrics, had just begun his work in the field of immunity. He was working under Rudolf Kraus at Paltauf's Serotherapeutic Institute in Vienna, on the immune precipitin reaction that Kraus had recently described.[19] He had also been appointed clinical assistant at the University Children's Clinic.

16   Bernhard Naunyn, "Die Entwicklung der inneren Medicin mit Hygiene und Bakteriologie im XIX Jahrhundert," *Verh. d. Gesellschaft Deutscher Naturforscher und Aerzte,* 72 Versammlung, zu Aachen (Leipzig: Vogel, 1900), pt. 1, 59–70. Naunyn uses "ontology" and "aetiology" almost synonymously.

17   Knud Faber, *Nosography: The Evolution of Clinical Medicine in Modern Times, with an Introductory Note by Rufus Cole* (New York: Hoeber, 1923), 84–85; see also Owsei Temkin, "The scientific approach to disease: specific entity and individual sickness, " in A.C. Crombie, ed., *Scientific Change* (London: Heinemann, 1963) 629–647.

18   Clemens von Pirquet to Gruber, letter d. June 2, 1903, published in Richard Wagner, *Clemens von Pirquet, His Life and Work* (Baltimore, Md.: Johns Hopkins University Press, 1968), 30. No originals of these letters are available. The von Pirquet papers at the National Library of Medicine, Bethesda, Md., show a gap in the holdings, which, it seems from the catalogue, may have covered these letters. The translations I have quoted are those of Wagner; no further details are obtainable as Dr. Wagner died in 1974.

19   Rudolf Kraus, "Ueber specifische Reactionen in keimfreien Filtraten aus Cholera, Typhus und Pestbouillon erzeugt durch homologes Serum," *Wiener klin. Wschr.* 10(1897):736–738. This paper is in part an answer to Gruber's contention that the agglutination of bacteria was due to a preliminary swelling of their bodies. Kraus shows that agglutination, that is, precipitation, can take place without bodies, in a

On receiving this letter, Gruber immediately invited von Pirquet to Munich, where Gruber was able to look at his graphic representation and recalculation of the results published by the young Danish serologist Thorvald Madsen in his paper on the neutralization of tetanus toxin by antitoxin.[20]

Madsen plotted toxin neutralized against antitoxin added; the amount of toxin neutralized was calculated from its equivalent in cubic centimetres of blood lysed by the remaining un-neutralized toxin. The result was, as Ehrlich had taught him, the stepped *Giftspektrumschema.* Gruber and von Pirquet replotted it more simply and directly, by putting percentage haemolysis against antitoxin added. The result of this was a smooth continuous curve, with intervals as close together as the doses of antitoxin permitted: there were no discontinuities.

Von Pirquet's notes contain many sheets of paper on which he worked out this idea: he plotted Madsen's results as logarithmic curves, as straight lines, as added toxin against toxin neutralized, as added toxin against toxin not neutralized, as a proportion of toxin neutralized with successive amounts of antitoxin added. Some of these curves are illustrated in Figure 6.2. In one version, he has drawn the smooth curve over Madsen's stepped one: this is the idea in its essence, the criticism that Gruber and von Pirquet were making – that the stepped *Giftspektrum* is an artefact, an attempt to describe a continuous change in terms of discontinuous separate specificities.[21]

Gruber was delighted. He wrote to von Pirquet a few days after his return to Vienna sending him further charts of his own experiments on toxin neutralization plotted in this new way and including one of Ehrlich's *Giftspektra,* "just for fun."[22] The next day he again wrote to von Pirquet – "I performed a new experiment today by a still more refined technique which shows the ridiculousness of Ehrlich's toxin theory" – and proposed that von Pirquet drop everything he was doing in Vienna and come at once to collaborate with him.[23] The day after this, he sent von Pirquet another postcard, telling him it was no longer necessary. Madsen himself had taken the step that they had planned: Gruber had just come across the paper by Madsen and the Swedish

cell-free solution containing substances secreted by the bacteria. Kraus's subsequent papers – for example, "Ueber Hämolysine und Antihämolysine," *Wiener klin. Wschr.* 13(1900):49–56 – show him to have been interested in a positive sense in Ehrlich's work, as was Paltauf the institute director.

20   Thorvald Madsen, "Ueber Tetanolysin," *Z. f. Hyg. u. Infektionskr.* 32(1899):214–238.
21   Clemens, Freiherr Pirquet von Cesenatico, manuscripts in the possession of the National Library of Medicine, Bethesda, Md., ms 47, Box 3, pt. 6.
22   Wagner, *von Pirquet* (1968) (n. 18), 33.
23   Wagner, *von Pirquet* (1968) (n. 18), 33.

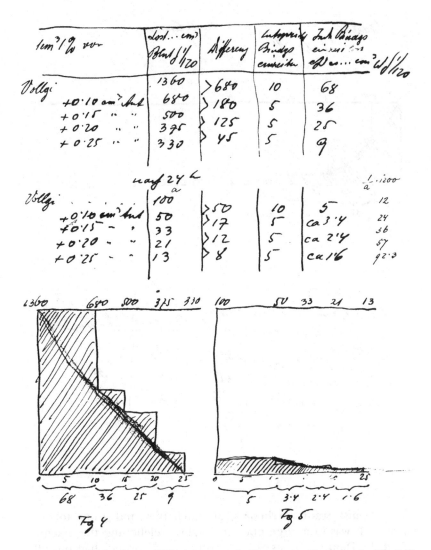

Figure 6.2 (*above and following two pages*). Clemens von Pirquet replots Ehrlich's *Giftspektrum.*

Three pages from von Pirquet's notes in which he is working over Thorvald Madsen's published paper, "Ueber Tetanolysin" (1899) (n. 20). He is testing the idea that the relationship can be described by a smooth curve rather that a series of discrete steps. From von Pirquet Papers, Box 3, pt. 6 (n. 21). (Ms. and photograph courtesy of National Library of Medicine, Bethesda, Md.)

physical chemist Svante Arrhenius, in which they, too, came to this conclusion. It was in a large book in English celebrating the inauguration of the Danish Statens Seruminstitut, which Gruber had put off reading until he had a vacation to do it in.[24] He still wanted to publish

24   Svante Arrhenius and Thorvald Madsen, "Physical chemistry applied to toxins and antitoxins," in Carl J. Salomonsen, ed., *Festskrift ved indvielsen af Staatens Seruminstitut 1902: Contributions from the University Laboratory for Medical Bacteriology to Celebrate the Inauguration of the State Serum Institute* (Copenhagen: Olsen, 1902), chap. 3, 1–111. Articles separately paginated (p. 71); published in English. On the State Serum Institute, see Thorvald Madsen, *Statens Seruminstitut: Institutets Udvikling 1902–1940* (Copenhagen: Luno, 1940).

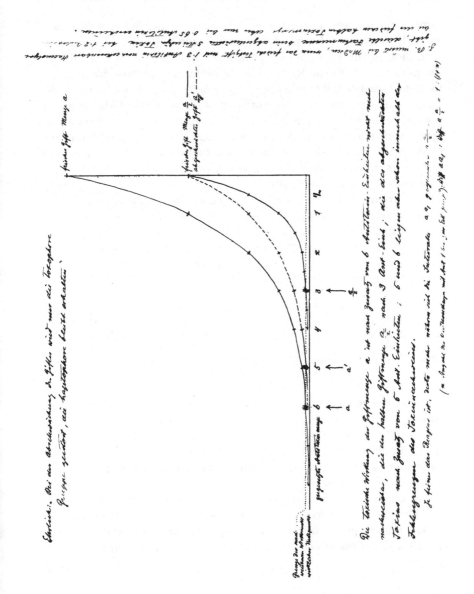

something with von Pirquet, however. Madsen, he said, out of loyalty to Ehrlich, had not drawn all the possible conclusions.

In fact, in the paper finally published by Gruber and von Pirquet, the examples of the similarity of the neutralization of toxin by antitoxin to that of the reversible reaction of weak acids with weak bases are very close to those given by Arrhenius and Madsen in the *Festskrift* of the Statens Seruminstitut. They quote the comparison of the immune reaction with that of ammonia and boric acid and add, "Arrhenius and Madsen are also obviously perfectly clear that Ehrlich's theory has been laid in its grave by their investigations, but they do not draw all the conclusions that are justifiable."[25]

In a footnote full of delighted *Schadenfreude* they write (that is, Gruber wrote, I am sure), "The sunrise in Copenhagen as I have lately noticed is also producing a faint glimmer in Frankfurt [he quotes a paper by Morgenroth and Sachs] though full daylight does not seem to have dawned there yet."[26] The words that caught his eye in Morgenroth's paper were

The conditions, however, are entirely different if the affinity of the complementophile group of the anchored amboceptor is very slight. In other words, we are dealing with an easily dissociated combination in a reversible process. In that case, in accordance with a well known chemical law, the more of one of the elements is in excess, the more of the completed combination will remain intact.[27]

This was not the kind of firm chemical combination that Ehrlich had had in mind. But Morgenroth was not disloyal. Further down the page he adds: "A most conspicuous role, however, is played by the fact that the immune serum is not a simple substance, but is made up of partial amboceptors to which various dominant complements of the sera correspond."[28]

Not content with having Arrhenius and Madsen bury Ehrlich's theory, Gruber wanted to dance upon the grave. The week before this critical paper appeared in the *Münchener medizinische Wochenschrift*, the *Wiener medizinische Wochenschrift* had printed an article entitled "New

25 Max von Gruber and Clemens, Freiherr von Pirquet, "Toxin und Antitoxin," *Münch. med. Wschr.* 50(1903):1193–1263 (p. 1259).
26 Gruber and v. Pirquet, "Toxin und Antitoxin" (1903) (n. 25), 1259.
27 Julius Morgenroth and Hans Sachs, "Ueber die quantitativen Beziehungen von Ambozeptor, Komplement und Antikomplement," *Berl. klin. Wschr.* (1902):817–822 (p. 819); and in translation in Charles Bolduan, trans. and ed., *Studies in Immunity by Professor Paul Ehrlich and His Collaborators* (New York: Wiley, 1st ed. 1906, 2d ed. 1910). Reference is to 2d ed., 250–266 (p. 257).
28 Morgenroth and Sachs, "Quantitative Beziehungen" (1902) (n. 27), 819 (Ger.), 257 (Engl.).

Fruits of the Ehrlich Toxin-Doctrine," which purported to be a defence of Ehrlich's theory in a letter addressed to Gruber by "Dr. Peter Phantasus, by the Grace of God Chemist," which Gruber explains means either that the writer is quite potty, or that he uses this heavenly qualification as he has no earthly one. "Dr. Phantasus" writes:

> The Ehrlich side-chain theory has triumphed throughout the world, and you, *Herr Professor*, are the only one to have called this theory in print, with astonishing shortsightedness and almost unbelievable arrogance, "completely, worthless, unbridled hypothesis-spinning," and "dangerous numbing with words," – a verdict all the more out of place, as you – as you yourself said publicly – "only looked over the material facts for a few weeks during the holidays." With that lack of respect which is characteristic of the demeanour of the proletariat to their superiors, whose greatness they are unable to understand, you were so bold as to have stood up against Ehrlich at the *Naturforscherversammlung* in Hamburg, if the chairman of the session had not told you just in time that the whole point of the session was to make Ehrlich's brilliant views accessible to a wider public, and that they should not be disturbed with criticism.[29]

In this cunning, funny, and bitter piece of polemic, Gruber demonstrates that water, which acts as a toxin – that is, a haemolysin – to red cells and can be titred out with added salt, also consists of a whole series of different toxins: a water-protoxoid, water-prototoxin, water-deuterotoxin, water-tritotoxin, and so on. The same Ehrlich *Giftspectrum* can also be made for sulphuric acid. The findings can be extended by analogy to compressed aether, too, and since radium sends out rays, why should there not be reciprocal elements that suck them in? The surface of the molecule must be covered by a thick layer of aether-receptors that reach out and trap the passing ray. In this way we can explain magnetism and electricity, as well as immunity.

This neat satire completed the ruin of Gruber's reputation in Vienna. In 1902, on the death of Hans Buchner, Gruber left to succeed him in the chair of Hygiene in Munich.

29 Max von Gruber, "Neue Früchte der Ehrlich'schen Toxinlehre," *Wiener klin. Wschr.* *16*(1903):791–793 (p. 791).

7

Max von Gruber and Karl Landsteiner

As a young man of twenty-eight in his first job, Karl Landsteiner was appointed second assistant to Max von Gruber at the Institute for Hygiene in Vienna. He joined the institute in January 1896 and stayed there until September 1897. He then transferred to the Institute of Pathological Anatomy, becoming an assistant under its director, the pathologist Anton Weichselbaum. He stayed at the Institute of Pathological Anatomy until 1907.[1]

Landsteiner (Figure 7.1) was born in June 1868 in the little spa town of Baden-bei-Wien, outside Vienna. He was the son of Leopold Landsteiner (born in 1818 in Vienna), who was an expert on French history and culture and a well-known political and economic journalist. Leopold Landsteiner was a member of that assimilated, liberal Jewish intelligentsia, the cultural elite so important in the professional and intellectual life of Vienna. He was the first editor of *Die Presse*, which was started in May 1848, and in the early fifties, he founded a paper of his own, *Die Morgenpost*. It dealt with current political and economic events in a style digestible (*mundgerecht*) for the lower classes.[2] Leopold Landsteiner printed and edited *Die Morgenpost* for twenty years, until his death at the age of fifty-seven in 1875, when his son Karl was only six. Karl's mother Franziska (Fanni), née Hess (Figure 7.2), was from Prossnitz in Moravia, from one of the many German-Jewish merchant families who had come to the capital from the German-speaking towns of Bohemia and Moravia. She was born in 1837 and was twenty years

1 Paul Speiser and Ferdinand G. Smekal, *Karl Landsteiner the Discoverer of the Blood-Groups and a Pioneer in the Field of Immunology: Biography of a Nobel Prize-Winner of the Vienna Medical School*, trans. Richard Rickett (Vienna: Hollinek, 2d ed., 1975), 24–32; Peyton Rous, "Karl Landsteiner," *Obituary Notices of Fellows of the Royal Society of London* 5(1947):295–312. Other articles dealing with Landsteiner's early life are based on these two sources.
2 Rudolf Till, "Erhebungen durch Archiverei der Gemeinde Wien," material gathered at the request of George Mackenzie, 1949–1951; American Philosophical Society, Landsteiner-Mackenzie Papers, B L32m, Box 2.

136

Figure 7.1. Karl Landsteiner at the age of about five, c. 1873, posing in a Hussar riding costume on the photographer's papier-maché rocks. (Photograph from George Mackenzie's collection, courtesy of the American Philosophical Society)

Figure 7.2. Franziska (Fanni) Hess, Mrs. Leopold Landsteiner, Karl's mother. After her death in 1908, he kept this picture with him for the rest of his life. (Photograph from George Mackenzie's collection, courtesy of the American Philosophical Society)

younger than her husband. She had been married only seven years when he died: she was to live with her son in a close and affectionate relationship until her own death in 1908.

Karl Landsteiner went to the gymnasium, the higher secondary school of the professional classes, in the IX district, one of the three politically liberal inner-city districts, where Jewish students made up more than half of the gymnasiasts.[3] Landsteiner went on to the university in 1885, reading medicine, though most children of journalists as a group tended to go into law.[4] Landsteiner's father had evidently died too early to affect his choice. The university, like the gymnasia that fed it, had a large proportion of Jewish students, about one-third on average. In spite of this strong presence, anti-Semitism was prevalent among the student body and its organizations from the 1880s onwards. On the faculty, Jews remained in the lower, unpaid, academic ranks of *Privat-dozenten* unless they converted. Professorships were subject to the same regulations as the other civil service posts under the Austro-Hungarian regime and were permitted only to Catholics.[5] In December 1890, just before his final examination, Landsteiner and his mother converted to Catholicism: he was baptized with the name of Karl Otto and received into the Roman church at the Schottenkirche in Vienna. He took the degree of doctor of medicine in February 1891.[6]

Landsteiner (Figure 7.3) came to Max von Gruber with no experience in immunology, but a good deal of it in organic chemistry. He had spent the years 1891 to 1893, after his final medical examination, studying chemistry with the three outstanding organic chemists of the time, Emil Fischer, Eugen von Bamberger, and Arthur Hantzsch. (Their influence on Landsteiner's thinking is discussed in Chapter 8.)

On joining Gruber at the Institute for Hygiene, Landsteiner immediately became involved in Gruber's controversy with Richard Pfeiffer concerning immune specificity. His first immunological paper dealt in part with the problem Gruber set for him dealing with the quantitative assessment of the reaction of an immune serum with a large group of bacteria similar to that used to produce it.[7] The serum was raised by

3 Steven Beller, *Vienna and the Jews 1867–1938: A Cultural History* (Cambridge: Cambridge University Press, 1989), 43–70 (p. 53).
4 Beller, *Vienna and the Jews* (1989) (n. 3), 60–61.
5 Beller, *Vienna and the Jews* (1989) (n. 3), 34–35.
6 Speiser and Smekal, *Landsteiner* (1975) (n. 1), 17.
7 Karl Landsteiner, "Ueber die Folgen der Einverleibung sterilisirten Bakterienculturen," *Wiener klin. Wschr.* 10(1897):439–444, Chase no. 8. Reference is to the 346-item Landsteiner bibliography prepared by Merrill W. Chase, first published in *J. Immunol.* 48(1944):1–16, and subsequently in Speiser and Smekal, *Landsteiner* (1975) (n. 1), 167–181. Henceforth these bibliographic references will be designated in the form C. 8.

Figure 7.3. Karl Landsteiner as a young man, c. 1896. (Photograph from George Mackenzie's collection, courtesy of the American Philosophical Society)

injecting a guinea-pig with *Bacillus typhimurium*, and it was tested against some *B. typhi* and *B. coli* types, as well as a number of more distantly related ones, including several that were included in Gruber's demonstration of non-specificity at Wiesbaden in April 1896: *Vibrio cholerae, V. metschnikovi, V. berolinensis, V. finkler-prior,* and *B. enteritidis* (Gärtner). Landsteiner found agglutination in many of the tubes, but it was complete only in those containing the matching organism. He wrote: "This then is specificity, in the sense that the phenomenon appears in traces with different species, but never as intensely as when matching [*gleich-nämig*] serum and bacteria act on each other."[8]

It was Gruber's problem. They were Gruber's organisms and it was Gruber's conclusion: but it was one that Landsteiner never repudiated. In his second paper, of 1898, he took up another Gruber problem, which had been touched on already in the first. Here, too, the argument was between Gruber and Pfeiffer and was about whether the agglutinins of an immune serum are directly involved in the defence of the immune animal against the immunizing organism: Gruber said yes, they are; Pfeiffer said no. Landsteiner tested their effectiveness on bacteria outside the body and found an inverse relationship between the concentration of antiserum and the number of surviving colonies of bacteria. He concluded for Gruber, but carefully pointed out that immune serum may not be the only protective weapon of the body.[9]

In a third one of these early immunological papers (1899), he began to expand upon the ideas already given him by Gruber. He had raised some sera against red blood cells and would have reported this at length, but Jules Bordet had just published a description of the same thing. He begins the paper:

In producing sera specific for vibrios, Gruber observed that the production of specific substances took place even if the immunizing material was a dead culture with no toxicity to speak of. This observation leads to the conception of the process of immunization in the animal body not as a reaction to harmful influences, but to the presence of certain materials, for which a complete characterization, perhaps in terms of their chemical nature, is not yet available. Such reactions to apparently indifferent materials cannot really be called immunizing, they should be seen as a special case of a general law-like regularity.[10]

Gruber was pleased with this development of his idea:

The first person to have understood the biological problem in its wider sense seems to have been my colleague Landsteiner who, from the fact that I had managed to obtain specific bactericidal and agglutinating sera by the injection of completely harmless saprophytes drew the right conclusion, that here was no protective measure against the danger of infection, or reaction product of the sick organism. From this starting point he then attempted to raise antisera to materials that had never been used till then.[11]

8 Landsteiner, "Einverleibung sterilisierten Bakterienculturen" (1897) (n. 7), 442.
9 Landsteiner, "Ueber die Wirkung des Choleraserums ausserhalb des Thierkörpers," *Zbl. f. Bakt.* (1 Abt. Orig.), *23*(1898):847–852, C. 11.
10 Karl Landsteiner, "Zur Kenntniss der specifisch auf Blutkörperchen wirkenden Sera," *Zbl. f. Bakt.* *25*(1899):546–596, C. 12 (pp. 546, 547).
11 Max von Gruber, "Zur Theorie der Antikörper: I. Ueber Antitoxin-immunität," *Münch. med. Wschr.* *48*(1901):1827–1830 (p. 1827).

One can see here the idea originating in Gruber's work and later being passed back to him in a more developed form, a form that he himself approved of and admired: that of a general law underlying the special cases of multifarious experience. It is also the first indication of the split that was developing in the field of immunology, between the bacteriological tradition of immunology as defence (which was to go on refining the multifariousness of experience in the diagnosis of different organisms and disease by immunological means) and the search for general laws that might explain the nature of specificity. It was already inherent in the controversies between the Koch bacteriologists and the Nägeli group, but within the next ten years the field itself split in two. In 1913, we find Ulrich Friedemann of Berlin, in an essay entitled "Infection and immunity" – written for a *Handbook* of which Gruber was one of the editors – making the same suggestion, that these two parts of the field should have different names. Immunology as a discipline has two parts, he says: the older (*schon Uralt*, prehistoric already) part is the study of resistance to disease, as seen in Jenner's and Pasteur's work on vaccinations. The other, more modern part is the study of specificity per se, which he names "serology."

Immunology sees all reactions only in terms of their usefulness to the organism, while *serology* without taking up any teleological position tries to explain them by physicochemical principles. From this point of view the formation of specific protective substances becomes but a special case of the law of specific antibody production, and it may seem as if . . . the teleological way of looking at the immune process is unjustified.[12]

A few days after Gruber's commendation of Landsteiner's broadening of the problem of immunity into a general biological process, with no teleological implications, the *Wiener medizinische Wochenschrift* printed

12   Ulrich Friedemann, "Infektion und Immunität," in Max Rubner, Max von Gruber, and Martin Ficker, eds., *Handbuch der Hygiene* (Leipzig: Hirzel, 1913), v. 3, pt. 1, "Die Infektionskrankheiten," 661–810 (pp. 663, 664); Isidor Fischer, "Friedemann, Ulrich," in *Biographisches Lexikon der hervorragenden Aerzte der letzten fünfzig Jahre* 2 v. (Vienna: Urban, 1933), v. 1, 450. Friedemann was born in 1877; his professional appointments ranged over almost the whole of the Koch–Ehrlich group. He was assistant to Max Neisser in Stettin, and also at Ehrlich's Institute at Frankfurt, where he worked on bacterial agglutination and its relationship with colloid precipitation, along with Neisser and Heinrich Bechhold. He was assistant in bacteriology at the Institute for Hygiene in Berlin while Max Rubner was its director, and at the institute he supervised Ludwik Hirszfeld's thesis *Haemagglutination und ibre physikalischen Grundlagen* (Berlin, 1907), discussed in Chapter 14, and Ludwik Hirszfeld, in Hanna Hirszfeldowa, ed., *Historia Jednego Zycia* (Warsaw: Institut Wydawniczy Pax, 2d ed., 1967). In 1920 he became a member of the Robert Koch Institute for Infectious Disease in Berlin, and medical director of the Department for Infectious Disease in the Moabit Hospital, Berlin.

the paper by Landsteiner explaining the isoagglutinins in human blood in a manner that cut through the tangle of teleological explanations that had grown up around the assumed protective function of the serum antibodies, and that obscured the relationship of these normal human antibodies to each other and to the red cells. His paper describes a simple reciprocal relationship between two cell antigens, A and B, and two antibodies, anti-A and anti-B. Human red cells might have either of the antigens, or none. Sera, too, might contain either of the antibodies, or both. No antibody is ever present that reacts with its own cells.[13]

This simple and neat pattern accounted for all the contradictory findings of earlier workers who had been guided by the assumption that these agglutinins were a result of the human body's defence against an attack of some past or present disease. Malaria and typhoid had been among those suggested. But there was no need to postulate a disease; the antibodies were a normal expression of immunological individuality.[14] They had nothing to do with defence, and they could not be cured.

This solution depended to some extent on the recent publication of an experiment by Ehrlich and his co-worker Morgenroth in which they immunized a group of goats with each other's blood. It appeared in their third *Mittheilung über Hämolysine* of 1900, in which they introduced into the literature the idea of immunological differences in the bloods of individuals of the same species. They called this their pluralistic conception of immunity.[15] Its extreme pluralism appears in the citation from their fourth *Mittheilung*, which they themselves quoted in their sixth:

We have obtained a whole series of *different* isolysins by injecting goats with goat blood. At present they number twelve. In the red blood cells not merely a single group but a large number of different groups must be considered, which, provided there are fitting receptors, can produce a corresponding series of immune bodies. All of these immune bodies again will be anchored by the blood cells employed in immunisation. We may assume that when an animal of Species A is immunised with blood cells of Species B, a haemolytic serum will be produced which contains a whole host [*ein ganze Schaar*] of immune bodies.

13 Karl Landsteiner, "Ueber Agglutinationserscheinungen normalen menschlichen Blutes," *Wiener klin. Wschr.* 14(1901):1132–1134 (C. 17).

14 Pauline M. H. Mazumdar, "The purpose of immunity: Landsteiner's explanation of the human isoantibodies," *J. Hist. Biol.* 8(1975):115–134; Peter Keating, "The problem of natural antibodies, 1894–1905," *J. Hist. Biol.* 24(1991):245–263.

15 Paul Ehrlich and Julius Morgenroth, "Ueber Hämolysine: dritte Mittheilung," *Berl. klin. Wschr.* 37(1900):453–458; and in Felix Himmelweit, Martha Marquard, and Sir Henry Dale, eds., *Collected Papers of Paul Ehrlich*, 4 v. (London: Pergamon, 1957), v. 2, 192–204 (Ger.); the English translation is cited where there is one.

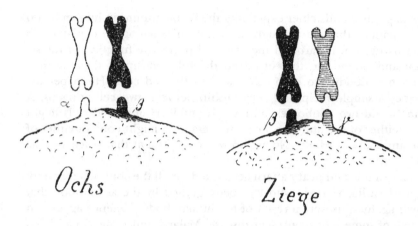

Figure 7.4. Ehrlich's conception of specificity. A rabbit immunized with ox [*Ochs*] cells, produces a serum that lyses both ox and goat [*Ziege*] cells. Absorption with ox cells removes both anti-ox and anti-goat, but goat cells only remove anti-goat activity, and leave anti-ox behind in the serum. Ehrlich supposes that each cell type has one unique, specific receptor and one that is common to both types. This technique and this interpretation became the basis for the study of blood groups and their genetics from the 1920s onwards. From Ehrlich and Morgenroth, "Ueber Hämolysine: sechste Mittheilung" (1901) (n. 16).

The immune bodies in their entirety are anchored by the blood cells of Species A.[16]

In the sixth *Mittheilung* itself, they add that "the pluralistic standpoint adopted by us raises numerous difficulties for thorough analytical work in this field but it leads to a deeper insight into the complicated problems . . . [of] immunity."[17]

Using the pluralistic conception, they explain what happens when a serum that is capable of agglutinating two different cell types – say, that of ox and goat – is absorbed with one of them (see Figure 7.4).

If a rabbit is injected with ox blood the amboceptors (immune bodies) corresponding to *alpha* and *beta* will be formed. Ox blood cells, by means of their *alpha* and *beta* groups, will be able to anchor all the immune bodies, whereas goat blood cells will anchor only the immune body of portion *beta*, leaving behind in the fluid the immune body of portion *alpha*.[18]

16 Paul Ehrlich and Julius Morgenroth, "Ueber Hämolysine: sechste Mittheilung," *Berl. klin. Wschr. 38*(1901):569–574, 598–604 (p. 570); and in *Coll. Papers* (n. 15), v. 2, 256–277 (Ger.), and 278–297 (Engl.) (p. 278–279).
17 Ehrlich and Morgenroth, "Sechste Mittheilung" (1901) (n. 16), 279.
18 Ehrlich and Morgenroth, "Sechste Mittheilung" (1901) (n. 16), 283.

And, indeed, this seems to explain what happens in practice. The anti-ox serum, after it has been absorbed by goat cells, still agglutinates ox cells: it still contains the anti-ox fraction. The explanation fits the phenomena.

The pluralistic conception was expressed even more neatly and clearly by G. M. Malkoff, a Russian student who was working under August von Wassermann at the Institute for Infectious Disease in Berlin. His research was undertaken at von Wassermann's instigation and done under his direction. He used the same method as Ehrlich, that of absorbing the serum with a given cell type, and testing for activity left unabsorbed. He concluded:

The agglutinin has to the morphological element that it agglutinates a specific binding affinity in that it is bound only by that and no other. In a normal serum that agglutinates different cells at the same time, there exist as many different specific agglutinins as the serum agglutinates different cells.[19]

There is *ein ganze Schaar* of them, a whole army, as Ehrlich had said. This quotation shows the typical sharply defined specificity, the one-to-one relationship of antigen and antibody, that is typical of the Koch tradition. In the eighties, the bacteria were sharply specific; in 1900 the same style of thought was applied to the relationship of antibodies and cells. There is no trace of *quantitative Abstufung* here.

Landsteiner also felt it to be typical, and he chose to quote it as an expression of the kind of opinion he wished to controvert. His co-worker on blood groups was Adriano Sturli, who was then a student and later became a clinical assistant at the Second Medical Clinic. Landsteiner himself had done some of his post-graduate clinical practice there.[20] They worked together at the Institute of Pathology, where Sturli spent any free time he could obtain from his clinical responsibilities. Although Landsteiner had left Gruber's laboratory in 1897, when he joined the staff of the Institute of Pathological Anatomy in 1902 Gruber's spirit was still with him:

A single kind of red cell is supposed to have an enormous number of different substances on it, and in the same way there are substances in the serum to react with many different animal cells. In addition the substances which match each kind of cell different in each kind of serum.

19 G. M. Malkoff, "Beiträge zur Frage der Agglutination der rothen Blutkörperchen," *Deutsche med. Wschr. 29*(1900):229–231 (p. 231).
20 Speiser and Smekal, *Landsteiner* (1975) (n. 1), 17, 149–150, 155. In 1902 Sturli and Alfred von Decastello-Rechtwehr, who was also an assistant at the II Medical Clinic, described the fourth and least common of the ABO blood groups, now called AB, but which they called "without type," since the serum contained no antibody.

The number of hypothetical different substances postulated makes this conception so uneconomical that the question must be asked whether it is the only one possible.

Indications of doubts in this direction can be found in Bordet, Gruber, Pfeiffer, Exner. We ourselves hold that another, simpler, explanation is possible.[21]

The alternative, the simpler more economical explanation that Landsteiner and Sturli preferred, was that the active materials were a series of homologous or otherwise chemically similar substances. Or, as Landsteiner suggested in a paper he sent to the *Münchener medizinische Wochenschrift* later in the same year (Gruber had just succeeded Hans Buchner in the chair of hygiene in Munich), the explanation of *seeming* specificity might lie in a summation of effects in themselves non-specific:

The work of Landsteiner and Sturli came to the conclusion that the existence of numerous specific substances in the serum has not been proved by anything that has been done up to now. However, it is not unlikely that the active materials in the serum are a series of non-specific substances which even in such simple components of the animal body as the fats have a not inconsiderable chemical multiplicity.[22]

Up to this time, Landsteiner had disliked Ehrlich's pluralistic hypothesis on methodological grounds alone. He had no evidence against it. His real objection to it was intuitive: he was looking for a means to show that nature does not consist of innumerable unrelated substances. He called Ehrlich's hypothesis "uneconomical," and tried to find another, a "simpler" one, that would unite the innumerable substances of the pluralistic conception under a single law.

The *Malkoff'sche Phänomen*, as Landsteiner later called it, depended on the procedure of testing for the activity remaining in a serum after it had been absorbed with certain cells. Landsteiner now reversed this procedure, and tested the material subtracted by the cells. He worked out a technique with which to free the absorbed antibody from the cells by a change in temperature, and to test this eluate for its activity.[23] Agglutination tests on the eluate now showed that the antibodies eluted from a single cell type would agglutinate a whole range of different cells more or less strongly.[24] This "specificity" seemed anything but

21   Karl Landsteiner and Adriano Sturli, "Häemagglutinine normaler Sera," *Wiener klin. Wschr.* 15(1902):38–40, C. 19 (p. 39).
22   Karl Landsteiner, "Ueber Serumagglutinine," *Münchener med. Wschr.* 49(1902):1905–1908, C. 26.
23   Karl Landsteiner, "Beobachtungen über Hämagglutination," *Wiener klin. Rundschau* 16(1902):774, C. 25.
24   Landsteiner, "Ueber Serumagglutinine" (n. 22) (1902), 1906, 1907.

Eluate from Rabbit Cells and Ox Serum:	vs. Rabbit Cells:	vs. Goose Cells:	vs. Pigeon Cells:	vs. Guineapig Cells:	vs. Mouse Cells:
	+++	++	+	+	+

Figure 7.5. Landsteiner's quantitative conception of specificity. Landsteiner tests the material taken up by the cells: an eluate from rabbit cells agglutinates a series of other cells more or less strongly. There is no sharp-edged cut-off. Specificity seems to be a matter of more or less good fit. From Landsteiner, "Ueber Serumagglutinine" (1902) (n. 22).

sharp edged. It could be explained, Landsteiner thought, by the assumption of few different substances not in themselves specific, with greater or less, but not absolute, affinity for the different cells (Figure 7.5).

By 1902, then, Landsteiner had evidence that could be interpreted to support the position he had held since his earliest work on the problem of specificity. His first papers were done in direct contact with Gruber and took up Gruber's view of specificity as a series of smooth quantitative transitions between species. He then moved to the hypothesis that seeming specificity could be explained by the resultant effect of a small number of substances in themselves non-specific. He suggested that these substances might be related to each other in the same way as the members of a homologous series in organic chemistry. The elution experiment represented an attempt to find experimental support for this position: the position itself long antedates the support.

Landsteiner's criticism of Ehrlich also originates in Gruber's position. He adopted Gruber's antagonism, but the criticism was his own: the feeling that Ehrlich's theory did not provide a good explanation because it postulated too many unknown entities: the huge numbers of different specific substances Ehrlich supposed to be present in the serum was an "uneconomical" hypothesis, and therefore implausible. A "simpler" explanation would be a better one.

This criticism, applied to a different aspect of Ehrlich's theory, had also been made by Buchner in 1900, in one of his last pieces of work.[25] It had been quoted and endorsed by Gruber.[26] "It obviously contradicts

25 Hans Buchner, "Zur Kenntniss der Alexine sowie der specifisch-bactericiden und specifisch-hämolytischen Wirkungen," *Münch. med. Wschr.* 47(1900):277–283, 279.
26 Max von Gruber, "Zur Theorie der Antikörper: II. Ueber Bakteriolyse und Hämolyse," *Münch. med. Wschr.* 48(1901):1924–1927, 1965–1967 (1965).

all economy of thought," wrote Buchner, to which Ehrlich replied that
his conclusions were not the result of speculations, but the inescapable
consequence of observation, and that nature might be even more complicated than this pluralistic theory suggests: he rejected without hesi-
tation the idea of economy as a guiding principle.[27]

The idea that the large number of different unrelated substances
involved make Ehrlich's theory unlikely appears as a much more important part of Landsteiner's criticism than of either Buchner's or
Gruber's, though both predate his use of it. It is also to be found, rather
unexpectedly, in a paper by Richard Pfeiffer, whose views on absolute
specificity seemed to have softened with time, or even to have come
round to Gruber's way of thinking:

If the specificity of the amboceptors of normal serum is complete, since the
animal body is naturally immune to an uncountable number of saprophytic and
pathological bacteria, . . . the blood must contain an absolutely inconceivable
number [*eine geradezu unabsehbare Menge*] of different specific amboceptors, a
concept which can only be admitted if every other explanation has been excluded.[28]

Landsteiner did not fail to notice this change and to quote Pfeiffer's
criticism in his own support. It was, of course, a criticism to which
Ehrlich was quite indifferent: he himself, after all, called his theory
"pluralistic." But its importance for Landsteiner is shown by his continued appeal to it. Several years later in a review paper of 1909, he
made it the basis of his scientific methodology.

In this essay, we can trace the transition between Landsteiner's studies of the *Malkoff'sche phänomen* and his later interpretation of specificity
as a resultant of non-specific forces, and its explanation in terms of the
physical chemistry of colloids. It is an essay in the history and philosophy of his science, as Landsteiner saw it, and it gives an important
insight into his thinking.[29]

The history begins with an attempt to explain how the animal body
is able to produce an antibody that matches any one of thousands of
possible antigens. It was at first suggested that the antibodies were de-

27 Paul Ehrlich, "Ueber Hämolysine: Vierte Mittheilung," *Berl. klin. Wschr.* 37(1900):
    681–687; *Coll. Papers* (n. 15), v. 2, 213–223 (Ger.), 224–237 (Engl.). As mentioned
    earlier, the Dale translation misses this idea: it is to be found only in the German
    text, p. 217.
28 Richard Pfeiffer and Ernst Friedberger, "Ueber die im normalen Ziegenserum en-
    thaltenen bacteriolytischen Stoffe (*Ambiceptoren* Ehrlichs)," *Deutsche. med. Wschr.*
    27(1901):834–836 (p. 836).
29 Karl Landsteiner, "Die Theorien der Antikörperbildung," *Wiener Klin. Wschr.* 22
    (1909):1623–1631; and in *Ergebnisse wiss. Med.* 1(1909-1910):185–207, C. 102.

rived from the antigen, but "as is well known, this simple hypothesis could not be made to harmonise well with the phenomena."[30] Although it was simple, and therefore attractive, it was dropped. Then "Ehrlich tried to solve the problem in a simple way that seemed very plausible; his hypothesis was that all immune substances were already present in normal serum."[31] This, too, was simple and therefore plausible, but the simplicity was only apparent:

Actually, the hypothesis made things no easier to understand since it offered no principle that was any simpler than the facts themselves. The riddle of the production of specific substances during immunization was replaced by something no more comprehensible, namely the physiological presence of innumerable substances.[32]

In other words, the hypothesis failed in its office, which was to simplify the phenomena. Malkoff's experiment, too, in Landsteiner's opinion, "in spite of its simplicity and clearness," was not well thought out and was easy to refute, although it was simple, and therefore attractive.[33] In Landsteiner's own elution experiment, "The findings can be interpreted quite simply, if in normal serum a number of agglutinins of low specificity are present, and this assumption made the basis for the explanation of the *Malkoff'sche phänomen.*"[34] This hypothesis has the virtue of diminishing the number of different substances postulated: there are now fewer substances than there are "phenomena," that is, agglutination reactions.

Earlier, both Landsteiner and Bordet had held the "simple and plausible hypothesis" that these few substances of low specificity could form a large number of more highly specific ones "by adaptive qualitative and quantitative combination."[35] Although this idea was not itself supported by the findings, it could be maintained in a modified form:

According to the older view, for every single effect of a serum, there was a separate substance, or at least a particular chemical group and . . . a normal serum contained as many different haemagglutinins as it agglutinated different cells. The situation was undoubtedly made much simpler if, to use the Ehrlich terminology . . . the separate haptophore groups can combine with an extremely large number of receptors in stepwise differing quantities [*in quantitativ abgestufter Weise*] as a stain does with numerous animal tissues, though not always with the same intensity. A normal serum would therefore visibly affect such a

30 Landsteiner, "Theorien der Antikörperbildung" (1909) (n. 29), 5.
31 Landsteiner, "Theorien der Antikörperbildung" (1909) (n. 29), 6.
32 Landsteiner, "Theorien der Antikörperbildung" (1909) (n. 29), 6.
33 Landsteiner, "Theorien der Antikörperbildung" (1909) (n. 29), 7.
34 Landsteiner, "Theorien der Antikörperbildung" (1909) (n. 29), 8.
35 Landsteiner, "Theorien der Antikörperbildung" (1909) (n. 29), 10.

large number of different blood cells . . . not because it contained countless special substances, but because of the colloids of the serum, and therefore of the agglutinins by reason of their chemical constitution and the electrochemical properties resulting from it. That this manner of representation [*diese Darstellungsweise*] as a considerable simplification [*Vereinfachung*, reduction] is clear; it also opens the way to direct experimental testing by the methods of structural chemistry.[36]

It has been objected, that the relation between the behaviour of antibody and that of a colloid is one of analogy only: but apart from the fact that the search for analogies is well known to be one of the most important means to knowledge, this is actually not just an analogy, as the immune bodies really are colloids, and the two classes of substances have many common properties. . . . It is by no means insignificant that on the basis of this colloid-chemical point of view, one can imitate the phenomena of agglutination with relatively simple compounds.[37]

The analogy leads to a simple model that can be used to test the *Darstellungsweise*, the model, which has been set up as a hypothesis.

The main thoughts of this essay, as I have interpreted it, are these: First, a simple conception is inherently plausible, but it must be made to accord with the phenomena. Second, the office of a hypothesis is to suggest a principle that makes our understanding of the phenomena easier, by simplifying them. If it does not do this, it is a failure. A field of phenomena is more simply described if they are related to each other by stepwise differences (*quantitative Abstufungen*) than if large numbers of separate unrelated substances are postulated. And third, the analogies between the behaviour of immune substances and those of inorganic colloids make it possible to set up a simple model that is accessible to testing.

This, then, is Landsteiner's *methodologische Einleitung*. Nowhere else in his writing does he give anything approaching a theory of scientific knowledge. Indeed, he does not do so here either; his views have to be disentangled from what is intended as a discussion of past and present theories of antibody production, where his methodological statements make themselves visible in that each one contains the word "simple," *einfach*.

One of the sources of this *erkenntnisstheoretisch* viewpoint is easy enough to find: Landsteiner defines himself by contrasting his view with those of Ehrlich. Where Ehrlich calls his theory *plurimistisch*, Landsteiner's is *einheitlich*; where Ehrlich postulates a whole host of separate substances and absolute specificity, Landsteiner sees *quantitative Abstufung*. Here he stands in continuity with Gruber: the objections that

36 Landsteiner, "Theorien der Antikörperbildung" (1909) (n. 29), 15.
37 Landsteiner, "Theorien der Antikörperbildung" (1909) (n. 29), 11.

Gruber raised to the work of Koch and Pfeiffer, and that he grasped at so eagerly in von Pirquet's criticisms of Ehrlich, are present in spirit in Landsteiner's discussion of Ehrlich's theory. The form is the same, only the content has altered: behind Landsteiner stands Gruber, who did not believe in specificity; Nägeli, with his *quantitative Abstufung;* and Schleiden, with his heuristic maxim of unity and his criticism of the divisions and subdivisions of the Linnaean botanists. As Schleiden wrote in 1838, "The basic principle of human reason, its inescapable striving for unity in its knowledge, is as valid for the study of the organism as it is elsewhere in science."[38]

38 Matthias J. Schleiden, "Beiträge zur Phytogenese," *Arch. f. Anat. u. Physiol.* 2(1838): 137–176; my translation.

8

Unity, Simplicity, Continuity: The Philosophy of Ernst Mach

Landsteiner's heuristic maxims, as expressed in his essay on the theories of antibody formation of 1909, are simplicity, continuity, and analogy.[1] He represents for his generation the line of unitarian thinkers that I have traced back in this book to Matthias Schleiden, for whom the unity of nature was always more important that its diversity. But in 1902, in his criticism of Ehrlich's pluralistic theory, Landsteiner had written: "The number of hypothetical substances postulated makes this conception so uneconomical that the question must be asked whether it is the only one possible."[2]

The linking of the idea of simplicity with that of economy immediately suggests the influence of the physicist and philosopher Ernst Mach.[3] His was an influence that could be found in a wide area of thought at this time, especially in Vienna, where Mach was professor of philosophy, specializing in the history and theory of the inductive sciences. The effect of Mach's teaching on scientific and other writing

---

1 Karl Landsteiner, "Die Theorien der Antikörperbildung," *Wiener klin. Wschr.* 22(1909):1623–1631, C. 102.
2 Karl Landsteiner and Adriano Sturli, "Ueber die Hämagglutinine normaler Sera," *Wieiner klin. Wschr.* 15(1902):38–40, C. 19.
3 The current biographical and critical sources for Mach are at present Karl Daniel Heller, *Ernst Mach Wegbereiter der modernen Physik, mit ausgewählten Kapiteln aus seinem Werk* (Vienna: Springer, 1964); Erna Lesky, *Die Wiener medizinische Schule im XIX Jahrhundert* (Graz: Bohlaus, 1965), 533–535; John T. Blackmore, *Ernst Mach: His Life Work and Influence* (Berkeley, Calif.: University of California Press, 1972); Alan Janik and Steven Toulmin, *Wittgenstein's Vienna* (New York: Simon, 1973), 132 ff.; Rudolf Haller und Friedrich Stadler, *Ernst Mach – Werk und Wirkung* (Vienna: Hölder-Pichler-Tempsky, 1988). A contemporary physiologist's account is Alois Kreidel, "Ernst Mach," *Wiener klin. Wschr.* 29(1916):394–396. Other discussions of Mach's thought and influence will be referred to as they occur in the text. His own summing up is Ernst Mach, *"Die Leitegedanken meiner naturwissenschaftliche Erkenntnislehre und ihre Aufnahme durch die Zeitgenossen,"* und *"Sinnliche Elemente und naturwissenschaftliche,"* zwei *Aufsätze* (Leipzig: Barth, 1919); and in Mach, *The Analysis of Sensations and the Relation of the Psychical to the Physical,* translated by C. M. Williams, from the 1st (1885), revised and supplemented from the 5th (1906) German edition by Sydney Waterlow (Chicago, Ill.: Open Court, 1914), "How my views have been received," 354–371.

at the end of the nineteenth century can be compared with that of Kant a hundred years earlier.

Mach's views on the economy and simplicity of scientific thought were put most neatly by himself in the course of a polemic of 1910, though they had by that time been in circulation for nearly thirty years. In this statement, Mach wrote that influenced by the Graz economist Emmanuel Herrmann he began to see the activity of the scientist as an effort to economize in thought:

Every abstract concept, every general expression of the relations of facts to each other, every replacement of a table of figures by a formula or a rule by which they might be reproduced, and the law that covers it, every explanation of a new fact by means of another, better-known one, can be looked at as the result of an economy. The further the analysis of the systematising, ordering simplifying mathematical structure is taken, the clearer it becomes, that what science actually does is to economise.[4]

Mach combined the economizing function with the Darwinistic idea of survival of the fittest in what he called a biologico-economic view of scientific knowledge: the adaptation of thoughts to facts is the adaptation of the human animal to its environment: "Every beneficial biological process is a self-preserving or adaptive one, and as such is more economical than one which is not beneficial to the individual. All useful knowledge is a special case of a biologically useful process."[5]

Mach's biographer calls this the shortest and most pregnant statement that Mach made of his theory of economy of thought.[6] It was also quoted as a nutshell version by Alois Kreidl, the Viennese physiologist who wrote a necrology of Mach in the *Wiener klinische Wochenschrift* on Mach's death in 1916.[7]

Mach first stated this theory of knowledge as economy of thought in a lecture in Vienna in 1882. This presentation became one of his most famous essays, "The Economical Nature of Scientific Research."[8] It was

4 Mach, *Leitgedanken* (1919) (n. 3), 3; Rudolf Haller, "Poetische Phantasie und Sparsamkeit – Ernst Mach als Wissenschaftstheoretiker," in Haller and Stadler, *Ernst Mach* (1988) (n. 3), 342–355; Haller (p. 345), cites Emmanuel Herrmann, *Allgemeine Wirtschaftslehre* (Graz: 1868), v. 1, pt. 1, 21: "It is the economy of nature . . . that is the best and most beautiful part of economics. What are our labour-saving devices, our building techniques, our machines, compared to the labour-saving of the organs which transport fluids in plants, or the organs of the senses and the will in animals."
5 Mach, *Leitgedanken* (1919) (n. 3).
6 Heller, *Ernst Mach* (1964) (n. 3), 134.
7 Kreidl, *Ernst Mach* (1916) (n. 3), 396.
8 Ernst Mach, "Die ökonomische Natur der physikalischen Forschung," Vortrag, gehalten in der feierlichen Sitzung der k. Akad. der Wissenschaften zu Wien, 25 Mai 1882; reprinted in *Popular Scientific Lectures*, trans. Thomas J. McCormack (Chicago, Ill.: Open Court, 1895); and in *Populär-wissenschaftliche Vorlesungen* (Leipzig: Barth, 1st ed. 1896, 5th ed. 1923). Quotations from 5th ed.

reprinted in a collected edition of Mach's work entitled *Popular Scientific Lectures*, which appeared first in English, published by Mach's friend Paul Carus at the Open Court in Chicago in 1895, then in German in 1896. The fifth edition, edited by his son Ludwig Mach, appeared in 1923, after Mach's death.[9]

In this often-quoted essay, Mach's description of the process of scientific understanding, like Landsteiner's, equates "economy" with "simplicity." Science takes the manifoldness of nature and tries to find in it an underlying unity; as long as the manifoldness remains, the field has not been truly explained: science has made no progress. This is the criticism that Landsteiner was making of Ehrlich's theory when he refused to accept that each receptor on a cell differed absolutely from every other, and that each had a matching specific antibody. The manifoldness of nature has not been reduced: the explanation is no more economical than the phenomena themselves. Mach brings out the same point in his essay:

When we look over a province of facts for the first time, it appears to us diversified, irregular, confused and full of contradictions . . . By and by we discover the simple, permanent elements of the mosaic when we have reached the point when we can recognise everywhere in this manifoldness the same facts, we no longer feel strange in that province: we comprehend it without effort, it is *explained* for us . . .

This goal, that of surveying a province with the least possible expenditure of thought, and of representing all the facts as parts of a single thought process, may justly be called an economical one.[10]

In 1895 Mach was appointed to the chair of philosophy in Vienna, "with special reference to the history and theory of the inductive sciences."[11] His first course of lectures was on the "psychology and logic of scientific investigation" and was given over the winter semester of 1895–1896. This course was worked up and published ten years later as a series of linked essays, *Knowledge and Error: Sketches in the Psychology of Scientific Research*, which appeared in 1905.[12]

9   Joachim Thiele, "Ernst Mach: Bibliographie," *Centaurus, 8*(1963):189–237.
10 Mach, *Popular Scientific Lectures* (1895) (n. 8); translation is that of McCormack, 5th ed., 225. McCormack does not bring out the "manifoldness of nature," however.
11 Blackmore, *Ernst Mach* (1972) (n. 3), 154.
12 Ernst Mach, *Erkenntnis und Irrtum: Skizzen zur Psychologie der Forschung* (Leipzig: Barth, 1st ed. 1905, 3d ed. 1917); Mach, *Knowledge and Error: Sketches on the Psychology of Enquiry*, translated from 5th ed. by Thomas J. McCormack and Paul Foulkes, edited by Brian McGuinness, with an introduction by Erwin N. Hiebert (Dordrecht: Reidel, 1976), v. 3. In series Vienna Circle Collection. My quotations are from 3d German edition, and translations are mine. Interestingly, Hiebert says nothing about analogy, simplicity, or unity as characteristics of Mach's thought.

Historians and contemporaries have stressed the enormous popular success of this lecture series, which was attended by many people outside the fields of philosophy and science.[13] These lectures contain possible prototypes for further elements of Landsteiner's methodology. For Mach, as for Landsteiner, there is a connection between simplicity and continuity:

Quantitative dependence is a particularly simple form of qualitative relationship ... In quantitative dependence, there is a clearly visible continuum of cases, while in the case of qualitative relations there is only a number of separate individuals. It is natural, wherever possible, to try to achieve the simplicity, regularity and clearness of quantitative handling. It becomes possible as soon as quantitatively identical markers can be discovered which fully describe the qualitatively dissimilar.[14]

The pattern followed by developing science, says Mach, is to find similarities in the apparently different, to relate things that seem at first to be poles apart: "A change in the object in question which takes place continuously or in small steps, leads to the understanding of the relationship between distant members of a series, and to the realization of what it is that remains constant through this change."[15]

Both of these citations reflect the mistrust of discontinuity and qualitative differentiation that was part of Landsteiner's mental furniture. Like Landsteiner, Mach feels he has understood something when he can see it as part of a continuous series, a *quantitative Abstufung*: it is simplified for him, as for Landsteiner, when seemingly qualitatively different elements are found to differ only quantitatively. Mach does *not* suggest that the function of science might include the attempt to discriminate among the apparently similar. If he had been a bacteriologist, one feels, he would have been an admirer of Nägeli.

A further correspondence with Landsteiner's thinking lies in Mach's emphasis on analogy as a tool of scientific thought. He quotes the seventeenth-century astronomer Johannes Kepler as having said, "Plurimum namque amo analogias, fidelissimos meos magistros, omnium naturae arcanorum conscios," he writes, "there is no lack of examples of the importance of analogy. It can hardly be too highly valued in natural science."[16] This remark occurs in the middle of a whole chapter

13 Blackmore, *Ernst Mach* (1972) (n. 3), 154ff.; Janik and Toulmin, *Wittgenstein's Vienna* (1973) (n. 3), 113, 133.
14 Mach, *Erkenntnis und Irrtum* (1905/1917) (n. 12), 201–219.
15 Mach, *Erkenntnis und Irrtum* (1905/1917) (n. 12), 220–231: "Die Ähnlichkeit und die Analogie als Leitmotive der Forschung," 224.
16 Mach, *Erkenntnis und Irrtum* (1905/1917) (n. 12), 220–231: "Ähnlichkeit und Analogie," 223.

of *Knowledge and Error* that is devoted to "similarity and analogy as leit-motifs for investigation."[17] One of the "popular scientific lectures" is also on this theme, that of analogy or comparison as "one of the most powerful of the inner elements of the life of a science."[18] We can find out by example from history, he says, what methods have been most successful in the past: "The methods of physical and mental experiment, of analogy, the principles of simplicity and continuity – it is in these that we may quite simply put our trust."[19]

In analogy, simplicity and continuity: these are the principles in which we may quite simply put our trust. We could hardly be closer to Landsteiner's methodology.

This does not mean, however, that Landsteiner was actually influenced by Mach. Landsteiner's own footnotes, for example, show no direct references to Mach. Those who share his feeling that Ehrlich's theory is uneconomical, he says, are Jules Bordet, Max von Gruber, and Richard Pfeiffer, all immunologists, and Siegmund Exner the physiologist. Hans Buchner, it will be remembered, had used this phrase himself in referring to Ehrlich's theory.[20] Nor is there evidence that Landsteiner was himself in personal contact with Mach, either as a friend or through his publications. The closest connection they appear to have had was through their mutual friend Wolfgang Pauli, whose relationship with both Mach and Landsteiner is discussed further on. Although Pauli was close to both of them, he does not provide any evidence of *direct* contact.[21]

At the same time, there is plenty of evidence to show that Mach had a powerful indirect influence that affected the thinking of many biological scientists, especially in Vienna. The writing of such men provides a number of examples of biological thinkers who were Landsteiner's

17 Mach, *Erkenntnis und Irrtum* (1905/1917) (n. 12), 220–231: "Ähnlichkeit und Analogie," 227.
18 Ernst Mach, "Ueber das Prinzip der Vergleichung in der Physik," Vortrag, gehalten auf der 66 Versammlung deutscher Naturforscher und Aerzte, Vienna, 1894; and in *Populäre-wissenschaftliche Vorlesungen* (1896/1932) (n. 8), 266–289 (p. 269).
19 Mach, *Erkenntnis und Irrtum* (1905/1917) (n. 12), 220–231, 220–231; "Ähnlichkeit und Analogie," 224.
20 Landsteiner and Sturli, "Hämagglutinin normaler Sera" (1902) (n. 2), 40. References to Ehrlich's theory as uneconomical are also to be found in Max von Gruber, "Zur Theorie der Antikörper: II. Ueber Bakterioloyse und Hämolyse," *Wiener klin. Wschr. 14*(1901):1214–1215; Hans Buchner, "Zur Kenntniss der Alexine, sowie der specifisch-bacterociden und specifisch-haemolytischen Wirkungen," *Münch. med. Wschr. 47*(1900):277–283 (p. 279); Richard Pfeiffer and Ernst Friedberger, "Ueber die normalen Ziegenserum enthaltenen bakteriolytischen Stoffe (*Ambiceptoren* Ehrlichs)," *Deutsch. med. Wschr. 27*(1901):834–836 (p. 836).
21 František Smutny, "Ernst Mach and Wolfgang Pauli's ancestors in Prague," *Gesnerus 46*(1989):183–194; and see Chapter 11, esp. n. 7.

contemporaries, and who actually did adopt ideas from Mach. An examination of which ideas it was they adopted will provide something of a contemporary perspective on Mach and his influence, and a context within which to assess Landsteiner's thinking.

Mach was closely connected with the Vienna medical school and with many physiologists. His life-long friendship with the physiologist Ewald Hering has been called "the intellectual forcefield around which Vienna medicine of the last third of the XIX century was oriented."[22] Hering and Mach had worked together in Prague, where Mach was professor of experimental physics, and Hering had succeeded Jan Purkyně in the chair of physiology in 1870. Hering had studied under Ernst Heinrich Weber and Gustav Theodor Fechner, eponymous owners of the Weber–Fechner law, which relates psychological events to physical by means of a mathematical function. Hering's interests in physiology, like those of Mach himself, lay in this same field – namely, the physiology of sensation, especially sound, vision, and the sense of movement. It was here in Prague that Mach's classical work on physics, on the physiology of sensation, and on the historico-critical analysis of mechanics, as well as the formulation of his theory of knowledge, all took place.[23]

The close association of his theory of knowledge with his physiology is evident throughout his first "philosophical" work, *The Analysis of Sensations*, which appeared during this period. In the author's preface, Mach writes: "My natural bent for the study of these questions received its strongest stimulus twenty-five years ago from Fechner's *Elemente der Psychophysik* (Leipzig 1860) but my greatest assistance was derived from Hering's solution of the two problems referred to [two problems of vision]."[24]

In this book Mach sets out to analyze sensation not through psychology alone, like Johannes Müller, nor by the laws of physics alone, like Hermann Helmholtz, but through the relation between the two: his goal is to understand the connection of psychologically observed data with the physical process that produces them.[25] His guiding principle, he states, is the complete parallelism of the psychical and the physical; not, however, as Fechner called them, two aspects of the same

22 Lesky, *Die Wiener medizinischen Schule* (1965) (n. 3), 533.
23 Heller, *Ernst Mach* (1964) (n. 3), p. 19.
24 Ernst Mach, *Beiträge zur Analyse der Empfindungen* (Jena: 1st ed. 1886, 9th ed. 1922); translated from 1st German ed. by C. M. Williams, and revised from 5th (1906) and supplemented by Sydney Waterlow, *The Analysis of Sensations and the Relation of the Physical to the Psychical* (Chicago, Ill · Open Court, 1914), viii. Quotations are from this 1914 edition.
25 Mach, *Analysis of Sensations* (1885/1914) (n. 24), 39.

reality, which Mach feels to be a metaphysical position, but as an effort to find a common physiologico-physical process underlying a common group of sensations. If, for example, the limitless multiplicity of colours can be reduced to a few elements, the same simplification is to be expected in the nerve processes.[26] Mach's own investigation on the mechanism and responses of the semicircular canals of the middle ear and the detection of movement gave him another example of psychophysical parallelism of this positive kind.[27]

Mach's theory of knowledge is developed from this physiological basis. Knowledge is a grouping of the elements of sensation formed in the sense organs into the complex called an object. The elements of colour, sound, and pressure of the outside world are connected together solely by association into objects, which in turn are associated by the mind in the relationship called cause and effect.

This rather British philosophy is given a biological turn by Mach in that the associations between elements are part of the Darwinian survival apparatus of the animal species. The ego itself is only an economical unity of sets of elementary sensations and has no real or permanent existence. In the words of Mach's famous *Satz*, *"das Ich ist unrettbar,"* the "I" is a phenomenon that cannot be saved. Like the *Ding-an-sich* of Kantian philosophy, the "I" consists only of a bundle of associated sensations.[28] The "thing" and the "I," the object and the subject, are not separate entities.[29] For the sober scientific thinker – the positivist, that is – they are metaphysical nonentities, the source of factitious philosophical pseudo-problems.[30] As Mach himself wrote in the preface to *Knowledge and Error*: "I have to say with Schuppe, the land of the transcendental is closed to me. . . . *I am no philosopher, only a scientist.*"[31] And he calls himself a Sunday driver (*ein Sonntagsjäger*, a Sunday hunter) in the field of philosophy.

The system of thought that has been called empiriocriticism, a family name that marks its British and its Kantian ancestry, could equally well in Mach's hands be called biological positivism.[32] Besides Hume and Kant, its ancestors included Johannes Müller, Hermann Helmholtz,

26 Mach, *Analysis of Sensations* (1885/1914) (n. 24), 266.
27 Mach, *Analysis of Sensations* (1885/1914) (n. 24), 147ff.
28 Mach, *Analysis of Sensations* (1885/1914) (n. 24), 1–37: "Introductory remarks: antimetaphysical." The famous phrase "Das Ich ist unrettbar," translated a little colourlessly as "The ego must be given up," is on p. 24, bottom.
29 Mach, *Erkenntnis und Irrtum* (1905/1917) (n. 12), 15.
30 Mach, *Erkenntnis und Irrtum* (1905/1917) (n. 12), 12.
31 Mach, *Erkenntnis und Irrtum* (1905/1917) (n. 12), viii.
32 M. Capek, "Ernst Mach's biological theory of knowledge," *Synthèse* 18(1968):171–191.

and Charles Darwin. As Hans Henning, who wrote an extremely partisan monograph on Mach in 1915, said, "Whoever rejects Mach's principle of economy, for him Darwin has never lived."[33] It is no accident that a philosophy so largely formed of biological ideas should have been attractive to biologists: the philosopher in this case was their own *frère de lait.*

There are at least three examples of biological thinkers for whom there is good evidence of a direct relationship to Mach. Their experience of Mach as a contemporary can be used to build up a picture of the parts of his thought that seemed important and interesting to his own time. The three examples I have chosen are the Viennese pathologist Samuel von Basch, and the physiologists Max Verworn, also of Vienna, and Jacques Loeb, who worked mainly in America. The names of Loeb and Verworn appear in the list of letters to Mach in the archives of the Ernst-Mach Institute.[34] Von Basch's relation with Mach has been discussed by Erna Lesky.[35]

Samuel von Basch was a student of the Leipzig physiologist Carl Ludwig and was himself director of the Laboratory for Experimental Pathology in Vienna. In his case, the exact lecture of Mach's that attracted his interest can be identified from his paper; it is the one that appears in Mach's *Popular Scientific Lectures* under the title of "On the Principle of Comparison in Physics"; it was first given in Vienna in 1894.[36] Von Basch quotes Mach's empiricist objections to the causal relationship and says that if he had understood Mach rightly, what he said was that the conditions that link together sensory phenomena are all that we can know or describe. We can never say that one is a cause and one an effect. We are drawn into calling them this by a kind of spiritual thirst, a need, for causality, or what von Basch calls a *causalitäts-durstigen Phantasie.*

We should attempt to formulate a law which will show that the relationship of conditions to conditioned is a necessary one, and does not depend on an un-

33 Hans Henning, *Ernst Mach als Philosoph Physiker und Psycholog* (Leipzig: Barth, 1915), 135.

34 H. Scharden, Director, "Briefe an Ernst Mach im Besitz des Ernst Mach Institut der Fraunhofer Gesellschaft, Freiburg im Breisgau." My thanks are due to the Institut für Geschichte der Medizin der Universität Wien for bringing this source to my attention.

35 Erna Lesky, "Kompensationslehre und denkokonomisches Prinzip," *Gesnerus* 23(1966):97–108. I was indebted to Professor Lesky for personally drawing my attention to this paper.

36 Samuel von Basch, "Die Compensationslehre von Erkenntniss theoretisch Standpunkte," *Verhandl.d.XIII Congr. f. inn. Med.* (Wiesbaden, 1895), 433–447.

derlying accident. The description or representation [*Schilderung*, that is, of these conditions] in the most concise, or, as Mach puts it, the most economical form, must be regarded as the goal of all scientific investigation.[37]

Von Basch uses this notion to attack the current interpretation of the enlarged heart of valvular heart disease as a compensatory enlargement. The enlarged heart had been assumed to be compensating for a slowing of the blood flow, regarded as the cause of the dyspnoea. The causal explanation, says von Basch, was allowed to take the place of detailed investigation of the process taking place; the facts were made to conform with the thought. The discovery of the actual conditions involved in the process makes this explanation in terms of the body's attempt to compensate for its losses unnecessary, and even shows it to have been misleading. Von Basch uses Mach's *Erkenntnis-theoretisch* viewpoint to attack the cause and effect, and the teleologic, interpretation of the symptoms of heart disease, which had served to conceal the fact that the processes were not known.

Landsteiner's interpretation of the human blood groups can be put into this same framework. Previous interpretations had all assumed that the antibodies were there for the purpose of defence, and that therefore a causal disease must be found. Landsteiner's interpretation, which simply describes their relations to each other and by-passes the causal and teleological assumptions, can be seen, like von Basch's, as an example of Mach's principles in action. But it is harder to prove that it is an example of Mach's influence. Landsteiner may have been aware of Mach's principle of economy, or he may have been aware of it in time to use the phrase in a paper that appeared eight weeks later.[38] In that paper it was applied to Ehrlich's theory, with its thousands of postulated substances, not to the teleological interpretations of the blood group antibodies. The assumption of a protective function for the immune substances, as he wrote in 1899, might stand in the way of the formulation of more general laws.[39] It was to the teleology, not to the lack of economy, that he objected here, and there is no direct evidence to connect that with Mach's work. But in both cases, that of the antibodies to red cells and that of Ehrlich's theory, Landsteiner appeals to simplicity as a criterion of a good explanation.[40]

37 Von Basch, "Compensationslehre" (1895) (n. 36), 436.
38 The paper on the interpretation of the blood groups (C. 17) appeared on 14 November 1901 and that in which he criticized Ehrlich's theory as uneconomical (C. 19) on 9 January 1902.
39 Karl Landsteiner, "Zur Kenntniss der spezifisch auf Blutkörperchen wirkenden Sera," *Zbl. f. Bakt.* 25(1899):546–549, C. 12 (p. 547); see Chapter 7 for discussion.
40 Landsteiner, "Spezifisch auf Blutkörperchen wirkenden Sera" (1899) (n. 39), 548.

Like von Basch, Max Verworn emphasizes conditions rather than causes in his explanation of physiological processes. Verworn introduces his textbook of physiology, which first appeared in 1894, with a *methodologische Einleitung*, a long essay entitled "Ends and Means in Physiological Research." Verworn closely follows the features of Mach's sensationalist epistemology. Representations (*Vorstellungen*) are nothing but memories of sensory experiences put together as *Vorstellungsassociationen* or thoughts, which are selected and cultivated (gezüchtet) or eliminated according to whether they match with sensory experience. The goal of natural-scientific explanation is the recognition of lawfulness, that is, the knowledge of the conditions that affect a state or process. When these conditions are known, the goal has been reached. Verworn's *Lehrsatz* was "Natural science must ever strive to eliminate the concept of cause from exact thinking; it is not causes we should seek, but conditions – *Nicht kausalismus sondern Konditionismus.*[41]

He goes on to eliminate from his epistemology the Kantian *Ding-an-sich*, the origin of pseudo-problems, as well as the materialist metaphysic of a common material or atomic substrate of things. Verworn wrote in the second edition of his book, in a passage that he reworked and then cut from later editions:

My own individuality is only a representation [*Vorstellung*] of my psyche, and so I can no longer say, *Die Welt sei meine Vorstellung*. All I can say is that the world is a *Vorstellung* or a sum of *Vorstellungen*, and what seems to me to be my individuality is only a part of this complex of *Vorstellungen*, like the individuality of other men, and the whole physical world.[42]

As Mach had said, *Das Ich ist unrettbar*: the "I" is a phenomenon that cannot be saved. Dualism of body and soul is thus eliminated. It must be remembered that Verworn was a student of Ernst Haeckel, and though he refers this view in his second edition to Mach's monistic *Beiträge zur Analyse der Empfindungen* of 1886, his mind must have been prepared for its reception by his contact with Haeckel's radical monism.

The main source of Verworn's epistemology is quite clear. He tells us in a footnote to the later editions that the views he developed are close to those of Ernst Mach, and for further discussion of these questions he directs the reader to *The Analysis of Sensations* and *Knowledge and Error*, along with the works of several other writers on the philos-

41   Max Verworn, *Allgemeine Physiologie eine Grundriss der Lehre vom Leben* (Jena: 1st ed., 1895, 5th ed., 1909), 1–62 (5th ed.): "Erkenntniss-theoretischer Konditionismus," 37.
42   Verworn, *Allgemeine Physiologie* (n. 41), 2d ed., 38.

ophy of natural science.[43] Verworn, it should be pointed out, was able
to give a complete *Mach'sche* epistemology without ever mentioning
economy of thought, analogy, simplicity, or continuity. His emphasis is
on Mach's empiricism, and on his elimination of the causal explanation.

Another physiologist in direct contact with Mach was Jacques Loeb.
Loeb's connection with Mach's thinking is not immediately obvious,
and it has not been of particular concern to historians who have written
about him.[44] Brailsford Robertson, a student and colleague of Loeb,
has remarked on his epiphenomenalism; but a disconnected parallelism between consciousness and the material world, which was Brailsford
Robertson's definition of epiphenomenalism, is not a particularly good
characterization of Loeb's style, nor of Mach's.[45] Blackmore, to whom
Loeb's letters to Mach were available, dismissed Loeb's philosophy with
his usual undervaluation of serious human effort.[46]

The first line of evidence for Loeb's interest in Mach's thinking is
that of his own statements in these letters. In one of them, dated 1890
and written from the Naples Zoological Station where Loeb was working on heliotropisms in marine animals, he wrote to Mach: "So far as
I have arrived at any particular view of nature, I owe it to reading your
Analysis of Sensations, and your other main works, and I see it as my
goal to orient the phenomena with reference to this point of view."[47]

Loeb's early work was done in close association with the senior botanist Julius von Sachs at Würzburg. Sachs was working on heliotropic
and geotropic movements in plants, and Loeb's work was a parallel
exposition of heliotropism in animals. One of his conclusions was that
"the dependence of animal movements upon light is point for point
the same as the dependence of plant movement upon that stimulus."[48]

The response is greater to the more refractile end of the spectrum,

---

43 Verworn, *Allgemeine Physiologie* (n. 41), 5th ed., 41.

44 The standard biographical sources on Jacques Loeb are Philip J. Pauly, *Controlling
    Life: Jacques Loeb and the Engineering Ideal in Biology* (Berkeley, Calif.: University of
    California Press, 1987): on Loeb and Mach, 41–45, 72–74; Donald Fleming, Intro-
    duction to Loeb, *The Mechanistic Conception of Life* (Cambridge, Mass.: Harvard Uni-
    versity Press, 1964), vii–xiii; Winthrop J. V. Osterhout, "Jacques Loeb," *J. Gen. Physiol.*
    (Jacques Loeb Memorial Volume), *8*(1928):ix– .

45 T. Brailsford Robertson, "The life and work of a mechanistic philosopher: Jacques
    Loeb," *Science Progress in XXth Century 21*(1926):114–129 (p. 128).

46 Blackmore, *Ernst Mach* (1972) (n. 3), 129–131.

47 Jacques Loeb to Ernst Mach, letter dated 26 February 1890, from collection of twenty
    letters to Mach, 1885–1905. Copies of these letters were made available to me
    through the courtesy of Dr. H. Schardin, director, Ernst Mach Institut der Fraun-
    hofer Gesellschaft, Freiburg-im-Breisgau.

48 Jacques Loeb, *Der Heliotropismus der Thiere und seine Uebereinstimmung mit dem Heliotro-
    pismus der Pflanzen* (Würzburg: Hertz, 1890), 109.

he says; it occurs only within a certain range of intensity and only within a certain temperature range. It is different for the upper and lower sides of some of the animals studied, and is present even in eyeless forms. He concludes that these heliotropic phenomena cannot be referred to specific properties of a nervous system, such as instincts or reflexes, since they occur in plants as well as animals.

The psycho-physical relationship here is, like Mach's, no metaphysical proposition. The psychological side, the behaviour of the animal towards light, is defined for it by the physical conditions in which it finds itself. In a subsequent paper Loeb wrote:

> In physics the conception that Mach has expressed so forcefully has come more and more to be accepted, that the explanation of a phenomenon consists only in the description of the conditions under which it takes place, a description which should be as simple and as complete as possible. The opposite standpoint from that of Mach, Kirchhoff and Ostwald, can be called the metaphysical or *naturphilosophisch*. . . . One field of biology, that of the instinctive reactions of animals, has, in the hands of most of the workers dealing with it, remained stuck at the *naturphilosophisch* or theological standpoint.[49]

As an example of a metaphysical interpretation, Loeb cites the suggestion that a moth flies towards a candle because it instinctively prefers the light or because it is curious about it. But if the conditions are manipulated, positively heliotropic animals can be made negatively heliotropic and vice versa; the same conditions affect the responses of plants. Sensation itself becomes a superfluous notion in this highly reductionist system, which is none the less closely related to Mach's sensationalist epistemology. The elementary sensations have been broken down further, into a group of general physico-chemical conditions.

Mach himself, though he admired Loeb's work and often quoted it, was not so sure that it was necessary to free oneself of all teleological explanation in biology. In the fourth edition of the *Analysis of Sensations*, he added a chapter on the problem of teleology, in which he admits it as a useful device where causal (or conditional) theories are not yet available.[50] Loeb's work represents a

> happily conceived and important effort to shake off the trammels of unnecessarily complicated assumptions impregnated with metaphysics. But I cannot agree with Loeb when he treats Darwin's phylogenetic research on the instincts

49 Jacques Loeb, "Zur Theorie der physiologischen Licht und Schwerkraftwirkungen," *Arch. f. die ges. Physiologie, 66*(1897):439–459 (p. 456).
50 Mach, *Analysis of Sensations* (1885/1914) (n. 24), 83–101 (p. 85).

as a fallacious and one-sided proceeding which ought to be dropped and re-placed by physico-chemical investigations.''[51]

It is the attempt to approach psychic phenomena from both sides that Mach approves of in the work of the physiologist Sigmund Exner, where the instincts, the anatomy of the nervous system, and the effects of stimuli are all considered in turn, and where the connection between physiological conditions and psychological phenomena is made in full.[52]

What, then, can be said in general about the relationship of the physiologists to Ernst Mach? These three individuals were clearly in direct contact with Mach. They heard him lecture, or referred to him in footnotes, or wrote to him. It is quite likely that others, too, did the same, particularly those who were working in the field of physiology of sensation, where Mach's own physiological interests lay, and where a sensationalist epistemology might act as a unifying link, bringing a world view into consonance with scientific thinking in their own field. For it was this aspect of Mach's work that was of most interest to his contemporaries. In each of these three figures, one finds the same features emphasized: the prescription for today's science, the good ad-vice that Mach offers, is that the goal of science is to describe condi-tions, the "functional dependence of the elements on each other."[53] Representations, *Vorstellungen,* are grouped sensational elements re-lated to each other in this way. We need not bring in any goal beyond the Darwinian one of the adaptation of thoughts to facts, and no re-lationships beyond association. Verworn's *Nicht Kausalismus, sondern Konditionismus* is a *Satz* that speaks for all three physiologists, and for Mach, too. It is the positivists' rejection of metaphysics, at bottom, to which they all subscribe so enthusiastically, which extends to every cat-egory of thought except that of quantitative relationship. Not one of the three has picked up the emphasis on analogy, simplicity, and con-tinuity found in Landsteiner's methodology and so evident in Mach's writing. The Mach by whom these three were influenced does not stand for simplicity and continuity, but for a sensationalist epistemology of which there is no trace in Landsteiner's writing.

The connection between these two parts of Mach's thought, the economy and simplicity side and the sensationalism, as it appeared to

51 Mach, *Analysis of Sensations* (1885/1914) (n. 24), 241.
52 Mach, *Analysis of Sensations* (1885/1914) (n. 24), 369–370, referring to Siegmund Exner, *Entwurf zu einer physiologischen Erklärung der psychischen Erscheinungen* (Leipzig: Deuticke, 1894). Exner begins with a chapter on the anatomy of the spinal cord, brain stem, and cerebral cortex, then discusses the physiology of the nervous system and voluntary movement; he then goes on to sensation, perception, *Vorstellung,* and finally intelligence, the instincts and the forms of thought.
53 Mach, *Analysis of Sensations* (1885/1914) (n. 24), 369.

another of Mach's contemporaries, is made by the Viennese writer Robert Musil.[54]

Musil was born in 1880 in Klagenfurt, and after an unpleasant education at the k. u. k. Militär-Oberrealschule at Mährisch-Weisskirchen, he read engineering at the k. k. Technische Hochschule in Brunn (Brno), where his father was teaching, and afterwards worked in the laboratory for machine technology at the Technische Hochschule in Stuttgart. But in 1903 he transferred into the humanities and in 1908 took a doctorate in the department of philosophy of Berlin University, with a thesis on Mach.[55] By the time his thesis was accepted, he had already published his first novel, and his subsequent career was a literary one.

Musil's thesis is a rather hostile critique of Mach's philosophy. The first third or so is devoted to a general exposition of Mach's thought, mainly on economy, continuity, and adaptation as criteria for the truth of a proposition. The next section provides a discussion of Mach's polemic against causality, and it is here that Musil finds what he considers distinctive about Mach's thought. Both sections give an important insight into Mach's place in the context of contemporary thought.

In his first general section, Musil criticizes the use of economy, continuity, and adaptation as criteria for truth: they themselves depend on truth. As he puts it, continuity can give rise to either knowledge or error – the difference being, according to Mach, that a "true" proposition would be better adapted to the facts. But adaptation need not be complete, for that would be uneconomical; it need only be enough [*zureichend*]. It is adapted enough when it leads to no contradiction, that is, when it is true. Only then can it be called economical, for the untrue, which leads to error, cannot be economical.[56] So, says Musil, economy is a criterion that is secondary to truth. The view of science as a biological adaptation of the most economical kind, although it has accorded well with the sceptical temper of the times and has therefore been so attractive, is an inadequate theory of knowledge.[57]

54 For bio-bibliographic data on Musil, see Robert L. Roseberry, *Robert Musil, ein Forschungsbericht* (Frankfurt-am-M.: Athenäum/Fischer, 1974); Karl Dinklage, ed., *Robert Musil: Leben, Werke, Wirkung* (Vienna: Amalthea, 1960), 133–142. The following is taken mainly from Roseberry, and from the *Vita* attached to Musil's thesis, n. 55 below.

55 Robert Musil, *Beitrag zur Beurteilung der Lehren Machs: Inaugural-Dissertation zur Erlangung der Doktorwrde genehmigt von der philosophischen Fakultät der Friedrich-Wilhelms-Universität zu Berlin* (Berlin: Dissertations-Verlag Carl Arnold, 1908); translated by Kevin Mulligan as *On Mach's Theories*, introduction by Georg Henrik von Wright (Washington, D.C.: Catholic University Press, 1982). This thesis is discussed by Johannes von Allesch, "Robert Musil in der geistigen Bewegung seiner Zeit," in Dinklage, *Robert Musil* (1960) (n. 54), 133–142.

56 Musil, *Beurteilung der Lehren Machs* (1908) (n. 55), 28, 29.

57 Musil, *Beurteilung der Lehren Machs* (1908) (n. 55), 13.

This stage-by-stage analysis by ex-engineer Musil is reminiscent of the stage-by-stage analysis of criteria for scientific theories of the physicist H. R. Hertz, which appears in the "Introduction" to his *Science of Mechanics.* The work of Hertz, which was part of the professional education of every engineer, comprised a three-stage consideration of the goodness of a scientific theory.[58] First, Hertz said, it must be logically permissible, and it must be correct: it must conflict neither with thoughts nor with things. But two theories that meet these criteria may still be of unequal value: they may differ in appropriateness (*Zweckmässigkeit*); one may reflect the properties of the object more clearly than the other. If both, finally, are equally clear, the best one will be that which contains the fewest superfluous relations, that is, the simpler of the two. Simplicity is thus a final criterion, which is of importance only after all the others have been met. This would seem to be the criticism that Musil is making of Mach's criterion of economy.

This discussion covers the first third of Musil's thesis. But it is in the following section, in which he discusses Mach's polemic against causality and its replacement by the concept of functional relations, that Musil finds what he feels to be specific to Mach.[59] It has been commonplace among physicists to believe that all phenomena should be presented as a function of other phenomena. Musil cites Kirchhoff's remark that the goal of the science of mechanics is the "clearest, simplest description" of the movements of bodies, and he finds the same attitude in the work of H. R. Hertz. For them, the word "force" is not a metaphysical first cause but only a name for certain algebraic expressions. Mach, however, expels causality from science, not only at this superficial level but utterly and completely.

Musil, then, contrasts Mach with these two other philosophers of science – or better, physicists writing on theory of knowledge – as examples of the main line in scientific thinking. Musil's analysis suggests that to a contemporary Mach appeared to view scientific theory as the simplest sensory *Vorstellung* of the world that is effective for survival and to see science as economical adaptation, as well as an attack on causation. To this we can add the features taken up by the physiologists, which in many cases were the same ones: the economical description of conditions, the functional dependence of elements on each other,

58 Janik and Toulmin, *Wittgenstein's Vienna* (1973) (n. 3), 174–175; Heinrich Rudolph Hertz, *Die Principien der Mechanik in neuem Zusammenhang dargestellt*, preface by Hermann von Helmholtz in *Gesammelte Werke*, v. 3 (Leipzig: 1894, Barth), 2; translated by Daniel E. Jones and John Thomas Walley, introduction by Robert S. Cohen, as *Principles of Mechanics Presented in a New Form*, preface by Hermann von Helmholtz (1899) (Dover, N.Y.: 1956).
59 Musil, *Beurteilung der Lehren Machs* (1908) (n. 55), 56, 57.

and the representation, the *Vorstellung*, as a group of sensational elements related in this way.

These ideas were not present in Landsteiner's 1909 essay on the theories of antibody production.[60] His leading principles – analogy, simplicity, and continuity – while they are to be found clearly stated, emphasized, and discussed at length in Mach, are not the ones his contemporaries found to be Mach-specific. Following Musil's suggestion, if we too take Kirchhoff and Hertz as examples of the main line in the philosophy of science, the non-specific contemporary background, we may be able to place Landsteiner's methodology in a better-fitting context.

It is by no means impossible to read Mach, especially *Knowledge and Error*, and come away with very little that is "Mach-specific" in Musil's sense. Mach quotes both Kirchhoff and Hertz rather as Musil does, as examples of the generally acceptable, as a means of tying his own thought to its background. He does so in this citation, from the beginning of his chapter on examples of science in action:

In general, the activity of scientists, their procedure in any given case, and the goal which they are pleased if they can reach, can be put in a few words by saying that they try to bring thoughts into the best possible agreement with the facts and with other thoughts. "Simple and complete description" (Kirchhoff, 1874), "economical presentation of the facts" (Mach, 1872), "Agreement between thought and being, and of thought processes with each other" (Grassmann, 1844) are all expressions, with slight variation, of the same thought.[61]

Mach is aware of the common ground between himself and other thinkers, and indeed he emphasizes it, as part of his position as one who is trying to tell us what science has done in the past, rather than as a philosopher setting up a new system. That same single sentence from Kirchhoff is quoted by both Mach and Musil, as it was by the physicist-physiologist Hermann Helmholtz. Helmholtz wrote: "Gustav Kirchhoff designated the task of mechanics, the most abstract of the sciences, as the description of motion in nature in the most complete and simplest way possible."[62]

Moritz Schlick in his commentary of 1921 on this essay calls Kirchhoff's phrase, "these famous words of Kirchhoff, which have become programmatic, so to speak, for any epistemologically-oriented phys-

60 Landsteiner, "Theorien der Antikörperbildung" (1909) (n. 1).
61 Mach, *Erkenntnis und Irrtum* (1905/1917) (n. 12), 287.
62 Hermann von Helmholtz, "Die Tatsachen in der Wahrnehmung," Rede, gehalten zur Stiftungsfeier der Friedrich-Wilhelm-Universität zu Berlin 1878, in Paul Hertz and Moritz Schlick, eds. and comm., *Hermann von Helmholtz Schriften zur Erkenntnistheorie* (Berlin: Springer, 1921), 109–175 (p. 132)

ics."[63] Schlick, it may be remarked, succeeded Mach as full-time professor of the philosophy of science in Vienna, and together with Rudolf Carnap in the early 1920s formed the positivist group that became known as the Vienna Circle.[64]

Like Kirchhoff, Hertz stresses simplicity as a criterion of good scientific theory. Hertz opens the preface of his *Principles of Mechanics* with the words, "All physicists agree that the problem of physics consists in tracing the phenomena of nature back to the simple laws of mechanics."[65] And further on, he writes,

It is true that we cannot *a priori* demand from nature simplicity, nor can we judge what in her opinion is simple. But with regard to images of our own creation we can lay down requirements. We are justified in deciding that if our images are well adapted to the things, the actual relations of the things must be represented by simple relations. And if the actual relations between the things can only be represented by complicated relations... we decide that these images are not sufficiently well adapted to the things. Hence our requirement of simplicity does not apply to nature, but to the images thereof which we fashion; and our repugnance to a complicated statement as a fundamental law only expresses the conviction that if the contents of the statement are correct and comprehensive, it can be stated in a simpler form by a more suitable choice of the fundamental conceptions.[66]

For both Kirchhoff and Hertz, then, simplicity is one of the most important properties of a law of nature. Mach's concept of economy may be seen, as he himself would probably agree, as a special expression of this generally held view.

In one important part of his thought, however, Hertz does not represent a general non-specific background for Mach, but instead offers a direct contradiction to him, a contrast which has been pointed out both by Robert Cohen and by Janik and Toulmin. Cohen writes that Hertz breaks clearly with Mach, for Hertz's laws of nature are less descriptive shorthand for experientially correlated perceptions than prescriptive interpretive symbolic systems.[67] Janik and Toulmin note that Hertz consistently uses the word *Darstellungen* rather than *Vorstel-*

63 Schlick, comments on *Helmholtz Schriften zur Erkenntnistheorie* (1921) (n. 62), 170.
64 For an account of these later developments, see Philipp Frank, *Between Physics and Philosophy* (Cambridge, Mass.: Harvard University Press, 1941); see also Friedrich Stadler, "Ernst Mach, zu Leben, Werk und Wirkung" in Haller and Stadler, *Ernst Mach* (1988) (n. 3), 11–63 (pp. 29–33).
65 Hertz, *Principles of Mechanics* (1899/1956) (n. 58), xxiii.
66 Hertz, *Principles of Mechanics* (1899/1956) (n. 58), 23.
67 Cohen, in Hertz, *Principles of Mechanics* (1899/1956) (n. 58), Introduction, sec. 3: "Hertz and Mach," final paragraphs (not paginated).

lungen: his mathematical models are constructed schemes for knowing, rather than mental images or bundles of sense impressions.[68]

This is an important point for understanding Landsteiner's methodology. Not only, as we have seen, does he *not* adopt the Mach view of empiricist *Vorstellungen* – as emphasized, for example, by Verworn – but his use of the alternative term *Darstellung* points in the opposite direction.

Landsteiner calls his analogy of the charge properties of an inorganic colloid and those of the serum agglutinin a *Darstellungsweise*, a kind of simple chemical model of the more complex live substances: "That this *Darstellungsweise* offers a considerable simplification is clear. Apart from this it gives an opportunity for direct experimental testing by the methods of structural chemistry."[69]

Landsteiner's *Darstellung* is a constructed scheme, a model, whose properties may predict and hence explain those of the agglutinin. It is not a *Vorstellung*, a mental image. Like Hertz's, it is a working model of the natural situation, with predictive rather than descriptive value. Here, then, we have reached a conclusion. Landsteiner's use of the term "uneconomical" in 1902 suggests that he was probably aware of Mach's thinking. His essay of 1909 shows that by that time he was not involved in it to any great extent: not enough, that is, to have made use of any of the concepts that his contemporaries saw as being peculiar to Mach. Landsteiner's *Leitmotiv*, the idea of simplicity, is to be found in Kirchhoff and Hertz, as well as in Mach.

But about this time the idea of simplicity also appeared in other quarters, where its presence could be seen as part of the outpouring of the ideas of science into the general culture in Vienna, an outpouring that may itself have been related to the popularity of Mach's teaching.

Mach's lectures of the winter semester of 1895–1896 were attended by a great many intellectuals from outside the fields of either philosophy or science.[70] Hugo von Hofmannsthal, at this time a lyric poet, along with Hermann Broch, a journalist, and Hermann Bahr, a playwright and *belleslettrist*, were all attracted to Mach's biological positivism and to his view of knowledge as *Vorstellung* of sense impressions. Seldom had a scientist "exerted such an influence upon his culture as Ernst Mach. . . . From poetry to philosophy of law, from physics to so-

68 Janik and Toulmin, *Wittgenstein's Vienna* (1973) (n. 3), 139.
69 Landsteiner, "Theorien der Antikörperbildung" (1909) (n. 1), 1628.
70 Janik and Toulmin, *Wittgenstein's Vienna* (1973) (n. 3), 113; Blackmore, *Ernst Mach* (1972) (n. 3), 154–157.

cial theory. Mach's influence was all-pervasive, in Austria and else-where."[71]

Hermann Bahr, writing in 1916, the year of Mach's death, remembered what he himself had experienced:

The influence of Mach, especially on the young people, was at the time very great, and it worked through one particular *Satz*. Mach had stated, *das Ich ist unrettbar*. The I was dethroned, the last idols seemed to be smashed, the last barriers down, the highest freedom was won, the work of negation complete. There was nothing left.[72]

As Karl Marx said in 1842, speaking of the philosophy of Ludwig Feuerbach, this was the river of fire through which they all must pass, and from which they emerged purified of illusions, in this case the illusions of metaphysics.[73] In 1903, Bahr himself described his feelings on the problem of the "I" and his meeting with Mach's *Analysis of Sensations* in an essay called "Das unrettbar Ich."[74] In the essay on Mach of 1921 he traces the thread of philosophical fashion after this collapse into utter skepticism: first it leads to the pragmatism of William James, for whom the truth is whatever is useful, and thence to the philosophy of *als ob* of Hans Vaihinger, for whom there is no truth at all, only conscious fictions that are needed for thinking.

Robert Musil's enormous novel *The Man without Qualities* was called "a philosopher's novel," based on the philosophy of Mach.[75] In a chapter entitled "Science Smiling into Its Beard, or First Full-dress Encounter with Evil," he wrote in terms of Mach's earthbound biological

71   Janik and Toulmin, *Wittgenstein's Vienna* (1973) (n. 3), 133.
72   Hermann Bahr, "Mach," in *Bilderbuch* (Vienna: Wila, 1921), 35–41 (p. 37). This very striking passage has also been quoted, in a different translation, by Blackmore, *Ernst Mach* (1972) (n. 3), 155.
73   Karl Marx, "Luther als Schiedsrichter zwischen Strauss und Feuerbach," in *Karl Marx-Friedrich Engels Werke* (Berlin: Dietz, 1964), v. 1, 26–27 (p. 27); Marx is punning on the name "Feuerbach," which means "stream of fire."
74   Joachim Thiele, "Zur Wirkungsgeschichte der Methodenlehre Ernst Machs," in *Symposium aus Anlass des 50 Todestages von Ernst Mach, veranstaltet am 11–12 März 1966 vom Ernst Mach Institut Freiburg-i-Br.* Thiele subtitles his paper "Dokumente zum Einfluss der Methodenlehre Machs im *nicht*-naturwissenschaftlichen Bereich." It may be remarked that Blackmore, *Ernst Mach* (1972) (n. 3), 187, mistranslating Thiele's word *Skizze* as "skit," misses the point that Bahr's essay "Das unrettbar Ich" is an account of Bahr's deeply felt struggle – Blackmore again manages to downgrade a serious problem. Here, apart from discussing Bahr's essay without having read it, Blackmore has not even read Thiele's quotation from it. This essay does not seem to be available in North America: I have not read it either.
75   Von Wright, introduction to Musil, *On Mach's Theories* (1982) (n. 55), 9. Von Wright says that Mach was the only source of Musil's abortive foray into philosophy, but that Nietzsche was more significant in his later work.

realism, of the historical rise of science as an "out and out intoxication, a very fire, of matter-of-factness [*Nüchternheit*]":

> In the struggle for existence, there are no philosophical sentimentalities, but only the will to kill off one's opponent by the shortest and most practical method. There, everyone is a positivist. . . . If one investigates what qualities it is that lead to discoveries, what one finds is freedom from traditional scruples and inhibitions, as much initiative as destructive spirit, the exclusion of moral considerations . . . nothing but the old hunters', soldiers' and merchants' vices, reinterpreted as virtues. And though by this means they are raised above the urge for personal and comparatively vulgar advantage, yet the element of evil is something they do not lose. . . . It is something indestructible and eternal, or at least as eternal as everything humanly sublime, since it consists in nothing less, nothing other, than the pleasure of tripping that sublimity up and watching it fall flat on its face.[76]

The new feeling for starkness and simplicity, the reaction against the decorative applied art of the Viennese Sezession and earlier, appears at its wildest in the work of the architect Adolf Loos. Loos arrived in Vienna in 1896. His early work, and his polemical writing, were too radical to be popular among patrons of architecture, so his first complete house was not built until 1910.[77] His early work was on the interiors of flats, including his own.

It is easy to see Loos's buildings and his writing as the expression of positivism in architecture, and his campaign for functionalism and against ornament as a parallel in architecture of Mach's economy of thought and biological positivism. Loos's essay of 1908, entitled "Ornament and Crime," pursues that idea: "Ornament is wasted labour and therefore wasted health. It has always been. But today it also represents wasted material and both represent wasted capital. Ornament is no longer organically related to our culture: it is therefore no longer the expression of our culture."[78]

Adolf Loos's buildings themselves, the earlier ones at least, come as rather a surprise after these fighting words. His use of classical detail

76 Robert Musil, *Der Mann ohne Eigenschaften* (Berlin: Rowoholt, 1931), translated by Eithne Wilkins and Ernst Kaiser (London: 1953), 360. The phrase is "ein Rausch und Feuer der Nchternheit," an intoxication, a very fire of sobriety (p. 481, Ger.), not, as might be supposed from the translation, "Sachlichkeit," or objectivity.

77 Ludwig Münz and Gustav Künstler, *Adolf Loos, Pioneer of Modern Architecture* (1964), translated by Harold Meek, introduction by Nikolaus Pevsner and appreciation by Oskar Kokoschka (New York: Praeger 1966), 13.

78 Adolf Loos, "Ornament und Verbrechen," in *Trotzdem* (1900–1930), reprinted in Franz Glück, *Adolf Loos sämtliche Schriften in Zwei Bänden* (Vienna: 1962), v. 1, 276–288 (p. 282); and translated in Münz and Künstler, *Adolf Loos* (1966) (n. 77); 226–231; Reyner Banham, *Theory and Design in the First Machine Age* (London: Architectural Press, 1960), 88–97, "Adolf Loos and the problem of ornament."

in columns, even though they are very plain columns – for example, in his 1910 shopfront on the Michaelerplatz in Vienna – seems to contradicts his own statements, until one sees that it faces the grand baroque gateway of the Hofburg on the other side of the Platz. The frightful plainness of the design produced an uproar in the press.[79] As late as 1930, Loos's buildings were still causing this kind of commotion among those who were unaware of the "modern movement" or were wishing it away, although by that time such houses were quite common throughout Europe.[80] Loos himself had moved up to the position suggested in his own earlier writing, and when he built the Müller House in Prague in 1930, he produced at least a façade that was the epitome of positivism and economy. The house had walls, a roof, doors, and windows – and nothing more. It had no ornament, no detail, no curves, no self-indulgence (Figure 8.1) It could well have stood for both Mach's description of the methods and function of science, and for Landsteiner's programmatic aspirations. It was *einfach*.

It is not quite correct to call this "*neue Sachlichkeit*," as this phrase was not introduced until 1925, and then in a different context.[81] But soon after its introduction, it was applied retrospectively to what had gone before. It was applied, for example, to the work of Loos by the writer of an article that appeared in the *Prager Tageblatt* in 1929 and that was quoted by Franz Glück in his collected edition of Loos's essays. Its occasion was one of the many times that the essay "Ornament and Crime" was reprinted during the battle over the effect of the Müller house in Prague on local real estate values. That writer says: "The so-called new realism [*neue Sachlichkeit*] is not so new as it would seem. It grows from the nineties of the last century, and was foretold by Adolf Loos."[82]

Foretellings of *neue Sachlichkeit* of this period were not confined to

79   Münz and Künstler, *Adolf Loos* (1966) (n. 78), 111.
80   Willy Hofmann, architect, in *Prager Tageblatt*, 22 February 1930, quoted by Münz and Künstler, *Adolf Loos* (1966) (n. 77), 149–154; Münz and Künstler include floor plans and photographs of the outside and inside of the Müller House.
81   Fritz Schmalenbach, "Der Name 'neue Sachlichkeit,'" in *Kunsthistorische Studien* (Basel: Schudel, 1941), 22–32; Schmalenbach, *Die Malerei der 'Neuen Sachlichkeit'* (Berlin: Mann, 1973). The name was introduced by Gustav Friedrich Hartlaub as the title of an exhibition at the Stadtischen Kunsthalle in Mannheim, "Ausstellung 'Neue Sachlichkeit,' " in summer 1925; it covered German painting since expressionism. The style lasted only until about 1937, when it was suppressed and replaced by Nazi painting. The painters involved included Otto Dix, Georg Schrimpf, and Alexander Kanoldt. These painters were not minimalists, like the architects, but realists. Minimalism in painting dates from about 1962, but it has also been linked to Mach, as in Peter Mahr, "Gestaltwahrnehmungen, Minimal Art," in Haller and Stadler, *Ernst Mach* (1988) (n. 3), 404–431 (pp. 421–431).
82   Article in *Prager Tageblatt*, 1929; quoted by Glück, *Loos sämtliche Schriften* (1962) (n. 78), v. 1, 457.

Figure 8.1. Adolf Loos, the Müller House in Prague (1929). Loos's purified design is an example of architectural positivism, which finally lives up to his ideal. There is no ornament at all, not even a door casing or a window surround. As Reyner Banham wrote, to build without decoration is to build like an engineer, in a manner proper to a Machine Age (Banham, *Theory and Design* [1960] [n. 77], 97). Pen drawing by Friedrich Kurrent, from *Adolf Loos: vierzig Photographien von Walter Zednicek mit einem Essay von Friedrich Kurrent* (Vienna: Tusch, 1984), opp. p. 12, #9.

visual art. "Music is not to be decorative, it is to be true," wrote Arnold Schoenberg in 1911. He turned away from the expansiveness in time and richness in orchestration of the romantic Mahler–Wagner style, a style that he had himself used in his early group of songs, the *Gurrelieder*, of 1900.[83] The thickly woven *Leitmotiv* structure, the lyr-

83 Arnold Schoenberg, *Gurrelieder*, settings of "Gurresange" (1868) by the Danish poet and novelist Jens Peter Jacobsen; music composed 1900–1901, final section orches-

icism, and the elaborate orchestral setting of the first sections of the *Gurrelieder* summed up for Schoenberg the musical past and gave him a *point d'appui* for his swing over to what he considered its opposite. Criticism of the superfluous in music and in all art is the theme of his twenty-three-minute opera *Die glückliche Hand* of 1910–1913.[84] The hero of the opera, the "prophet of the new objectivity," clearly represents Schoenberg himself.[85] In the central scene, the hero walks into a workshop where men are busy at benches, filing, hammering, working at machines. He watches them work for a minute, until an idea occurs to him, and he says

Das kann man einfacher – "That can be done more simply."
   Radiant (a lighting effect), swelled with the sense of power, he strikes a piece of gold a single hammer-blow, and produces a perfect diadem set with stones. *So schafft man Schmuck –* he says, "That's how to make jewellery."[86]

Landsteiner's leading maxim, then, is not peculiar to himself, is not even confined to science, nor need it be an example of direct borrowing from Mach, although *Einfachheit*, along with *Kontinuität* and *Analogie*, is certainly to be found in Mach's writing. But as Mach himself was aware, these criteria for scientific theories were also to be found in the work of other philosophers of science, both his contemporaries and his predecessors. Mach's famous principle of economy, if equated only with simplicity, is the least specific, least original part of his thought. But in effect, although simplicity is an important part of the principle of economy, it is only a part of it. The physiological empiricism that accompanies it cannot be left out of account in deciding whether Landsteiner was, or was not, really influenced by Mach himself. It would seem that he was not.

   But during this period – from 1895, when Mach first lectured in Vienna, over the first decade or so of the new century – the philosophy of science and the positivistic way of thinking were poured into the broader culture of Vienna from this source. Mach's ideas, which were a part of this outpouring, can be found in every field; they became one

trated 1911; Pierre Boulez, conductor, M. Napier, T. Thomas, Yvonne Minton, et al., B.B.C. Symphony Orchestra and Chorus (Columbia M233303, 1975); Schoenberg, *Gurrelieder von Jens Peter Jacobsen, Deutsch von Robert Franz Arnold, für Soli Chor und Orchester*, Score (Vienna: Universal, 1920).

84 Arnold Schoenberg, *Die glückliche Hand* Scenario-libretto both by the composer, Op. 18 (1910–1913), R. Craft, Conductor, R. Oliver, Bass, Columbia Symphony Orchestra and Chorus in *Music of Arnold Schoenberg*, v. 1 (Columbia, M2s679, 1963); Schoenberg, *Die glückliche Hand*, score/libretto (Vienna: Universal, 1926).

85 Theodor Adorno, "Schoenberg and progress," in his *Philosophy of Modern Music*, trans. A. G. Mitchell and W. V. Blomster (New York: Seabury, 1973), 46.

86 Schoenberg, *Die glückliche Hand*, score/libretto (1926) (n. 84), 18, 20.

of the conditions of existence. Their traces, both direct, among those in contact with Mach, and indirect among those who caught only the reflected rays, can be found everywhere that the change to *die neue Sachlichkeit* was occurring.

Landsteiner's leading principles, then, are compounded of ideas that came to him from his nineteenth-century forebears in biology, and from the culture of his contemporary Vienna. The year 1896 brought his first appointment with Gruber, Nägeli's student, and his first contact with immunology; it was also the year of Mach's first lectures in Vienna, and that of Adolf Loos's first appearance there. Landsteiner stood at the focal point of unitarianism, where the influence of the past, through Gruber, and that of his contemporaries, as exemplified by Mach and the physicists, crossed. Long before he came to any solution of the problems on which he worked, his thinking had been formed. His criticisms of Ehrlich are understandable not in terms of an alternative solution to a problem, but in terms of a life's work guided by a particular set of principles, which were established quite independently of any empirical data. They determined in advance – twenty years in advance, as will be seen in Chapter 9 – what solution was going to be possible for him, and when he would know that he had reached it.

PART III

Chemical Affinity and Immune Specificity: The Argument in Chemical Terms

The very term, "elective affinity," must lead into error.

Claude-Louis Berthollet, *Récherches sur les Lois de l'Affinité* (1801), trans. M. Farrell (Baltimore, Md.: Nicklin, 1809), 146

Wolfg. Ostwald has actually said that . . . "specific chemical affinity" is a concept with no definite content.

Karl Landsteiner, "Zur Frage der Spezifizität der Immunreaktin, und ihre kolloidchemischen Erklärbarkeit," *Biochem Z.* 50(1913):176–184

9

Structural and Physical Chemistry in the Late Nineteenth Century

In February 1891 Karl Landsteiner received his medical degree from the University of Vienna. Between this time and January 1896, when he joined Max von Gruber at the Institute of Hygiene, Landsteiner continued his medical education: in internal medicine at the Second Medical Clinic of the University, and in surgery as "surgical apprentice" (*Operationszögling*) at the First Surgical Clinic in Vienna.[1]

During these years he also spent some time abroad, in the laboratories of three famous chemists: first in Germany, in Würzburg, with Emil Fischer, then during the winter semester of 1892–1893 at Munich University under Eugen Bamberger who was lecturing on benzene derivatives, and finally in Switzerland with Arthur Hantzsch at the Eidgenossische Züricher Polytechnikum. In each of these three laboratories, Landsteiner was assigned a problem from the research in progress and published a paper on the work he did. With Emil Fischer, he wrote on the preparation of glycolaldehyde, a substance of significance as the first member of the sugar series.[2] With Eugen Bamberger, he worked on the reactions of diazobenzene; the resulting paper became the second of Bamberger's series on the structure of diazobenzene.[3] With Roland Scholl of Hantzsch's laboratory, he worked on the structure of the *pseudo*-nitroles and oximes.[4]

Since the chemical rather than the clinical subjects that Landsteiner

1 Paul Speiser and Ferdinand G. Smekal, *Karl Landsteiner the Discoverer of the Blood Groups and a Pioneer in the Field of Immunology: Biography of a Nobel Prizewinner of the Vienna Medical School* (Vienna: Hollinek, 1st ed., 1961), 2d ed., translated by Richard Rickett, 1975, 17–24. This biography contains a complete bibliography of Landsteiner's work and that of his co-workers, by Merril W. Chase. Numbers attached to the Landsteiner papers in this book refer to the Chase bibliography.
2 Emil Fischer and Karl Landsteiner, "Ueber den Glycolaldehyde," *Ber. d. deutsch. chem. Ges.* 25(1882):2549–2554, C. 1.
3 Eugen von Bamberger and Karl Landsteiner, "Das Verhalten des Diazobenzol gegen Kaliumpermanganat," *Ber. d. deutsch. chem. Ges.*, 26(1893):482–495, C. 3.
4 Roland Scholl and Karl Landsteiner, "Reduction der Pseudonitrol zu Ketoximen," *Ber. d. deutsch. chem. Ges.* 29(1896):87–90, C. 7.

studied during these five years played a large part in his later thinking on the problem of specificity, it is worth inquiring what kind of chemistry these three men represented, and what other kind was possible at this time. All three were at work on problems of organic structure, particularly on its most recent branch, stereochemistry. The peculiar nature of this kind of chemistry was sensitively expressed by the organic chemist Victor Meyer, who was professor at Heidelberg from 1889 to 1897.[5] In September 1889, the Naturforscherversammlung met in Heidelberg and Meyer gave the opening lecture of the general sessions, entitled "The Problems of Chemistry today":

In today's chemistry, imagination and intuition play a greater role than they do in other sciences. In working in it, as well as a purely scientific satisfaction, there is an enjoyment which is in a sense like that of art. . . . In experimental work in organic chemistry, the sensing of things for which no law has yet been found that can be stated in words can be extraordinarily successful. Something which we may call a "feel for chemistry" [*chemisches Gefühl*] comes to the aid of thought – an expression which will disappear as soon as chemistry gets close to being a mathematico-physical discipline.[6]

Meyer's chemistry, and that of the other structural chemists, had in it much of the sense of organic form, the visual imagination of the earlier comparative anatomists. His *chemisches Gefühl*, the intuition of variations on a basic ground plan, of series of ascending complexity in which a simple starting point is the clue to the understanding of the higher members, has a parallel in the intuition of the transcendental anatomist. This chemistry represents the morphology and systematics of science.

The visual nature of the ideas of structural chemistry and its problems makes it difficult to speak about them without diagrams, or indeed, after the beginning of stereochemistry, without three-dimensional models. Victor Meyer, giving a long review of past and present stereochemistry to the Deutsche chemische Gesellschaft in 1890, is showing his audience models as he talks: a carbon atom in the middle of a tetrahedron, for ex-

---

5   Richard E. Meyer, *Victor Meyer Leben und Wirken eines deutschen Chemikers und Natur-forschers 1848–1897*, in Wilhelm Ostwald's series "Grosse Männer: Studien zur Biologie des Genies," v. 4 (Leipzig: Akademische Verlag, 1917), 250–251.
6   Victor Meyer, "Chemische Probleme der Gegenwart," in *Deutsche Naturf. u. Aerzte, Tageblatt der 62 Versammlung,* Heidelberg (Heidelberg: Hörning, 1890), 126–134 (p. 127); and quoted in part in Meyer (n. 5), 239–241; see n. 46 for Ostwald's complementary view of the development of physical chemistry, which Ostwald feels has now reached that stage of mathematical description. Meyer is describing what Ostwald refers to as the second stage of development of a science, that of *systematische Ordnung,* or systematics as it is called in biology.

ample, or Kekulé's big models made of wooden balls and brass wires, which he says are good for demonstration, but too clumsy and expensive for thinking with.[7] He has a new kind, developed by his friend Professor Paul Friedländer of Karlsruhe. They are made of thin rubber tubing about the size of match sticks, with coloured tips like match heads to stick in the tubes to represent atoms. He shows his audience how easy it is to answer the question of the form of a row of three carbon atoms by just seeing a model. Many of the problems of the relationship of different groups to each other on the molecule can be posed by making a model of this kind, and can be addressed by experiments involving condensations between neighbouring groups: the formation, for example, of anhydrides. The specific affinities of the atoms and radicals attached to the carbon atoms play an important part in determining what structures are possible, what are likely, and what are excluded.

Meyer gives a good picture of the way organic chemistry looked from inside in 1890. His detailed discussion of its history and its present problems omits one important detail, however: the close connection of the problems of organic structure and organic synthesis with the chemical industry, particularly the coal-tar dye industry.[8] He provides an intellectual context in which to set the work of the three chemists mentioned earlier, the structural chemists with whom Landsteiner trained.

Meyer's review was entitled "Aims and Results in Stereochemistry." He himself had introduced the term "stereochemistry" two years earlier. He begins by tracing the history of the subject back to the publication of Jacobus Henricus van't Hoff's little book, *La Chimie dans l'Espace*, and to an article by Joseph-Achille LeBel that appeared a few months later and that contained almost the same idea.[9] Both van't Hoff

---

7 Victor Meyer, "Ergebnisse und Ziele der stereochemischen Forschung," *Ber. d. deutsch. chem. Ges.* 23(1890):567–619, 571–572; for a discussion of the part played by the models in the genesis and acceptance of the theories, see O. Bertrand Ramsay, "Molecular models in the early development of stereochemistry: I. The van't Hoff model; II. The Kekulé models and the Baeyer strain theory," in O. Bertrand Ramsay, ed., *Van't Hoff-LeBel Centenniel: Symposium Arranged by Division of the History of Chemistry of the American Chemical Society,* 11–12 September 1974 (Washington, D.C.: American Chemical Society, 1975), 74–96.

8 Ernst Bäumler, *Ein Jahrhundert Chemie* (Dusseldorf: Econ, 1975), 21–70, "Am Anfang war der Teer" (In the beginning was tar).

9 Jacobus Henricus van't Hoff, *La Chimie dans l'Espace* (Rotterdam: Bazendijk, 1875), translated by F. Hermann, *Die Lagerung der Atome im Raume, mit einem Vorwort von Dr Johannes Wislicenus* (Braunschweig: Vieweg, 1877; 2d ed., 1894); Joseph-Achille LeBel, "Sur les relations qui existent entre les formules atomiques des corps organiques et le pouvoir rotatoire de leurs dissolutions," *Bull. de la Soc. Chim.* (Paris), 22, n.s. (1874): 337–347. The suggestion that the structure is not flat is common to both authors, but

and LeBel had suggested that certain cases of isomerism could be explained by assuming that the carbon atom and its valencies were not flat but occupied a volume in space. The carbon atom was imagined as lying at the centre of a tetrahedron, with its four valencies pointing to the corners. It followed that if four different groups were attached to a single carbon atom, the structure could take two possible forms that were mirror images of each other; these substances, when in solution, had the power of rotating polarized light, one isomer rotating it to the left and one to the right. The classical example quoted by both van't Hoff and LeBel was tartaric acid and its three isomers. Up to that time (1874), this isomerism had been impossible to explain on structural grounds, but the idea in *Chemie dans l'Espace* of an "asymmetric carbon atom," an atom to which were attached four different substituent groups, provided an attractive explanatory model.

This spatial model led Adolf von Baeyer to suggest that the valencies of the carbon atom might protrude in particular directions. Under this so-called strain theory, the most natural position for a chain of carbon atoms was a C-shaped ring, and these rings were easily closed by anhydride formation.[10] This would explain the commonness in nature, and the great stability, of the closed benzene ring.

With the popularization of van't Hoff's ideas, many chemists had tried out similar speculations based on *Chemie dans l'Espace.* Arthur Hantzsch and his co-worker Alfred Werner, for example, had suggested a stereochemistry of nitrogen: Hantzsch regarded the nitrogen atom not as a flat object with three planar valencies that could be written $-\text{N}-$ but as an atom sitting on top of a figure formed by its three

arms:

the "asymmetric carbon atom," though mentioned by LeBel at one place, is much more elaborately treated by van't Hoff. Meyer's account does not differentiate between them, but in fact van't Hoff's was much the more influential of the two, perhaps because van't Hoff was the more active thinker and worker in later years. See John W. Servos, *Physical Chemistry from Ostwald to Pauling: the Making of a Science in America* (Princeton, N.J.: Princeton University Press, 1990), 24–33.

10 Ramsay, "Molecular models in the early development of stereochemistry" (1974) (n. 7). Van't Hoff's coloured cardboard models (now in the Deutsches Museum, Munich), which he sent to many prominent chemists along with his book, probably contributed to its effectiveness. Ramsay also illustrates the Kekulé–von Baeyer wire and ball models and quotes the British dye-chemist William H. Perkin, a student of Baeyer's in Munich in the 1880s, as saying that it was the models that gave rise to the *Spannungs-theorie* (p. 88).

which implied that the compounds of nitrogen such as oximes could have stereoisomers:

$$
\begin{array}{c}
\quad\ \ \text{OH} \\
\backslash\quad / \\
\text{C} = \text{N} \\
/
\end{array}
$$

$$
\begin{array}{c}
\backslash \\
\text{C} = \text{N} \\
/\quad\ \ \backslash \\
\qquad\text{OH.}
\end{array}
$$

Meyer felt that stereoisomerism of atoms other than carbon was not very well supported at the time. Stereoisomerism rested mainly on the possibility of free rotation between carbon atoms; this would be prevented if a double bond was present to hold the molecule rigid, or if groups had a similar charge and so repelled each other, giving them no choice of position. Apart from this, if Arthur Hantzsch's theory of the stereochemistry of nitrogen were true, there should have been, in practice, a number of different oximes of phenanthrene. There should have been three different dioximes:

$$
\begin{array}{l}
\qquad\ \text{OH} \\
\qquad / \\
\text{C}_6\text{H}_4 - \text{C} = \text{N} \\
|\qquad\quad | \\
|\qquad\quad | \\
\text{C}_6\text{H}_4 - \text{C} = \text{N} \\
\qquad\qquad\quad \backslash \\
\qquad\qquad\quad\ \text{OH}
\end{array}
\qquad
\begin{array}{l}
\text{C}_6\text{H}_4 - \text{C} = \text{N} \\
|\qquad\quad |\quad\ \backslash \\
|\qquad\quad |\qquad \text{OH} \\
\qquad\qquad\quad\ \text{OH} \\
\qquad\qquad\quad / \\
\text{C}_6\text{H}_4 - \text{C} = \text{N}
\end{array}
\qquad
\begin{array}{l}
\text{C}_6\text{H}_4 - \text{C} = \text{N} \\
|\qquad\quad |\quad\ \backslash \\
|\qquad\quad |\qquad \text{OH} \\
\text{C}_6\text{H}_4 - \text{C} = \text{N} \\
\qquad\qquad\quad \backslash \\
\qquad\qquad\quad\ \text{OH}
\end{array}
$$

and two monoximes:

$$
\begin{array}{l}
\qquad\ \ \text{OH} \\
\qquad\ / \\
\text{C}_6\text{H}_4 - \text{C} = \text{N} \\
|\qquad\quad | \\
|\qquad\quad | \\
\text{C}_6\text{H}_4 - \text{C} = \text{O}
\end{array}
\qquad
\begin{array}{l}
\text{C}_6\text{H}_4 - \text{C} = \text{N} \\
|\qquad\quad |\quad\ \backslash \\
|\qquad\quad |\qquad \text{OH} \\
\text{C}_6\text{H}_4 - \text{C} = \text{O}
\end{array}
$$

But, says Meyer, there are no isomeric oximes of phenanthrene at all.

Landsteiner's three teachers can be placed in this tradition of late nineteenth-century argument about the form of organic molecules. But within it, each is his own man. Bamberger deviates perhaps the least from the mainstream of structural organic chemistry.

The winter semester of 1892–1893, which Landsteiner spent in Munich, was Eugen von Bamberger's last term there as *ausserordentlicher* professor of organic chemistry. In 1893 he moved on to the Eidgenossische Technische Hochschule in Zurich as *Ordinarius*. He had gone to Munich in 1883 as an assistant to Adolf von Baeyer, who succeeded Justus von Liebig at the Ludwig Maximilian University in 1875. Von Baeyer began by putting up a large new institute building containing two floors, one for organic and one for inorganic chemistry. This was the setting for one of the most important large-scale research training schools of German chemistry: at the height of its operation, there were about 230 students working in the inorganic section and 270 in the organic section. Some of them were students of medicine or pharmacy, but about 560 chemists, including 395 Ph.D.'s, were trained in von Baeyer's institute between 1875 and 1915, the year he retired. Many of these trained research chemists were absorbed by the growing chemical industry. This institute was "the principal seed-bed for the next generation of leaders in the development of organic chemistry in Germany."[11]

The students were mainly supervised by one of von Baeyer's succession of *Privatassistenten*. Few of them had any contact with the man himself, whose Prussian presence was occasionally seen striding through the halls. Eugen Bamberger, Landsteiner's supervisor of 1892–1893, was one of the *Assistenten*. He had worked first in the analytical laboratory for beginners, then in the organic laboratory, and finally in the advanced one, for a total of eleven years, 1882 to 1893, after which he was appointed to a professorship in Zurich.

Louis Blangey, in his memorial to Bamberger, speaks of Bamberger's lively chemical imagination, his *Lust zum formulieren*, as he himself expressed it. What really interested him, says Blangey, were subtle structural problems and the mechanism of complicated transformations, which he would try to break down into separate stages, each of which could be paralleled by a well-known reaction. He possessed an especially keen sense of smell and a memory for smells, and he could identify each typically smelling substance. Bamberger's teaching was as good as his research: his lectures, says Blangey, were well-made, *formvollendet*, as well as clear and lively, and he took great pains with the young men who, like Landsteiner, carried out a project under his direction. Bamberger himself first showed the student how do all the operations, melt-

11 Joseph S. Fruton, *Contrasts in Scientific Style: Research Groups in the Chemical and Biochemical Sciences* (Philadelphia, Pa.: American Philosophical Society, 1990), 128–129.

ing point determinations, crystallizations, identification reactions, and so on, and then had him do them himself in Bamberger's presence. Blangey writes of his teaching in Zurich of a later date:

Bamberger never gave his explanations at the writing table. The results of an experiment were discussed and confirmed at the laboratory bench, theoretical conclusions drawn, and new experiments discussed or carried out. In these daily discussions the student not only got to know the reactions, he learned above all to think chemically, draw the conclusions himself and ask the next question.[12]

The diazobenzene problem on which Landsteiner worked with him was one that had arisen from some work of Victor Meyer's of 1877, in which he had investigated the reaction of aliphatic ketones with diazocompounds.[13] This reaction gave a yellow product thought by Meyer to be a diazocompound, and by Bamberger to be a hydrazone:

$$CH_3CO - CH_2COOC_2H_5 \quad + \quad \langle\bigcirc\rangle - N = NOH$$

$\Rightarrow$

$$CH_3 - COCHCOOC_2H_5 \qquad \text{or} \qquad CH_2 \; COCHCOOC_2H_5$$
$$\underset{N \, = \, N \, - \, \langle\bigcirc\rangle}{|} \qquad\qquad HO - N - NH - \langle\bigcirc\rangle$$

diazocompound   hydrazone

For Bamberger, the question was whether diazobenzene reacted as an oximide:

$$\langle\bigcirc\rangle - N = NOH$$

or as a nitroso-compound:

$$\langle\bigcirc\rangle - \underset{H}{N} - N = O$$

12 Louis Blangey, "Eugen Bamberger 1857–1932," *Helv. chem. Acta 16*(1933):644–676, followed by a bibliography, 676–685 (p. 647).
13 Victor Meyer, "Einführung Stickstoffhaltiger Radicale in Fettkorper," *Ber. d. Deutsch. chem. Ges. 10*(1877):2075–2078.

The oxidation and reduction reaction might be written most simply as:

but it might also take place through an intermediary stage:

Bamberger and Landsteiner concluded that their investigations had not been decisive for either of the formulae, and that it was possible that diazobenzene was a tautomeric substance that existed in both forms.[14] But though it was inconclusive, this paper provided an excellent exercise in chemical methods and thinking. The argument was as follows: diazobenzene was allowed to react in alkaline solution with the oxidizing agent potassium permanganate. Along with a strong and unpleasant smell, and a series of little explosions, a number of end-products were produced: nitrosobenzene, nitrobenzene, azobenzene, phenyl isonitrile (the source of the smell), and a substance called diazobenzenic acid. Starting from aniline, the diazobenzene was first prepared, then oxidized, then the end-products separated and identified, by quantitative analysis and by physical methods. The formula of the derived diazobenzenic acid was then investigated by examining its reactions. Two of these seemed significant and suggested that it should have a phenyl nitramine formula: its ease of splitting into aniline and nitric acid, and its ease of rearrangement into *ortho* – and *para*-nitraniline:

phenyl nitramine formula
for diazobenzenic acid

p-nitraniline

Its formation could be explained most easily by assuming the nitrosamine formula for diazobenzene:

14 Bamberger and Landsteiner, "Diazobenzol u. Kaliumpermanganat" (1893) (n. 3). This paper is discussed in the context of the controversy between Hantzsch and Bamberger, by James R. Partington, *History of Chemistry* (London: Macmillan, 1964), v. 4, 844; Partington does not mention Landsteiner!

But this compound too (that is, benzenic acid) might have had tauto-meric forms:

The experiments did not lead to a final decision, but they provided solid training in methods of analysis, preparation, and chemical think-ing.

Landsteiner's third chemist, Arthur Hantzsch, began his career as *Assistent* under Gustav Wiedemann at the Institute for Physical Chem-istry in Leipzig. Wiedemann, who was Wilhelm Ostwald's predecessor at Leipzig, was more of a physicist than a physical chemist, although according to Ostwald he was then the "world's only professor of phys-ical chemistry."[15] Hantzsch was primarily interested in structural or-ganic chemistry, but because of this background, he often brought in the methods of physical chemistry to solve his structural problems.[16] In 1885, at the age of twenty-eight, he succeeded Victor Meyer as professor at the Eidgenossische Züricher Polytechnikum. In 1893 he succeeded Emil Fischer at Würzburg, and in 1903 he returned to Leipzig to suc-ceed Johannes Wislicenus in the more senior of the two Leipzig chem-istry chairs. The second chair was now held by Wilhelm Ostwald, who had replaced Wiedemann in 1897.[17]

In 1890 Hantzsch and his Zurich student Alfred Werner published the first of their papers on the stereoisomers of nitrogen compounds, whose existence, as already mentioned, Victor Meyer was inclined to doubt. In 1884 Bamberger had found what he thought was good evi-dence for a series of iso-diazocompounds:

15 Wilhelm Ostwald, *Lebenslinien eine Selbstbiographie* (Berlin: Klasing, 1933), v. 2, 192. Ostwald himself became professor in Leipzig in 1887.
16 Friedrich Hein, "A. Hantzsch," *Z. für Elektrochemie*, 42(1936):1–4 (p. 2).
17 Albert B. Costa, "Hantzsch, Arthur Rudolf," in Charles C. Gillispie, ed., *Dictionary of Scientific Biography* (New York: Scribner's, 1972), v. 6, 107–109; Servos, *Ostwald to Pauling* (1990) (n. 9), 21, 48.

p-nitro-diazo-benzene

iso-p-nitro-diazobenzene
(p-nitro-phenyl-nitrosamine)

In the same year, Hantzsch brought out an elaborate criticism (Partington's phrase) of this in which he asserted that there was no good evidence for iso-diazocompounds, and that as Bamberger himself had actually thought earlier, the difference in the behaviour of their derivatives might be due to tautomerism, in accordance with his own theory of the stereoisomers of nitrogen: the normal diazocompounds were in the "syn-" form, and the iso-diazocompounds in the "anti-" form:[18]

"syn" or ordinary
diazobenzene

"anti" or iso-
diazobenzene

This controversy began in 1894 and continued for many years, the arguments slowly evolving on both sides. In the end, it was resolved by Hantzch, using ultraviolet absorption spectra. It has been remarked that throughout the controversy, Bamberger argued from grounds of pure chemistry, while Hantzsch tended to use physico-chemical methods, such as spectrometry and conductivity measurements to distinguish between isomers.[19] Hantzsch began to use these methods from about 1894 onwards, and he argued from them more and more in his controversy with Bamberger.[20] At this time, few organic chemists could make use of these tools.

This controversy had not yet started, however, when Landsteiner was at the Zurich Polytechnic; and by 1893, when Hantzch left Zurich, he

18 Arthur Hantzsch, "Ueber Stereoisomerie bei Diazoverbindungen und die Natur der 'Iso-diazokorper,'" *Ber.d.Deutsch. chem. Ges.* 27(1894):1702–1731. This controversy is neatly explained by Partington, *History of Chemistry* (1964) (n. 14), v. 4, 842–847; and more recently by O. Bertrand Ramsay, *Stereochemistry* (London: Heyden, 1981), 129–131.
19 Costa, "Hantzsch" (1972) (n. 17), 108.
20 Partington, *History of Chemistry* (1964) (n. 14), 845.

does not seem to have leaned much toward physical chemistry, at least in his publications. His little book *Grundriss der Stereochemie* of 1893 gave it no prominence.[21] His reason for writing the book seems to have been to enlarge upon the stereoisomers of nitrogen.

The problem that Landsteiner worked on in Hantzsch's laboratory, together with Ronald Scholl, was not in any case one of Hantzsch's own; it was a purely structural one. Victor Meyer had discovered the *pseudo*-nitrole series twenty years before, in 1875.[22] Scholl had found in 1888 that the members of the series could be prepared from their ketoximes.[23] Meyer's old formula allowed for their relationship to the ketoximes, but so did a new one, also proposed by Meyer:[24]

$$
\text{old:} \quad \begin{array}{c} NO_2 \\ \diagdown\diagup \\ C \\ \diagup\diagdown \\ NO \end{array} \qquad\qquad \text{new:} \quad \begin{array}{c} \diagdown\diagup \\ C = NO.NO_2 \\ \diagup\diagdown \end{array}
$$

A reduction of propyl *pseudo*-nitrole gave acetoxime, which seemed to support the newer formula:

$$
\begin{array}{c} CH_3 \\ \diagdown \\ C = N.NO_2 \\ \diagup \\ CH_3 \end{array} \qquad\qquad \begin{array}{c} CH_3 \\ \diagdown \\ C = N.OH \\ \diagup \\ CH_3 \end{array}
$$

propyl-*ps* nitrole   acetoxime
(new formula)

But Scholl and Landsteiner lean towards the older form, with an interpolated intermediary product:[25]

21 Arthur Hantzsch, *Grundriss der Stereochemie* (Breslau: Trewendt, 1893, 1st ed.) (Leipzig, Barth, 1904, 2d ed.).
22 Victor Meyer, E. Demole, and W. Michler, "Ueber die Nitroverbindungen der Fettreihe, II. Abhandlung" (Liebigs), *Ann. d. Chem.* 175(1875):88–164; Victor Meyer, J. Tscherniak, J. Locher, and M. Lecco, "Untersuchung über die Verschiedenheiten der primären, secundären und tertiären Nitroverbindungen," *Ann. d. Chem.* 180(1876):111–206.
23 Roland Scholl, "Umwandlungen von Ketoximen in Pseudonitrole," *Ber. d. Deutsch. Chem. Ges.* 21(1888):506–510.
24 Victor Meyer and L. Oelkers, "Ueber die negative Natur organischer Radicale: Untersuchung des Desoxybenzoins," *Ber. d. Deutsch. chem. Ges.* 21 1295–1306 (1888).
25 Scholl and Landsteiner, "Reduction der Pseudonitrol" (1896) (n. 4).

$$
\begin{array}{ccccc}
\begin{array}{c}
CH_3 \ \ NO_2 \\
\diagdown \ \diagup \\
C \\
\diagup \ \diagdown \\
CH_3 \ \ NO
\end{array}
&
\Longrightarrow
&
\begin{array}{c}
CH_3 \\
\diagdown \\
C(NH.OH)_2 \\
\diagup \\
CH_3
\end{array}
&
\Longrightarrow
&
\begin{array}{c}
CH_3 \\
\diagdown \\
C = NOH \\
\diagup \\
CH_3
\end{array}
\end{array}
$$

The intermediary was *nicht fassbar*. It could not be isolated. But Scholl and Landsteiner were prepared to argue for it as a new bridge between the *pseudo*-nitroles and the ketoximes.

The third of the famous chemists, the first actually in point of time, with whom the young Landsteiner gained his laboratory experience was Emil Fischer.

Like Eugen Bamberger, Emil Fischer was a product of Adolf von Baeyer's training. He had been with von Baeyer in Strasbourg, where he had been one of von Baeyer's first students, and accompanied him to the new institute in Munich. There he reached the level of *ausserordentlicher* professor, going on to Erlangen as *Ordinarius*, and then to Würzburg, where Landsteiner worked with him. It was here that Fischer's research programme on the sugars began to get into its stride. In 1892 he went to Berlin to succeed the organic chemist August Hofmann. Hofmann had been the founder (in 1868) and for many years the president of the Deutsche chemische Gesellschaft.[26] On his death, the Berlin faculty "called" the three most highly respected organic chemists to replace him: August Kekulé, then aged sixty-three, and Adolf von Baeyer, aged fifty-seven, the famous men of the elder generation, and Emil Fischer, who was forty. It was understood, on account of their advanced age, that the first two would naturally refuse the position, and leave it to Fischer.[27]

Fischer's institute in Berlin was to rival von Baeyer's in size and productivity. Like von Baeyer, he found the existing facilities at the site to be inadequate for his large-scale plans, and with the help of funding from the chemical industry, by 1900 he had the old laboratory building with its small separate rooms recast as four large halls. Here he could have about 250 workers, as well as about 50 others, including 16 *Privatdozenten* and assistants.[28] Like von Baeyer, he was a powerful and distant figure to his juniors and ruled his kingdom with absolute authority; but unlike von Baeyer, he expected all the work in the laboratory to be focused on his own research problems.[29] He was awarded a Nobel prize in 1902 for his investigations of sugars and purines.

26   Meyer, *Victor Meyer* (1917) (n. 5), 441–443.
27   Emil Fischer, *Aus meinem Leben* (Berlin: Springer, 1922), 141.
28   Fruton, *Scientific Style* (1990) (n. 11), 198–199.
29   Fruton, *Scientific Style* (1990) (n. 11), 128, 163–229 (pp. 170–171).

Figure 9.1. Emil Fischer in the teaching laboratory, demonstrating on an organic series. (Photograph by Deutsche Illustrations-Gesellschaft, Berlin, n.d.; from the collection of the National Library of Medicine)

Fischer (Figure 9.1) stood out among the organic chemists in that his interests centred on the structure of the substances found in organisms, that is, on structural biochemistry rather than pure organic chemistry.[30] At the beginning of the nineties, Fischer was working, along with

30 Eduard Farber, "Fischer, Emil Hermann," in Charles C. Gillispie, ed., *Dictionary Scientific Biography* (n. 17), v. 5, 1–5.

his many students and co-workers, on the structure and stereochemistry of carbohydrates, especially on the synthesis of the sugars. In June 1890, while he was at Würzburg, he gave a lecture to the Deutsche chemische Gesellschaft on this subject. His was the third in a series that had started off with Victor Meyer's lecture on *Chimie dans l'Espace*, the whole series being organized by Hofmann to cover all of organic chemistry.[31] The idea for the series may have arisen from the twenty-fifth anniversary of the discovery by Kekulé of the ring structure of benzene, which was celebrated with extraordinary, and to Kekulé, embarrassing, pomp by the Gesellschaft on 11 April 1890, at its *Benzolfest* in the Berlin Rathaus. There were lectures, flowers, congratulations from the government in Berlin and from societies all over the world, and a banquet.[32] The various biographies, autobiographies, and memoirs indicate what an important part the Gesellschaft played in the social and intellectual lives of the organic chemists.

Fischer's lecture-review was on the synthesis of the sugars.[33] His approach was based on Heinrich Kiliani's method of 1885–1886.[34] Kiliani's synthesis depended on the reaction of ketones and aldehydes with hydrocyanic acid:

$$
\begin{array}{ccccc}
\begin{array}{c} H \quad O \\ \backslash\!\!\!/\!\!/ \\ C \\ | \\ (CH.OH)_n \\ | \\ CH_2OH \end{array}
& \xrightarrow{\text{HCN}} &
\begin{array}{c} C \equiv N \\ | \\ H.C.OH \\ | \\ (CH.OH)_n \\ | \\ CH_2OH \end{array}
& \Longrightarrow &
\begin{array}{c} H \quad O \\ \backslash\!\!\!/\!\!/ \\ C \\ | \\ H.C.OH \\ | \\ (CH.OH)_n \\ | \\ CH_2OH \end{array}
\end{array}
$$

31 August W. Hofmann, report of session of 28 January 1890, *Ber. d. Deutsch. chem. Ges. 23*(1890):97–99 (p. 99).

32 G. Schultz, "Bericht über den Feier der Deutschen chemischen Gesellschaft zu Ehren August Kekulé's," *Ber. d. Deutsch. chem. Ges. 23*(1890):1265–1312; this account of the *Benzolfest* (Hofmann's expression, p. 1269) includes remarks by Hofmann on the history of benzene itself, by Adolf von Baeyer on the theories of benzene structure and the part they played in the creation of the science of organic chemistry, and a reply by Kekulé.

33 Emil Fischer, "Synthesen in der Zuckergruppe, I.," *Ber. d. deutsch. chem. Ges. 23*(1890):2114–2141, and in *Untersuchungen über Kohlenhydrate und Fermente 1884–1908* (Berlin: Springer, 1909), 1–29.

34 Heinrich Kiliani, "Ueber das Cyanhydrin der Laevulose, I. Mittheilung," *Ber. d. deutsch. Chem. Ges. 18*(1885):3066–3072; Kiliani, "Ue. das Cyanhydrin der Laevulose, II. Mittheilung," *Ber. d. deutsch. Chem. Ges. 19* (1886):221–227; Kiliani, "Ueber die Einwirkung von Blausäure auf Dextrose," *Ber. d. deutsch. Chem. Ges. 19*(1886):767–772; Kiliani, "Ueber die Constitution der Dextrose-carbon-säure," *Ber. d. deutsch. Chem. Ges. 19*(1886):1128–1130.

This could be used to add successive C-atoms to the sugar chain, and thus synthesize the whole series, a series rich in stereoisomers, of which only two – dextrose and laevulose, with left- and right-handed optical activity – had been known earlier. These now took their place, in the new terminology proposed by Fischer, as *d*, *l* and *dl*-glucose (dextrose) and *d*, *l*, and *dl*-fructose (laevulose), 6-C sugars of the aldehyde and ketone, or aldose and ketose series, respectively.

Fischer's new key to the structure of the sugars lay in the compounds they formed with phenylhydrazine, compounds with bright yellow colours and distinctive crystal forms, and with sharply defined melting points, unlike the sugars themselves. Glucose and fructose, one an aldehyde and one a ketone, formed identical ozazones, on which he could base his structural arguments:[35]

```
    H    O                    H                          H    H
     \  //                     \                          \  /
      C                         C = N.NH                  HO.C
      |                         |                          |
    H.C.OH                      C = N.NH                   C = O
      |                         |                          |
    (CH.OH)n                  (CH.OH)n                   (CH.OH)n
      |                         |                          |
    CH2OH                      CH2OH                      CH2OH


    Aldose                    Ozazone                    Ketose
```

Fischer ended his review with a startling biological speculation, the kind of chemical-biological idea that was quite Fischer-specific. If the new and unnatural sugars that he had synthesized were fed to an animal in place of the naturally occurring ones, might its liver not make a new and unnatural glycogen? Might the proteins and fats of the animal's body not be different? And might changes in the building blocks not lead to changes in the architecture? Perhaps changes in the carbohydrates of the food might lead even to changes in animal form.

The paper on which Landsteiner worked with Fischer, and which was published in 1892, was one of this group on the structure of the sugars.[36] Glycolaldehyde, a 2-C substance, which could be regarded as a biose, the simplest aldose, was prepared from bromacetaldehyde and, as predicted, it did behave as a simple sugar. It made an ozazone:

35 Kurt Hoesch, *Emil Fischer sein Leben und sein Werk im Auftrage der Deutschen chemischen Gesellschaft dargestellt* (Berlin: Verlag Chemie, 1921), 297–304.
36 Fischer and Landsteiner, "Glycolaldehyde" (1882) (n. 2).

$$
\begin{array}{c}
\text{H}\quad\text{O} \\
\backslash\,// \\
\text{C} \\
| \\
\text{CH}_2\text{Br}
\end{array}
\qquad
\begin{array}{c}
\text{H}\quad\text{O} \\
\backslash\,// \\
\text{C} \\
| \\
\text{CH}_2\text{OH}
\end{array}
\qquad
\begin{array}{c}
\text{HC} = \text{N.NH} - \\
| \\
\text{HC} = \text{N.NH} -
\end{array}
$$

Bromacetaldehyde   Glycolaldehyde   Glyoxalosan

The published report of this project shows that it was, like the later papers that Landsteiner produced with Bamberger and with Scholl, a good exercise in chemical methods. But it also appears, from a remark by Fischer elsewhere, that they had tried to establish a biological relationship for their simple sugar: to test the activity of a live yeast with the glycolaldehyde. It did not work, however, as the bromine compound killed the yeast.[37]

This attempt is not mentioned in the paper itself, but it is of some significance. It was far from being a completely new idea, of course. Pasteur had used a yeast to separate tartaric acid isomers: the *l*-isomer was less easily metabolized. But it establishes a connection between a most important aspect of Fischer's work and what later became an important part of Landsteiner's: that is, the attempt to link a chemical structure with a biological phenomenon. Fischer found the idea highly exciting. He noted that a yeast will ferment indifferently 3-C, 6-C, and 9-C sugars, but in each case only the isomer of the *d* series:

We have before us a quite new and I might even say astonishing fact, that the most common function of a living being depends more on the molecular geometry than on the composition of its food material. The explanation of this is a task of the greatest importance for biology. . . . The most important components of a living cell are its proteins. They too are optically active, and have molecular asymmetry. . . . If a sugar comes into contact with the proteins of a yeast cell, it is conceivable that it will only be attacked and fermented if the geometrical structure of its molecule is not too far from that of the protein.[38]

It was an idea that was quickly followed up: in 1894, Fischer and a co-worker Hans Thierfelder produced a study of the effects of twelve different yeasts on a series of carbohydrates of different chain lengths, isomeric structures, and optical activities. The fermentations were car-

37 Emil Fischer, "Synthesen in der Zuckergruppe, II.," *Ber. d. Deutsch. chem. Ges.* 27(1894):3189–3232, and in Fischer, *Kohlenhydrate und Fermente* (1909) (n. 33), 30–75 (p. 39).
38 Emil Fischer, "Die Chemie der Kohlenhydrate und iher Bedeutung für die Physiologie: Rede gehalten zur Feier des Stiftungstages der Militärärztlichen Bildungsanstalten am 2 August 1894," in Fischer, *Kohlenhydrate u. Fermente* (1909) (n. 33), 96–115 (p. 108).

ried out under standard conditions, and after eight days they were tested with Fehling's solution. Fischer and Thierfelder reported the results semi-quantitatively. Scores were graded from +++, in which Fehling's test was negative, showing that all the sugar had been used up by the yeast, through ++, +, and −, in which the test was maximally positive, showing that the sugar was untouched.[39]

The results showed that *d* glucose, *d* mannose, and *d* fructose, which had the same three asymmetric carbon atoms, were affected similarly by the yeast; *d* galactose, which differed at one of them, was metabolized more slowly and in some cases not at all. Any further changes made them quite unfermentable (Figure 9.2).[40]

In explaining this striking influence of configuration on enzyme activity, Fischer wrote, in a phrase that has become famous: "To make use of an image, I shall say that enzymes and glycoside must fit each other like a lock and key in order to have any chemical effect on each other."[41]

Landsteiner's contact with Emil Fischer, then, did not simply give him competence in chemical methods. In this and in the later papers by Fischer, a pattern is visible which can be found again in Landsteiner's work on the chemistry of the antigens. Fischer's series of natural and synthetic sugars differing from each other by small structural modulations, and tested against the fermenting enzymes, is exactly paralleled by the pattern of Landsteiner's tests on series of synthetic chemical antigens of twenty-five years later, as is his method of semi-quantitative scoring. Here structural chemistry made contact with a biological effect, an ideal that Landsteiner was to work towards for so many years.

Fischer and von Baeyer, however, may both have contributed something else to Landsteiner. They may have provided him with a role model for the dominating figure of the laboratory director, the powerful and distant *Geheimrat* who controlled the activities of his group of juniors and assistants. It was a role that others perceived Landsteiner as playing to the end of his life, even though he was never to achieve the status of director of his own institute.

39 Emil Fischer and Hans Thierfelder, "Verhalten der Verschiedenen Zucker gegen reine Hefen," *Ber. d. Deutsch. chem. Ges.* 27(1894):2031–2037, and in Fischer, *Kohlenhydrate u. Fermente* (1909) (n. 33), 829–835.

40 Emil Fischer, "Bedeutung der Stereochemie für die Physiologie," *Z. f. physiol. Chem.* (Hoppe-Seylers) 26(1898):60–87, and in Fischer, *Kohlenhydrate u. Fermente* (1909) (n. 33), 116–137 (p. 118).

41 Emil Fischer, "Einfluss der Konfiguration auf die Wirkung der Enzyme, I.," *Ber. d. Deutsch. chem. Ges.* 27(1894):2985–2993, and in Fischer, *Kohlenhydrate u. Fermente* (1909) (n. 33), 836–844 (p. 843).

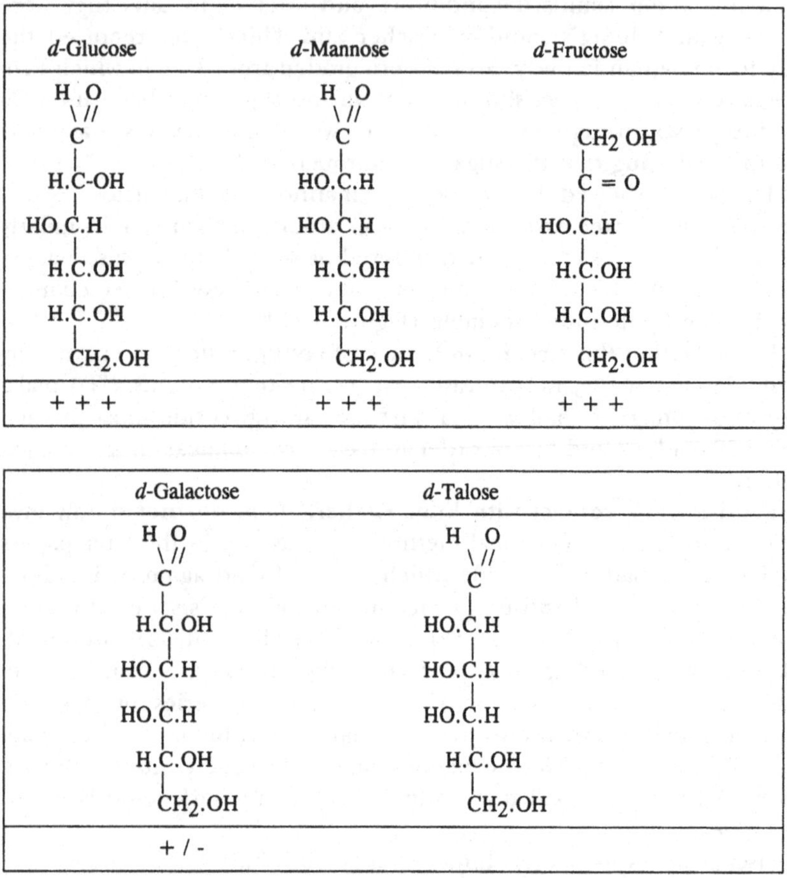

d-Glucose	*d*-Mannose	*d*-Fructose
H O \\ // C | H.C-OH | HO.C.H | H.C.OH | H.C.OH | CH₂.OH	H O \\ // C | HO.C.H | HO.C.H | H.C.OH | H.C.OH | CH₂.OH	CH₂ OH | C = O | HO.C.H | H.C.OH | H.C.OH | CH₂.OH
+ + +	+ + +	+ + +

d-Galactose	*d*-Talose
H O \\ // C | H.C.OH | HO.C.H | HO.C.H | H.C.OH | CH₂.OH	H O \\ // C | HO.C.H | HO.C.H | HO.C.H | H.C.OH | CH₂.OH
+ / -	

Figure 9.2. Emil Fischer's experiment on the fermentation of stereo-isomeric hexoses by a yeast. *d*-glucose, *d*-mannose, and *d*-fructose all have the same three asymmetric C atoms, and are all fermented by the yeast; *d*-galactose, which differs at one of them, is metabolized more slowly, and any further changes make the sugar quite unfermentable. This technique and the presentation of the results may be compared with Landsteiner's experiment of 1917: Landsteiner and Lampl, "XI Mitteilung über Antigene" (1917) (see Chapter 12 of this volume, n. 50, Figure 12.3). Modified from Emil Fischer, "Bedeutung der Stereochemie für die Physiologie" (1898) (n. 40); and Fischer and Thierfelder, "Verhalten der verschieden Zucker gegen reine Hefen" (1894) (n. 39).

This account of the organic chemistry of the 1890s, the chemistry in which Landsteiner received his training, gives only half the picture, however. There was another chemistry, not represented in the *Berichte der Deutschen Chemischen Gesellschaft*, whose development was not covered by Victor Meyer's review. Unlike organic chemistry, physical chemistry, or general chemistry – as it was called by its most active proponent, Wilhelm Ostwald – had a strong mathematical component. Ostwald saw it as a vindication of Kant's reproach that chemistry was not a science because it was impossible to handle mathematically.[42]

Wilhelm Ostwald had held the chair in chemistry in Riga, Latvia, the town where he was born, but was called to Leipzig in 1887 to the chair in physical chemistry, which he held until 1905. In Leipzig, he became the leader and inspiration for the small group of "Ioner," the electro-chemists Svante Arrhenius and J. H. van't Hoff, later joined by Walther Nernst.[43]

Ostwald's view of the history of chemistry is of a science that originated in a primitive conception of specificity. Goethe's novel *Die Wahl-verwandschaften* is his example, and he quotes the passage in which one of the protagonists explains the theory of specific or elective affinity with great sympathy for the desire of the substances in the reaction to unite with each other.[44] This primitive anthropomorphic specificity becomes more and more generalized until all the phenomena with which it deals can be seen as special cases under general, mathematical laws, from which all specificity, in fact all material basis, has been expunged – a *Chemie ohne Stoff* as he calls it. It is a system of general concepts and quantitative relationships or natural laws, which apply to all matter whatever its nature. From this point of view, the properties of all individual substances take their place as examples of those general laws that could be deduced by means of given constants by inserting their quantities into those equations.[45]

This was Ostwald's programme for the series of lectures on the history of chemistry that he gave at the Massachusetts Institute of Technology in the autumn of 1905. The sixth lecture, entitled "Affinität," gives his picture of this development taking place through the course of the nineteenth century.[46] The threshold point, at which earlier gropings became self-conscious mastery of the problem, was the phase rule

42 Wilhelm Ostwald, *Der Werdegang einer Wissenschaft: sieben gemeinverständliche Vorträge aus der Geschichte der Chemie* (Leipzig: Akademische Verlag, 2d ed. 1908), 220; 1st ed. (1906), as *Leitlinien der Chemie*, lecture series of 1905, Massachussetts Institute of Technology.
43 Erwin N. Hiebert and Hans-Günther Körber, "Ostwald, Friedrich Wilhelm," in Charles C. Gillispie, ed., *Dictionary of Scientific Biography* (New York: Scribner's, 1978), v. 15, suppl. 1, 455–469.
44 Ostwald, *Werdegang einer Wissenschaft* (1906) (n. 42), p. 207. He quotes Johann Wolfgang Goethe, *Die Walhlverwandschaften* (1808) (Munich: D.T.V.-Gesamtausgabe, 1972), v. 19, 34.
45 Ostwald, *Lebenslinien* (1933) (n. 15), v. 2, 387–388.
46 Ostwald, *Werdegang einer Wissenschaft* (1906) (n. 42), 207–252.

of the American Willard Gibbs, which appeared in print between 1875 and 1878 and which Ostwald himself translated into German in 1892.[47] In Gibbs's work, mathematical chemistry attains the level of exactness and universality that mathematical physics had reached a hundred years earlier.[48] Ostwald adopted thermodynamics or "energetics," as he called it, as an alternative to chemistry, and he went on to organize and institutionalize this alternative.

One paragraph sums up Ostwald's history of chemical concepts:

The principle of shifts in equilibrium gives us information on all the processes that are bound up with the shift in equilibrium conditions, even to the constants of the law of mass action. This law, that the effect of any material is proportional to its concentration, was first vaguely stated by Wenzel for the speed of reactions, and by Berthollet for the equilibria, and then shown experimentally much later by Guldberg and Waage, and by Julius Thomsen and his successors. The application of thermodynamics to chemical equilibria between gases gave the same laws, as was shown first by Horstmann, then more exhaustively by Willard Gibbs. By van't Hoff's discovery that the gas laws applied unchanged to substances in solution and to their osmotic pressure, this theoretical demonstration was widened to cover an enormous area.[49] Instead of being true only of a few gases, the law of mass action now became valid for numberless substances in solution.[50]

This was the position according to Ostwald, when in 1887 the full statement of Svante Arrhenius's theory of electrolytic dissociation ap-

47 Josiah Willard Gibbs, "On the equilibrium of heterogeneous substances," *Trans. Connecticut Acad.* 3(1875–76):108–248; 3(1877–78): 353–524, and in H. A. Bumstead and R. G. Van Name eds., *The Scientific Papers of J. Willard Gibbs* (1928) (New York: Dover, 1961). Translated by Wilhelm Ostwald as *Thermodynamische Studien von J. Willard Gibbs unter Mitwirkung des Verf. aus dem englischen übersetzt* (Leipzig: Engelmann, 1892).

48 Ostwald, *Werdegang einer Wissenschaft* (1906) (n. 42), 236.

49 The dates of the chemists mentioned by Ostwald in this history of increasingly general laws in chemistry are: Carl Friedrich Wenzel (1740–1793), Claude-Louis Berthollet (1748–1822), Julius Thomsen (1826–1909), Peter Waage (1833–1900), Cato M. Guldberg (1836–1902), J. Willard Gibbs (1839–1903), August F. Horstmann (1842–1929), and Jacobus Henricus van't Hoff (1852–1911).

50 Ostwald, *Werdegang einer Wissenschaft* (1906) (n. 42), 243. Ostwald's version of the history of chemistry as passing through three stages, "Kennenlernen der Objekte," "systematische Ordnung," and "Ermittlung der allgemeinen Gesetze," is expressed in several other places, for example, in "Die Aufgaben der physikalischen Chemie," of 1887 and "Altes und neues in der Chemie," of 1890. Both these lectures are reprinted in his *Abhandlungen und Vorträge allgemeinen Inhaltes 1887–1903* (Leipzig: Veit, 1904); cf. Meyer, "Chemische Probleme der Gegenwart" (1890) (n. 6), 560, who describes for organic chemistry what Ostwald calls "systematische Ordnung." Ostwald's feeling is that it is the highest stage, in which mathematics has appeared in chemistry, that is represented by physical chemistry. Only at this stage does chemistry match Kant's 1786 definition of a real science. This early history has been dealt with by John W. Servos, *Physical Chemistry from Ostwald to Pauling: The Making of a Science in America* (Princeton, N.J.: Princeton University Press, 1990), 11–39.

peared.[51] With this the equilibrium concept grew even more important: if a substance in solution dissociates into ions, the ions can be regarded as independent substances, so that by the law of mass action the equilibrium between the ions and the undissociated material could be covered by a single equation.

According to Ostwald, Arrhenius's work began a new period of electrochemistry.[52] Van't Hoff had already derived a set of equations that drew an analogy between the behaviour of substances in solution and the laws governing pressure and temperature relationships in gases.[53] Arrhenius's theory explained the failure of van't Hoff's equations to apply to electrolytes in solution, where the physical measurements, that is, the osmotic pressure, boiling and freezing point changes, were all too high: in the case of KCl, twice as high as van't Hoff expected, in that of K_2SO_4 three times as high. Van't Hoff had incorporated an irrational coefficient i in his equation to take into account this irrational behavior, so the "gas-law" equation for osmotic pressure read $PV = iRT$. Ostwald called it *die ominose Koeffizient.* Arrhenius showed that i always and only appeared if the substance in solution was an electrolyte, and that it depended on dilution. At high dilution, where the substances were completely dissociated, it reached its maximum. In short, says Ostwald, all apparent contradictions of the theory disappeared with the assumption of electrolytic dissociation, and became so many confirmations.

Ostwald devoted himself with terrific energy to the promotion of the new science. It needed both a text and a journal: Ostwald's textbook *Lehrbuch der allgemeinen Chemie* appeared in two parts, the first in 1885 and the second in 1887.[54] Together with van't Hoff, in 1887 he founded the *Zeitschrift für physikalische Chemie, Stoichiometrie u. Verwandtschaftslehre.* The same year, he became professor of chemistry at Leipzig, taking over the chair held by Gustav Wiedemann, the "world's only chair of physical chemistry."[55] In 1887, Arrhenius's solution theory appeared.

51 Svante Arrhenius, "Ueber die Dissociation der in Wasser gelösten Stoffe," *Z. f. physik. Chemie* 1(1887):631–648; on Arrhenius and solution theory, see Servos, *Ostwald to Pauling* (1990) (n. 48), 33–45.
52 Ostwald, *Werdegang einer Wissenschaft* (1906) (n. 40), 185.
53 Jacobus Henricus van't Hoff, "L'équilibre chimique dans les systèmes gazaux ou dissous à l'état dilué," *Arch. neerlandaises des Sciences Exactes et Naturelles* 20(1886): 239–302. On the relationship between Arrhenius's and van't Hoff's theories, see Robert Scott Bernstein, "Svante Arrhenius and ionic dissociation: a revaluation," in George Dubpernell and J. H. Westbrook, eds., *Proceedings of the Symposium on Selected Topics in the History of Electrochemistry, Proceedings of the Electrochemical Society* 78(1977): 201–212.
54 Wilhelm Ostwald, *Lehrbuch der allgemeinen Chemie* (Leipzig: Engelmann, 1885–1887), v. 1, Stoichiometrie; v. 2, Verwandtschaftslehre.
55 Ostwald, *Lebenslinien* (1933) (n. 15), v. 2, 19.

Figure 9.3. A postcard from the Ostwalds at their country house, "Landhaus Energie," dated 12 September 1908. Svante Arrhenius is among the guests. (Photograph courtesy of the Library of Congress, Washington, D.C.)

Ostwald calls 1887 "*ein Wendepunkt der Wissenschaft,*" a turning point for both him and his science. Van't Hoff, Arrhenius, and Ostwald were the *Coryphäen,* the stars, of the new science; Ostwald, as he himself says, was the organizing principle.[56] In 1906 he resigned from his Leipzig chair and retired to his country house, "Landhaus Energie," to write on pacificism, monism, and "cultural energetics" (Figure 9.3).[57]

In spite of Ostwald's boosting, physical chemistry gained status as a legitimate part of chemistry rather slowly.[58] In 1890, it could not compare in power and influence with organic chemistry, with the large numbers of chairs held by organic chemists, the large-scale institutes for training and research, funded in part by the chemical industry, and with the influence of their organization, the Deutsche chemische Gesellschaft, which, with its journal, was founded in 1868.

It is not surprising that Landsteiner, as a young man seeking training in chemistry in 1891, should have looked for it with the great research schools of the organic chemists. The major teaching institutes, in Munich and later in Berlin, were institutes of structural organic chemistry, devoted to what Ostwald called the stage of "*systematische Ordnung,*" the synthesis and analysis of series of organic structures, the attempt to deduce structure from synthetic reactions and end-products, and, in Fischer's case, the attempt to relate chemical structure to biological behaviour. The source from which Landsteiner carried away the most seems to have been Emil Fischer: Fischerian habits reappear in recognizable ways in Landsteiner's later work. Indeed, unless he had gone to Leipzig, and come across Ostwald and his influence secondarily, as some chemists, Hantzsch among them, were just beginning to do, he would probably have heard of no other chemistry but the organic kind.

56   Ostwald, *Lebenslinien* (1933) (n. 15), v. 2, 20.
57   Fruton, *Contrasts in Scientific Style* (1990) (n. 11), 241–262 (p. 246).
58   W. Yost, "The first 45 years of physical chemistry in Germany," *Ann. Rev. Phys. Chem.* 17(1966):1–14.

10

Ehrlich's Chemistry and Its Opponents: i. The Dissociation Theory of Arrhenius and Madsen

The earliest theory of the relationship of antigen to antibody was that of Paul Ehrlich. It was, as one might expect, a theory that explained antibody specificity as a special case of chemical affinity, and in which the reaction of antigen with antibody forms a compound as stable as those of organic chemistry. Physical properties, or what Ehrlich considered to be physical properties, are not involved.

Ehrlich makes this plain in a paper on staining and pharmacological effects that he first read in 1898 and published in full in 1902.[1] In it he directly discusses the problem of whether, in "Ostwald's terminology," the biological properties of substances are *additive*, that is, physical, or *constitutive*, meaning chemical:

To which group, then, do the properties of affinity, i.e., the power of elements to effect chemical reactions, belong? Evidently to the constitutive, for daily experience teaches us that the nature as well as the arrangement of the elements is a factor. . . . Butyric acid and acetic ester are not only composed of the same materials, but have the same molecular weight, yet their affinities are different. There is probably no doubt that those properties of organic substances which interest us as therapeutists are constitutive in nature.[2]

Although he refers this example to Ostwald's *Grundriss der allgemeinen Chemie*, this is not really what Ostwald said there. In Ostwald's view, the

1 Paul Ehrlich, "Ueber die Beziehungen von chemische Constitution, Vertheilung und pharmakologischer Wirkung," lecture to Verein für innere Medizin, Berlin 12 December 1898, and reported in *Münch. med. Wschr.* 45(1898):1654–1655. First published in full in *Internationale Beiträge zur innere Medicin: Ernst von Leyden zur Feier seines 70-jährigen Geburtstages, am 20 April 1902, gewidmet von seinen Freunden und seinen Schülern* (Berlin: Hirschwald, 1902), v. 1, 645–679; also in Felix Himmelweit, Martha Marquardt, and Sir Henry Dale, eds., *Collected Papers of Paul Ehrlich* (London: Pergamon 1956), v. 1, 570–595 (Ger.), 596–618 (Engl.), translated by Charles Bolduan from Bolduan, ed., *Collected Studies in Immunity by Professor Paul Ehrlich and His Collaborators* (New York, N.Y.: Wiley, 1910), 404–442 (referred to as *Ehrlich-Bolduan*).
2 Ehrlich, "Beziehungen von chemische Constitution, Vertheilung und pharmakologischer Wirkung" (1898) (n. 1), 600 (translation is slightly modified). All page references are to the English versions in *Collected Papers* (1956) (n. 1), where they exist.

concept of affinity, of *Wahlverwandschaften*, might be entirely replaced by a system of equations built up out of the law of mass action, the phase rule, and the gas laws as applied to solutions – in other words, by a quantitative *allgemeine Chemie* that left no room for specificity.[3] For Ehrlich, the only truly additive property was mass, as Ostwald said at the beginning of his chapter.[4] Ehrlich did not follow him to the end, where he had reduced everything, even the constitutive characters, to the additive form.

Although the properties of poisons, drugs, dyes, and other "unnatural" substances depend on their chemical constitution, their union with the protoplasm is not, says Ehrlich, a chemical union. He distinguishes these substances carefully from the ones he calls "natural," those that are foodstuffs capable of assimilation, whose union with the body is chemical or synthetic. Chemical union at this time meant what was later to be classed as covalent linkage, as found in the firmly bound compounds of organic chemistry.[5] The substances that are natural to the body are linked to it by this kind of firm union. The sugar molecule is Ehrlich's example: it cannot be washed out of a cell by water but must be split off by acids. Its chemical union, as in every organic synthesis, implies

the presence of two combining groups of maximal chemical affinity which are fitted to one another. Those groups in the cell which anchor foodstuffs I term *side-chains* or *receptors*, and the combining groups of the food molecule, the *haptophore groups*. Hence I assume that the living protoplasm possesses a large number of such side-chains and that these by virtue of their chemical constitution are able to anchor the greatest variety of foodstuffs. In this way cell metabolism is made possible.[6]

3   Wilhelm Ostwald, *Grundriss der allgemeinen Chemie* (Leipzig: Engelmann, 1889), translated by J. Walker as *Outlines of General Chemistry* (London: Macmillan, 1890), 372–386 (p. 372). Ostwald begins his chapter "The nature, composition and constitution of substances and their affinity" by saying, "What are these 'essential' properties still to be found in the elements? . . . They are the additive properties. No strictly additive property is known save mass." But after discussing the relationship between constitution and various other physical properties, such as electrical conductivity, he derives an affinity constant, then finds it to be proportional to the heat of dissociation on decomposition (of an acid) into ions. He ends his chapter with the words: "We see the essentially constitutive constants of affinity resolved into the additive form" (p. 386).

4   Ehrlich, "Beziehungen von chemische Constitution, Vertheilung und pharmakologischer Wirkung" (1898) (n. 1), 614.

5   John Parascandola and Roland Jasensky, "Origins of the receptor theory of drug action," *Bull. Hist. Med.* 48(1974):199–220 (p. 208); Parascandola, "The development of receptor theory," in Michael J. Parnham and Jacques Bruinvels, eds., *Discoveries in Pharmacology,* v. 3, *Pharmacological Methods, Receptors and Chemotherapy* (Amsterdam: Elsevier Science, 1986), 129–156.

6   Ehrlich, "Beziehungen von chemische Constitution, Vertheilung und pharmakologischer Wirkung" (1898) (n. 1), 613.

Only substances with haptophore groups that fit the receptors on the protoplasm can enter into this kind of synthetic combination with it. Antigens and toxins are like foodstuffs, and their haptophore groups unite firmly with the cell receptors. Only substances with haptophore groups can act as antigens. But when a toxin unites with a cell, it damages the cell instead of nourishing it: the receptor is torn away, and the cell, in its effort at repair, hyper-regenerates the receptors or side-chains, and frees them into the serum. The free side-chains in the serum are the serum antibodies.

The union of antigen and antibody must therefore be a firm and irreversible one, a synthetic union, as Ehrlich calls it, that is, like those of organic synthesis. The kind of union that *un*natural substances like drugs and dyes make with protoplasm is different, though it, too, depends on chemical constitution. Ehrlich dislikes the physical interpretation: "The purely mechanical conception which refers it entirely to physical processes such as surface attraction and adsorption can probably be discarded for the staining of substances in general."[7]

The possibilities that are left are two, the same two that he discussed in 1878 in his thesis on staining.[8] The first is the formation of lakes, insoluble salt-like compounds of dye and fibre; the second is the formation of solid solutions of dye in fibre, "à la van't Hoff."[9] These are loose combinations that can be easily dissociated, in which no chemical union takes place. But even without a true chemical union, a *chemisch-synthetische Veränkerung*, the conditions necessary for selection of fibres by dye are present: the chemistry involved in the case of unnatural foreign substances like dyes and drugs is that of salt formation and the

---

7   Ehrlich, "Beziehungen von chemische Constitution, Vertheilung und pharmakologischer Wirkung" (1898) (n. 1), 614.

8   Paul Ehrlich, "Beiträge zur Theorie und Praxis der histologischen Färbung," Inaugural Dissertation, University of Leipzig, 1878; *Coll. Papers* (n. 1) 29–64 (Ger.), 65–98 (Engl.) (p. 734). In this thesis Ehrlich had rejected the physical explanation of dyeing, but his discussion of the lake theory states that every dyestuff must have a salt-forming group and a chromophore group, e.g., the azo-group – N N -. Here the haptophore does not give a "*chemisch-synthetische Veränkerung*," but a salt linkage.

9   Jacobus Henricus van't Hoff, "Ueber feste Lösungen und Moleculargewichtsbestimmung an festen Körpern," *Z. für physik. Chem.* 5(1890):322–339. Van't Hoff conceives of the idea of a "solid solution" as an explanation of the existence of mixed crystals, which though homogeneous are composed of more than one crystalline substance. He sees them as a third level, comparable to mixtures in gas and in fluid media. Gases are perfectly miscible, solutions in fluids are limited, in some cases insoluble – while solubility in solids is usually less than in liquids, though there are plenty of examples of it. In this way he brings solid substances of variable composition into continuity with solutions in all three phases.

preferential partition of these between different solvents, a preference that must be constitutional.[10]

In this early paper on drug action, Ehrlich is drawing a sharp distinction between the action of natural and unnatural substances: drugs and dyes have *no* haptophore groups and so do not take part in true synthetic reactions. It is only to the union of receptor and haptophore that the metaphor of lock and key, symbolizing true specificity, applies.[11]

This paper provides several pieces of information: one is that Ehrlich rejects physico-chemical explanations, but that he does not include salt formation or even partition between different solvents in the "physical" category. The second is that the union of antigen with antibody is like the unions of organic synthesis: a true, or to use an anachronism, a "covalent" linkage, not a salt-like linkage, and most particularly not a physical relationship. Unlike the dye or the drug, the substance that has a haptophore group is irreversibly linked and cannot be washed out. It depends on chemical affinity, *Wahlverwandschaft*, in its choice of the receptor to which it is linked. For Ehrlich, antibody specificity is chemical affinity and both are sharply discontinuous: the antibody fits its antigen, in Fischer's famous simile, like a key fits a lock. There are no master keys. Ehrlich's later papers on drug action, however, move towards assimilating it to chemical specificity and true "synthetic union" through receptors and ascribe the same haptophore and toxophore structure to dyes and drugs as to antibodies.

The entry of physical chemistry into immunology in 1902 was a product of the contact between the Swedish physical chemist Svante Arrhenius and Thorvald Madsen at the Statens Seruminstitut in Copenhagen.[12] Madsen's dissertation on diphtheria toxin was presented in Copenhagen in 1896, while he was assistant to Carl Julius Salomonsen at the University Laboratory for Medical Bacteriology.[13] He later took charge of the department which prepared diphtheria serum for

10 Ehrlich, "Beziehungen von chemische Constitution, Vertheilung und pharmakologischer Wirkung" (1898) (n. 1), 616.
11 Parascandola and Jasensky, "Origins of the receptor theory of drug action" (1974) (n. 5), 207.
12 E. Schelde-Møller, *Thorvald Madsen: i Videnskabens og Menneskehedens Tjeneste* (Copenhagen, Nyt Nordisk Forlag, 1970). Madsen was to become director of the Statens Seruminstitut in 1909 and to lead the League of Nations Health Organisation in its work on the standardization of antisera between the two World Wars.
13 Thorvald Madsen, "Experimentelle Undersøgelser over Difteriegiften," diss. (Copenhagen, 1896), and as "Ueber Messung der Stärke des antidiphtherischen Serums," *Z. f. Hyg. u. Infectionskr.* 24(1897):425–442.

clinical use. When the Statens Seruminstitut was inaugurated in 1902, he was able to make use of Salomonsen's contacts in both France and Germany: he spent some time with Ehrlich, where he did assays of diphtheria and tetanus toxin in the Ehrlich manner, and some time at the Institut Pasteur, where he was in touch with Jules Bordet and Jean Danysz, both of whom were opposed to Ehrlich's theories.[14]

In Copenhagen, with the collaboration of Svante Arrhenius, Madsen set up quantitative experiments to investigate the neutralization of toxin by antitoxin from a physico-chemical point of view. Since Madsen was working with Ehrlich in the late nineties, it seems quite possible that Ehrlich's early emphasis on salt-formation by dyes and drugs had something to do with Madsen's selection of salt-formation as his model for the antigen-antibody reaction. But to Arrhenius, salt-formation meant a physico-chemical process rather than a specific affinity-based one, and so that is what it became in the hands of the two collaborators.

Madsen also experimented with the haemolytic effect of bases, as Jean Danysz had been doing in Paris.[15] He discovered that the haemolysis produced by ammonia was blocked by the addition of ammonium salts. This suggested to Arrhenius and Madsen that the neutralisation of toxin by antitoxin was analogous to that of acid by base, and to the action of ammonium salts on the electrolytic dissociation of ammonia. They saw that the curve of neutralisation of toxin and antitoxin closely resembled that of the equilibrium between a substance in partial dissociation and its products of association: a reversible equilibrium reaction, according to the Arrhenius solution theory, in which the equilibrium depended upon the quantities of the reactants present. By the mass-action equation,

$$\frac{\text{Free toxin}}{\text{Vol.}} \cdot \frac{\text{Free antitoxin}}{\text{Vol.}} \; \rightleftharpoons \; K \cdot \frac{(\text{combined toxin-antitoxin})^2}{\text{Vol.}}$$

The tetanus toxin had an obvious haemolytic effect on red cells, so that free toxin could be estimated in terms of haemolysis. Ammonia, which like the toxin was haemolytic, was neutralized by boric acid in an exactly similar way, and the curve also expressed by the mass-action equation.

Unlike Ehrlich's stepped *Giftspektrum*, these neutralization curves are smooth exponentials, but they have the same form as Ehrlich's (Figure

14 Thorvald Madsen, "Ueber Tetanolysin," *Z. f. Hyg. u. Infectionskr. 32*(1899):214–238.
15 Jean Danysz, "Contribution a l'étude de l'immunité: Propriétés des mélanges des toxines avec leurs antitoxines. Constitution des toxines," *Ann. de l'Inst. Pasteur 13*(1899): 581–595.

10.1). Madsen did not regard himself as an opponent of Ehrlich and did not show unholy triumph at the weakening of Ehrlich's theory and the vindication of his own earlier suggestion that the neutralization curve was not stepped but continuous.[16] He courteously said that the stepped *Giftspektrum* was the result of using the complex deteriorated diphtheria toxin, and that his own theory was built upon this foundation. Ehrlich was interested, and not antagonized, and invited Arrhenius to work for some time in his Institute in Frankfurt.[17]

We have seen the mixture of joy and envy with which Gruber, and following his example, von Pirquet, welcomed Madsen's work as "disproving" Ehrlich's theory.[18] Gruber's two students, Roland Grassberger and Arthur Schattenfroh, also saw Madsen's curve in this light, but they were among the very few to do so. Others, as will be seen, saw it as an offshoot of Ehrlich's theory, as based on affinity chemistry and not on physical chemistry. Schattenfroh succeeded Gruber as *Ordinarius* and director of the Institute for Hygiene in Vienna in 1905, and when he died in 1927 at the early age of fifty-four he was succeeded by Grassberger.[19] The foreword to the monograph *The Relationship of Toxin to Antitoxin* published in 1904 gives a spirited picture of immunological thinking:

The relations between toxin and antitoxin are the subject at present of lively controversy. About three years ago in a series of lectures and publications, Gruber took an open stand against the ruling Ehrlich hypothesis, and since then research workers have been divided between the two camps. On the one side, Ehrlich and his school have shown both brilliance and obstinacy in defending their position. On the other side, Ehrlich's opponents have found reinforcement in the sensational [*Aufsehen erregend*] work of Arrhenius and Madsen.

Madsen, the erstwhile pupil of Ehrlich, has recently carried the campaign into Ehrlich's own territory. Instead of the complicated relations suggested by Ehrlich for the binding of diphtheria toxin and antitoxin, depending on a large number of substances in the toxin solution, he has tried to show that the course of the reaction suggests the binding of substances with weak affinities.

16 Madsen, "Ueber Tetanolysin" (1899) (n. 14), 226; Lewis P. Rubin, "Styles in scientific explanation: Paul Ehrlich and Svante Arrhenius," *J. Hist. Med. 35*(1980):397–425.

17 Svante Arrhenius, *Immunochemistry: The Application of the Principles of Physical Chemistry to the Study of the Biological Antibodies* (1907) (New York, N.Y.: Macmillan, 1907), Preface, vii.

18 Richard Wagner, *Clemens von Pirquet, His Life and Work* (Baltimore, Md.: Johns Hopkins University Press, 1968), 32 ff.

19 "Grassberger, Roland," in Isidor Fischer, *Biographisches Lexikon der hervorragenden Ärzte der letzten fünfzig Jahre* 2 v. (Vienna: Urban, 1933), v. 1, 529; "Schattenfroh, Arthur," *Biographisches Lexikon*, v. 2, 1375; Erna Lesky, *Die Wiener medizinische Schule im XIX Jahrhundert* (Graz: Bohlaus, 1965), 602.

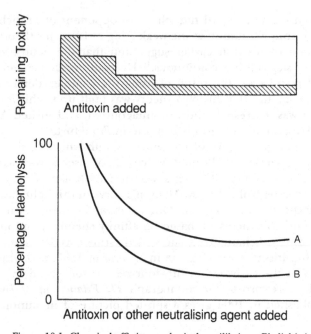

Figure 10.1. Chemical affinity or physical equilibrium: Ehrlich's interpretation of the relationship of toxin and antitoxin compared with that of Arrhenius and Madsen. Ordinates: remaining toxicity. Abscissae: neutralizing agent added. Top: Ehrlich's stepped *Giftspektrum*. Irreversible organic syntheses occurring in order of their chemical affinity. Bottom: Arrhenius and Madsen's neutralization curves. A. The model reversible reaction, neutralization of ammonia by boric acid:

$$\frac{\text{Free NH}_3}{\text{Vol}} \cdot \frac{\text{Free H}_3\text{O}_3\text{B}}{\text{Vol.}} \quad \begin{array}{c}\Diamond\\\Diamond\end{array} \quad K.\frac{(\text{NH}_4\text{H}_2\text{O}_3\text{B})^2}{\text{Vol.}}$$

B. Neutralization of toxin by antitoxin, also a reversible reaction.

The events follow each other fast. It has hardly appeared that Madsen has gained a decisive victory over Ehrlich, when Ehrlich announces triumphantly that Arrhenius, Madsen's distinguished co-author, again believes in the multiplicity of active substances in the diphtheria toxin solution.

In this battle, opinion has recently gone far beyond the material necessary for proving anything.[20]

Grassberger and Schattenfroh thought that Arrhenius and Madsen's theory was a sensational refutation of Ehrlich's. Their military description of immunological thought, their imagery of armed camps, battle

---

20   Roland Grassberger and Arthur Schattenfroh, *Ueber die Beziehungen von Toxin und Antitoxin* (Vienna: Deuticke, 1904), Foreword.

and victory, positions defended, and reinforcements brought up was based upon the violent dialectic carried on by Gruber and Ehrlich.

But not everyone saw this theory as either sensational or a refutation, or even as physico-chemical in nature. As early as 1896, Jules Bordet had written, "It is probable that serum acts on bacteria by changing the relations of molecular attraction between the bacteria and the surrounding fluid."[21] And he maintained this more truly physical theory in opposition to Ehrlich's affinity chemistry to the end of his life. In a review written in 1909, Bordet contrasts his theory with those based on chemical affinity, which can call either for strong affinities and complete reactions, like Ehrlich's, or for weak affinities and reversible reactions like Arrhenius and Madsen's (Figure 10.1).[22] For Bordet as for Ehrlich, salt formation was not a physico-chemical matter, though Arrhenius and Madsen thought of it as such. For Bordet, Arrhenius's theory was only a minor offshoot of Ehrlich's.

Arrhenius's theory, in fact, seems to have attracted little support. Though it appeared to its authors to be a new and different interpretation – using the new physical chemistry, that is, the Arrhenius solution-theory – it was overtaken almost at birth by the much more powerful physico-chemical theories based on adsorption and colloid chemistry, which had Bordet's support. The only extensive treatment of Arrhenius's theory seems to have been Arrhenius's own, in the lectures he gave in California in 1904 and published under the title of *Immunochemistry* in 1907.[23]

Comments on the idea of a reversible equilibrium were almost all negative. Landsteiner had shown that the antigen–antibody reaction was not really reversible, and so not a true equilibrium.[24] The physical chemist Walther Nernst claimed that Arrhenius was careless in his experimental technique and in his choice of equation.[25] Nernst was

21 Jules Bordet, "Sur le mode d'action des sérums préventifs," *Ann. de l'Inst. Pasteur* 10(1896):193–219, and in *Studies in Immunity by Professor Jules Bordet and His Collaborators*, collected and translated by Frederick P. Gay (New York, N.Y.: Wiley, 1909), 93 (referred to as *Bordet-Gay*).
22 Jules Bordet, "A general résumé of immunity," in *Bordet-Gay* (1909) (n. 21), 496–530 (p. 516). This essay was written especially for Gay's Bordet anthology, corresponding to the essay written by Ehrlich, "A general review of the recent work in immunity," which appeared in *Studies in Immunity by Professor Paul Ehrlich and His Collaborators*, collected and translated by Charles Bolduan (New York, N.Y.: Wiley, 1906), 577–586. An advertiser's leaflet found inside a copy of *Bordet-Gay* at the Royal Society of Medicine of London urges readers to buy both of these books so that they, too, can join in the controversy between these two famous men.
23 Arrhenius, *Immunochemistry* (1907) (n. 17).
24 Karl Landsteiner and Nikolaus von Jagić, "Ueber die Verbindungen und die Entstehung von Immunkörpern," *Münch. med. Wschr.* 50(1903):764–768, C. 31.
25 H. Walther Nernst, "Ueber die Anwendbarkeit der Gesetze des chemischen Gleich-

against him, even Nernst, who had been Ostwald's assistant in 1887 when Ostwald first took up his chair in Leipzig and who must have remembered the heroic days when van't Hoff, Ostwald, and Arrhenius were clearing the jungle and planting the first seeds of physical chemistry.[26] None the less, he wrote:

Reading this publication of Arrhenius it is easy to get the impression that here is a new conception of the field of serum-therapy, one which basically contradicts the older views of von Behring and Ehrlich. Looked at more closely, however, it easy to see that Arrhenius and Madsen's investigation deals only with a question of rather minor importance, and not with the central problems of serum therapy. For example, Ehrlich's fundamental account of the formation of antitoxins in the organism is absolutely untouched.[27]

Arrhenius's theory is not only wrong but also trivial. This very damaging criticism, from a man whose support Arrhenius must have expected, led to an estrangement between them.[28] Nernst, now in the chair of physical chemistry at the Chemical Institute in Berlin, had taken over from Ostwald, the founder, as the leader of the physical chemistry school. It was Nernst's textbook *Theoretical Chemistry* (the first edition came out in 1893) that set the pattern for twentieth-century thinking.[29] It is very likely that Nernst's bad review contributed a good deal to the marked lack of success of Arrhenius's theory. It is probably this criticism, for example, that Gideon Wells of the University of Chicago had in mind when he wrote in a review of Arrhenius's book:

Already many of the conclusions expressed in this book, and previously published in special periodicals, have been sharply attacked; and in the opinion of the reviewer the attacks have justified, for there is no question that among the

gewichts auf Gemische von Toxin und Antitoxin," *Z. f. Elektrochem. 10*(1904):377–380.
26 Wolfgang Ostwald, *Lebenslinien eine Selbstbiographie,* 2 v. (Berlin: Klasing, Volksausgabe, 1933), v. 2, 36. On Nernst, see Erwin Hiebert, "Nernst and electrochemistry," in George Dubpernell and J. H. Westbrook, eds., *Proceedings of the Symposium on Selected Topics in the History of Electrochemistry,* in *Proc. Electrochem. Soc.* (Philadelphia), 78(1977): 180–200; Hiebert, "Nernst, Hermann Walther," in Charles C. Gillispie, ed., *Dictionary of Scientific Biography* (New York, N.Y.: Scribners, 1978), v.15, suppl. 1, 432–453.
27 Nernst, "Anwendbarkeit der Gesetze des chemischen Gleichgewichts" (1904) (n. 25), 377.
28 James R. Partington, *A History of Chemistry,* 4 v. (London: Macmillan, 1964), v. 4, 674. Partington himself spent some time in Nernst's laboratory, and saw Arrhenius once or twice, though he "never had the honour to speak to him."
29 Hiebert, "Nernst and electrochemistry" (1977) (n. 26), 187; Nernst, *Theoretische Chemie von Standpunkt der Avogadroschen Regel und der Thermodynamik* (Stuttgart: Enke, 1893), 15th ed., 1926.

many investigations recorded are to be found serious errors of experiment and interpretation.[30]

As Gideon Wells said it would be, the word "immunochemistry" was adopted into the vocabulary of the "immunologist" – Wells places quotation marks around this new word – but the Arrhenius theory itself was without any long-term following. The writers of general reviews covering the field of immunology tended more or less to ignore it. In Bordet's essay of 1909, it is mentioned only as an unimportant modification of Ehrlich's chemical-affinity theory. In a massive review of the biochemistry of antigens that appeared in 1912, Ernst Peter Pick of the State Serotherapeutic Institute in Vienna lists Arrhenius's book in his bibliography but does not appear to discuss it anywhere in his text.[31] In Ulrich Friedemann's historical treatment of the subject (1913), it is only mentioned to be dismissed.[32] Gruber and his students seem to have been alone in finding it sensational.

In Ehrlich's as in Madsen's toxin–antitoxin system, the first aliquots of antitoxin added were absorbed without affecting toxicity, the next few reduced it steeply, and the final few had less effect. This could be ascribed to a series of discrete chemical reactions occurring in order of chemical affinity, and conforming to the law of definite proportions, as in Ehrlich's model, or as an equilibrium conforming to the law of mass action, as in Arrhenius and Madsen's. The phenomenon as a whole was sometimes called *das Ehrlich'sche Phänomen.*

At the Institut Pasteur, Jean Danysz, working on *le phénomène d'Ehrlich* with the plant toxin ricin, which agglutinates red cells, discovered a similar phenomenon, which came to be called *le phénomène de Danysz.* He found that to get Ehrlich's results the series of toxin-antitoxin mixtures had to be made by pouring the ricin into the serum in a single dose. If, instead, a series was set up by preparing a neutralized mixture, separating it into aliquots, and twenty-four hours later, adding the different quantities of excess ricin to each tube, a different curve was produced.[33] More toxin was neutralized by a given amount of antitoxin when toxin was added all at once than when it was added in divided doses.

30 Harry Gideon Wells, Review of *Immunochemistry* (n. 17), in *J. Am. Chem. Soc. 30*(1908): 650–652, 651.

31 Ernst P. Pick, "Biochemie der Antigene, mit besonderer Berücksichtigung der chemischen Grundlagen der Antigenspezifizität," in Wilhelm Kolle and August von Wassermann, eds., *Handbuch der pathogenen Mikroorganismen*, 5 v., 2d ed. (Jena: Fischer, 1912), v. 1, 685–868.

32 Ulrich Friedemann, "Infektion und Immunität," in Max Rubner, Max von Gruber, M. Ficker, eds., *Handbuch der Hygiene* (Leipzig, Hirzel, 1913), v. 3, pt. 1, 661–810 (p. 777).

33 Jean Danysz, "Contribution a l'étude des propriétés et de la nature des toxines avec leurs antitoxines," *Ann. de l'Inst. Pasteur, 16*(1902):331–345.

Jules Bordet had described the same thing in 1900: if he added a batch of red cells all at once to a haemolytic serum, twice as much could be lysed than if he added it in divided doses.[34] To Bordet, this suggested an analogy with dyeing: to illustrate the variable proportions in which dye, or antibody, can be taken up, he describes what would happen if a large piece of filter paper were cut in bits and dipped piece by piece in a dilute dye solution: the first few pieces would be deeply stained, the last few not at all. If the whole piece is put in at once it is all uniformly stained.[35]

At the State Serotherapeutic Institute in Vienna, Philipp Eisenberg and Richard Volk obtained the same effect again, this time with typhoid bacilli and anti-typhoid serum. Like Danysz, they began from Ehrlich's quantitative relationship between antigen and antibody and then found that "the capacity of the agglutinable substance is not a constant, but changes with the changing concentration of agglutinin. We were unable to obtain complete saturation of agglutinable substance with agglutinin."[36]

For Bordet, this was the outstanding fact: the agglutinable substance absorbed amounts of agglutinin that varied according to the relative proportions of the reacting substances. Commenting on Eisenberg and Volk's paper, he wrote:

It would seem legitimate, then, to assume that the law of fixed proportions is not applicable [here]. . . . The conditions are very unlike those met with in straight chemistry which depend on equations and equivalents. It is *simply for the purpose of expressing this idea* more emphatically that we have *compared* these phenomena with those of dyeing . . . [it has been] supposed that we overlooked the "chemical nature" of the combination . . . that we regard this fixation as depending entirely on mechanical causes (surface adhesion, etc.) to the exclusion of any elective or specific affinity. . . . We have never committed ourselves to the intimate nature of the reaction. . . . The point of importance is that these

34 Jules Bordet, "Les sérums hémolytiques, leurs anticorps et les théories des sérums cytolytiques," *Ann. de l'Inst. Pasteur, 14*(1900):257–296, and in *Bordet-Gay* (1909) (n. 22), 186–216 (p. 195).

35 Bordet, "Sur le mode d'action des sérums" (1896) (n. 21), 195; Bordet, "General résumé" (1909) (n. 22), 521.

36 Philipp Eisenberg and Richard Volk, "Untersuchungen über die Agglutination: vorläufige Mittheilungen," *Wiener klin. Wschr. 14*(1901):1221–1223 (pp. 1221–1222). Eisenberg and Volk published their results later *in extenso* in a paper entitled "Untersuchungen über Agglutination," *Z. f. Hyg. u. Infectionskr. 40*(1902):155–195. This contains a review of the problem, the quantitative results, and a long bibliography. This is the paper that was mainly referred to by other writers who replotted Eisenberg and Volk's results to fit in with their own interpretations – as was done by, among others, Svante Arrhenius (1903), Karl Landsteiner (1905) (C. 45), Jean Billitzer (1905), Wilhelm Biltz (1904), and Clemens von Pirquet (manuscript notes). These papers are listed in the bibliography and discussed in Chapter 11.

reactions differ from those of ordinary chemistry in that they are not expressed by equations: the proportions in which the substances unite vary according to the conditions of the experiment.[37]

Bordet says elsewhere that some bacteriologist has argued that antigen and antibody must combine in definite proportions because twice as much serum is needed to combine with two than with one dose of bacterial emulsion. That, Bordet says scornfully, is like claiming that paint must combine in definite proportions with a wall.[38]

37 Jules Bordet, "Sur le mode d'action des antitoxines sur les toxines" *Ann. de l'Inst. Pasteur* 17(1903):161–186; translated in *Bordet-Gay* (1909) (n. 21), 259–279 (p. 262).
38 Bordet, "Sur le mode d'action des antitoxines" (1903) (n. 37); in *Bordet-Gay* (n. 21), 264.

11

Ehrlich's Chemistry and Its Opponents: ii. The Colloid Theory of Landsteiner and Pauli

In November 1897, Landsteiner left Gruber's laboratory and joined the staff of the Institute of Pathological Anatomy, first as a voluntary assistant, and from October 1898, as *Assistent*. The director was the pathologist Anton Weichselbaum, whose own work was mainly in the field of bacteriology and infections.

The institute (Figure 11.1) had two functions: it was the pathology department of both the Allgemeines Krankenhaus, and the university, where its task was to teach the medical students. The daily routine there was to have the assistants (Figure 11.2) work through the autopsies in the mornings, watched by the students, and demonstrated them to the director at 11:45 a.m. Between 12:15 and 1 p.m., the director lectured to the students on some of the specimens. The assistants were then free to do their own research for the rest of the day.[1] Landsteiner's publications from this period include a mixture of reports on pathological specimens, bacteriological problems, and serology. His serological papers are the most important for this discussion. Among them are the series of papers that defined the human blood groups.[2]

Between 1891 and 1893, when Landsteiner studied under the structural chemists, and 1903, when his earliest immunochemical paper appeared, there is an interval of nearly ten years, a gap in which his thinking in immunology had been deeply influenced by his two years with Gruber at the Institute for Hygiene, and by Gruber's opposition to Ehrlich. Landsteiner's chemical training had been in structural chemistry. He had experienced at first hand and at a high level the

---

1 Ernest Gold, interview with George M. Mackenzie, 11 February 1948; Gold graduated in Vienna in 1915, worked for three years at the Institute for Pathological Anatomy. American Philosophical Society, Landsteiner–Mackenzie Papers, B L23m, Box 2, no. 11.

2 Pauline M. H. Mazumdar, "The purpose of immunity: Landsteiner's interpretation of the human isoantibodies," *J. Hist. Biol.* 8(1975):115–133; Arthur M. Silverstein, "The Donath–Landsteiner autoantibody: the incommensurate languages of immunological dispute," in his *A History of Immunology* (New York: Academic Press, 1989), 190–213.

Figure 11.1. The Institute of Pathological Anatomy, University of Vienna. From the collection of George Mackenzie, American Philosophical Society, Philadelphia.

thinking and the laboratory methodology of the structural organic chemists of the last quarter of the twentieth century, the chemistry represented by Victor Meyer's review, which had followed on the work of Adolf von Baeyer and J. H. van't Hoff; and he had also experienced the newer structural chemistry of Emil Fischer. He does not seem to have had any contact with physical chemistry of the kind represented by Ostwald, and by Arrhenius's solution theory. Nevertheless, when Landsteiner entered the field of immunochemistry it was not as a proponent of structural or stereochemical explanations, but of explanations based upon physical chemistry.

The first of his papers to make use of a chemical model to explain an immunological phenomenon is unmistakably a product of this influence, and not that of the structural chemists. The position he takes up is that of Ehrlich's opponent.[3] For Ehrlich, he says, the immune reaction is one of true chemical compound formation; the side-chains of his theory are the side-chains of structural chemistry. Some of the other reactions that take place in the body may be absorption phenomena, but not these. For Bordet, on the other hand, serological reactions are analogous to those of dyes, which he also holds to be different from the reactions of "real" chemistry. Following Bordet's lead, Landsteiner points out that Eisenberg and Volk have shown that there is, to say the least, no evidence of the formation of compounds of constant composition.

3 Karl Landsteiner and Nikolaus von Jagić, "Ueber die Verbindungen und die Entstehung von Immunkörpern," *Münch. med. Wschr.* 50(1903):764–768, C. 31 (p. 766).

Figure 11.2. Anton Weichselbaum's Assistenten and their assistants outside the Institute of Pathological Anatomy, dated to early 1903: Wiesner joined in 1903, and Milan Sachs died on 5 May 1903 of a laboratory plague-pneumonia infection, picked up in Berlin at Koch's institute. From left to right: 1. Pletschnig 2. Schneider (assistant) 3. Kryle 4. Milan Sachs 5. Oskar Stoerk 6. Anton Ghon 7. Richard Wiesner (who sent the photograph to Mackenzie) 8. Karl Landsteiner 9. Julius Bartel 10. An assistant. The *Assistenten* as pathologists wore black coats.

Oskar Stoerk spent nineteen years as *Assistent*, from 1893 to 1912; he reached *a.o. Professor* in 1909, and *o. Professor* in 1915. In 1912, he became director of the University Institute for Histopathology. Anton Ghon was *Assistent* from 1894 to 1910, and became professor of Pathological Anatomy at the German University, Prague, in 1910. Richard Wiesner was *Assistent* from 1903 to 1920, and reached *a.o. Professor* in 1917; he succeeded Landsteiner as Prosektor at the Wilhelminenspital. Landsteiner was at the Institute of Pathological Anatomy first as unpaid *Assistent*, then in a regularized position, from 1897 to 1908; in 1909, he went to the Wilhelminenspital as *Prosektor*. He was not promoted to *a. o. Professor* until 1911 and then had no further promotions. Julius Bartel joined the institute in 1898, and remained there as *Assistent* until he died in 1925. Pletschnig and his assistant, Zaritsch (not shown) and Stoerk were among the colleagues who gave blood samples for Landsteiner's first blood-grouping experiment in 1901. From Fischer, *Biographisches Lexikon* (1933) (Bibl. Sect. 1). (From the collection of George Mackenzie, American Philosophical Society, Philadelphia)

The type of reaction that this suggests to Landsteiner is that of colloid chemistry. It is a new idea, so he explains what it means.[4] Colloid

4   The best contemporary account of the history of colloid chemistry is that of Wilhelm Ostwald's son, Wolfgang Ostwald, in his *Grundriss der Kolloidchemie* · (Dresden: Steinkopf, 1909), pt. 1, Geschichte der Kolloidchemie, pp. 3–76. This was translated into

phenomena are both physical and chemical: the adsorption products
form when salts, for example, taken up by colloidal silica have a composition that is dependent on the physical parameters of concentration
and temperature; but the adsorption coefficient is different for each
substance adsorbed. He quotes the colloid chemist Jacob van Bemmelen, professor of chemistry at Leyden, who regards these "adhesion
phenomena" as the effects of a loose chemical binding, an intermediate stage between the purely physical and the chemical, between reactions in which the proportions are continuously variable, and those
with simple fixed proportions. He quotes Walther Nernst, too: "The
difference between a physical mixture and a chemical compound is
gradual and stepwise, and all stages [*alle Abstufungen*] between the two
are found in nature."[5]

These reactions are chemical, says Landsteiner, in that they are influenced by chemical constitution, but in their quantitative relationships they are like physical phenomena such as solubility. "This being
the case, as no one is in a position to give a clear-cut distinction between so-called physical and chemical compounds, it would be a good
thing if this discussion of names were dropped, and people contented
themselves with describing the processes instead."[6]

Landsteiner has picked out from the writings of the colloid chemists
quotations that emphasize the blending of the two poles of the field,
the continuity of nature's transition between the physical and the
chemical. This transitionalness of the colloid state was also emphasized

English by Martin H. Fischer in 1917. According to Partington (p. 729), it was the
first systematic text on the subject: curiously, for a physical chemistry, it is completely
non-mathematical; Martin H. Fischer, "Wolfgang Ostwalds Weg zur Kolloidchemie,"
Kolloid Z. 145(1956):1–2; Ernst v. Meyer, *Geschichte der Chemie von den ältesten Zeiten bis
zur Gegenwart, zugleich Einführung in das Studium der Chemie* (Leipzig: 1st ed., 1888; 4th
ed., 1914), 468–470 (4th). This is an extremely exiguous treatment of it: von Meyer
feels in 1914 that the history of colloid chemistry lies mainly in the future (p. 469).
For historians' discussions of colloid chemistry see: James R. Partington, *A History of
Chemistry*, 4 v. (London: Macmillan, 1964), v. 4, 729–740. Joseph Fruton has called the
colloid movement a "source of confusion" as compared to its conceptual opposite,
structural chemistry. J. S. Fruton, "Early theories of protein structure," in P. K. Sri-
nivasan, Joseph S. Fruton, and John T. Edsall, eds., *Origins of Modern Biochemistry: A
Retrospect on Proteins* in *Ann. N.Y. Acad. Med.* 325(1979):53–73 (pp. 1–15, 8–9). For a
more detailed recent account, see John W. Servos, *Physical Chemistry from Ostwald to
Pauling: The Making of a Science in America* (Princeton, N.J.: Princeton University Press,
1990), esp. 11–39, on the early years of colloid chemistry in Germany.

5 Landsteiner and v. Jagić, "Verbindungen und Entstehung der Immunkörpern"
(1903) (n. 3), 767, quoting H. Walther Nernst, *Theoretische Chemie vom Standpunkt der
Avogadro'schen Regel und der Thermodynanik* (Stuttgart: Enke, 1st ed., 1893, 3d ed. 1900),
32. There were many later editions of this leading textbook of physical chemistry.
6 Landsteiner and v. Jagić, "Verbindungen und Entstehung von Immunkörpern"
(1903) (n. 3), 767.

by Wolfgang Ostwald, Wilhelm Ostwald's son and successor as chief propagandist for physical chemistry.[7] Ostwald's stepwise transition was from the large particle size characteristic of a suspension of a solid in a fluid, through the stages in which the "solution" was a colloidal one, a sol or a gel, to the smallest particle size where the result was a true solution.[8]

The most important of the transitions covered by the field of colloid chemistry, and the one mainly responsible for the enormous interest in colloids in the first quarter of the twentieth century, was that between living and non-living. The Viennese Wolfgang Pauli, another chemist cited by Landsteiner in his discussion of colloid reactions, was interested in colloids mainly from the physiological point of view, as intermediates between the organic and the inorganic, between matter and life (Figure 11.3).[9] His principal concern, he says, is "the extensive parallelism between the laws which govern changes in the colloid state in vitro and in the living organism."[10]

This "extensive parallelism" becomes for Pauli the basis of a kind of "mechanistic vitalism" – the phrase is from the Viennese physiologist

7 Wolfgang Ostwald was the son of Wilhelm Ostwald and a physical chemist like his father. He took up colloid chemistry, the new physical chemistry of the younger generation, and spent some time in Jacques Loeb's laboratory in California while writing what was the first textbook of colloid chemistry: Ostwald, *Grundriss der Kolloidchemie* (1909) (n. 4), v. See also Philip J. Pauly, *Controlling Life: Jacques Loeb and the Engineering Ideal in Biology* (Berkeley: University of California Press, 1987), 113, 151–152.

8 Wolfgang Ostwald, *Die neuere Entwicklung der Kolloidchemie: Vortrag gehalten auf der 84 Versammlung deutscher Naturforscher und Aerzte zu Münster-i-W. 1912* (Dresden: Steinkopf, 1912).

9 Isidor Fischer, "Pauli, Wolfgang," in *Biographisches Lexikon der hervorragenden Aerzte der letzten fünfzig Jahre*, 2 v. (Vienna: Urban, 1933), v. 2, 1182. Pauli was born in 1869 in Prague and studied there at the Deutsche Universität, where he was a student of the physiologist Franz Hofmeister, who carried the ideas of physical chemistry into the new areas of physiology, the physical properties of proteins, and the effect of salts on precipitation (Hofmeister ion series). At this time too Ernst Mach was professor of experimental physics, and was twice rector of the German University. Pauli was there in Prague during the nationalist disturbances of Mach's second rectorate, and he *promovierte* in 1884, the year Mach resigned the Rectorate (on Mach, see John T. Blackmore, *Ernst Mach: His Life Work and Influence* (Berkeley: University of California Press, 1972), 73–83. Pauli *habilitierte* again in internal medicine in Vienna but in 1913 he became *a.o.* and in 1919 *o. Professor* of bio-physical chemistry and director of the Institute for Colloid Chemistry in Vienna. Pauli was an admirer of Mach and of the physiologist Ewald Hering, also of Prague, Mach's associate, whom he often cites. According to Blackmore (l.c., 315), Pauli asked Mach to be godfather to his son, born 1900, later to become a well-known physicist and also called Wolfgang. Mach thanks Pauli senior for reading his proofs, those for *Erkenntnis und Irrtum*, as well as for the fourth (1902) edition of *Analysis of Sensations*. At the time of writing the essays referred to here, he had recently been doing this.

10 Wolfgang Pauli, *Physical Chemistry in the Service of Medicine: Seven Addresses* translated by Martin H. Fischer (New York: Wiley, 1907), iii. These translations have no footnotes or citations, of which there are a great many in the original publications.

Figure 11.3. Wolfgang Pauli, portrait photograph, n.d. From Speiser & Smekal, *Karl Landsteiner* (1975) (n. 60). (Courtesy of Gebrüder Hollinek, Vienna)

Max Verworn – which is not uncommon among the colloid chemists.[11] Verworn defines it as

the view that the life processes rest basically on physico-chemical factors but that the conditions in the living organism are so complex that they have up to now not been elucidated. These complex conditions which are peculiar to living

11   Pauline M. H. Mazumdar, "The antigen–antibody reaction and the physics and chemistry of life," *Bull. Hist. Med. 48*(1974):1–21.

organisms in contrast to inorganic nature, one may for the present call the life force.[12]

In an essay of 1899, Pauli writes, "The methods of physical chemistry ... unquestionably enlarge that territory which the organic and the inorganic world have in common. The last barriers between the two cannot as yet be broken down, however. ... There always remains an unsolved portion, the kernel, as it were, of vital phenomena."[13] The unwillingness of the proteins of the protoplasm to submit to the gas laws, or to any of the laws governing chemical reactions, to the law of definite proportions, or the law of mass action led to the highly stimulating idea that the life processes might belong to a different universe, whose laws were those of colloid chemistry.[14] Landsteiner writes, "It seems then, that this extraordinary type of reaction plays a particularly large part in living organisms; living substance is mostly made up of colloids, whose reactions often conform to no fixed proportions."[15]

It is interesting to note that at this time the theory of the structure of colloidal gels was still that of Carl von Nägeli, who had suggested that they were composed of micelles or giant molecules united together into a ramifying gel.[16] This interpretation was supported by Jacob van Bemmelen against the alternative theory of a regular honeycomb structure, proposed by Otto Bütschli of Heidelberg. Nägeli's theory was finally confirmed by X-ray diffraction studies.[17]

The micellar structure of protein crystals ("crystalloids," Nägeli called them, using the word in quite the opposite sense ascribed to it by Thomas Graham in his founding document)[18] – these micellar structures, or gels, were the first stage of Nägeli's scheme of the evolution of living protoplasm from the non-living. During that gradual

---

12   Max Verworn, *Allgemeine Physiologie: ein Grundriss der Lehre vom Leben* (Jena: Fischer, 1st ed. 1894; 5th ed. 1909), 50 (5th).

13   Wolfgang Pauli, *Ueber physikalisch-chemische Methoden und Probleme in der Medicin: Vortrag, gehalten den 10 November 1899 in der Gesellschaft der Aerzte in Wien* (Vienna: Perles, 1900), and in Pauli, *Physical Chemistry in the Service of Medicine* (1907) (n. 10), 1–23 (p. 23).

14   Mazumdar, "Antigen–antibody reaction" (1974) (n. 11), 16.

15   Landsteiner & v. Jagić, "Verbindungen und Entstehung der Immunkörpern," (1903) (n. 3) 767.

16   Partington, *History of Chemistry* (1964) (n. 4), 738–739.

17   John S. Wilkie, "Nägeli's work on the fine structure of living matter, III. Studies tending to confirm, correct and develop Nägeli's views. i. Optical studies," *Ann. Sci.* 16(1960):209–239. Wilkie does not give any attention to the colloid chemists, but, as he says, concentrates on the botanists, Nägeli's direct successors.

18   Thomas H. Graham, "On the diffusion of liquids: Bakerian Lecture to the Royal Society of London December, 1849," *Phil. Trans.* 140(1850):1–46; Graham, "Liquid diffusion applied to analysis," *Phil. Trans.* 151(1861):183–199, 183. These two papers founded the science of colloid chemistry by distinguishing the properties of crystalloids from those of colloids.

transition, the inorganic molecule became ever more elaborate and complex until it formed a simple unicellular organism. As mentioned in Chapter 1, this theory was a development of Schleiden's cell theory, in which cells crystallized out in a nitrogenous mother-liquor. For Nägeli, the protein that formed not true crystals but "crystalloids" was a stage in the *quantitative Abstufung* between crystal and protoplasm, between matter and life.

The two essays of Wolfgang Pauli, cited by Landsteiner in his paper of 1903, are both very critical of Otto Bütschli's theory of the honeycomb structure of gels and conclude that physico-chemically there is no evidence of a two-phase system.[19] But Pauli does not mention Nägeli by name: he refers only to "a view expressed years ago and based on studies of the way in which water is held in gels."[20] Instead he refers to the separation of opposing reactions in living tissue, with the Hering–Mach interpretation of it. Landsteiner picks up this reference to Hering, but does not mention Mach.

It is this new, highly exciting colloid chemistry that Landsteiner now began to apply to the problems of immunology. He seems to have met with Pauli and with colloid chemistry at almost the same time: at least, he mentions neither before 1903. Pauli, in his essay of 1902, refers to "the beautiful investigations of Bordet, Ehrlich, and Landsteiner."[21] The following year, he gave a "festival address" to the k. k. Gesellschaft der Aerzte, which celebrated Landsteiner's colloid-chemical interpretation of the antigen–antibody reaction, and his re-interpretation of Ehrlich's results.[22] Landsteiner's first colloid-chemical paper describes what was for him a new revelation. But he discusses it with the cool impersonality, the *sachliche Kühle* that contrasts, in Erna Lesky's phrase, with the excitement of his *temperamentvoller einstiger Lehrer*, Max Gruber, when he and von Pirquet had got hold of a stick with which to beat Ehrlich.[23]

Landsteiner nevertheless manages to suggest that a new era is beginning for him. In colloid chemistry it is clear that he felt he had found

19　Wolfgang Pauli, "Der kolloidale Zustand und die Vorgänge der lebenden Substanz," Address to Morphologisch-physiologische Gesellschaft in Vienna, 13 May 1902; in his *Physical Chemistry in the Service of Medicine* (1907) (n. 10), 44–71 (p. 57).

20　Pauli, "Kolloide Zustand und lebende Substanz" (1902) (n. 20) (p. 57, below).

21　Wolfgang Pauli, "Allgemeine Physiko-chemie der Zellen und Gewebe," in *Ergebn.d. Physiol.* 1 1 Abt. (Biochemie) (1902):1–14; and in *Physical Chemistry in the Service of Medicine* (1907) (n. 10), 23–44 (p. 38). This essay and that referred to in n. 19 are the two cited by Landsteiner.

22　Wolfgang Pauli, "Wandlungen in der Pathologie durch die Fortschritte der allgemeinin Chemie," Festival Address at the 3d Annual meeting of the k. k. Ges. d. Aerzte in Vienna, 24 March 1905; in *Physical Chemistry in the Service of Medicine* (1907) (n. 10), 101–137.

23　Erna Lesky, *Die Wiener medizinische Schule im XIX Jahrhundert* (Graz: Bohlaus, 1965), 600.

the tool that would best fit his hand. The *quantitative Abgestufte* nature of the field itself and its processes, the generality of its laws, as much as their applicability to the peculiar case of the antigen–antibody reaction – all made it the ideal alternative to the affinity chemistry of the Ehrlich theory. Immune specificity, he suggested, could be regarded as the sum of a number of component reactions, in themselves nonspecific.[24]

Landsteiner's next step was to investigate the reactions of the "immune bodies," by which he means agglutinins and precipitins, in terms of a colloid-chemical analogy. The paper appeared in a short preliminary form in the Vienna medical weekly in January of 1904, obviously to secure the priority for what he knew was a great idea. A second version, more fully worked out but still highly speculative, appeared in July 1904 in the Munich weekly.[25] We quickly find him too, writing to a journal that had published a paper by the Viennese chemist Jean Billitzer on the subject, and had omitted the footnote acknowledging Landsteiner's priority.[26]

Landsteiner's model of the immune reaction was the precipitation of protein, and the agglutination and haemolysis of bacteria and red cells, by silica gel. He found a parallel for Eisenberg and Volk's quantitative results in the relationship between the concentration of the silicate, and the amount of it absorbed by the cells or proteins. The reaction, like the toxin–antitoxin reaction, was only partly reversible, and there was even a certain degree of specificity, in that bacteria were not agglutinated while spermatozoa were very sensitive to even minute amounts of colloid.

Unlike Arrhenius's theory, which had appeared at almost the same time, this physico-chemical alternative to Ehrlich's specific affinities found immediate support: the physical chemists working in the new field of colloid chemistry at once adopted it, as did Walther Nernst, in some ways Wilhelm Ostwald's successor as leader of the field. Landsteiner was able to cite many influential names: Heinrich Zangger, Wilhelm Biltz, and Georg Bredig – all important figures in Wolfgang Ostwald's history of colloid chemistry of 1909, particularly Biltz and

24 Landsteiner and v. Jagić, "Verbindungen und Enstehung der Immunkörpern" (1903) (n. 3), 768.
25 Karl Landsteiner and Nikolaus von Jagić, "Ueber Analogien der Wirkung kolloidaler Kieselsäure mit der Reaktionen der Immunkörper und verwandte Stoffe," *Wiener klin. Wschr.* 17(1904):63–64, C. 34. Karl Landsteiner and Nikolaus von Jagić, "Ueber Reaktionen anorganischer Kolloide und Immunkörper," *Münch. med. Wschr.* 51(1904):1185–1189, C. 39.
26 Karl Landsteiner, "Bemerkung zur Mitteilung von Jean Billitzer, 'Theorie der Kolloide, II,'" *Z. f. physik. Chem.* 51(1905):741–742, C. 45; see n. 44 for ref. to Billitzer's paper.

Bredig.[27] Nernst suggested a colloidal model for the immune reaction without referring to Landsteiner, and probably independently of him, in the same paper in which he did so much to destroy Arrhenius's theory.[28] He referred only to the chemist Wilhelm Biltz. Biltz, however, in turn, did Landsteiner the courtesy of this footnote:

Historically, the following should be noted: the first to go into the question of how far serotherapeutic reactions form "compounds of variable proportions" was, as far as I know, Bordet. . . . The first to recognise that there was a far-reaching analogy between immunological and similar phenomena, and those of adsorption in inorganic colloids, were Landsteiner . . . and Zangger. . . . Lively interest has lately been shown in this relationship with colloid chemistry by Neisser, Friedemann and Bechhold. . . . [29] Of all these authors, Landsteiner has undoubtedly expressed himself most clearly and as a result of these views he has also produced much new research material.[30]

Biltz now replotted the quantitative results of Eisenberg and Volk in the form of an adsorption curve, and to make the parallel as clear as possible placed by the side of the curve of typhoid serum against bacteria, some curves of the adsorption process in colloids (Figure 11.4). Biltz's curves included one for the adsorption of (crystalloidal) hydrochloric acid by colloidal stannic acid, one for the adsorption of an organic colloid, benzopurpurin, by another colloid, aluminum hydroxide; and one for the adsorption of an inorganic colloid, molybdenum blue, by silk, which Biltz interprets as adsorption of an inorganic by organic colloid; it is also a dyeing process.

Biltz plots the first of these, the curve of hydrochloric against colloidal stannic acid, from Jacob van Bemmelen's data. One of the ear-

27 Wolfgang Ostwald, *Grundriss der Kolloidchemie* (1909) (n. 4), Biltz has twenty-three
    page references in Ostwald's index, and Bredig twenty-two; Zangger has seven ref-
    erences only, but figures in Ostwald's list of the most important chemists interested
    in biological and physiological applications of colloid chemistry, along with himself
    and Wolfgang Pauli (p. 71). His list of those making important applications of the
    ideas of colloid chemistry to immunity included Bordet, Zangger, Biltz, and Land-
    steiner and Pauli, with some others (p. 71).
28 H. Walther Nernst, "Ueber die Anwendbarkeit der Gesetze des chemischen Gleich-
    gewicht auf Gemische von Toxin und Antitoxin," *Z. f. Elektrochem.* *10*(1904):377–380
    (p. 379).
29 Max Neisser was head of Bacteriology and Hygiene under Paul Ehrlich at Frankfurt;
    Heinrich Bechhold was a chemist also with Ehrlich at Frankfurt; Ulrich Friedemann
    was assistant under Neisser at Stettin, and later with Ehrlich in Frankfurt; he wrote
    many papers on immunology. All three of these men appear in Ehrlich's Festschrift
    (Hugo Apolant, et al., *Paul Ehrlich eine Darstellung seines wissenschaftlichen Wirkens: Fest-
    schrift zum 60 Geburtstage des Forschers* (Jena: Fischer, 1914). It is significant these three
    close associates of Ehrlich were, as Biltz says, interested in colloid models.
30 Wilhelm Biltz, "Ein Versuch zur Deutung der Agglutinierungs-vorgange," *Z. f. physik.*
    Chem. 48(1904):615–623 (p. 1, n. 4).

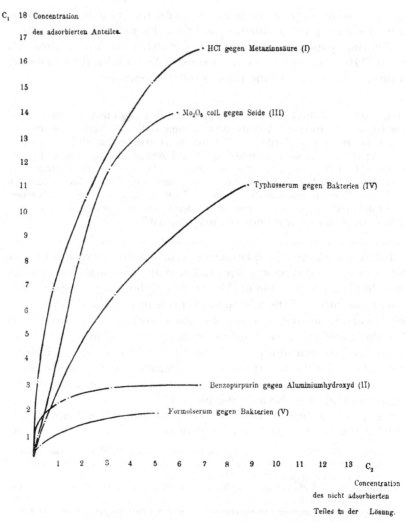

Figure 11.4. The analogy between colloid phenomena and those of immunity (1904). Wilhelm Biltz compares a plot of the published results of Eisenberg and Volk for typhoid bacteria agglutinated by their antiserum with those for the precipitation of inorganic colloids. From Biltz, "Versuch zur Deutung der Agglutinierungs-vorgange," (1904) (n. 30), 619. (Photograph by Instructional Media Services, University of Toronto)

liest examples of this type of curve was produced by Wilhelm Ostwald in 1885 for the adsorption of hydrochloric acid by animal charcoal.[31] Ostwald found that the adsorption was regular and quantitative, and that the process was completely reversible; he therefore supposed it to be a purely physical process, a *mechanische Affinität*, and he likened it to dyeing. Van Bemmelen on the other hand inclined at first to a chemical theory. The controversy that was taking place here, as to the chemical or physical nature of the process, and the analogy with dyeing, was the same in form as the controversies of the later period, about the nature of the immune reaction.

Landsteiner's argument now moved forward rapidly: he and his co-workers quickly found that it was the charged colloids, the acidic and basic ones, that precipitated each other, and the proteins and cells upon which they tested them. Salt formation was clearly of some importance. The inorganic colloids also moved in an electric field towards the pole in the direction suggested by their acidity and basicity. He was able to cite Jean Billitzer on this. Colloids that moved in opposite directions consisted of particles with opposite charge. If this picture was transferred to amphoteric organic colloids, two of these, if charged in opposite senses, or even just to different degrees, could influence each other's electrical behaviour and form salt-like compounds at the same time.

Arguing like this, he came to the conclusion that

From this point of view there is no sharp boundary line separating the reactions of the immune bodies from chemical processes between crystalloids, just as in nature there exists every stage between crystalloid and colloid. The nearer the colloid particle approximates to that of the normal electrolyte, the nearer its compounds must obviously come to conforming to the law of simple stoichiometric proportions, and the compounds themselves to simple chemical compounds. At this point it should be recalled that Arrhenius has shown that the quantitative relationship between toxin and antitoxin is very similar to that between acid and base.[32]

Landsteiner, unlike Nerst, found Arrhenius's parallel with salt formation very attractive. He reminds us more than once of Arrhenius's suggestion that the quantitative relationship between antigen and an-

31 Partington, *History of Chemistry* (1964) (n. 4), 740; Wolfgang Ostwald, *Grundriss der Kolloidchemie* (1909) (n. 4), 409–417, talks about this problem and its history; as usual for him, he uses neither mathematics nor curves. He quotes from his father's textbook (2d ed., 1890), and says that the data were obtained in 1885 and had remained unpublished till 1890. Charcoal in the form of biscuits was a long-standing remedy for gastric problems, now seen to be acting by the adsorption of excess stomach acid.
32 Landsteiner and v. Jagić, "Ueber Reaktionen anorganischer Kolloide und Immunkörper" (1904) (n. 25), 1188–1189.

tibody could be compared with the relationship between acid and base.[33] In this respect, as in many others, Landsteiner, like Grassberger and Schattenfroh, is a student of Gruber.

His final conclusion in this almost entirely speculative paper leaves no doubt that his intention is to provide a complete electrochemical explanation for immunological specificity: "It is conceivable that quantitative changes [*Abstufungen*] in acidity and basicity may be important in the relations between the immune bodies and the relations of protein substances with each other, just as they are in elective or specific staining of animal tissues."[34]

As early as 1897 Arrhenius suggested that "chemical affinity depends on the electric properties of atoms, and their attraction is due to their electric charges."[35] Billitzer's connection of the acidity and basicity of inorganic colloids with their electrical behaviour made it possible for Landsteiner to apply Arrhenius's words to immune specificity.

Landsteiner's extension of colloid chemistry into the field of immunology found an immediate response among the colloid chemists, many of whom were already working around the edges of this idea themselves. The conception of immune affinity as a resultant of electrochemical charge was being approached from all sides. Landsteiner's first colloid paper of May 1903 was a preliminary communication written specially to secure the priority.[36] It was an important priority that he was later very careful to insist on.[37] His second colloid paper appeared in January 1904.[38] The next came out in July.[39] Bredig in 1901 in his book *Inorganic Ferments* had already called the precipitation of colloids a *kapillärelektrische Phänomen*.[40] In 1904, Biltz suggested that colloids with opposite charge precipitated each other by neutralizing each other's charge.[41] Billitzer suggested that the ions of added electrolytes precipitated the charged colloid particles in the same way, by neutral-

33 Landsteiner and v. Jagić, idem (n. 27), 1189, as well as in Karl Landsteiner and M. Reich, "Ueber die Verbindungen der Immunkörper," *Centr. Bakt. Orig.* *39*(1905): 83–93, 84; and other places.

34 Landsteiner and v. Jagić, "Ueber Reaktionen anorganischer Kolloide und Immunkörper" (1904) (n. 27), 1189.

35 Svante Arrhenius, *Textbook of Electrochemistry* (London: Longmans, 1902), 19; the English translation by J. McCrae was from the German translation of 1901, which in turn was a revised version of lectures given in Swedish in 1897 (from preface by the author for the English edition).

36 Landsteiner and v. Jagić, "Verbindungen und Entstehung von Immunkörpern" (1902) (n. 3).

37 Landsteiner, "Bemerkung zur Mitteilung von Billitzer" (1905) (n. 26).

38 Landsteiner and v. Jagić, "Analogien der Wirkung kolloidaler Kieselsäure" (1904) (n. 25).

39 Landsteiner and v. Jagić, "Reaktionen anorganischer Kolloide und Immunkörper" (1904) (n. 25).

40 Georg Bredig, *Anorganische Fermente* (Leipzig: Engelmann, 1901).

41 Biltz, "Deutung der Agglutinierungs-vorgange" (1904) (n. 30).

ization.[42] Landsteiner's speculations were supported by this work, and accepted in turn by these men, chemists working outside Landsteiner's field.

The Ehrlich group too plunged, though rather hesitantly, into colloid chemistry: Bechhold, Neisser, and Friedemann from Ehrlich's Institute in Frankfurt presented a joint paper to the *Naturforscherversammlung* in Kassel in September 1903 demonstrating that the "salting-out" (*Ausflockung*) of colloids and of bacteria depended on the valency of the ions used.[43] The members of the Gesellschaft Deutscher Naturforscher und Aerzte, as its name suggests, included both scientists and physicians, and both must have responded to this paper. Because of the different professional interests involved, says Bechhold, they decided to split their work into two long papers: one for the *Zeitschrift für physikalische Chemie*, Ostwald and van't Hoff's journal; the other for a medical journal, the *Münchener medizinische Wochenschrift*.[44]

But the Ehrlich group did not go on to a complete acceptance of physico-chemical explanations of immune specificity. Rather, after this brave start they seemed to withdraw from the field. The next episode, and it is tempting to see it at least in part as the cause of the withdrawal, was Pauli's "Festival lecture" to the k. k. Gesellschaft der Aerzte in Wien of 1905, in which he enthusiastically presented colloid chemistry as the new explanation of almost everything in medicine and biology. Ehrlich's structural chemistry was outdated: all that was new and brilliant in modern pathology was a result of the application of the methods of colloid chemistry. Pauli pictured Landsteiner as providing the solution to the problems first crudely perceived by Ehrlich, a solution vaguely and intuitively approached by Bordet in his dye analogy, but now made plain by Landsteiner: "Landsteiner was the first who, independently and recognizing the goal toward which he was travelling, studied the connection between immunity and colloidal reactions experimentally."[45]

Specific substances serve only to give colloidal particles the charge

42 Jean Billitzer, "Theorie der Kolloide, II.," *Z. f. physik. Chem.* 51(1905):129–166; see n. 28 for Landsteiner's remarks on this paper.

43 Heinrich Bechhold, "Die Bakterienagglutination ein physikalisch-chemisches Phänomen: Vortrag, in Gemeinschaft mit Prof. Dr Max Neisser und Dr Friedemann," *Verh. d. Ges. Deutsch. Naturf. u. Aerzte,* 75 Versammlung, Cassel, 1903 (Leipzig, Vogel 1904), pt. 1, 487–488.

44 Heinrich Bechhold, "Die Ausflocung von Suspensionen bzw. Kolloiden und die Bakterienagglutination," *Z. f. physik. Chem.* 48(1904):385–423 (p. 385, n. 1). Max Neisser and Ulrich Friedemann, "Studien über Ausflockungs-erscheinungen, I.," *Münch. med. Wschr.* 51(1904):466–469; Neisser and Friedemann, "Studien über Ausflockungs-erscheinungen, II. Beziehungen zur Bakterienagglutination," *Münch. med. Wschr.* 51(1904):827–891.

45 Pauli, "Wandlungen in der Pathologie durch allgemeinen Chemie" (1905) (n. 22), 128.

and size necessary for precipitation. Apparently, says Pauli, all immune sensitizing reactions can be explained in the same way. Even bones can be considered to be colloids: he and Ludwig Mach, Ernst Mach's son, had performed an experiment in which a protein packed into a steel tube and heated at high pressure, became hard as a bone.[46] Pauli is extravagant, enthusiastic, vitalistic, and delightful to read. He calls Bordet "brilliant" and several people's experiments "beautiful" (including both Ehrlich's and Landsteiner's, in another paper),[47] and he makes Gruber and even the sober *sachlicher* Landsteiner have intuitions of vitalism, too.[48] This must have been a splendid lecture to listen to: one can hardly resist clapping on reading it to the end.

Pauli's "Festival lecture" celebrating the sunrise of colloid chemistry and of Landsteiner, and the *Untergang* of structural chemistry and Ehrlich, was answered from Frankfurt by Heinrich Bechhold, with buckets of cold water and common sense. Bechhold pointed out that organic chemistry, not physical chemistry, had been dominating the field and that modern pathology and bacteriology had been dependent on the chemistry of stains, that is, on Ehrlich's chemistry and Ehrlich's stains. He called Pauli's a one-sided explanation of a whole string of different processes, which in the end would only discredit the cause of physical chemistry, and he compares Pauli himself to a physician who thinks he can cure all diseases with a single drug:

Above all, I do not think it a very happy idea to try to explain immune specificity solely in terms of variations [*Abstufungen*] in charge. . . . I do not claim that organic chemistry with its structural or configurational formulae has found the most exquisite expression of all the phenomena shown by a substance in its reactions, but it was certainly a good idea of Ehrlich's to apply this thinking in the organic world. Where would it get us if we were to classify all the innumerable phenols, alcohols, acids, amines etc., only in terms of their conductivity? It would explain very little about them.[49]

Pauli's reply reveals the extent to which this division into opposing camps was overt and self-conscious. It shows the importance of this dialectic to the participants, and the degree to which they themselves saw their attitudes being determined by the group to which they belonged.

46 Pauli, "Wandlungen in der Pathologie durch allgemeinen Chemie" (1905) (n. 22), 123–133.
47 Pauli, "Allgemeine Physiko-chemie der Zellen" (1902) (n. 10), 38.
48 Pauli, "Wandlungen in der Pathologie durch allgemeinen Chemie" (1905) (n. 22), 135.
49 Heinrich Bechhold, Review of Pauli, "Wandlungen in der Pathologie durch allgemeinen Chemie" (1905) (n. 22), in *Wiener klin. Wschr.* 18(1905):550–551, 551.

Like Pfeiffer's remark at Wiesbaden in 1896, that Gruber was Nägeli's student, and so could not be expected to believe in specificity,[50] or Grassberger and Schattenfroh's remark that Arrhenius's work gave succour to Ehrlich's enemies,[51] Pauli's response to Bechhold's defense of Ehrlich demonstrates his participation in the *agon*:

> In my lecture . . . I tried to bring together the work of a few investigators especially that of Landsteiner and Biltz, which shows how colloid-chemistry touches the experience we have of the behaviour of the immune bodies. This line of approach is necessarily in conflict with the thinking of the Ehrlich school, which puts down the peculiar qualities of toxins and antibodies exclusively to their structure, that of groups combining different ways. It is not surprising, then, that a critique of my lecture by Herr Bechhold, a member of the Frankfurt Institute, should be an expression of this conflict.[52]

Pauli followed this with a more overt and less restrained attack on Ehrlich than he had presented in the original lecture. Bechhold now answered with an essay called, "Unsolved Questions on the Part Played by Colloid Chemistry in the Problems of Immunity."[53] According to Pauli, and later Landsteiner, Ehrlich had failed to show the existence of his supposed chemical groups, which were only a kind of fiction.[54] Bechhold saw this as an unfair criticism: "Emil Fischer explained the working of enzymes on glycosides through their stereochemical form, in that they matched each other like 'lock and key.' This is what is meant by Ehrlich's 'haptophore' groups: no-one can accuse Emil Fischer of messing about with 'fictions.' "[55]

Just as Bechhold and Pauli were now engaged in combat, Landsteiner attacked Bechhold's Frankfurt colleague, Ulrich Friedemann. Landsteiner summarized a paper by Friedemann for the literature survey that appeared in the *Zentralblatt für Physiologie*. He presented it as a simple confirmation of his own results on the precipitation of inorganic colloids and of the amphoteric character of proteins, and he took the opportunity, rather unusual in such summaries, to point out an error

50 Richard Pfeiffer, in discussion with Max von Gruber, "Ueber active und passive Immunität gegen Cholera und Typhus," *Verh. d. Kongr. f. inn. Med.* (XIII Kongress, Wiesbaden, 1896):207–220 (p. 220).

51 Roland Grassberger and Arthur Schattenfroh, *Ueber die Beziehungen von Toxin und Antitoxin* (Leipzig: Deuticke 1904), Vorwort, v.

52 Pauli, "Ueber den Anteil der Kolloidchemie an der Immunitätsforschung," *Wiener klin. Wschr. 18*(1905):665–666.

53 Heinrich Bechhold, "Ungelöste Fragen über den Anteil der Kolloid-chemie an der Immunitätsforschung," *Wiener klin. Wschr. 18*(1905):666–667.

54 Karl Landsteiner and R. Stanković, "Ueber die Adsorption von Eiweisskörpern und ber Agglutinin-verbindungen," *Zentralbl. f. Bakt. 41*(1906):108–117, (p. 115).

55 Bechhold, "Ungelöste Fragen" (1905) (n. 60), 667.

in Friedemann's interpretation of his, Landsteiner's, own work.[56] Ulrich Friedemann and his co-worker Hans Friedenthal returned the compliment by attacking Landsteiner:

L. sees the immune reactions as salt-type linkages with amphoteric colloids. The comparison of colloidal with ionic reaction (salt-formation) seems inappropriate to us, since the sole similarity between them is that both imply equalization of electric charge. . . . Further, we do not agree with Landsteiner that the immune reactions are those of amphoteric colloids.[57]

Landsteiner answered this once. But after the next reply, he simply dismissed the matter, saying that they had failed to understand it.[58]

Thus, though the Ehrlich group in Frankfurt were not untouched by the rage for colloid chemistry, loyalty and the attacks of the opposition combined to put them in a position hostile to it. Where Pauli claimed that almost everything in nature could be explained by electrochemistry, and Landsteiner, that electrochemical forces were responsible for specific affinity, the Ehrlich group, though they did not reject colloid chemistry as a whole, saw it as an explanation for only the second stage, the precipitation, involved in the immune reaction, and not for the primary selection of antigen by antibody. The primary problem was for them still unsolved, as Bechhold had said, but it was likely to be a matter of chemical structure.[59]

One of the most interesting points in Bechhold's review of Pauli's lecture, is his interpretation of a new piece of evidence, the work of a Viennese dye-chemist, Wilhelm Suida. Suida was a student of the chemist Ernst Ludwig, who was *Ordinarius* for Applied Medical Chemistry in Vienna from 1874. Landsteiner had also studied under Ludwig during his medical training.[60]

56 Ulrich Friedemann, "Ueber die Fallung von Eiweiss durch andere Colloide und ihre Beziehungen zu den Serumkörperreaktionen," *Arch. f. Hyg.* 55(1906):361–389; summarized (*referiert*) in *Zbl. f. Physiol.* 20(1906):171, by Karl Landsteiner. The *Zentralblatt für Physiologie* ran a review of the literature that appeared fortnightly. Authors who wanted their papers to be dealt with were asked to send their reprints to the editors, Alois Kreidl and Otto van Fürth of Vienna and René du Bois-Reymond of Berlin. Kreidl was a Vienna physiologist associated with Ernst Mach, who wrote an obituary for Mach in 1916.
57 Ulrich Friedemann and Hans Friedenthal, "Beziehung der Kernstoffe zu den Immunkörpern," *Zbl. f. Physiol.* 20(1906):585–587.
58 Karl Landsteiner, "Bemerkung zu der Mitteilung von U. Friedemann und H. Friedenthal, 'Beziehungen der Kernstoffe zu den Immunkörpern,' " *Zbl. f. Physiol.* 20(1907):657–658, C. 67. Karl Landsteiner, "Zu der Erwiderung von Friedemann und Friedenthal," *Zbl. f. Physiol.* 20(1907):806, C. 69.
59 Bechhold, "Ungelöste Fragen" (1905) (n. 53), 667.
60 Paul Speiser and Ferdinand G. Smekal, *Karl Landsteiner, Discoverer of the Blood Groups and Pioneer in the Field of Immunology: Biography of a Nobel Prize Winner from the Vienna Medical School,* trans. Richard Rickett (Vienna: Hollinek, 2d ed., 1975), 17, 139, fig. 93.

According to Bechhold, Suida acylated or alkylated wool and silk fibres. The colloidal character of the fibre was unchanged, but the acidic groups were altered chemically, so that whereas the wool and silk usually took up colour well, they now resembled cotton. That is, they took up the dye, thanks, says Bechhold, to their colloidal nature – but owing to the chemical change in the fibre it was not held fast and was easily washed out again. The change could be reversed and the fibres returned to their original easily-dyed state by soaping out the acetyl or acyl groups.[61] Here, says Bechhold, we can see how important the chemical groups are for the binding of the two colloids: the fastness of the dye depends on the chemical reaction, which is blocked by acylation.[62]

Landsteiner, too, was interested in Suida's experiment, but for him it seemed to lead to the opposite conclusion.[63] For Landsteiner, "chemical affinity depend[ed] on the electrical properties of atoms, and their attraction [was] due to their electric charges," as Arrhenius had taught.[64] An electrical explanation therefore did not conflict with a chemical one: they amounted to the same thing: "The dependence of adsorption on the chemical nature of the adsorbing substance . . . leads to the conclusion that chemical reactions are involved in these processes. But this is in no way incompatible with the idea that there is an electrical attraction operating."[65]

And Landsteiner adduces Suida's work in support of this notion. Suida had blocked the acid groups that took up basic dyes, so that his fibres were no longer coloured by them. Landsteiner repeated Suida's experiment using the protein casein instead of silk and wool fibres, and found that where it was originally coloured by basic crystal violet, and only slightly by acid Ponceau red, after its acid groups were blocked by acylation and alkylation, the position was reversed. The

61 Wilhelm Suida, "Ueber das Verhalten von Teerfarbstoffen gegenber Stärke, Kieselsäure und Silikaten," lecture to the Royal Academy of Sciences, Vienna, Mathematical-Scientific Section, 16 June 1904, and published in *Monatsh. f. Chem.* (Vienna), 25(1904):1107–1143; Suida, "Ueber den Einfluss der aktiven Atomgruppen in den Textilfasern auf das Zustandgekommen von Färbungen," lecture to the same, 12 January 1905, and in *Monatsh. f. Chem.* (Vienna), 26(1905):413–427; P. Gelmo and Wilhelm Suida, "Studien über die Vorgänge beim Färben animalischer Textilfasern," lecture, 25 October 1906, and in *Monatsh. f. Chem.* (Vienna), 27(1906):1193–1198.
62 Bechhold, Review of Pauli, "Wandlungen in der Pathologie durch allgemeinen Chemie" (1905, n. 22) (1905) (n. 49), 551.
63 Landsteiner and Stanković, "Adsorption und Agglutiniverbindungen" (1906) (n. 54), C. 55.
64 Arrhenius, *Textbook of Electrochemistry* (1902) (n. 35), 19.
65 Landsteiner and Stanković, "Adsorption und Agglutinin-verbindungen" (1906) (n. 61), 108.

same applied to the uptake of the plant toxin abrin, Landsteiner wrote triumphantly:

It can therefore hardly be doubted that the reaction between abrin and casein is a process which is quite similar to that of the uptake of abrin and other agglutinins by blood cells. These experiments constitute an important new support for the concept that we have put forward, that the combination of the immune bodies is of a salt-like nature, and is dependent on the amphoteric character of these substances. As we have said before, these reactions are similar to those of inorganic colloidal acids and bases, and to many dyeing processes (cf. Bordet). The difference between the view expressed here and previous hypotheses on the reactions of the immune bodies, is that in ours, the actual nature of the chemical processes that occur is given a definite, experimentally testable form, while earlier the actual nature of the supposed chemical process was never even discussed.[66]

It was, of course, Ehrlich who never discussed the actual nature of the supposed chemical process.

The first experimental test to be made of this hypothesis as to the nature of the supposed chemical process was Landsteiner and Pauli's investigation of the charge properties of the immune substances. Wolfgang Pauli was a year younger than Landsteiner; he had his *Habilitation* in 1899, three years before Landsteiner in 1903. He reached acting *ausser-ordentlicher* professor in 1908, while Landsteiner reached *a.o.* in 1909. But in 1919 their paths diverged: Pauli became full Professor and Director of the Institute for Medical Colloid Chemistry in Vienna, and Landsteiner, despairing of any further advancement, began planning to leave Vienna. In 1902 the enthusiastic and charismatic Pauli was already on the way to becoming a leading authority on the physical chemistry of biological processes. He had contributed the opening essay on the foundation of the new journal *Ergebnisse der Physiologie*, on the general physical chemistry of cells and tissues. The first words of this essay, and the first reference in the literature he cites – which oddly, begins instead of ends the article – pay homage to Ernst Mach. Mach's fundamental work on the theory of knowledge has freed natural science from the ballast of pseudo-problems and extreme positions, such as the conflict between vitalism and mechanism, and Pauli states what appears to be the manifesto of the new journal:

In the realm of general physiology the physico-chemical method of looking at things has been the first to make it possible to ask many questions in a general way and to answer them according to the present status of physico-chemical

66 Landsteiner and Stanković, "Adsorption und Agglutinin-verbindungen" (1906) (n. 54), 115.

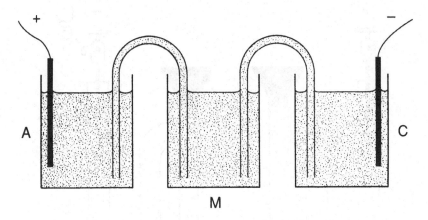

Figure 11.5. Landsteiner and Pauli's apparatus for the investigation of charge in proteins. First version, 1906. A is the anode (+) and C the cathode (-); the protein solution goes in the middle beaker, M. Negatively charged material moves towards the positive pole when the current is passed. Illustration from Pauli, *Kolloidchemie der Eiweisskörper* (1920) (n. 78), 19. (Re-drawn and photographed by Instructional Media Services, University of Toronto)

investigation. New analogies and transitions between phenomena in living and dead matter have been discovered; and it has often proved no small task to discover that side of a phenomenon which characterizes it as a specifically biological one.[67]

Pauli himself had been working on the movement of proteins in an electric field with a simple apparatus of three beakers connected by siphons (Figure 11.5). The protein solution was put in the middle beaker and the current passed for some time. The siphons were then taken out and the protein concentration in the anode and cathode cells determined by a nitrogen estimation using Kjeldahl's method. After all electrolytes were removed by dialysis against distilled water for several weeks, albumin moved only slightly, on prolonged electrophoresis. But the addition of a trace of acid to the fluid made it move quickly to the negative pole, and vice versa if alkali was added.

There were still some practical problems, however. When the protein came into contact with the electrical poles, it took on the charge of the pole and started to move back in the opposite direction. This made

67 Wolfgang Pauli, "Allgemeine Physiko-chemie der Zellen" (1902) (n. 21, p. 24), n. 1 in Pauli's list is Ernst Mach, *Analyse der Empfindungen* (3d ed., 1902); Martin Fischer's translation of this paper in Pauli, *Physical Chemistry in the Service of Medicine* (1907) (n. 10), lacks the literature citations.

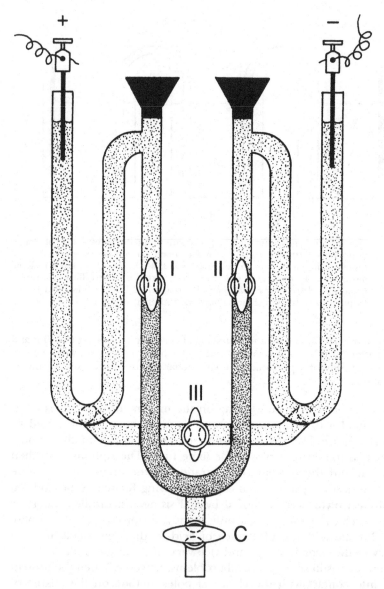

Figure 11.6. Landsteiner and Pauli's apparatus for the investigation of charge in proteins. Improved version of 1908. In this version, materials are introduced at C. Negatively charged substances move towards the positive pole and positively towards the negative, as before. But now test substances do not come in contact with the poles, and only the boundary at I and II actually moves. Illustration from Pauli, *Kolloidchemie der Eiweisskörper* (1920) (n. 78), 22. (Redrawn and photographed by Instructional Media Services, University of Toronto)

it impossible to investigate anything that was weakly charged and moved slowly, as the return flow spoiled the results.[68]

Landsteiner and Pauli then improved the apparatus, on Pauli's behalf, and used it to investigate the charge properties of immune substances, on Landsteiner's. In the new version, the test substances were kept away from the poles, and only a boundary between the protein and the electrolyte moved (Figure 11.6). In this version, materials are introduced at C. Negatively charged substances move towards the positive pole and positively towards the negative, as before (Figure 11.5). But now test substances do not come touch the poles, and only the boundary at I and II actually moves.

They presented this version at the Congress for Internal Medicine in Vienna in 1908, and used it to show that the plant agglutinins ricin and abrin and a chicken serum agglutinin were amphoteric colloids, with very slight inherent charge. Colloids with strong inherent charges such as silica gel and the metallic hydroxides naturally precipitated any kind of protein or cell non-specifically, while these subtly charged amphoteric substances were able, they suggested, to precipitate selectively and specifically.[69]

This interpretation naturally did not convince everyone: a champion arose from the opposition to attack it. The Berlin physical chemist Leonor Michaelis could not accept that electrical properties were sufficient to account for specificity. He reviewed the problem in a new *Handbuch* on physical chemistry and medicine, of 1908, which, unlike the huge *Handbücher* on more ancient subjects such as bacteriology, had only two

68 Wolfgang Pauli, "Untersuchung über physikalische Zustandsänderungen der Kolloide. Fünfte Mitteilung: Die Elektrische Ladung von Eiweiss," *(Hofmeisters) Beiträge z. chim. Physiol. u. Path.* 7(1906):531–547; Pauli, *Kolloidchemie der Eiweisskörper* (Dresden: Steinkopf, 1920), translated by P. C. L. Thorne as *Colloid Chemistry of Proteins* (Philadelphia, Pa.: Blackiston's, 1922), figs. 2 and 3 (pp. 19, 22).

69 Karl Landsteiner and Wolfgang Pauli, "Elektrische Wanderung der Immunstoffe," *Verh. d. Kongr. f. inn. Med.* (XX Congress, Vienna, 1908), 571–574, C. 85. Moving boundary electrophoresis, using this improved apparatus, was taken up by Leonor Michaelis to distinguish bio-colloids from each other by means of their iso-electric points, sc., that point at which no movement occurs when a current passes (see n. 79). Michaelis was to become the most influential of the proponents of electrical potential or pH as a determining factor in colloid behaviour, a line of thought that gradually replaced the purely mechanical in colloid chemistry. See Pauline M. H. Mazumdar, "The template theory of antibody formation and the chemical synthesis of the twenties," in Mazumdar, ed., *Immunology 1930–1980: Essays on the History of Immunology* (Toronto: Wall, 1989), 13–32. Arne Tiselius used an apparatus similar to Landsteiner and Pauli's to fractionate protein components in mixtures such as serum by means of their pH-mobility relationships. Arne Tiselius, *The Moving Boundary Method of Studying the Electrophoresis of Proteins*, Inaugural Dissertation (Uppsala: 1930); Tiselius, "A new apparatus for electrophoretic analysis of colloidal mixtures," *Trans. Faraday Soc. 33*(1937):524–531.

slim volumes. There was no clear evidence, he said, that toxin and antitoxin moved in opposite directions in an electric field, nor was the toxin–antitoxin "compound" split up by electrolysis. It seemed that little progress had been made since Ehrlich first proposed that specific affinity should be explained as chemical affinity:

Everything we know points to the conclusion that specific affinity is chemical affinity in the purest sense, and is quite independent of electrical affinity, as Ehrlich has maintained from the beginning in spite of all the criticism. . . . Organic chemistry offers innumerable examples of direct chemical reactions in which no electrical phenomena are involved at all. It is a step backwards to try to make all reactions depend on opposite electrical charges, as Berzelius did.

Berzelius's antiquity makes Arrhenius and the modern electrochemists seem both less original and less advanced than they claimed to be. Michaelis concludes his article, as Bechhold had done before him, by calling on Emil Fischer, the man with the key:

As the cause of the specific affinity of toxin and antitoxin we must appeal to purely chemical forces, and we understand these in the sense originally meant by Emil Fischer for fermentations, by his image of "lock" and "key." The "keys" in the living organism are reaction products of the "lock" and can only be understood in terms of Ehrlich's side-chain theory.

Ehrlich's theory explains the origin of antibodies, their specificity, and the antigen-antibody reaction. It is still the most powerful theory.[70]

70 Leonor Michaelis, "Physikalische Chemie der Kolloide," in Alexander von Korányi and Paul Friedrich Richter, eds., *Physikalische Chemie und Medizin: ein Handbuch*, 2 v. (Leipzig: Thieme, 1908), v. 2, 341–453 (pp. 452–453). Michaelis was later to use the Landsteiner–Pauli apparatus to distinguish proteins from each other by their iso-electric points. See Michaelis, "Elektrische Ueberführung von Fermenten," *Biochem. Z. 16*(1908):81–86.

12
Ehrlich's Chemistry and Its Opponents: iii. The New Structural Chemistry of Landsteiner and Pick

In spite of the growing popularity of physical, and especially colloid, chemistry in the first decades of the century, the heuristic effectiveness of the Ehrlich theory was not exhausted. Ehrlich himself, from about 1902 onwards, had turned his attention away from research in immunology to the new area of chemotherapy. Immunology at the Frankfurt Institute was now carried on by the group, but no longer by its leader. The testing and standardization of sera on a large scale was the institute's most important function. Here, Ehrlich's serological methods were of the essence.

In Vienna, the policy of the State Serotherapeutic Institute stayed close to the Koch–Ehrlich research line, and to Ehrlich's methods of serum production and testing. Richard Paltauf was director of both the serum institute and the Institute for General and Experimental Pathology.[1] His invited contribution as Austrian representative to the *Koch-Festschrift* number of the *Deutsche medizinische Wochenschrift* of 1903, produced by Koch's "grateful students," is mainly a review of the work carried out by his institute staff on problems of agglutination and precipitation.[2] The group includes Philipp Eisenberg and Richard Volk,[3]

1 Erna Lesky, *Die Wiener medizinische Schule im XIX Jahrhundert* (Graz: Böhlaus, 1965), 577– 578, calls Paltauf "one of the great school-builders." In 1893, he was appointed to the *Prosektur*, the senior post in pathology, at the Vienna Rudolfspital, where he started the State Serotherapeutic Institute; from this nucleus, he built up an empire of experimental pathology in Vienna over the next fifteen years. In 1900 he became, in addition, professor and director of the Institute for General and Experimental Pathology. In 1908, when the Institute for Hygiene moved out of the old munitions factory into its new building, two institutes came under the same roof. Most of Paltauf's own work was on pathological anatomy, but as director of the serum institute, he was a supporter of the Ehrlich theory and of Ehrlich's practical methods. These methods were used generally in serum institutes across Europe.
2 Richard Paltauf, "Ueber Agglutination und Präcipitation," *Deutsche med. Wschr.* 29(1903):946–950.
3 Philipp Eisenberg and Richard Volk, "Untersuchung über die Agglutination," *Wiener klin. Wschr.* 14(1901):1121–1123; Eisenberg and Volk, Untersuchung über die Agglutination," *Z. f. Hygiene u. Infektionskr.* 40 155–195 (1902). These papers were the source

Rudolf Kraus,[4] and Friedrich Obermayer and Ernst Peter Pick.[5] He
even includes the two serum horses, Zoroaster and Elsa, who appear
also in Eisenberg and Volk's paper. The article might be regarded as
the homage of Paltauf and his students to Koch. The Ehrlich theory
was the working hypothesis of the entire Paltauf school.[6]

In 1906, this group produced a striking and literal justification of the
chemical basis of the side-chain theory. The clinical chemist Friedrich
Obermayer and the younger Ernst Peter Pick, under Paltauf's *aegis*, set
the perspective for their paper by reviewing the problem of species
specificity.[7] The extreme sensitivity of Kraus's precipitin reaction made
it possible to distinguish between the serum proteins of even very
closely related species, such as the hare and rabbit. Paul Uhlenhuth, a
graduate of Koch's Institute for Infectious Disease in Berlin, had described the procedure in his contribution to the *Festschrift* for Koch.[8]
Obermayer and Pick had been working together on the chemistry of
specificity for some time. In 1902 they had read a paper on it to the k.
k. Gesellschaft der Aerzte in Wien in which they were raising antisera
in rabbits by injecting them with bovine serum albumin altered in various ways.[9] At first they altered them by physico-chemical methods, such

of much basic information on the relation of antigen and antibody and were cited,
discussed, and replotted by many people. See Chapter 10, n. 36.

4   Rudolf Kraus, "Ueber spezifische Reaktionen in Keimfreien Filtraten aus Cholera-,
Typhus – und Pest-bouillon-culturen durch homologes Serum," *Wiener klin Wschr.*
10(1897):736–738; Kraus and L. Löw, "Ueber Agglutination, *Wiener klin. Wschr.*
12(1899):95–98, "Ueber Fadenbildung," *Wiener klin.Wschr. 12*(1899):761–764; Kraus
and P. Clairmont, "Ueber Hämolysme und Antihämolysine," *Wiener klin. Wschr.*
13(1900):49–56. These papers describe the discovery of immune precipitation *in vitro*.
5   Friedrich Obermayer and Ernst Peter Pick, "Ueber den Einfluss physikalischer und
chemischer Zustandsänderungen präcipitogener Substanzen auf die Bildung von Im-
munpraecipitinen," lecture to the k.k. Gesellschaft der Aerzte in Vienna, May 1903,
reported in *Wiener klin. Wschr. 16*(1903):659–660; Obermayer and Pick, "Biologisch-
chemische Studie über das Eiklar. Ein Beitrag zur Immunitätslehre," *Wiener klin. Rundschau 16*(1902):277–279. Their work on the chemistry of specificity is discussed later
in this chapter.
6   Lesky, *Die Wiener medizinische Schule* (1965) (n. 1), 599.
7   Friedrich Obermayer and Ernst Peter Pick, "Ueber die chemischen Grundlagen der
Arteigenschaften der Eiweisskörper. Bildung von Immunpräzipitinen durch chemisch
veränderte Eiweisskörper," *Wiener klin. Wschr. 19*(1906):327–333.
8   Isidor Fischer, "Uhlenhuth, Paul," *Biographisches Lexikon der hervorragenden Aerzte der
letzten fünfzig Jahre,* 2 v. (Vienna: Urban, 1933), v. 2, 1594. Uhlenhuth was assistant and
Oberarzt at the Berlin Institute, and from 1906, director of the bacteriological section
of the Reichsgesundheitsamt. He described methods for distinguishing human from
animal blood, for detecting adulteration in meat, and for identifying tubercle and
lepra bacilli. Paul Uhlenhuth, "Zur Lehre von der Unterscheidung verschiedener
Eiweissarten mit Hilfe spezifischer Sera," in *Festschrift zum sechsigsten Geburtstage von
Robert Koch, herausgeben von seinen dankbaren Schülern* (Jena: Fischer 1903), 49–74.
9   Obermayer and Pick, "Ueber den Einfluss physikalischer und chemischer Zustandsänderungen präcipitogener Substanzen auf die Bildung von Immunpräcipitinen"
(1903) (n.5).

as heating. The antibodies raised against heated albumin still reacted with untreated albumin, but they also reacted with a number of its protein degradation products as well. Physico-chemical changes did not change species specificity, they concluded, though they added to it.

Chemical alteration had a different effect. They resuscitated an old method "long since forgotten," reported by Obermayer in the *Berichte der Deutschen chemischen Gesellschaft* in 1894.[10] Obermayer and Pick changed the protein molecule chemically by introducing a diazo-group, and then used the reactivity of the diazo-group to couple other groups with the protein, such as phenols and amines. In this way they were able to create "new" proteins with a large series of different added side chains. Some had added iodine, and some had modified diazo-groups, as in the so-called xantho-proteic reaction, which could be used as a test for protein.[11] The striking result of their experiment was that the introduction of new side chains completely changed the species specificity of the proteins: an antiserum raised against iodo-bovine albumin reacted with iodo-albumin from any other animal or bird, or even from a plant, and not with either native bovine serum albumin or albumins with any other side chain: "The experiments show, that with the entry of the iodo – nitro – or diazo-group into the protein, a change is produced such that with a single stroke the species specificity of the protein vanishes."[12]

The *experimentum crucis*, as they called it, that showed the species specificity to have been totally expunged by the addition of the new side chain, was the production of anti-rabbit nitro-albumin antibody by a rabbit: the animal produced antibody to protein from its own species.

Obermayer and Pick interpreted this extraordinary and highly significant (*höchst eigenartig und merkwürdig*) result in terms of structural chemistry. The new side chains, they thought, probably entered the molecule by substitution in a benzene ring, for example the ring of the amino acid tyrosine:

HO $-\left\langle\bigcirc\right\rangle-$ H$_2$CH(NH$_2$)COOH

10 Friedrich Obermayer, "Farben thierischer Fasern und Gewebe unter Erzeugung von Azoderivaten ihrer Eiweissartigen Bestandtheil," *Ber. d. Deutsch. chem. Ges.* 27 (IV), Referate (1894):354–355.

11 Friedrich Obermayer, "Ueber Xanthoprotein (Vorläufige Mittheilung)," *Zbl. für Physiol.* 6(1892):300–301; Obermayer had been interested at this time in clinico-chemical tests for protein in urine – see Obermayer, "Nucleoalbumin-ausscheidung im Harn," *Wiener klin. Wschr.* 4(1891):966–967.

12 Obermayer and Pick, "Chemische Grundlagen der Arteigenschaften" (1906) (n. 7), 331.

It was reasonable to assume, if specificity was changed by introducing a substituent side chain into a ring, that it was the ring and its substituents that were responsible for specificity. The side-chain theory was literally vindicated: "the aromatic complex [that is, the benzene ring] gives a sort of central point for the actual species specific grouping of the side-chains."[13] Ehrlich's figurative side chains were the actual side chains of structural chemistry.

Pick had been assistant in Paltauf's serum institute since 1899. Though he does not seem to have worked directly with Paltauf – in fact, very little of Paltauf's own work was in immunology – the stamp of Paltauf's loyalties is plain to see in Pick's writing. Besides using structural chemistry, Obermayer and Pick weight their references on specificity with citations from the work of the Koch–Ehrlich group on the immunological diagnosis of species. They list, for example, the work of Max Neisser and Hans Sachs on forensic tests to distinguish different bloods, Paul Uhlenhuth's work on the same subject and on distinguishing between proteins from different species (a contribution to the *Koch Festschrift*), Hermann Pfeiffer's effort to distinguish spermatic proteins from others of the same species, and Friedenthal's studies on blood relationship, as well as the work of Paltauf himself and Kraus.[14]

But Pick left the serum institute in 1906 to join the Institute for Pharmacology, and when the following year he wrote a review of meth-

---

13 Obermayer and Pick, "Chemische Grundlagen der Arteigenschaften" (1906) (n. 7), 332.
14 Max Neisser and Hans Sachs, "Ein Verfahren zum forensischen Nachweis der Herkunft des Blutes," *Berl. klin. Wschr.* 42(1905):1388–1389. See Chapter 11, nn. 29, 44, for institutional affiliations of Neisser (and of Ulrich Friedemann); also Chapter 14, nn. 3–6, for their relationship with the chemist Heinrich Bechhold and with Ludwik Hirszfeld. Isidor Fischer, "Sachs, Hans" (n. 8), v. 2, 1349–1350. Sachs became a member of Ehrlich's institute in 1905, and in 1915 *stellvertretender Direktor*, after Ehrlich's death in August of that year. Paul Uhlenhuth, "Ein Verfahren zur biologischen Unterscheidung von Blut verwandter Tier," *Deutsche med. Wschr.* 31(1905):1673–1678. Uhlenhuth, "Unterscheidung verschiedener Eiweissarten" (1903) (n. 9); see n. 8 for institutional affiliations of Uhlenhuth. Hermann Pfeiffer, "Beiträge zur Lösung des biologisch-forensischen problems der Unterscheidung von Spermaeiweiss gegenber den anderen Eiweissarten derselben Spezies durch die Präzipitinmethode," *Wiener klin. Wschr.* 18(1905):637–641. Pfeiffer also wrote on forensic tests for blood: Pfeiffer, "Der biologische Blutnachweis," in Emil Abderhalden, ed., *Handbuch der biologischen Arbeitsmethoden* (Berlin: Urban adn Scwarzenberg 1920–1939), sec. 4, *Angewandte chemische und physikalische Methoden* (1923), v. 12, pt. 1, 105–176. Isidor Fischer, "Friedenthal, Hans" (n. 8), v. 1, 450. Friedenthal was an anthropologist, who worked with Ulrich Friedemann on immunological problems; he had his own private laboratory at Nicolassee. Karl Landsteiner attacked their papers (Chapter 11, nn. 58). See Chapter 11, n. 29 for institutional affiliations of Ulrich Friedemann. Hans Friedenthal, "Weitere Versuche über die Reaktion auf Blutverwandtschaft," *Archiv f. Physiol.* (Engelmann's) (1904):387–388. Friedenthal worked on the blood relationship between man and apes.

ods of antigen preparation for Kraus and Levaditi's *Handbuch,* his perspective seems to have shifted markedly:

The only thing that we know about the physico-chemical nature of antigens that is accepted by almost everyone, with few exceptions, is that they are colloids. . . .
    In our opinion nothing prevents us regarding the production of antibodies by antigens as the result of the formation of adsorption complexes between certain colloids and the toxins. A similar conception of the effect of toxin on antitoxin has gradually gained ground through the investigations of Landsteiner and Biltz (see also Pauli and Zangger).[15]

Pick was stating an alternative in colloid terms for both parts of the Ehrlich theory: the side-chain interpretation of the formation of antibody, as well as the structural-chemical conception of the union of antigen and antibody. His references for this are Landsteiner, Biltz, Pauli, and Zangger, not, as one might have expected from his earlier paper and his earlier associates, Ehrlich, or Bechhold, or Michaelis.
    The results of Obermayer and Pick's experiments on antigenic specificity were generally accepted and often quoted. Landsteiner wove them into his own theoretical structure:

Immune chemistry . . . [is] the chemistry of the amphoteric colloids. But this is not an explanation of specificity as the theory of the influence of amphoteric colloids on each other is not fully worked out. . . .
    . . . [A]cidic or basic groups may be important in the fine adjustment of their electrical behaviour. It is particularly noteworthy that, as in the case of dyes, the specificity in precipitation is greatly influenced by the substituents of the aromatic nucleus of the protein, for example $-NO^2$ and Cl (Pick and Obermayer) . . . the peculiarities in the reactivity of the individual immune substances are related to peculiarities of chemical constitution, so that for example, each haemagglutinin binds to very many species of red cell which are sensitive to it in different degrees. For a single agglutinin the number of species of cell to which it is sensitive is greater the lower the grade of reaction one takes into consideration.[16]

As Landsteiner says, "colloid" is not an explanation in itself. It is only a key word that indicates a rejection of structural organic chemistry and its implied firm bindings, and that allows him to replace one-to-one specificity by *quantitative Abstufung* and charge-properties. Nor is there a contradiction between electrical behaviour, acidity and basicity,

15 Ernst Peter Pick, "Darstellung der Antigene mit chemischen und physikalischen Methoden," in Rudolf Kraus and Constantin Levaditi, eds., *Handbuch der Technik und Methodik der Immunitätsforschung* (Jena: Fischer, 1908), v. 1: *Antigene* 331–586 (p. 332).
16 Karl Landsteiner, "Die Theorien der Antikörperbildung," *Wiener klin. Wschr.* 22(1909):1623–1631.

and chemical constitution. *Quantitative Abstufung* in specificity can be related to any of these levels of discussion. The serum agglutinins are colloids whose electrochemical behaviour is a resultant of their chemical constitution.

In 1912, the chemistry of antigens was the subject of an enormous review by Ernst Peter Pick that appeared in the second edition of Kolle and von Wassermann's *Handbuch*. Pick, too, agreed that whatever else they were, all antigens were colloids and all were proteins, and he coined the *Satz, Kein Antigen ohne Eiweiss* – "No antigen without albumin," or, in more modern terminology, only proteins can be antigens.[17] It would seem, says Pick,

that, in general, in the case of antigens we are not dealing with chemical individuality in the usual organic-chemistry sense, but with large colloid complexes, in which the meaning of the elementary composition and the constitutive structure is often pushed into the background by the physicochemical properties and the energies developed through them, though obviously the purely chemical texture of the substance plays a similar role, partly as the basis and partly as the result of the physico-chemical process.[18]

He then suggests that the chemical composition of antigens may be inconstant, altering with the cultural conditions, as he says is found to happen with bacteria! Substances that are not normally antigenic can perhaps become antigenic by the formation of more or less irreversible adsorption compounds:

All conceptions of specificity which reduce it either solely to structural chemistry, like Ehrlich's side-chain theory, or to the purely physical condition, cannot be made to accommodate all the facts. Only those which like Landsteiner's take into account both chemical and physical properties can do so. Landsteiner considers that specificity of immune substances like the elective behaviour of dyes is dependent upon chemical constitution, and that further the electrical properties of these substances which are amphoteric colloids is affected by their acid and basic groups. This last is supported by the experiments of Landsteiner and Pauli, in which they managed to show that abrin, ricin and a chick serum agglutinin acted as amphoteric substances in the electric current.[19]

17 Ernst Peter Pick, "Biochemie der Antigene, mit besonderer Berücksichtigung der chemischen Grundlagen der Antigen spezifizität," in Wilhelm Kolle and August von Wassermann, eds., *Handbuch der pathogenen Mikroorganismen* (Jena: Fischer, 2d ed., 1912), v. 1, 685–868 (p. 687).
18 Pick, "Biochemie der Antigene" (1912) (n. 17), 694; Pauline M. H. Mazumdar, "The antigen–antibody reaction and the physics and chemistry of life," *Bull. Hist. Med.* 48(1974):1–21.
19 Pick, "Biochemie der Antigene" (1912) (n. 24), 702; Jsidor Traube, "Die Resonanztheorie, eine physikalische Theorie der Immunitätserscheinungen," *Z. f. Immunitätsf* 9(1911):246–274; Traube suggested that a unified explanation of immune

This nearly two-hundred-page review covered an enormous amount of literature: there are 695 references. It have carried great weight. Pick, almost in Landsteiner's own words, gives Landsteiner his official blessing as the author of the theory that succeeds Ehrlich's.

Pick's review of 1912, and doubtless his complimentary references to Landsteiner, seem to have drawn Landsteiner's attention more particularly to the whole series of Pick's papers on the chemistry of antigens. Landsteiner (Figure 12.1) had earlier mentioned the 1906 one, rather casually in footnotes, but in 1913, with the help of Emil Prášek (Figure 12.2), he repeated and confirmed all Pick's 1906 experiments in the light of Pick's 1912 amplifications and amendments, and also in the light of Wilhelm Suida's experiments on the acylation of textile fibres.[20] With this, Landsteiner's Pauli-period came to an end, and his Pick-period began.

Pick had written that the changes in specificity produced by physical treatment such as heating were less far-reaching in effect than those produced by chemical alterations. Species specificity should therefore be sought not in physico-chemical differences between antigens, but in structural-chemical differences, as his experiments of 1906 had attempted to show. He suggested a terminology to bring out this distinction: *Zustandsspezifizität*, specificity of state, and *konstitutive Spezifizität*, that which depends on structural chemical constitution.[21]

Landsteiner's experiments were done semi-quantitatively. Using physically and chemically altered albumins from different animals, he found that there were grades of loss of specificity, that the apparent loss for a given chemical change was different for two different antisera.[22] Species specificity was diminished but not completely destroyed by chem-

phenomena was provided by this "resonance" theory, a theory in which both specificity and aggregation were shown to depend on surface tension. Traube's theory was attacked by Landsteiner, "Bemerkungen zu der Abhandlung von Traube, etc.," *Z. f. Immunitätsf.* 9(1911):779–786, C. 122. Landsteiner claimed that he himself had first suggested a *quantitative Abgestufte* explanation for specificity (p. 785): "It was I who first suggested the possibility, and its consequences that the specificity of antigen and antibody depends on quantitatively graded *[graduellabstufbare]* (as I hypothesised, electrochemical) forces. These forces are so constituted that the affinity of a specific antibody for a particular antigen is especially high, i.e, that these materials are quantitatively related to each other." The difference between them is that Landsteiner brings in the chemical constitution as the basis for these forces, while Traube sees them as purely physical.

20 Landsteiner, "Theorien der Antikörperbildung" (1909) (n. 16), 1428 and elsewhere. Karl Landsteiner and Emil Prášek, "Ueber die Aufhebung der Artspezifizität von Serumeiweiss: IV. Mitteilung über Antigene," *Z. f. Immunitätsf.* 20(1913):211–237, C. 142.
21 Pick, "Biochemie der Antigene" (1912) (n. 17), 705.
22 Landsteiner and Prášek, "Aufhebung der Artspezifizität" (1913) (n. 28), 218.

Figure 12.1. Karl Landsteiner, portrait photograph, n.d. (From the collection of the National Library of Medicine, Bethesda, Md.)

Figure 12.2. Karl Landsteiner and his co-worker, Emil Prášek from Belgrade, photograph 14 December 1913. The two worked together on the chemical manipulation of the specificity of serum albumin. Landsteiner and Prášek, "Aufhebung der Artspezifizität" (1913) (n. 20). (From the collection of George Mackenzie, American Philosophical Society, Philadelphia, Pa.)

ical changes: it did *not* vanish at one stroke, as Obermayer and Pick had said in 1906.[23] The three anti-horse nitro-albumin antisera he used still precipitated the horse nitro-albumin more strongly than bovine or chicken nitro-albumin, and they still reacted, though weakly, with native horse albumin. In addition, species specificity was also affected, though to a lesser degree. In place of Pick's sharp dichotomy, Landsteiner offers a series of smooth transitions: all changes are continuous, stepwise, *graduelle abgestufte*. A single antibody raised against a nitro-

23 Obermayer and Pick, "Grundlagen der Arteigenschaft" (1906) (n. 7), 331.

246 *Chemical Affinity and Immune Specificity*

albumin would both precipitate other nitro-albumins more or less strongly, and its native albumin weakly.[24]

The new experiments, Landsteiner's own, that he carried out in addition to repeating those of E. P. Pick, in fact owed as much to Suida's dye chemistry as to Obermayer and Pick's methods. Like Suida, he now introduced into the albumin substitute groups whose effect was to change the charge properties of the molecule. Instead of adding nitro- and diazo-groups, which were thought to enter the tyrosine rings, he treated the protein with alcohol and sulphuric and hydrochloric acids, as Suida had done, to alkylate the protein and block its acid groups by the formation of esters.[25]

Technically these preparations were very difficult to handle. The treated albumin was insoluble and it was difficult to get a good antibody response to it from the animals. It was also impossible to use a precipitation test for the reaction with antibody because the material was already precipitated. The alternative, the complement-binding test of Bordet and Gengou, was much more difficult to handle and to interpret. In some cases, too, the acylation procedure did not seem to work.

But the results were clear-cut enough to be significant. The treatment of the albumin with alcohol and sulphuric acid, apparently with the formation of an ester, produced a change that extended across species boundaries: the anti-horse-alkyl-albumin serum reacted with chick- and rabbit-treated albumins, as did the sera against nitro- and diazo-albumins in Pick's experiments.[26]

Here, then, there was no suggestion of anything affecting the benzene rings; only the charge-bearing groups were involved in the formation of esters. Landsteiner concluded that his artificial structural specificity and the original species specificity were opposite ends of the continuum; as one increased, the other diminished.[27] As Obermayer and Pick had done, he performed the *experimentum crucis*: where species specificity was lost, the altered albumin was antigenic in its species of origin.[28] As Suida had done, he tried treating the antigen with acetic

24 Landsteiner and Prášek, "Aufhebung der Artspezifizität" (1913) (n. 20), 219.24 Landsteiner and Prášek, "Aufhebung der Artspezifizität" (1913) (n. 20), 219.
25 Wilhelm Suida, "Ueber den Einfluss der aktiven Atomgruppen in den Textilfasern auf das Zustandekommen von Färbungen," *Monatsh. f. Chem.* (Wien) 26(1905):413–427.
26 Karl Landsteiner and Hans Lampl, "Ueber die Abhängigkeit der serologischen Spezifizität von der chemischen Struktur (Darstellung von Antigenen mit bekannter chemischer Konstitution der spezifischen Gruppen): XII. Mitteilung über Antigene," *Biochem. Z.* 86(1918):343–394, C. 169, 351.
27 Landsteiner and Prášek, "Aufhebung der Artspezifizität" (1913) (n. 20), 234.
28 Karl Landsteiner and B. Jablons, "Ueber die Bildung von Antikörpern gegen verändertes arteigenes Serumeiweiss: (V. Mitteilung uber Antigene)" *Z. f. Immunitätsf.* 20(1914):618–621, C. 145.

anhydride and acetyl chloride, to produce what was (probably) an acetyl derivative: the effect was the same. The acetyl-albumin showed a new species specificity.[29] So did a methyl-albumin.[30]

The immunology was clear, in spite of the practical difficulties. But the exact nature of the changes produced in the protein molecule was not clear at all. Landsteiner and his co-workers had reached the boundary of known structural chemistry. The action of the acid anhydrides on the protein was thought to introduce an acyl group. With a new co-worker, Hans Lampl, he tried a whole series of different acid anhydrides, acetyl, chloracetyl, dichloracetyl, trichloracetyl, butyryl, valeryl: the serum raised against the chloracetyl derivative, whatever it was, cross-reacted with acetyl, dichloracetyl, and trichloracetyl and reacted very weakly with the higher radicals.[31] They tried other preparative methods: the Schotten–Baumann preparation, using an acyl chloride in alkaline solution, and a method from Suida in which acylation was carried out in the presence of quinoline.[32]

Landsteiner writes in conclusion:

It seems difficult to us to explain these results in any way other than by the influence of the chemical constitution of the acyl groups. The only other possibility is, in our opinion, the very artificial one that these two different acylation methods, the one with water-free anhydrides, the other with bicarbonate in watery solution, place the acid radicals selectively at exactly the same spot on the protein molecule, or produce exactly the same degree of acylation. . . .

However, it is extremely likely that the type and grade of acylation, as well as the chemical nature of the acid groups, can also create serological differences.[33]

The paper is very long. The amount of work, both chemical and immunological, that must have gone into it is enormous; hundreds of compounds were synthesized and tested, hundreds of animals immunized, and this was all done during the war, when it was extremely difficult in Vienna to work at all. There are scattered remarks in the papers from the time suggesting the difficulties: Landsteiner notes that it was difficult to obtain a good response to the immunizing injections, perhaps because the animals were weak. That was most likely from the lack of warmth and the lack of proper food, since the same was true of the investigators themselves, who tried to continue working as the

29 Karl Landsteiner and B. Jablons, "Ueber die Antigen-eigenschaften von acetyliertem Eiweiss: VI. Mitteilung über Antigene," *Z. f. Immunitätsf, 21*(1914):193–201, C. 146.
30 Karl Landsteiner, "Ueber die Antigeneigenschaften von methyliertem Eiweiss: VII. Mitteilung über Antigene," *Z. f. Immunitätsf. 26*(1917):122–133, C. 162.
31 Karl Landsteiner and Hans Lampl, "Ueber Antigene mit verschiedenartigen Acylgruppen: X. Mitteilung über Antigene," *Z. f. Immunitätsf. 26*(1917):258–276,.C. 166.
32 Suida, "Einfluss der aktiven Atomgruppen in den Textilfasern" (1904) (n. 25), 418.
33 Landsteiner and Lampl, "X. Mitteilung über Antigene" (1917) (n. 31), 275.

life of the city foundered and the empire came to an end.[34] The experimenters also had to contend with technical difficulties concerning both the chemistry and the immunology.

The report has no true conclusion; it is full of phrases suggesting uncertainty and frustration. It starts off by acknowledging that the problem is insoluble: "The investigation of antigen specificity is now prevented by the insufficiency of our knowledge of the chemical structure of proteins, to which the main body of antigens belong. It is therefore not possible to say upon what chemical differences the known serological differences depend." And further on: "So far no satisfactory method for the solution of the problem has been found."[35]

The problem is not solved: the results are "extremely probable," the alternative explanation "very artificial." The form of the problem has been defined, but the solution itself has not been reached. There is no means of knowing what structures have been created during the chemical manipulations of the protein molecule. There is no means of knowing how chemical structure determines specificity.

But there is no doubt now in Landsteiner's mind that the solution will be found in the area of structural chemistry. For the first time in the twenty-five years or so since his training in organic chemistry, Landsteiner is making use of its methods. All these different preparative methods for the introduction of alkyl and acyl groups – the Schotten–Baumann reaction as well as the anhydride and the acid chloride procedures – were the standard methods of structural organic chemistry. The description of the Schotten–Baumann procedure, the mention of the disappearance of the sharp smell of the chlorine freed from the acid radical as the acyl group enters the protein, the series of preparations of ascending complexity, from the simple acetyl chloride through dichloracetyl, trichloracetyl, palmityl, valeryl, benzoyl, anisoyl, and phenyl-acetyl chlorides – all this has the sound of classical organic-structural methodology.[36]

In a purely chemical paper on the methylation of albumin, in which Landsteiner was second author along with the chemist Josef Herzig, there is even a reference to a method, first published in 1891, for estimating methyl groups. It was new at the time of Landsteiner's chemical training, twenty-four years before.[37]

34 Arthur Koestler, *The Case of the Midwife-Toad* (London: Hutchinson, 1971), 73 ff. Koestler describes the difficulties in keeping such projects, with live animals, going during the harsh wartime conditions in Vienna. Landsteiner and Lampl, "XI Mitteilung über Antigene" (1917) (n. 41), 365.

35 Landsteiner and Lampl, "X. Mitteilung über Antigene" (1917) (n.. 31), 258–259.

36 Landsteiner and Lampl, "X. Mitteilung über Antigene" (1917) (n. 40), 265–266.

37 Josef Herzig and Karl Landsteiner, "Ueber die Methylierung von Eiweisstoffen,"

It is hard to recognize now the Landsteiner who once rejected structural chemistry as an explanation of immune specificity. The dying flare of Ehrlich's chemistry in the hands of E. P. Pick has scored its last heuristic success. Although Landsteiner still emphasizes that charge effects are the basis of specificity, his methods are those of structural chemistry. The voice, one might say, is Jacob's voice, but the hands are the hands of Esau.[38]

The final episode in this series of papers was made possible by a new impulse from structural chemistry. The chemist Hermann Pauly from the Institute for Physiology at Heidelberg had been working on the diazo-reaction of proteins for many years. It was his paper of 1904 that suggested to Obermayer and Pick in 1906 that the diazo-groups entered the protein molecule at the tyrosine or histidine rings. In 1915 Pauly found that by preparing an insoluble diazo-compound using diazo-benzene arsinic acid (Atoxyl), instead of the usual diazo-benzene sulphonic acid, which gave a soluble product, he was able to analyze the compound and show that the tyrosine ring took on two molecules of the diazo-compound, and was a *bis*-diazo-tyrosine, with the structure shown. The same was true of the ring in the amino acid histidine.[39]

$$CH_2CH(NH_2)COOH$$

$$H_2O_3\,As \quad N = N \quad \underset{OH}{} \quad N = N \quad AsO_3H_2$$

Biochem. Z. 61(1914):458–463, C. 148; Rudolf Benedikt and Max Bamberger, "Ueber die Einwirkung von Jodwasserstoffsäure auf schwefel-haltige Substanzen," *Monatsh.f. Chem.* (Wien), 12(1891):1–4. Note that this is Max Bamberger, not Landsteiner's teacher, Eugen von Bamberger.

38   Gen. 27.5.22.

39   Hermann Pauly, "Ueber die Konstitution des Histidins: I. Mitteilung," *Z. f. Physiolog. Chem.* (Hoppe-Seylers) 42(1904):508–518. The diazo-reaction was well known as Ehrlich's urine test of 1882. Paul Ehrlich, "Ueber eine neue Harnprobe," *Z.f.Klin. Med.* 5(1882):285–288, and in Felix Himmelweit, Martha Marquardt, and Sir Henry Dale, eds., *Collected Papers of Paul Ehrlich*, 4 v. (London: Pergamon, 1956), v. 1, 619–629. The addition of diazo-benzene sulphonic acid to urine produced a beautiful red colour in certain cases of fever, owing to the presence of a "urochromogen" whose nature was then unknown. Diazo-benzene arsinic or "Atoxyl" came on the market in 1902 and was tested by Paul Ehrlich and his associate Kyoshi Shiga for activity against trypanosomes. It was found at this time to be inactive. Later it was taken up and tested again, after Ehrlich had proposed a new structure for it, and was found to be active. This substance formed the basis of Ehrlich's 606 modifications that were tested for therapeutic activity by Ehrlich and Sahashiro Hata. See Alfred Bertheim, "Chemie der Arsenverbindungen," in Hugo Apolant et al., *Paul Ehrlich eine Darstellung seines wissenschaftlichen Wirkens* (Jena: Fischer, 1914), 447–476; Martha Marquardt, "Atoxyl and other arsenicals," in *Paul Ehrlich* (New York: Schuman, 1951),

Landsteiner seemingly did not become aware of Pauly's work until two or more years after it was published, but when he did, it provided him with exactly what was needed. Pauly's *bis*-diazo-tyrosine was an area on a protein molecule whose structure was precisely known, which was easily chemically altered, and which, if the diazo-benzene sulphonic form was used, was even soluble.[40]

The benzene rings could be substituted in endless ways; Landsteiner and Hans Lampl, beginning with the diazo-benzene sulphonate, prepared thirty-three other diazo-compounds, in some cases preparing the parent substance as well as diazotizing it. Horse serum was their source of protein, and from this they made up the series of azo-proteins they used as immunizing antigens, with a parallel series prepared from chick serum and from the plant protein edestin. Two series of immunizations were carried out with about a year's interval between them, the first acting as a pilot study, and the second using many more animals.

The results of this immense project – they used 33 antigens, 23 sera, and 759 test reactions – appear in several pages of tables. They show clearly the profound effect of chemical constitution on the specificity of antigens (Figure 12.3).

The fact that individual sera reacted only with those antigens that bore the homologous acid radical showed the overriding importance of the charged groups. Within these families of antigens, small chemical differences in structure added up to stepwise, gradual changes in specificity. The position of the acid groups was important. The sulphonic acid family contained two recognizable subsets: in one the acid was in the *ortho-* or *meta*-position relative to the diazo-group, and in the other in the *para*-position. The seven sera of the *ortho-* and *meta*-set gave twenty-eight reactions with antigens of this group and none with those of the *para*-set.

If the tests were read after a longer interval, or quantitatively more serum was used, "irregular" reactions were more common: Landsteiner suggested that there was a weak serum component directed to the diazo-group itself, analogous to the weak reactions one might get with an anti-human serum anti-serum if it was tested with the serum of other animals. This component crossed "species" boundaries and reacted with all azo-proteins.

That Landsteiner recognized this to be an apotheosis of all that he

141–157; Ernst Bäumler, *Paul Ehrlich, Scientist for Life*, trans. Grant Edwards (New York: Holmes, 1984), 110, 122–123.

40 Pauly's paper appeared in *Hoppe-Seylers Zeitschrift*, a German journal; it may have been delayed in publication or it may have been difficult to get the journals from abroad in Vienna. There are other indications in these papers that conditions in Vienna were hard. See n. 34.

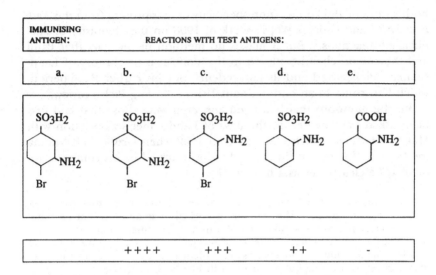

Figure 12.3. Chemical structure and immune specificity: Landsteiner and Hans Lampl's demonstration (1917–1918). The experiment showed a continuous spectrum of reactions to the benzene sulphonic acid family of antigens. The immunizing antigen is: (a) 3-bromo-, 2-amino-benzene sulphonic acid. The serum raised gives its strongest reaction with the homologous antigen, (b) but it also reacts, less strongly, with (c) 3-bromo-, 1-amino-benzene sulphonic acid and (d) 1-amino-benzene sulphonic. It does not react with (e) amino-benzoic acid. The pattern of the experiment and its semi-quantitative scoring recall Emil Fischer's experiment of 1894–1897 on the fermentation of the stereoisomeric hexoses. From Landsteiner and Lampl, "XI. Mitteilung über Antigene" (1917) (n. 41).

had been working for is proved by the triumphant, though cool, fanfare with which his published account opens (see Figure 12.3): "The elucidation of the chemistry of species difference and species relationship is of importance both for its diagnostic application and for biology. Apart from this, these reactions are also of great interest in themselves; they possess one unique property, that of specificity, which apart from a few fermentations, is different from all other known chemical phenomena."[41]

Landsteiner goes on to trace the history of the problem – a very whiggish history that leads up to his own crowning achievement, the elucidation of the chemistry of species specificity. Interestingly, it begins with Obermayer and Pick's paper of 1906 and includes Uhlen-

41 Karl Landsteiner and Hans Lampl, "Ueber die Antigen-eigenschaften von Azoprotein: XI. Mitteilung über Antigene," *Z. f. Immunitätsf.* 26(1917):293–304, C. 167 (p. 343).

huth's work of 1903 on the non-specificity of lens-protein from the *Koch Festschrift*[42] and Gideon Wells's work of 1908 on the chemistry of ana-phylactic reactions.[43] But it makes no mention of the part that Pauli and the colloid chemists played in the story. It is the history of Land-steiner's Pick-period and its antecedents, not his earlier Pauli-period, which had come to an end in about 1912, following Pick's review.

But the questions that had been answered were those that had first been asked even earlier, in the days of Landsteiner's association with Max von Gruber, which had ended in 1902 when Gruber left Vienna for Munich.[44] It was to this time that he returned in his summing up of what he clearly saw as a life's work:

We, like other authors (Bordet, Gruber, Pfeiffer) did not think that it was pos-sible to explain rationally the working of normal or immune serum on innu-merable cell-types by the supposition of innumerable different antibodies, and by analogy a similar fantastic number of receptors on each cell. We held, like Gruber, that the simpler concept was quite adequate, that an antibody can react with a variety of related, but not necessarily identical, antigens. . . .

In the opinion of Gruber and ourselves, the specificity of serum reactions appears to be the expression of graded quantitative affinity (*quantitative abges-tufte Affinität*) which reaches a maximum in certain combinations – those of antigen and homologous antibody. This implies that antibodies must exist with greater or smaller breadth of reaction, which would show up as lesser and greater specificity, while the Ehrlich theory admits only of a single absolute specificity.

Nor has he forgotten Arrhenius: it was he who suggested that the serum reactions might be salt-type linkages, an idea that was supported by the effect on them of traces of acid or alkali, and by the analogy with colloid reactions:

Though certain changes have taken place in our conception, we do not see in our present results any contradiction of this hypothesis [the charge-colloid hy-pothesis]. In the azo-protein reactions described here, so it seems, the influence of the acid groups speaks rather in its support, in the same way as the significant

42 Paul Uhlenhuth, "Unterscheidung verschiedener Eiweissarten" (1903) (n. 8), 67–70.
43 H. Gideon Wells, "Studies on the chemistry of anaphylaxis," *J. Infect. Disease*, 5(1903): 449–483. Harry Gideon Wells had his medical degree from Rush Medical College, Chicago, in 1898 and later a Ph.D. from the University of Chicago. At this time he was professor of pathology at the University of Chicago. He afterwards became pres-ident of the American Association for Immunology and one of Landsteiner's strong-est supporters in America.
44 These periods of Landsteiner's thinking may be dated approximately as follows: 1896–1902, immunity and continuity, Gruber; 1902–1912, colloid chemistry, Pauli; 1912–1920, structural chemistry, Pick. The boundaries are not sharp-edged, of course, and the thinking of the earliest period persists strongly in all the others.

changes in the antigenic properties after acylation and alkylation which we reported earlier.[45]

Thus, with this paper of 1917 Landsteiner answered the questions of 1896 in the way that Gruber, the student of Nägeli, had impressed on him that they would be answered. The pluralistic hypothesis of the one-to-one relationship of antigen to antibody was untenable, or as Nägeli had said, there are no absolute differences in Nature.[46] It was Nägeli's *unendlich Theilbarkeit* that Landsteiner had recognized in the responses to his chemical antigens, a general property, with, as Nägeli put it, specific grades of intensity (*spezifische Abstufung in der Intensität*).[47]

At this point, Landsteiner knew that he had reached his goal. He had found what he and Gruber before him had known the truth would be. Nature is a continuum of smoothly graded stepwise differences, seamless in every direction, a single unified whole.

45 Landsteiner and Lampl, "XI. Mitteilung über Antigene" (1917) (n. 41), 389–390, 392. This reference to "Bordet, Gruber, Pfeiffer" and the fantastic numbers of different substances is from Landsteiner and Adriano Sturli, "Ueber die Hämagglutine normaler Sera," *Wiener klin. Wschr.* 15(1902):38–40, in which he called the Ehrlich theory "uneconomical" by reason of the number of hypothetical entities it requires. Although Landsteiner makes no direct reference to Emil Fischer here, his paper follows the same pattern as Fischer's investigation of the fermentation of the stereo-isomeric hexoses, begun in 1894, just after Landsteiner's period of apprenticeship with him. See Chapter 9, Figure 9.2.

46 Carl von Nägeli, "Die Schranken der Naturwissenschaftlichen Erkenntniss," lecture to *50 Versammlung deutscher Naturforscher u. Aerzte,* Munich, 1877. For full reference and discussion, see Chapter 1, nn. 33, 65.

47 Carl von Nägeli, "Schranken der Naturwissenschaftlichen Erkenntniss" (1877) (n. 62); Foreword to this in *Mechanisch-physiologische Theorie der Abstammungslehre* (Munich: Oldenbourg, 1884); Nägeli, *Theorie der Gärung: ein Beitrag der molecular Physiologie* (Munich: Oldenbourg, 1879), 116.

13

The Decline and Persistence of Ehrlich's Theory: Landsteiner Surrenders Vienna

The Ehrlich chemical theory was not destroyed by its adversaries: it was slowly and gradually abandoned by its chief proponents, including Ehrlich himself. This abandonment did not mean that the Ehrlich group adopted the rival immunochemical theory, but rather that they lost interest in immunochemistry altogether and turned to other kinds of immunology, where multiple discrete specificities and the idea of a one-to-one relationship between antigen and antibody still appeared to be tenable.

This process had begun many years before Landsteiner's demonstration and by 1918 was already more or less complete.[1] There were no attacks on Landsteiner's later papers from the Ehrlich side, because there was no one left there who was working on the chemical problem. It seems there were not even any replies to E. P. Pick's review of 1912, in which he declared Ehrlich's theory to have been superseded by Landsteiner's colloid-chemical one.[2] But perhaps its very appearance, in a *Handbuch* edited by Wilhelm Kolle and August von Wassermann, both members of the Koch–Ehrlich group from Berlin, was an indication that Ehrlich's affinity-chemistry was no longer convincing even to his colleagues in immunology.[3] Pick's slogan *Kein Antigen ohne Eiweiss*,

1   Karl Landsteiner and Hans Lampl, "Ueber die Abhängigkeit der serologischen Spezifizität von der chemischen Struktur. (Darstellung von Antigenen mit bekannter chemischer Konstitution der spezifischen Gruppen), XII Mitt. über Antigene," *Biochem. Z. 86*(1918):343–394, C. 169.

2   Ernst Peter Pick, "Biochemie der Antigene mit besondere Rücksichtigung der chemischen Grundlagen der Antigen-spezifizität," in Wilhelm Kolle and August von Wassermann, ed., *Handbuch der pathogenen Mikro-organismen* 2d ed. (Jena: Fischer, 1912), v. 1, 685–869.

3   Isidor Fischer, "Kolle, Wilhelm," in *Biographisches Lexikon der hervorragenden Aerzte der letzten fünfzig Jahre*, 2 v. (Vienna: Urban, 1933), v. 1, 798. Kolle was born in 1868, and was assistant at the Koch Institut für Infektionskrankheiten in Berlin from 1893 to 1897. He took part in two of the expeditions in pursuit of tropical diseases and in 1901 became section-director at the Koch Institute, in 1906 *Ordinarius* for hygiene and bacteriology in Bern. In 1917 he became director of the Staatsinstitut für experimentelle Therapie and of the Georg-Speyer-Haus in Frankfurt, replacing Hans Sachs,

that all antigens were proteins and depended on the laws of colloid chemistry, was widely accepted.[4] It had become a dogma of immunology, to be stated in the first line or two of any discussion, then assumed to underlie all the rest. By 1912 the colloid theory had superseded Ehrlich's, although in the practice of the serum institutes the old assumption of clear-cut, one-to-one specificity was essentially unchanged.

In 1905, however, Ehrlich's chemical theory appeared to have survived all the attacks: those of Gruber, of Arrhenius, and of Bordet. Ehrlich's closest collaborator, Julius Morgenroth, was writing confidently:

The concept of chemical binding between toxin and antitoxin, an idea borrowed from organic and particularly stereochemistry, is the actual foundation of the theoretical structure built up by Ehrlich and his school. Upon this conception of the binding of antigen and antibody rests, by analogy, the relationship of ferment to anti-ferment, amboceptor to anti-amboceptor, complement to anti-complement, and finally that of toxin, agglutinin and amboceptor to the receptors on cells. No less central, it is well known, is the connection of this basic idea of chemical binding with the biological concept by which the side-chain theory seeks to solve the problem of the origin of the antibodies within the organism.[5]

who had held the position briefly as *Stellvertretender Direktor* after Ehrlich's death. Kolle maintained Ehrlich's methods in the preparation, testing, and distribution of sera and serum standards. As mentioned earlier (Chapter 3, n. 56), he had a specially close relationship with Koch. See Chapter 3 (n. 56), and Kurt Kolle, ed., *Robert Koch, Briefe an Wilhelm Kolle* (Stuttgart: Thieme, 1959). Kolle edited the *Handbuch* first with von Wassermann, then with Rudolf Kraus of the Paltauf Institute in Vienna, and Paul Uhlenhuth of Frankfurt.

August von Wassermann was born in 1866 and in 1890 became assistant at the Robert Koch Institut; from 1906 he was director of the department of experimental therapy and biochemistry. He was, as Fischer says (n. 3, v. 2, 1645), first student then friend of Ehrlich, and he worked especially on the problem of the diagnosis of syphilis by immunological means; he is the author of the eponymous "Wassermann Reaction." In 1917 he became director of his own new Institute for Experimental Therapy at Berlin-Dahlem.

4 Both these assumptions were challenged in the early twenties. See Michael Heidelberger and Oswald T. Avery, "The soluble specific substances of *Pneumococcus*," *J. Exper. Med.* 40(1923):301–306, where he found that the antigen in this case was a carbohydrate. Jacques Loeb, *Proteins and the Theory of Colloidal Behaviour* (New York: McGraw-Hill, 1922), demonstrated that proteins do undergo ordinary chemical reactions and that adsorption and precipitation depend not on special colloid laws but on ionization and the mass-action equation. See Pauline M. H. Mazumdar, "The antigen–antibody reaction and the physics and chemistry of life," *Bull. Hist. Med.* 48(1974):1–21.

5 Isidor Fischer, "Morgenroth, Julius" (n. 3), v. 2, 1068. He was for eight years Ehrlich's closest collaborator: from 1905 he was director of the Department of Bacteriology at the Pathological Institute in Berlin, and from 1919 director of the department of chemotherapy of the Robert Koch Institut für Infektionskrankheiten. Julius Morgenroth, "Ue. die Wiedergewinnung von Toxin aus seiner Antitoxinverbindungen," *Berl. klin. Wschr.* 42(1905):1550–1554, 1550.

256   *Chemical Affinity and Immune Specificity*

Ehrlich's chemists did not stick absolutely to the party line. Max Neisser had spoken for the group in 1904 in defending the Ehrlich theory against Arrhenius's alternative.[6] But at the same time Neisser himself was quite interested in colloid chemistry.[7] Heinrich Bechhold, too, did several studies on the details of the analogy between the agglutination of bacteria and the physico-chemical process of colloid flocculation.[8] But he still vigorously answered Pauli's "festival lecture," in which Pauli enthusiastically presented Landsteiner's colloid-chemical theory as superseding Ehrlich's.[9] As Bechhold said, Pauli claimed that there was a colloid-chemical explanation for every single phenomenon in biology and medicine.[10]

In 1904 Ehrlich himself was utterly certain, so he said, that his theory was correct, and that his position was "like that of a chess player who, even though his game is won, is forced by the obstinacy of his opponent to carry it on move by move till the final mate."[11] His loss of interest in immunochemistry can be dated from the foundation of the Georg-Speyer-Haus in the spring of 1906, and the beginning of his collaboration with the chemist Alfred Bertheim on the chemistry of Atoxyl.[12]

6 Max Neisser, "Kritische Bemerkungen zur Arrhenius'schen Agglutinin Verteilungsformel," *Zbl. f. Bakt. 36*(1904):671–676.
7 Max Neisser and Ulrich Friedemann, "Studien über Ausflockungserscheinungen," *Münch. med. Wschr. 51*(1904):464–469; Neisser and Friedemann, "Studien zur Ausflockungserscheinungen, II. Beziehungen zur Bakterienagglutination," *Münch. med. Wschr. 51*(1904):827–831.
8 Heinrich Bechhold, "Die Ausflockung von Suspensionen bzw. Kolloiden und die Bakterienagglutination," *Z. f. physik. Chem. 48*(1904):385–423; Bechhold, "Strukturbildung in Gallerten," *Z. f. physik. Chem. 52*(1905):185–199; Bechhold, "Kolloidstudien mit der Filtrations-method," *Z. f. physik. Chem. 60*(1907):257–318; Bechhold, "Die elektrische Ladung von Toxin und Antitoxin," *Münch. med. Wschr. 54*(1907): 1921–1922.
9 Wolfgang Pauli, "Wandlungen in der Pathologie durch die Fortschritte der allgemeinen Chemie," Festival Address at the 3d annual meeting of the k.k. Gesellschaft der Aerzte in Wien, *Wiener klin. Wschr. 18*(1905):550–551; translation in M. H. Fischer, trans. and ed., *Physical Chemistry in the Service of Medicine* (New York: Wiley, 1907), 101–137. (See chap. 11, n. 22).
10 Heinrich Bechhold, "Ungelöste Fragen über den Anteil der Kolloidchemie an der Immunitätsforschung," *Wien. klin. Wschr. 18*(1905);666–668 (see chap. 11); Bechhold, review of Pauli (n. 10), *Wiener klin. Wschr. 18*(1905):550–551. (See chap. 11, n. 53).
11 Paul Ehrlich, "Preface to the German Edition," in Charles Bolduan, trans. and ed., *Studies in Immunity by Professor Paul Ehrlich and His Collaborators* (New York: Wiley, 1st ed., 1906; 2d ed., 1910), viii.
12 Martha Marquardt, *Paul Ehrlich* (New York: Schumann, 1951), 127–128, 141–144; Ernst Bäumler, *Paul Ehrlich, Scientist for Life* (1984), trans. Grant Edwards (New York: Holmes, 1984), 83–84; 118–130. The Georg-Speyer-Haus was an institute for research on chemotherapy, built for Ehrlich in 1906 with funds donated by Franziska Speyer, widow of one of Frankfurt's leading bankers, in memory of her husband. Speyer himself had already donated funds for the establishment of a cancer research department at the Royal Institute for Experimental Therapy.

In 1906, however, he could still be persuaded by his American admirer Charles Bolduan to write a new essay for the collection of his papers on immunity, which Bolduan had translated and which appeared first in 1906 then in 1910 in a second edition. This final summing up of Ehrlich's immunological period served in translation to transmit his turn-of-the-century views engraved in stone to later generations in the English-speaking world. In it, Ehrlich could still say that "the stereo-chemical conception of the immunity reaction, despite numerous attacks has proven itself able to dominate every phase of the subject."[13] He could still adopt the work of Obermayer and Pick as giving support to the receptor theory of sharply defined specificity.

We see therefore that the introduction of a certain chemical group into the albumin molecule completely alters the latter's power to excite the production of antibodies. This certainly corresponds entirely to the view that the production of antibodies is dependent on the chemical constitution of the exciting agent, a view which finds expression in my receptor theory.[14]

Indeed, in 1906, Obermayer and Pick themselves thought so too.[15] But, Ehrlich continues significantly,

The heuristic value of the receptor idea, the idea which underlies my side-chain theory, can best be appreciated by studying the development of our knowledge concerning the cytotoxins of blood serum. As prototype of these substances, the haemolysins occupy a prominent place in this volume.[16]

As Ehrlich said, the receptor idea was an important part of the work on the cytotoxins of blood serum. It had already provided the theoretical explanation behind the famous "six communications on haemolysis" of Ehrlich and Morgenroth, which included the paper on the individuality of goat bloods.[17]

13 Paul Ehrlich, "A general review of the recent work on immunity" (1906) (n. 11), 577–586 (p. 577).
14 Ehrlich, "General review" (1906) (n. 11), 579.
15 Friedrich Obermayer and Ernst Peter Pick, "Ue. die chemischen Grundlagen der Arteigenschaften der Eiweisskörper," *Wiener klin. Wschr.* *19*(1906):327–334.
16 Ehrlich, "General review" (1906) (n. 11), 579.
17 Paul Ehrlich and Julius Morgenroth, "Zur Theorie der Lysinwirkung," *Berl. klin. Wschr.* *36*(1899):6–9; Ehrlich and Morgenroth, "Ueber Haemolysine: Zweite Mittheilung," *Berl. klin. Wschr.* *36*(1900):481–486; Ehrlich and Morgenroth, "Ueber Haemolysine: dritte Mittheilung," *Berl. klin. Wschr.* *37*(1900):453–458; Ehrlich and Morgenroth, "Ueber Haemolysine: vierte Mittheilung," *Berl. klin. Wschr.* *37*(1900):681–687; Ehrlich and Morgenroth, "Ueber Haemolysine: fünfte Mittheilung," *Berl. klin. Wschr.* *38*(1901):251–257; Ehrlich and Morgenroth, "Ueber Haemolysine: sechste Mittheilung," *Berl. klin. Wschr.* *38*(1901):569–574, 598–604. The description of the experiment that showed the individuality of the goat bloods is in "Dritte Mittheilung."

These experiments and the receptor theory that went with them were used by Emil, Freiherr von Dungern, and his Polish co-worker Ludwik Hirszfeld in their investigation of blood groups of families of dogs, and then of the families of the Heidelberg faculty, in which they showed that blood group specificity was inherited according to the Mendelian rule.[18] It was the first example, except for the rather uncertain case of eye colour, of the Mendelian inheritance of a normal human character.[19]

It was von Dungern who wrote the article on "receptor specificity" for the *Ehrlich-Festschrift* of 1914.[20] In it, he traces the origin of his own

18 Emil von Dungern, "Ueber Nachweis und Vererbung biochemischer Strukturen und ihre forensiche Bedeutung," *Münch. med. Wschr.* 57(1910):293–295; v. Dungern and Ludwik Hirszfeld, "Ueber eine Methode, das Blut verschiedener Menschen serologisch zu unterschieden," *Münch. med. Wschr.* 57(1910):741–742; v. Dungern and Hirszfeld "Ueber Nachweis und Vererbung biochemischer Strukturen," *Z. f. Immunitätsf.* 4,(1910):531–546; v. Dungern and Hirszfeld, "Ueber Vererbung gruppenspezifischer Strukturen des Blutes II," *Z. f. Immunitätsf.* 6(1910):284–292; v. Dungern and Hirszfeld, "Ueber gruppenspezifische Strukturen des Blutes, III," *Z. f. Immunitätsf.* 8(1911):526–562. The work of von Dungern and Hirszfeld is discussed more fully in Chapter 15.

19 Emil von Dungern was at this time director of the scientific section of the Heidelberg Cancer Institute. He had been a visitor at Ehrlich's institute for some time around 1900, where he was working on the blocking of antigen–antibody reactions by added antigen: Emil von Dungern, "Beiträge zur Immunitätslehre, I. Neue Experimente zur Seitenkettentheorie," *Münch. med. Wschr.* 47(1900): 677–680; v. Dungern, "Beiträge zur Immunitätslehre, II. Rezeptoren und Antikörperbildung," *Münch. med. Wschr.* 47(1900):962–965. He also published *Die Antikörper* (Jena: Fischer, 1903). Isidor Fischer's biographical information in this case is not very helpful, as the contact with Ehrlich is not mentioned: it occurred during von Dungern's time as *Privatdozent* at the University of Freiburg. Ludwik Hirszfeld was von Dungern's assistant from 1907 to 1911 in Heidelberg. He had done his doctoral thesis under Ulrich Friedemann in Berlin. Ludwig Hirschfeld, "Untersuchung über Haemagglutination und ihre physikalische Grundlagen," *Arch. f. Hyg.* 63(1907):237–286. Hirszfeld was brought up on the side-chain theory and was so excited by it that he decided to become a serologist. According to Hirszfeld, the project on the inheritance of blood groups in dogs was well advanced, and he and von Dungern were discussing how to transfer the results to human beings. He writes, "during our chats over a glass of wine, we recalled the almost unnoticed research by Landsteiner." It is clear that this research was carried out under the inspiration of the Ehrlich receptor theory, and that Landsteiner came into it rather late. See Chapter 15 for detailed discussion. See also Ludwik Hirszfeld, *Historia Jednego Zycia* (Warsaw: Instytut Wydawniczy Pax, 1st ed., 1957; 2d ed., 1967; Hanna Hirszfeldowa, ed.). A translation was made available to me by Col. F. R. Camp, Jr., then commander/director, U.S. Army Medical Research Laboratory, Fort Knox, Ky.; it has been deposited in the National Library of Medicine, Bethesda, Md.

20 Emil von Dungern, "Rezeptorenspezifizität," in Hugo Apolant et al., *Paul Ehrlich: eine Darstellung seines wissenschaftlichen Wirkens Festschrift zum 60 Geburtstage des Forschers* (Jena: Fischer, 1914), 162–165. Apolant appears as "author" only because his name comes first. He was head of the cancer research section of Ehrlich's Royal Institute for Experimental Therapy in Frankfurt. Emil von Dungern had also written the

use of the idea of receptor specificity to the work he did in Ehrlich's Institute in 1900, and to Ehrlich and Morgenroth's classic articles on the goat bloods, through Ehrlich's elective absorption method, by which a cell could be shown to absorb a single specificity from a serum, leaving unaltered other test specificities. Von Dungern writes, "Specificity as we meet it in serological reactions does not depend on the whole protoplasmic structure of the species, but upon certain components, the receptors, which are characteristic of a species or a group of species."[21]

His article summarizes the work on receptor specificity at the Frankfurt serum institute. There had been projects on species specificity using the red cells of different animals, on organ specificity using the organs of the same individual, and on individual specificity using "homologous cell types within a species, which show an astonishingly large number of receptors." From Ehrlich and Morgenroth's goats of 1900, von Dungern moved to his own work on human blood groups of 1909. Landsteiner's name is mentioned, but only in passing. The inspiration clearly comes from Ehrlich: von Dungern is, after all, writing for Ehrlich's *Festschrift*. He discusses the biological phenomenon, the extraordinary diversity of the specific receptors. But he never discusses their chemical nature, even speculatively. Ehrlich in 1906 was already conscious that it was in this kind of immunological work that the future of his receptor theory lay, and not in attempts at chemical explanation.

It is striking that Bolduan's Ehrlich collection nowhere contains any structural chemistry *sensu stricto*. Not a single structural formula appears on its pages, though many of Ehrlich's visual-aid diagrams of antigen uniting with antibody are there, suggesting an image not so much of a complicated lock and key, as of the snapping together of two press-studs. Ehrlich's use of the words "chemical union" is as much a metaphor as was Fischer's use of "lock and key." As Ehrlich himself implied, the chief heuristic value of the receptor idea was not in the *chemistry* of species difference, but in the differentiation of species.

To the English-speaking reader, Ehrlich's chief opponent appeared to be Jules Bordet, a position he occupied by courtesy of the publisher John Wiley and of Frederick P. Gay, of Harvard, Bordet's friend and translator. Gay's collection and translation of Bordet's articles appeared as a volume uniform with Bolduan's Ehrlich translations, and in answer to them, in 1909. The papers produced by Bordet and his group at the

article on specificity for the *Festschrift zum sechsigsten Geburtstage von Robert Koch, harausgeben von seinen dankbaren Schülern* (Jena: Fischer, 1903), 1–16.
21 V. Dungern, "Rezeptorenspezifizität" (1914) (n. 20), 164.

260   *Chemical Affinity and Immune Specificity*

Institut Pasteur du Brabant of Brussels had kept up a continuous dialogue with those of Ehrlich and his school, as they answered each other point by point during Ehrlich's immunology period. This dialogue, however, was much less dramatic and passionate than the *agon* in Vienna. Bordet, though brilliant, was not as *temperamentvoll* as Gruber.[22]

From the beginning, Bordet rejected Ehrlich's chemistry, though he did not completely rule out the idea of specific affinities. The antigen and antibody did not unite in fixed proportions, as happened in "straight" chemistry. It was originally, said Bordet in 1903, "simply for the purpose of expression" that he had compared the phenomena with those of dyeing. "We have never committed ourselves on the intimate nature of the reaction."[23]

In these two famous opponents, then, we have the paradoxical situation that in both cases the physico-chemical "theory" that provided the underpinning for their immunology and that was the cause of so much controversy was in many ways not a hypothesis, but a metaphor. For Ehrlich, the metaphor originated in structural chemistry, for Bordet in dyeing. But Bordet and Octave Gengou in Brussels, and Gay at Harvard, did in fact take the problem out of the metaphorical realm and investigate the phenomena of adsorption and their relation to those of immunity.[24] It was one of Bordet's reproaches that Ehrlich's theory was not a stimulus to such investigation:

It attempted, at a stage when the data were still very limited, to interpret the mode of action of antibodies with antigens; here it seems to me that it has exercised a perturbing influence on the progress of knowledge, and has really hindered the free development of investigation. In offering explanations which seem definitive and schematic, which satisfy the experimenter and appease his

22 Erna Lesky, *Die Wiener medizinische Schule im XIX Jahrhundert* (Graz: Böhlaus, 1965), 600.
23 Jules Bordet, "Sur la mode d'action des antitoxines sur les toxines," *Ann. de l'Inst. Pasteur 17*(1903):161–186, and in Frederick P. Gay, ed., *Studies in Immunity by Professor Jules Bordet and His Collaborators, Collected and Translated by Frederick P. Gay, Including a Chapter Written Expressly for this Publication by Professor Bordet* (New York: Wiley, 1909), 259–279 (p. 263).
24 Octave Gengou, "Récherches sur l'agglutination des globules rouges par les précipités dans les mediaux colloidaux," *Ann. de l'Inst. Pasteur 18*(1904):678–700; *Bordet-Gay* (n. 26), 312–332; Jules Bordet and Frederick P. Gay, "L'absorption de l'alexine et le pouvoir antagonist des sérums normaux," *Ann. de l'Inst. Pasteur 22*(1908):625–643; *Bordet-Gay* (n. 26), 398–413; Octave Gengou, "Contribution a l'étude de l'adhésion moleculaire et de son intervention dans les phénomènes biologiques," *Arch. Internat. de Physiol. 7*(1908):1–210; resumé in *Bordet-Gay* (n. 26), 414–439; Jules Bordet and Oswald Streng, "Les phénomènes d'absorption et la conglutinine du sérum de boeuf," *Zbl. f. Bakt.* (1 Abt. (orig.), *49*(1909):260–276; *Bordet-Gay* (n. 26), 440–461.

curiosity, Ehrlich's theory has come to make certain problems which have scarcely been touched on, be regarded as worked out. . . . And the guiding thought, to my way of thinking fatal, which they [its partisans] have endeavored to enforce, is the constant attribution to the molecule of the antibody of separate atom groups for each of the phenomena to which the antibody gives rise . . . each of which appears as a resting-place of such and such a property. These groups are simply invoked by the theorist, as he wishes.[25]

Whatever the reason, it was true that Ehrlich's group used the *image* of "separate atom groups for each phenomenon" but did not investigate it chemically. Von Dungern's work on blood groups is a good example.

When Ehrlich did apply the methods of structural chemistry, it was not to the problems of immunology, but to those of chemotherapy. It was in this field that what Morgenroth calls Ehrlich's "ideal goal" – the investigation of biological phenomena by chemical means – came nearest to being fulfilled. In the earlier field of immunity, said Morgenroth in 1914, no one knew better than Ehrlich that this ideal had slipped away from him into the far distance. In the time allowed for one life's work, it was no longer within reach:

The drawing together of chemical and biological principles had reached a barrier. Sometimes it was the advancement of the biological side that came up against insoluble difficulties, sometimes chemical thinking and experimentation could not keep up with the increasing richness and development of the biological phenomena. It was the first that put an end to his dye studies, the second those in immunity.[26]

Morgenroth himself, who had collaborated with Ehrlich in the days of his enormously successful "studies on haemolysis" of 1899–1901, followed him into the field of chemotherapy.[27] In 1917, a year after Ehrlich's death and a few months before the appearance of Landsteiner's paper on the chemical explanation of specificity, Morgenroth wrote, in a paper on the relations between chemical constitution and chemotherapeutic effect,

The concept of specificity in chemotherapy has come to have a form, which in Ehrlich's theory of chemoceptors is closely analogous to that which his receptor theory had in immunology in the olden days [*vor Jahr und Tag*]. Above all he

25 Jules Bordet, "A general resumé of immunity," in *Bordet-Gay* (n. 25), 496–530 (p. 498–499). This paper was written especially for the collection, in 1909.
26 Julius Morgenroth, "Chemotherapeutische Studien," *Ehrlich-Festschrift* (1914) (n. 20), 541–582 (p. 541–542).
27 Ehrlich and Morgenroth, "Ue. Haemolysine, I.—VI." (1899–1901) (n. 17).

managed to free himself from the botanical and zoological concept of species [sic] and with these in both cases hypothetical substances, the receptors in immunology and the chemoceptors in chemotherapy he has made the first steps towards a chemical conception [that is, of specificity].[28]

But in a footnote to that same article, Morgenroth wrote that as yet, there existed "no bridge leading from chemistry to the field of immunology."[29] A year later, in 1918, Landsteiner triumphantly cited that remark as an expression of Ehrlich's final defeat.[30]

In 1917, then, the year after Ehrlich's death, Morgenroth, who had been his closest collaborator, believed that Ehrlich's theory was unable to link the biological phenomenon to its underlying chemistry, at least in the field of immunology. He had known this at least since 1914. Ehrlich himself had probably known it since 1906.

This failure of the Ehrlich chemical explanation is probably the cause of a phenomenon that I first noticed in 1973, but for which I was unable at that time to find an explanation.[31] This phenomenon is the virtual disappearance of chemistry from the writing of the German immunologists, that is, from the *Zeitschrift für Immunitätsforschung*, between the date of its foundation in 1909, when articles on chemistry amounted to about 7 percent of its output, and 1930, by which time they had almost completely disappeared (Figure 13.1).

The curves have been plotted by counting the papers that appeared in the *Zeitschrift* volumes and dividing them into three categories: those dealing with practical diagnostic problems, those dealing with problems of more basic theoretical importance, and those dealing with the chemical explanation of immune phenomena. For example, the first volume, which came out in 1909, contained forty-one papers in all, of which 61 percent were classed as "basic" and 39 percent as "diagnostic." Seven percent of the total fell into the "chemical" class: these were all "basic" papers. By way of example, the "basic immunology" group contains a large number of papers on anaphylaxis. One of them was Ernst Friedberger's "On Anaphylaxis: XVII Communication: the Meaning of Sessile Receptors for Anaphylaxis" of 1911. In this group are also classed a number of papers on blood grouping, such as those of von Dungern

28 Julius Morgenroth, "Zur Kenntnis der Beziehungen zwischen chemischer Konstitution und chemotherapeutischer Wirkung," *Berl. klin Wscher. 59*(1917):55–63 (p. 62).
29 Morgenroth, "Beziehungen zwischen chemischer Konstitution u. Wirkung" (1917) (n. 31), 59.
30 Landsteiner and Lampl, "Abhängigkeit der serologischen Spezifizität von chemischen Struktur, XII" (1918) (n. 1).
31 Pauline M. H. Mazumdar, "Karl Landsteiner in America," *Acta Congressus Internationalis XXIV Historiae Artis Medicinae*, 25–31 August 1974 Budapest (Budapest: Semmelweiss Museum, 1976), 1538–1542.

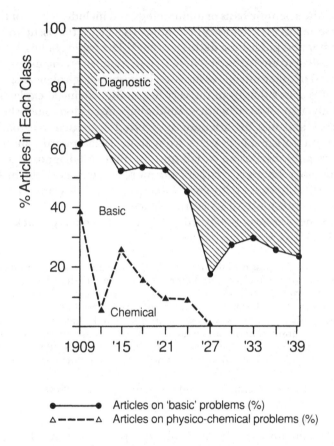

Figure 13.1. The *Zeitschrift für Immunitätsforschung,* 1909–1939:
Count of the types of papers published divided into two categories, "basic" or
theoretical research, and "diagnostic," dealing with laboratory methods for
clinical use. Chemical subjects are a subsection of the basic type. As theoretical
research declines over the period, interest in the physico-chemical approach,
never very strong, declines more steeply, reflecting the fading of the Ehrlich
chemical theory. Compare the papers published in the American *Journal of
Immunology* over the same period (Figure 13.3). (Drawing by Instructional Media Services, University of Toronto)

and Hirszfeld of 1909, 1910, and 1911, and a number on cancer immunity by Hugo Apolant from Ehrlich's institute.[32]

The papers classed as "chemical" make up a small subgroup of these

32 Ernst Friedberger and S. Girgolaff, "Ueber Anaphylaxie: XVII Mitteilung: die Bedeutung sessiler Rezeptoren für die Anaphylaxie," *Z. f. Immunitätsf.* 9(1911):575–
582. V. Dungern and Hirszfeld, "Nachweis u. Vererbung biochemischer Strukturen,
&c. I, II & III" (1910–1911) (n. 18). Hugo Apolant, "Ueber die Empfindlichkeit von

papers on the scientific basis of immunology and include eleven of Land-steiner's series on the chemistry of immune specificity, that is, up to 1917, when he stopped publishing in this journal.[33] The papers falling into the "practical-diagnostic" group include a large number on the diagnosis of syphilis, and of other infectious diseases, by immunological means.[34]

The curve shows that from the very beginning, chemical explanation played only a small part in the immunology discussions published in this journal, and that the attention it received decreased irregularly as time went on. In some later years, from about 1925 onwards, almost all the papers published came to fall into the practical-diagnostic group. As immunology became a larger and more established field in Germany, practical hospital pathology, rather than basic research, came to occupy more and more of its journal space.

In 1918 the *Zeitschrift* published a paper that explicitly attacked the chemical point of view. The author wrote:

The many efforts in past decades to explain the immune reaction in chemical terms have up to now had no very happy results. The fact that these reactions like chemical reactions can take place *in vitro*, that is, outside the animal body, has led scientists to seek for the chemical nature of these unknown materials, before enough is known of how they are produced and what they do. Often a chemical purification of antibodies is attempted, without asking whether today's chemistry is capable of it. The best that can be achieved would be only an analogy, an imitation of an immune reaction, with known chemical materials.[35]

He seems to be expressing a prevailing point of view, a view that echoes Morgenroth's *Satz* that there was no bridge from chemistry into immunology yet.[36]

It was the members of Ehrlich's old group who had thus lost confidence in the heart of the Ehrlich theory. The four editors of the *Zeitschrift*, the men whose interests were reflected in it, were Ernst Friedberger, Rudolf Kraus, Hans Sachs, and Paul Uhlenhuth. They were all members of the Koch–Ehrlich group, and belonged to the

Krebs-mausen gegen intraperitoneale Injektionen," *Z. f. Immunitätsf.* 3(1909):108–114; Apolant, "Ueber die Beziehungen der Milz zur aktiven Geschwülstimmunität," *Z. f. Immunitätsf.* 17(1913):219–232.

33 Karl Landsteiner and Hans Lampl, "Ueber die Antigen–eigenschaften von Azoproteinen, XI. Mitteilung über Antigene," *Z. f. Immunitätsf.* 26(1917):293–276.

34 L. Berczeller, "Soll die Wassermann'sche Reaktionen mit aktiven oder inaktivierten Patientenserum ausgeführt werden?" *Z. f. Immunitätsf.* 27(1918):305–325. Rudolf Kraus and J. M. de la Barrera, "Studien über Flecktyphus in Südamerika," *Z. f. Immunitätsf.* 34(1922):1–35.

35 G. Mansfeld, "Eine physiologische Erklärung der Agglutination," *Z. f. Immunitätsf.* 27(1918):197–212 (p. 197).

36 Morgenroth, "Beziehungen zwischen chemischer Konstitution u. Wirkung" (1917) (n. 31), 59.

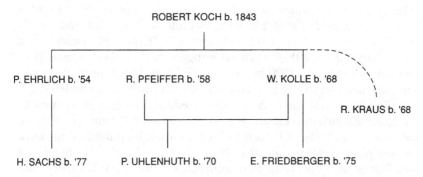

Figure 13.2. The four editors of the *Zeitschrift für Immunitätsforschung* at its founding in 1909: Ernst Friedberger, Rudolf Kraus, Hans Sachs, and Paul Uhlenhuth. All except Kraus are "grand-students" of Robert Koch; Kraus is a member of Paltauf's group in Vienna, and closely associated with the Koch-Ehrlich family.

second generation after Koch, his grand-students, as it were (Figure 13.2) except for Kraus, who was a little older and a member of Paltauf's serum institute in Vienna. Ernst Friedberger had studied in Berlin under Kolle at the Robert Koch Institute in Berlin and from 1901 to 1908 had been assistant at the Institute of Hygiene in Konigsberg, where Richard Pfeiffer was *Ordinarius*. In 1908 he returned to Berlin and in 1913 became director of the Department of Immunology and Experimental Therapy at the Pharmacological Institute there. In 1915 he succeeded to Loeffler's chair at the Institute of Hygiene at Greifswald and in 1926 became director of the Preussische Institut für Hygiene und Immunitätsforschung in Berlin-Dahlem. Friedberger was thus successively student, assistant, and successor of a Koch student. In 1919 he published a textbook of bacteriology with Richard Pfeiffer.[37]

Paul Uhlenhuth, who has been mentioned several times in this chapter, was assistant and *Oberarzt* at the Robert-Koch Institut in Berlin, and then at the Institute of Hygiene in Greifswald under Loeffler, where he published his paper on the immunological differentiation of proteins from different animals, which appeared in the *Koch-Festschrift* of 1903. Uhlenhuth regarded himself as one of Koch's students; he signed in the list of *dankbaren Schülern* who dedicated the *Festschrift* to Koch. In 1906 he became director of the Bacteriological Section of the Reichsgesundheitsamt; he held several chairs of hygiene including those at Strassburg, Marburg, and Freiburg.[38]

37 Isidor Fischer, "Friedberger, Ernst" (n. 3), v. 1, 449; "Pfeiffer, Richard," v. 2, 1206; "Loeffler, Friedrich," v. 1, 929.
38 Paul Uhlenhuth, "Zur Lehre von der Unterscheidung verschiedener Eiweissarten

Hans Sachs, the youngest of the editors, was born in 1877. In 1905 he became a member of Ehrlich's institute in Frankfurt and briefly after Ehrlich's death in 1915 *stellvertretender Direktor*. He published a number of papers with Ehrlich on antigens and antibodies, and later on chemotherapy. In 1920 he became *Ordinarius* for Immunology and Serology, and director of the Cancer Research Institute in Heidelberg.[39]

Rudolf Kraus was the only one of the editors not directly trained by the Koch–Ehrlich succession. He was born in 1868, and in 1893, he was assistant to Richard Paltauf in Vienna, when he published his work on precipitating antibodies, already discussed. As has also been pointed out, Paltauf and his group were adherents of the Ehrlich theory, and opponents of Gruber. Kraus cooperated with Uhlenhuth and with Kolle in editing two *Handbücher* and with Constantin Levaditi of the Institut Pasteur on a third, the *Handbuch der Technik und Methodik der Immunitätsforschung*.[40] In 1923, after ten years in Brazil, he came home to take up the leadership of the Serotherapeutic Institute in Vienna.

The editors of the *Zeitschrift für Immunitätsforschung* were all representatives of the Koch–Ehrlich school of immunology. This journal, like the earlier *Zeitschrift für Hygiene und Infektionskrankheiten* founded by Koch himself in 1885, had something of the character of a house-organ, the voice of the new generation of immunologists, students and grandstudents of the bacteriologists of Koch's generation.[41]

In this case, however, there was no German journal to represent non-Ehrlich immunology, as there had been in the 1880s in Pettenkofer's *Archiv für Hygiene*, which had represented the "other" hygiene of vital statistics against Koch's bacteriology. Following the enormous success and expansion of Koch's influence, there were few immunologists in the German-speaking world who were not part of this group. Landsteiner's scientific formation in an atmosphere of opposition to this tradition would have been difficult to repeat anywhere in Europe except perhaps in the Instituts Pasteur, in Brussels or perhaps in Paris.

Although American immunology in 1920 was in many ways very sim-

mit Hilfe spezifischer Sera," in *Koch Festschrift* (1903) (n. 20), 49–74, vi; Isidor Fischer, "Uhlenhuth, Paul" (n. 3), v. 2, 1594.

39 Isidor Fischer, "Sachs, Hans" (n. 3), v. 2, 1350.

40 On Kraus and the Paltauf group's attack on Gruber, see chap. 6, n. 3; on their homage to Koch, see chap.1 2, nn. 2–5. Isidor Fischer "Kraus, Rudolf" (n. 3), v. 1, 816; Rudolf Kraus and Paul Uhlenhuth, eds., *Handbuch der mikrobiologischen Technik, unter Mitarbeit hervorragender Fachgelehrten* (Berlin: Urban, 1923–23); Wilhelm Kolle, Rudolf Kraus and Paul Uhlenhuth, eds., *"Handbuch der pathogenen Mikro-organismen*, 3d ed. (Jena: Fischer, 1929). Rudolf Kraus and Constantin Levaditi, eds., *Handbuch der Technik und Methodik der Immunitätsforschung* (Jena: Fischer, 1908).

41 On the *Zeitschrift für Hygiene und Infektions krankheiten*, founded in 1885 by Koch and Flügge, see chap. 3, n. 31.

ilar to that in Germany, there was a subtle difference in the attitude to a possible chemical explanation for immunological phenomena. The papers that were published by the *Journal of Immunology* covered the same range of problems as those in the *Zeitschrift für Immunitätsforschung*. The journal was started in 1916 as the official organ of the American Association of Immunologists and the Association for Serology and Hematology. The president of the Association of Immunologists at that time was J. W. Jobling, who had been trained at Johns Hopkins and at Koch's Institute in Berlin. He had also been an associate of the Rockefeller Institute from 1904 to 1909.[42]

A count of the papers along the same lines as was done in the *Zeitschrift für Immunitätsforschung* brings out the rather striking difference between the two. In the American journal (Figure 13.3), there is no downwards trend in the numbers of papers in the "basic research" category, as there had been in the *Zeitschrift*. Instead, the proportion falling into this group increases irregularly between 1916 and 1939. The other difference is that the proportion falling into the "chemical" group climbs slowly. Immunochemistry, a subject of minimal and diminishing interest in the *Zeitschrift*, was growing in popularity in the journal.

An example of this popularity can be found in the presidential address to the American Association of Immunology of 1924, which was given by Harry Gideon Wells, professor of pathology at the University of Chicago, and which was confidently entitled "The Chemical Basis of Immunological Specificity." In this speech, Wells attacked Hans Zinsser, another of those who had doubted that any connection between protein chemistry and antigen specificity would be made in the near future. Wells uses Zinsser as a straw man so that he could quote dozens of examples of this connection, of parallels between the chemical properties of natural proteins, from milks, seeds, and so on, and their immunological properties.[43] He then discussed Landsteiner's work, which had been discussed but, says Wells, misunderstood, by Hans Zinsser.[44] One might almost feel that this amounts to a policy

42 George W. Corner, *A History of the Rockefeller Institute 1901–1953 Origins and Growth* (New York: Rockefeller Institute Press, 1964), 59, 109. Jopling had worked with Flexner at the Rockefeller Institute; they had discovered a transplantable rat tumour. He left in 1909 and went to the Michael Reese Hospital in Chicago and subsequently became professor of pathology there, and later at Columbia.
43 H. Gideon Wells, "The chemical basis of immunological specificity," presidential address to the American Association of Immunologists, 17 April 1924, *J. Immunol.* 9(1924):291–308. Wells's career has been mentioned in Chapter 12, n. 57.
44 Hans Zinsser, *Infection and Resistance: An Exposition of the Biological Phenomena Underlying the Occurrence of Infection and the Recovery of the Animal Body from Infectious Disease, with a Chapter on Colloids and Colloidal Reactions by Prof. S. W. Young,* 3d ed. (New York:

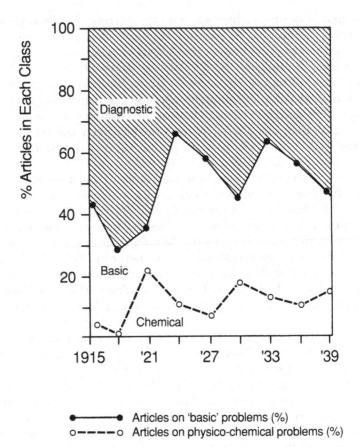

Figure 13.3. The *Journal of Immunology*, 1916–1939. Types of paper published, categories as in Figure 13.1. Interest in the physico-chemical approach shows an increase, in contrast to the pattern found in the *Zeitschrift für Immunitätsforschung* over the same period. In fact, the contrast is even greater (Arthur Silverstein, pers. comm., 1992). This was not the only journal to publish papers on immunochemistry. (Drawing by Instructional Media Services, University of Toronto)

Macmillan, 1923), 107. Zinsser, in the first edition of his textbook of 1914, leans heavily on Ehrlich's side-chain theory. He says, for example (p. 98), "the haemolysins (and others) ... react respectively only with the particular variety (of bacteria etc.) employed in their production. This indicates that each of these antigens – of almost unlimited number – must possess a chemical structure individually characteristic and different from all the others ... our knowledge of the chemical constitution of protein has not yet advanced to a point at which specificity can be based on definite variations in chemical structure, and the complexity of the problem is such that it does not seem likely that we can hope in the near future to attain such knowledge." In his third edition (107, 114), Zinsser discusses Obermayer and Pick's work, and

statement by the American association: indeed, in 1928, it was Landsteiner himself who came to give the presidential address.[45]

Gideon Wells, in fact, was a convinced anti-Ehrlich partisan, although
he had spent some time at Koch's Institute in Berlin. In 1924, the year
of his address to the American Association, Gideon Wells produced a
monograph, *The Chemical Aspects of Immunity*, in a series organized by
the American Chemical Society.[46] The editors intended this to be "a
serious attempt to found an American chemical literature without primary regard to commercial considerations."[47]

It was edited by William A. Noyes, and his cousin Arthur A. Noyes
was on the editorial board. Arthur Noyes, a student of Wilhelm Ostwald
in Leipzig in the early 1890s, ran what the historian John Servos calls
"an economy version of Ostwald's institute," at the Massachusetts Institute of Technology (MIT).[48] Noyes had astonished Ostwald in the
course of a visit to MIT in 1905 with the statement that he hoped in
time to shift the spiritual centre of gravity of the human race from
Europe to the other side of the Atlantic Ocean. Ostwald, in telling this
story in his autobiography, follows it with some serious reflections on
cultural development and decline, beginning with the Greeks.[49]

Such was the intention, then, of this series, and Gideon Wells was to
do it for immunology. In the preface to the first edition of his monograph, Wells wrote:

Originally the reactions of immunity were studied with the purpose of solving
urgent problems . . . of disease. After a time there came to be a growing recognition of their importance as general biological phenomena. . . . For the most

then Landsteiner's. He misinterprets Landsteiner's in three ways: (1) he interprets
it to show the importance of the benzene rings in specificity, in conformity with
Obermayer and Pick; (2) he does not notice the more-or-less-good-fit effect, and (3)
he repeats his remarks of 1914 that protein chemistry has not yet advanced sufficiently to make the correlation between protein structure and specificity possible. It
is this that Gideon Wells attacks. Isidor Fischer, "Zinsser, Hans" (n. 3), v. 2, 1726.
Zinsser was born in 1878, in New York, and trained at Columbia University. He was
a bacteriologist at Columbia from 1905 to 1906 and rose steadily from instructor to
associate professor; then to professor at Stanford University in 1911, at Columbia in
1913, and at Harvard in 1923.

45 Karl Landsteiner, "Cell antigens and individual specificity," *J. Immunol.* 15(1928):
589–600, presidential address to the American Association of Immunologists, 30 April
1928, C. 244.

46 H. Gideon Wells, *The Chemical Aspects of Immunity*, American Chemical Society, Monograph Series (New York: Chemical Catalog, 1924; 2d ed., 1929).

47 William A. and Arthur A. Noyes, "General introduction," on behalf of the American
Chemical Society (1929) (n. 61), 5.

48 John W. Servos, *Physical Chemistry from Ostwald to Pauling: the Making of a Science in
America* (Princeton, N.J.: Princeton University Press, 1990), 100–155 (p. 114).

49 Wilhelm Ostwald, *Lebenslinien: eine Selbstbiographie* (Berlin: Klasing, 1933), v. 3, 56–57
(p. 58).

part their chemical significance was less appreciated. . . . Perhaps the hypothetical presentation of the subject in the terms of the Ehrlich nomenclature, with pictorial conceptions which had no chemical significance, had some influence in satisfying many investigators that they understood the principles when they merely understood the hypothesis. As Dean has said in this connection, "Ignorance, however aptly veiled in an attractive terminology remains ignorance."[50]

Wells goes on to cite Bordet in his support, and his book contains a large section on Landsteiner's work, and on the physical chemistry of agglutination and precipitation.

The growing importance of immunochemistry in America was, however, only a part of the new interest in chemical explanations in all fields of biology. This interest was particularly strong at the Rockefeller Institute, where Jacques Loeb and his students formed a centre from which interest in a reductionist chemical physiology spread widely. Loeb was deeply admired by Simon Flexner, scientific director of the Rockefeller Institute.[51] The decline of the Ehrlich chemical theory in America did not leave a vacuum, as it had done in Germany, where the alternative to Ehrlich's chemistry had been no chemistry. A newer chemical physiology with a close relationship to the physical chemistry of the Ostwalds, and to Arrhenius's solution theory, as practised by Jacques Loeb, was already available. In Germany, the members of the Ehrlich group, as long as they remained part of the group, were unable to move very far in this direction. Wells said that it was the prevalence of the Ehrlich theory that had prevented the earlier growth of immunochemistry. But with the theory gone, the social power of the Ehrlich group was still directed against physico-chemical explanations, and so none were developed.

In one area of immunology, however, the Ehrlich theory did not decline. In the field of blood group serology, it continued to be a source of new work for a very long time to come. As Ehrlich himself had recognized, it was in the field of cytotoxins and haemolysins that the future of the receptor theory lay.

Paul Ehrlich died during the First World War, at the early age of sixty-one. Wilhelm Kolle, a fellow-protégé of Robert Koch, succeeded him as director of the Frankfurt institute. After the war, the League of Nations under Thorvald Madsen took over what had been Ehrlich's responsibility for the standardization of serum. Since Germany was not

---

50 H. Gideon Wells, *Chemical Aspects of Immunity* (1929) (n. 61), 9, "Author's preface to the first edition."

51 Mazumdar, "Landsteiner in America" (1975) (n. 31); Philip Pauly, *Controlling Life: Jacques Loeb and the Engineering Ideal in Biology* (Oxford: University Press, 1987), 164–169.

at first permitted to join the league, the German laboratory was excluded, and the leadership passed from Frankfurt to the Statens Seruminstitut in Copenhagen. The techniques and the units used, however, were still those that Ehrlich had introduced.[52] In practice, the side-chain theory was as firmly institutionalized as ever: Ehrlich still had the serum institutes on his side.

Productive though he was, Landsteiner had not prospered in Vienna. It is difficult to tell exactly why this should have happened, especially since in later life Landsteiner himself never talked about his Vienna years. There are at least three possible reasons, all of which may have had some bearing on his career.

The years from 1915 to 1919 saw some of Landsteiner's most important work, but they had been hard for him. Since about 1905, working with many different co-workers, he had produced a steady ten or twelve papers a year, but in 1915, the numbers fell: to one in that year, five in 1916, seven in 1917, three in 1918, and one in 1919. His mother had died in April 1908, just after he started work at the Wilhelminenspital. In November 1916, he married Helene Wlasto (Figures 13.4, 13.5), an actress whom he had known for many years, and in April 1917, when Landsteiner was nearly fifty years old, his son Karl Ernst was born.[53] He now had to look after two other people.

With the collapse of the Donau monarchy at the end of the 1914–1918 war, conditions in Austria became severe, especially in Vienna. The defeated city was without all municipal services, without heat, light, food, or transport. In his autobiography, Ludwik Hirszfeld wrote of visiting Vienna just after the end of the fighting in 1918:

I will not describe the hunger that prevailed in Vienna at that time nor the psychology of human strata who had once been the rulers and were now waking up the defeated ones. I would like to mention scientists only.

In Vienna, I met Paltauf, Pick, Landsteiner and Bacher. . . . Landsteiner made

52 League of Nations, Health Committee, *Report of the International Conference on the Standardisation of Sera and Serological Tests, London, 1921* (Geneva: League of Nations, Document no. C.533.M.378.1921.III); Klaus Jensen, "Report on international standards maintained at the Statens Seruminstitut, Copenhagen, Denmark, on behalf of the League of Nations," *League of Nations, Quarterly Bulletin of the Health Organisation,* Special Number on Biological Standardisation, II, November 1936, pp. 728–734; W. Charles Cockburn, "International contributions to the standardisation of biological substances: the League of Nations Health Commission," *Biologicals 19*(1991):161–169.

53 Paul Speiser and Ferdinand G. Smekal, *Karl Landsteiner Entdecker der Blutgruppen Biographie eines Nobelpreisträgers aus der Wiener medizinischen Schule* (Wien: Hollinek, 1961; 2d ed., 1975); translated by Richard Rickett as *Karl Landsteiner: The Discoverer of the Blood Groups and a Pioneer in the Field of Immunology: Biography of a Nobel Prize-Winner of the Vienna Medical School* (Vienna: Hollinek, 1975).

Figure 13.4. Helene Wlasto (photograph dated April 1897), as a young actress in *Romeo and Juliet*. (From the collection of George Mackenzie, American Philosophical Society, Philadelphia, Pa.)

Figure 13.5. Helene Wlasto (photograph dated December 1906). She married Karl Landsteiner in 1916. (From the collection of George Mackenzie, American Philosophical Society, Philadelphia, Pa.)

the biggest impression on me. I had met him at the Congress of Microbiologists in Berlin in 1913. At that time he was a very good-looking man: tall, a high forehead, a manly posture. In Vienna I saw a wreck. He was obviously starved.[54]

Landsteiner himself never spoke to anyone of this period of his life. The Swiss serologist Robert Doerr, who knew him before he left Vienna, felt that Landsteiner wanted to forget the period of misery and difficult working conditions during and after the war.[55]

Apart from the difficult political conditions at the end of the war, Landsteiner's personal career had not gone well. For someone so productive, he had moved ahead very slowly. At the end of 1897, after leaving the Hygiene Institute and Max von Gruber, Landsteiner moved to the Institute for Pathological Anatomy as *Assistent* under Anton Weichselbaum. Although the working conditions were good there, and he stayed for ten years (until December 1908), he did not get on particularly well with Weichselbaum, or with his fellow-*Assistenten*.[56] During these ten years he published 75 papers: 52 on the subject of serology, 12 on bacteriology or virology, and 11 on pathologico-anatomical problems. In January 1909 he left the institute and took over the *Prosektur* or Department of Morbid Anatomy at the Wilhelminenspital in Vienna (Figure 13.6), where he stayed for the next ten years, until 1919. In 1911 he was given the title of *ausserordenlicher Professor* and in 1917, in recognition of his war service as pathologist to the k. k. Kriegspital No. 1, he was given the title of *Regierungsrat*.[57]

This seems an extremely slow rate of professional advancement for a man who by 1919, at the age of fifty-one, had published 171 papers, many of them contributions of the highest calibre. It is certainly slow compared to the careers of the three German editors of the *Zeitschrift für Immunitätsforschung*: Friedberger, Uhlenhuth, and Sachs, who were made *Ordinarius* at the ages of 40, 41, and 43, respectively. Kraus, who had reached *ausserordentlicher Professor* in Paltauf's Serotherapeutic Institute in Vienna in 1910, one year ahead of Landsteiner (they were both born in 1868) had gone to Argentina as director of the Institute for Bacteriology in Buenos Aires in 1913. He returned to Vienna from 1923 to 1928 to take over the directorship of the Serotherapeutic Institute. The pathologists Oskar Stoerk and Anton Ghon, both productive men who had been *Assistenten* along with Landsteiner at the

54 Hirszfeld, *Historia* (1967) (n. 21), 67.
55 Robert Doerr to George Mackenzie, typescript ms., n.d.: Landsteiner-Mackenzie Papers, B L23m, Box 3, American Philosophical Society.
56 Ernst Peter Pick to George Mackenzie, interview, n.d., probably c. November 1951; Landsteiner-Mackenzie Papers, B L23m Box 1, American Philosophical Society.
57 Speiser and Smekal, *Landsteiner Entdecker der Blutgruppen* (1975) (n. 53), 25, 51.

Figure 13.6. The Wilhelminenspital in Vienna. A general view during the time Landsteiner was pathologist, 1908–1920. (From the collection of George Mackenzie, American Philosophical Society, Philadelphia, Pa.)

Institute for Pathological Anatomy, had their professorships and directorships by 1912 and 1910, respectively. Many of the less productive, of course, did not (see Figure 11.2).

The conflict between Gruber and Ehrlich offers one possible reason for Landsteiner's failure to thrive. The Serotherapeutic Institute group had been Ehrlich's partisan defenders in the conflict that ended in Max von Gruber's resignation and his leaving Vienna. Landsteiner's outspoken criticism of Ehrlich's theory, and perhaps his association with Gruber, may have locked him out of the serum institute. Max von Gruber himself interpreted it in that way: Speiser and Smekal quote a letter from Gruber, written in 1908, which appeared in the *Wiener medizinische Wochenschrift* in 1931 – a month or two after Landsteiner's Nobel prize became public. Gruber wrote, in part:

Do not take it ill of me, if I express my surprise that nobody yet in Vienna seems to be sure of Dr Landsteiner's quality. . . .

His recognition in Austria seems to be hindered by his not belonging to Paltauf, and in Germany by his having set himself up against Ehrlich. But in France, Belgium, England and Denmark his reputation is of the highest. Bordet once said to me, that in his opinion L. is the only brilliant mind among the Austrian bacteriologists.[58]

Several contemporaries, including the Hungarian immunologist Béla Schick, remembered that Landsteiner had been disappointed not to succeed Paltauf as director of the Serotherapeutic Institute.[59] It seems likely that this was the trigger that led Landsteiner to leave his homeland and settle elsewhere.

George Mackenzie's research for the biography that he began working on in 1943, soon after Landsteiner's death, uncovered another possible factor that may have forced Landsteiner to leave Vienna: the growing anti-Semitism of the period. Mackenzie felt that this was at the root of Landsteiner's constant personal discomfort with his surroundings, his chronic pessimism, and feeling of being persecuted. Mackenzie's attention was drawn to this problem by Landsteiner's refusal to be included in *Who's Who in American Jewry*, and his lawsuit of 1937 against

58 Max von Gruber to Prof. Dr Hermann von Schlesinger, dated Munich, 15 December 1908; in *Wiener med. Wschr. 81* 309 (1931). Speiser cites the letter not as evidence of a conflict, but as a testimonial to Landsteiner. Though he speculates as to the reason for Landsteiner's lack of professional success, he does not mention the conflict between Gruber and Ehrlich, or Paltauf's sympathy with the Ehrlich theory, as a possible cause.
59 Béla Schick, conversation with George Mackenzie, dated 17 November 1943; Landsteiner-Mackenzie Papers, B L23m, Box 1; William V. Berger, conversation with Mackenzie, n.d., but accompanied by letter Mackenzie to Berger, dated 16 November 1946, *idem,* Box 2: American Philosophical Society.

John Simons, its editor.[60] Landsteiner's parents were Jews by birth, but he was not a Jew by religion. In the 1880s, his widowed mother had converted and had herself and her son baptized as Catholics.

Conversion was not an unusual step. Between 1868, the year of Landsteiner's birth, and 1903, 9,000 Viennese Jews converted. Many did so to promote their career, since the state religion in Austria was Catholicism and practising Jews were not admitted to the upper bureaucracy, the army, or the diplomatic service. They were none the less numerous in Vienna among the highly educated, the professional, and intellectual classes. Between 1880 and 1936, 50–60 percent of lawyers were Jews although those of the Jewish *faith* were barred from the judiciary. During the same period, Jews made up 60 percent or more of doctors. In 1910 Jews by descent as well as religion together made up about 40 percent of the University Law Faculty and 60 percent of the Medical Faculty. In Vienna University, the *Ordinarius*, or full professorship, as a high government post, could not in theory be held by practising Jews. Conversion was often followed by promotion, although even converts could face difficulties.[61] Several of Mackenzie's Viennese sources, men who knew both Landsteiner and Vienna, say discreetly that Landsteiner was unlikely to have gained the directorship he wanted. But that was not obvious: Max von Gruber was Jewish, as were Ernst Peter Pick and Wolfgang Pauli, and the pathologist Oskar Stoerk, all of whom reached the level of *Ordinarius* and institute director, though many otherwise well-known names in medicine did not.[62] Landsteiner's co-worker, Michael von Eisler-Terramare, who became director of the State Serotherapeutic Institute in the thirties, was one of those who, like Pick and Pauli, fled Austria after the Nazi *Anschluss* of 1938.[63] E. P. Pick himself, responding to Mackenzie's suggestion, said that he did not think that Landsteiner's feeling of persecution could be put down to personal experience of anti-Semitism. Gruber, he thought, was inclined to be politically liberal,

Weichselbaum was certainly not antisemitic – he had Jewish Assistants, Oskar Stoerk, Erdheim, besides L. L. made in the University-titles and jobs the same

60 Grace Robinson, "Scientist sues to bar listing as notable Jew," *N.Y. Daily News*, 5 April 1937.
61 Steven Beller, *Vienna and the Jews 1867–1938 a Cultural History* (Cambridge: University Press, 1989), 188–189
62 Moshe Atlas, "Grosse Jüdische Aerzte Wiens im neunzehnten und zwanzigsten Jahrhundert," in Josef Fraenkel, ed., *The Jews of Austria: Essays on Their Life, History and Destruction* (London: Valentine, 1967), 41–65.
63 Hans-Peter Kröner, "Die Emigration Deutschsprachiger Mediziner im Nationalsozialismus," *Berichte zur Wissenschafts-geschichte*, Sonderheft 1989, on the occasion of the XVIII Congress of the International Union of the History of Science in Berlin (Weinheim: VCH Verlagsgesellschaft, 1989), 32, n. 83.

progress as his contemporaneous colleagues. The violent Viennese antisemitismus never influenced the life of L.; his departure from Vienna was more commanded by the fear of future events after the lost war than by some feeling of personal persecution.[64]

This ambiguity was typical of the complex climate of veiled personal intolerance accompanied by public statements rejecting political anti-Semitism that was common among the liberal bourgeoisie in Vienna in the twenties.[65] Political anti-Semitism was not, of course, therefore absent: the Christian Social or Catholic Populist party under Karl Lueger had campaigned since 1897 against a combined Marxist-*cum*-Jewish enemy. It was here, in an atmosphere of peculiarly virulent political anti-Semitism, that the young Hitler learned to fuse the two.[66]

Landsteiner, then, felt the pressure of several forces, all tending to increase his dissatisfaction with Vienna and to strengthen his resolve to leave. Living conditions there had become very hard indeed, as the political structure of Austro-Hungary collapsed, and the future seemed very uncertain. Landsteiner's opposition to the Ehrlich theory placed him outside the dominant group in immunology. His work on the chemical basis of specificity was part of a dwindling minority interest, unfashionable in the German-speaking world, which was still controlled by the Ehrlich group, who now rejected chemical explanations. He was not going to get to the highest professorial rank, nor was he ever going to succeed Paltauf as director of the Serotherapeutic Institute. The directorship went eventually to Rudolf Kraus, a member of the group. In the face of these disadvantages, the conversion to Catholicism that had served others was not enough on its own to ensure promotion.

In 1919 Landsteiner obtained an appointment in Holland as *Prosector* to the R. K. Ziekenhuis in The Hague. On 27 May 1920, he resigned from the *Prosektur* at the Wilhelminenspital and collected his pension. Like Gruber before him, he left Vienna for good.

64 Ernst Peter Pick to George Mackenzie, letter d. 28 February 1949. Landsteiner-Mackenzie Papers, B L23m Box 2, American Philosophical Society.
65 Sigurd Paul Scheichl, "The contexts and nuances of anti-Jewish language: were all the 'antisemites' antisemites?" in Ivaar Oxaal, Michael Pollack and Gerhard Botz, eds., *Jews, Antisemitism and Culture in Vienna* (London: Routledge, 1987), 89–110.
66 Robert S. Wistrich, "Social democracy, antisemitism and the Jews of Vienna," in Oxaal, Pollack, and Botz, *Jews, Antisemitism* (1987) (n. 79), 111–120.

PART IV

Absolute Specificity and Blood Group Genetics

Ehrlich währt am längsten

German proverb

14

Immunology and Genetics in the Early Twentieth Century: The Receptor and the Unit-Character

The transition from Ehrlich's work on the cytotoxins of animal serum to the work on human blood groups was made by Emil, Freiherr von Dungern, and his young Polish co-worker, Ludwik Hirszfeld (Figure 14.1). Von Dungern regarded himself as a student of both Koch and Ehrlich. He signed the list of Koch's students who dedicated the *Festschrift* of 1903 "presented by his grateful students" and wrote the first essay in it. He states his position in this way:

The phenomena can be adequately explained by the Ehrlich receptor-theory. Everything goes to show that the antibodies are reaction-products of the organism which unite chemically with certain binding molecular complexes of the substance involved. These binding groups are the bearers of specificity.[1]

At this time von Dungern was *ausserordentlicher* professor of bacteriology and hygiene at Freiburg. In 1906 he went to Heidelberg to head the scientific section at the Cancer Research Institute, and in 1913 he became its director. During his time at Freiburg, he was visiting the Ehrlich institute at Frankfurt, working on cytotoxins, and in 1914 he contributed an article on receptor-specificity to the *Ehrlich-Festschrift*, as he had to Koch's in 1903. In this *Festschrift*, eleven years after the one for Koch, he was still saying the same things of the receptor theory: "Ehrlich and Morgenroth introduced the concept of receptors in their first experiments with the binding of immune haemolysins. They defined them as the binding groups of the protoplasmic molecule."[2] Ludwik Hirszfeld, too, was brought up on the Ehrlich side-chain theory. In 1903 when he was nineteen, Hirszfeld went to Berlin to complete his

---

1 Emil von Dungern, "Spezifität der Antikörperbildung," *Festschrift zum 60 Geburtstage von Robert Koch, herausgeben von seinen dankbaren Schülern* (Jena: Fischer 1903), 1–16 (p. 1).
2 Emil von Dungern, "Rezeptorenspezifität," in Hugo Apolant, et al., eds., *Paul Ehrlich eine Darstellung seines wissenschaftlichen Wirkens: Festschrift zum 60 Geburtstage des Forschers* (Jena: Fischer 1914), 162–165 (p. 162).

Figure 14.1. Ludwik Hirszfeld, portrait photograph taken in 1953. (My thanks to Rune Grubb, Lund, for this picture)

medical training. The first scientific monograph he ever read, he says in his autobiography, was Oppenheimer's "Toxins and antitoxins."[3] Many years later he remembered reading it:

Those who lived at the time that this theory [the side-chain theory] was established, know how ravishing was the thought that the sensitivity of the cell to intoxication is also a guarantee of overcoming that intoxication. . . .
     It is the task of every theory to incite creative unrest, and to enrapture. None of the existing theories has accomplished that more forcefully than the side-chain theory.[4]

He sat over Oppenheimer's article all night, and in the morning decided to become a serologist. His thesis work was done at the Institute for Hygiene in Berlin, under Ulrich Friedemann, who was at that time *Assistent* in bacteriology. Friedemann had been working with the bacteriologist Max Neisser and the chemist Heinrich Bechhold at Ehrlich's institute in Frankfurt, on the relation between bacterial agglutination and colloid precipitation, during the brief period when this group was interested in colloid chemistry.[5] Hirszfeld says that Friedemann was often visited in the laboratory by Julius Morgenroth, and for hours they talked about resistance, with young Hirszfeld listening in, enraptured by the side-chain theory.

Hirszfeld's rather extraordinary thesis contains all these themes, including the sense of amazement and enthusiasm that he often evokes

3   He is probably referring here to Carl Oppenheimer, "Toxine und Schutzstoffe," *Biol. Zbl.* 19(1899):799–814. Oppenheimer explains sympathetically Ehrlich's contributions to immunology, including the side-chain theory and the assay of antitoxin.
4   Ludwik Hirszfeld, *Historia Jednego Życia,* 2d ed., Hanna Hirszfeldowa, ed. (Warsaw: Instytut Wydawniczy Pax, 1967), 91; translation courtesy of F. R. Camp and F. R. Ellis, eds., in series "Selected Contributions to the Literature of Blood Group Serology," from the Blood Transfusion Division, U.S. Army Medical Research Laboratory, Fort Knox, Ky. This translation was not actually published; it has been circulating in *samizdat* only. A copy has been deposited at the National Library of Medicine, Bethesda, Md. Although he is now dead, I should record my thanks to Alexander S. Wiener for bringing this series of translations to my notice. Ludwik Hirszfeld, *Historia* (1967), 9–10; he says this is cited from his book on immunology. Hirszfeld wrote two books on immunology, but I have not been able to find this passage in either of them. They are Hirszfeld, *Konstitutionsserologie und Blutgruppenforschung* (Berlin: Springer 1928), translated by the F. C. Farnham Co., Philadelphia, Pa., for F. R. Camp and F. R. Ellis, eds., "Contributions" (n. 3), v. 3, part i, *Constitutional Serology and Blood Group Research* (1969); Hirszfeld, trans. Hanna Hirszfeld, *Les Groupes Sanguines: leur Application à la Biologie à la Médecine et au Droit* (Paris: Masson, 1938).
5   Heinrich Bechhold, "Die Bakterienagglutination: ein physikalisch-chemisches Phänomen," Vortrag in Gemeinschaft mit Dr Max Neisser and Dr [Ulrich] Friedemann am kgl. Institute für experimentelle Therapie, Frankfurt-am-Main. *Verhandl. d. Ges. Deutsche Naturf. u. Aerzte* (75 Versammlung, Cassel, 20–26 September 1903) (Leipzig: Vogel 1904), pt. 1, 487–488.

in his writing.[6] It is important to this discussion in that it demonstrates the carryover of Ehrlich's absorption techniques and the *Malkoff'sche Phänomen* into the study of blood group specificity, where they became the basic procedures of investigation in the new field.

The *Malkoff'sche Phänomen* has been discussed before. It was a confirmation of one of the most important features of Ehrlich and Morgenroth's haemolysin experiments of 1899–1901, that a given cell could absorb from a serum one or more well-defined antibody specificities and leave untouched any other specificities present. Malkoff and Ehrlich cited it as a confirmation that there were multiple specific receptors on cells and multiple specific antibodies in serum, a demonstration *ad oculos* of the pluralistic theory.[7] Landsteiner had criticized this conclusion and shown that the material eluted from the absorbing cells still reacted to different degrees with several different cell types. For Landsteiner, it was a demonstration that antibodies were *not* sharply specific, but reacted more or less well with many antigens. They were not plural, but *quantitativ abgestufte*. These two interpretations mirror *in parvo* the clash between specificity and unity, which is the theme of this book and the problem that Hirszfeld tried to resolve in his thesis.

Hirszfeld found, to his surprise, that in the case of normal, non-immune serum from any animal he tried, the cells of a series of different species were always agglutinated in the same order. Chick cells always had the highest titre, and guinea-pig cells the lowest. Agglutination using organic colloids instead of serum gave the same result. He put it down to the physical properties of cells and serum: agglutination by non-immune serum, he thought, was a physico-chemical effect. Immune serum might be different. His conclusion works both against Ehrlich's hypothesis that normal and immune sera contained the same kind of side chains, and also against Landsteiner's conclusion from his eluates, that specificity was a matter of more or less good fit. It is rather funny that neither party seems to have paid any attention to this startling conclusion. The *Geheimrats* were not impressed.

In 1907, after finishing his thesis, Hirszfeld went to Heidelberg to the Cancer Research Institute. At the time, the American Arthur Coca was von Dungern's assistant in the serology department there.[8] When

---

6 Ludwig Hirschfeld (he spells it the German way at this time), *Untersuchung über die Haemagglutination und ihre physikalischen Grundlagen*, Inaugural Dissertation, Berlin 1907 (Munich 1908). Published as "Untersuchungen über die Hämagglutination und ihre physikalischen Grundlagen," *Arch. f. Hyg.* 63(1907):237–286.
7 For a discussion of this "phenomenon" and its implications see Chapter 7, nn. 19, 23; for its relation to later work on blood groups see n. 68, this chapter; and for its parallel in the work of R. A. Fisher, see Chapter 16, n. 53.
8 Isidor Fischer, "Coca, Arthur Fernandez," in his *Biographisches Lexikon hervorragenden*

he went back to the United States, the assistantship was given to Hirsz-feld, who had already become attached to von Dungern as a friend. Von Dungern – aristocratic, sophisticated, and inspired – inspired Hirszfeld, too.

Dungern was reproached for not being able to force himself to systematic work. . . . He himself did not realize that he had the greatest and most fervent creative force. . . . Von Dungern was a spiritual poet who had to fall in love with a problem in order to be able to work on it. . . . He was a flame burning from within.[9]

Their first project started as an attempt to investigate the response of the individual to tumour cells – it was a Cancer Research Institute – and von Dungern approached the problem through Ehrlich and Morgenroth's experiment showing the individuality of goat bloods. Using dogs instead of goats, they found that the dogs belonged to definite serological races, which had nothing to do with their appearance, but which were inherited.[10]

Their method was to inject several pairs of male and female dogs with each other's blood, so that they produced immune isoagglutinins. By differential absorptions of these sera with the dog cells, von Dungern and Hirszfeld were able to produce sera with which they could trace the inheritance of the red cell antigens.

They were, in effect, repeating and continuing the Ehrlich and Morgenroth goat experiment. In 1900 Ehrlich had written: "It is to be expected that a sufficient repetition of the experiments will finally lead us to recognise a certain cycle of constantly repeating types. The attainment of this goal however is rendered very tedious by the fact that in some cases, . . . no isolysin is formed."[11]

Von Dungern and Hirszfeld continued this experiment, which Ehr-

Aerzte der letzten fünfzig Jahre (Vienna: Urban, 1933), v. 1, 256. Coca graduated in medicine from the University of Pennsylvania in 1900 and spent some time in Heidelberg, where he was a student for four years, then assistant at the Heidelberg Cancer Research Institute from 1907 to 1909. From 1911 he was at Cornell University, until 1921 as instructor in pathology and immunology, then as assistant professor, and finally from 1924 as professor. He wrote on blood grouping and on hypersensitivity and was the author of one of the first student's textbooks of immunology, *Essentials of Immunology for Medical Students* (Baltimore, Md.: Williams, 1925).

9 Hirszfeld, *Historia* (1967) (n. 4), 18. Many passages in Hirszfeld's work sound like blood grouping through the eyes of Goethe's Werther.

10 Emil von Dungern and Ludwig Hirschfeld, "Ueber Nachweis und Vererbung biochemischer Strukturen," *Z. f. Immunitätsf.* 4(1910):531–546.

11 Paul Ehrlich and Julius Morgenroth, "Ueber Hämolysine: Dritte Mittheilung," *Berl. klin. Wschr.* 37(1900):453–458, and in Felix Himmelweit, Martha Marquardt, and Sir Henry Dale, eds., *Collected Papers of Paul Ehrlich*, 4 v. (London: Pergamon, 1956), v. 2, 196–204 (Ger.), 205–212 (Eng.) (p. 209). See chap. 5, nn. 30–33.

lich had found too tedious, and took it a stage further by showing that the repeating types were also inherited, something that had not struck Ehrlich. The methods they used – the immunizations of the animals with each other's blood, and the absorption of the sera with cells – to give sera with residual sharp specificity, were Ehrlich's methods.

It was not until they began to wonder how to transfer the results to human subjects that they remembered the Landsteiner paper of 1901 that showed them that in this case they did not need to immunize the families to get active antisera.[12] In the case of the human subjects, the isoagglutinins were already present in normal serum. Landsteiner's paper seems to have attracted little attention between its first appearance in 1901 and resurrection by von Dungern and Hirszfeld in about 1910.

The human families they tested were those of the Heidelberg dons: twenty-one professors, five Drs., a Dr. Dr., two *Rats*, and two directors. The results were stated as follows:

> The results show that structure A as well as structure B never appear in the children unless they are present in one of the parents ... the inheritance of the structures is discontinuous. It sometimes happens that a component is lacking in the children which is present in the parents. This is the kind of inheritance which fits the rule first observed by Mendel. The Mendelian rule is shown up most clearly if one keeps one's eye on one character at a time. . . . If two individuals of [these] two [different] races cross, discontinuous inheritance of the character in question can often be observed. The Mendelian rule is that, in the first generation of descendants, all the children are alike, but then in the second, some individuals appear which resemble the grandparents, and other the parents. . . . The character which appears in the first generation is called the dominant, and the other, the one that is not visible, the recessive.[13]

Accordingly, von Dungern and Hirszfeld call the characters A and not-A, B and not-B: A and B are dominant, not-A and not-B recessive.

The statement of the Mendelian rule in this form, with this emphasis on the dominance and recessiveness aspect of it, was a peculiarity of these early years of the "rediscovery" of Mendelism, especially of the original rediscoverers Hugo de Vries and Carl Correns.[14] De Vries even

---

12 Karl Landsteiner, "Ueber Agglutinations erscheinungen normalen menschlichen Blutes," *Wiener klin. Wschr.* 14(1901):1270–1271.

13 Emil von Dungern and Ludwig Hirschfeld, "Ueber Vererbung gruppenspezifischer Strukturen des Blutes II," *Z. f. Immunitätsf.* 6(1910):284–292 (pp. 286–288).

14 The history of Mendelism along with its "rediscovery" by Hugo de Vries, Carl Correns, and Emil von Tschermak-Seysenegg around 1900 and its subsequent spread has a very large literature. Leslie C. Dunn, *A Short History of Genetics: the Development of Some of the Main Lines of Thought, 1864–1939* (New York: McGraw-Hill, 1965), which gives a good summary of the important ideas. The argument has been well treated by Robert C. Olby, *Origins of Mendelism* (New York: Schocken, 1st ed., 1966; 2d ed., 1985), 124–144; Olby, "Mendel no Mendelian?" *Hist. Sci.* 17(1979):53–72. The re-

stated that "the hybrid always shows the character of one of its two parents, and that in its entirety. The character which is present in one parent and absent in the other never appears reduced to half."[15] Correns protested against this absolute statement.[16] In fact, he tried later to redefine dominance as a relative concept: he was, after all, a student of Carl von Nägeli.[17] However, the emphasis on dominance and recessiveness as the most immediately striking aspect of Mendelism persisted.

Von Dungern's explanation of the Mendelian rule also gives the rule this emphasis: "The regularity consists in this, that in the first mixed generation all the children are alike, but then in the next the characters of the grandparents again appear separately."[18] The example he gives to explain it is a curious one. If two axolotls, of pure black and pure white race, are paired, the first generation are all black. If these are now mated, the second generation shows the ratio of three-quarters black to one-quarter white. This odd example seems to come from the work of Valentin Haecker, professor of zoology at Halle, though von Dungern gives no direct reference. Haecker had been working on skin colour inheritance by crossing black axolotls with an albino variety. They gave a very good demonstration of Mendelian dominance and recessiveness, except that the segregating individuals of the F2 generation that should have been whites were not pure white, but had speck-

visionist view is summarized by Peter J. Bowler, *The Mendelian Revolution: The Emergence of Hereditarian Concepts in Modern Science and Society* (Baltimore, Md.: Johns Hopkins University Press, 1989), 113–119. William Bateson's part in the popularization of Mendelism is dealt with by Robert Olby, "William Bateson's introduction of Mendelism into England: a reappraisal," *Brit. J. Hist. Sci.* 20(1987):399–420. William B. Provine, *The Origins of Theoretical Population Genetics* (Chicago, Ill: University of Chicago Press, 1971). Provine creates a neatly patterned flow of ideas, though without mentioning blood groups. He also, I feel, underestimates the part played by the German-speaking geneticists, who were not involved in the controversy he sees as his nodal point. The reconciliation of Mendelism and biometry that is the subject of Provine's book could equally well be called, as Dunn calls it (p. 120), the effect of Wilhelm Johannsen's *Elememente der exakten Erblichkeitslehre* of 1909. Provine does not mention this work; for him, this "reconciliation" comes by way of R. A. Fisher's paper "Correlation between relatives" of 1918 and is one of the bases of his book's argument. It is possible that this neglect of the German scene by the English-language historians of the sixties and seventies is linked to the connection between genetics and Naziism. See Mikulas Teich, "The unmastered past of human genetics," in Mikulas Teich and Roy Porter, *Fin de Siècle and Its Legacy* (Cambridge: The University Press, 1990), 296–234.

15 Hugo de Vries, "Sur la loi de disjonction des hybrides," *C. R. de l'Acad. des Sciences* 130(1900):845–847.
16 Carl Correns, "G. Mendel's Regel über das Verhalten der Nachkommenschaft der Rassenbastarde," *Ber. d. Deutsche Bot. Ges.* 18(1900):158–161.
17 Carl Correns, "Ueber die dominierende Merkmale der Bastarde," *Ber. d. Deutsche Bot. Ges.* 21(1903):133–147.
18 Emil von Dungern, "Ueber Nachweis und Vererbung biochemischer Strukturen und ihre forensische Bedeutung," *Münch. med. Wschr.* 57(1910):293–295.

led heads. Haecker himself interpreted albinism as the absence of a character for colour that could be complete or partial giving all grades of colour from absence to full pigmentation.[19]

If this is indeed the paper on black and white axolotls that von Dungern read, which seems quite possible, it is interesting that he pays no attention at all to these subtleties, but uses Haecker's axolotls as an example illustrating sharp discontinuity. There are no *quantitative Abstufungen* between A and not-A. Von Dungern says that in the first generation,

apart from the visible character black, the character white is also present in the germ-plasm. . . . In the formation of germ-cells, the two characters . . . remain separate and never combine with each other, so that one germ cell contains the dominant character black and the other the recessive character white. . . . If a black germ cell comes together with another black one, we have a purebred black. If a white meets a white, we have a purebred white. But if a germ cell with the character black unites with a germ with the character white the result is a black *Bastard* which also contains the white recessive character.[20]

Repeatedly in this passage, von Dungern makes the assumption of a direct, one-to-one relationship, between the genetic determinant in the germ cell and the character as it appears on the axolotl's skin. "One germ cell contains the dominant character black," he says, and even "a black germ cell."

This assumption of the virtual equivalence of the two stages of the genetic process, the hypothetical genetic "unit" and its visible expression in the "character," was a typical feature of the early years of Mendelism. The Cambridge geneticist William Bateson was not among the original rediscoverers of Mendelism, but he was one of its most important promoters in English.[21] He was well aware that the earlier emphasis on dominance had been a temporary phase only. Those who first

---

19 Valentin Haecker, "Ueber Axolotlkreuzung," *Verh. d. Deutsche Zool. Ges.* (18 Jahresversammlung, Stuttgart 9–11 June 1908), 194–205; Haecker, *Allgemeine Vererbungslehre* (Braunschwieg: Vieweg, 1911). Although in his article on axolotl crossing, Haecker finds the albinos in the second filial generation to be tinged with colour, in this general textbook he uses the axolotl crosses in exactly the way von Dungern does, as a black and white example of dominance and recessiveness (p. 220), though he repeats his doubts (on p. 231), in small print. Haecker says in his *Vorwort* that the book was originally a series of lectures that he had been giving "seit einer Reihe von Jahren, zuerst an der Technische Hochschule in Stuttgart, an der landwirtschaftliche Hochschule in Hohenheim, später in Halle" (p. v). The material was therefore quite widely diffused.
20 V. Dungern, "Vererbung biochemischer Strukturen" (1910) (n. 18), 295.
21 William Coleman, "Bateson, William," in Charles C. Gillispie, ed., *Dictionary of Scientific Biography* (New York: Scribner's, 1971), v. 1, 505–506; Beatrice Bateson, *William Bateson, F.R.S., Naturalist: His Essays and Addresses together with a Short Account of His Life* (Cambridge: Cambridge University Press, 1928).

treated of Mendel's work, he wrote, fell into the error of enunciating a law of dominance.[22] He fully accepted, however, the concept of the unit-character.[23] The symbols used in the Mendelian terminology could mean either the genetic unit or the visible character, interchangeably. Where they were so closely identified, the absence of the character implied the absence of the genetic unit.[24] As Bateson wrote,

Mendel himself probably conceived of allelomorphism as depending on the separation of a definite something responsible for the dominant character from another something responsible for the production of the recessive character. It is however evidently simpler to imagine that the dominant character is due to the presence of something which in the case of the recessive is absent.[25]

This hypothesis fitted well with the data by Lucien Cuénot on the inheritance of coat colours in mice. In Cuénot's terminology, a determinant for colour C was an allele of albinism, A. The actual colours, *gris* G and *noir* N, were also a pair of alleles. Bateson added more letters, and made colour and no-colour, grey and no-grey, black and no-black the alleles. Cuénot had already suggested that the factor C produced, or actually was, an enzyme or *une différence matérielle d'ordre chimique* that was needed to convert a substrate into pigment.[26] The factor itself, the unit-character, might actually be the enzyme.[27] Bateson's best example was the pair of characters "round" and "wrinkled" in Mendel's peas: where the factor was absent, there was no enzyme that could convert the sugar to starch, a solid storage material; as a result, the pea became wrinkled as it aged.

Von Dungern and Hirszfeld's description of the inheritance of blood group "structures" as paired alleles, A and not-A, B and not-B, at first sounds as if it, too, might be an example of the presence and absence hypothesis. Von Dungern, however, is quite clear in his explanation of the axolotl colours, that white is not "nothing." There is a "white germ cell" even in the black *Bastard*. In their paper on the Heidelberg families, he and Hirszfeld wrote, "If two germ cells with the *Anlage* for

22 William Bateson, *Mendel's Principles of Heredity* (Cambridge: Cambridge University Press, 1909, 1st ed.; 1913, 2d ed.), 13 in both editions.
23 Bateson, *Mendel's Principles* (1909, 1913) (n. 22), 5, 15, in both editions.
24 Bateson's part, and that of the Americans in this, is discussed by R. C. Swinburne, "The presence and absence theory," *Ann. Sci. 18*(1962):131–145.
25 Bateson, *Mendel's Principles* (1909, 1913) (n. 22), 76, both editions.
26 Lucien Cuénot, "L'heredité de la pigmentation chez les souris," *Arch. de Zool. Expér. et Génetique* 4th ser., *1*(1903):33–41.
27 Haecker, *Allgemeine Vererbungslehre* (1911) (n. 19), 265, equates the terms in English, French, and German: Faktorenhypothese (hypothèse factorial, presence-and-absence theory), Faktor (Déterminante, Bestimmer, caractère-unité, Elementäreigenschaft, Erbeinheit, gene).

absence [*Nichtvorhandensein*] of the character [*Eigenschaft*] meet, the individual which develops does not have the character."[28]

The *Anlage* for absence is not, that is, itself absence. But if "A and not-A" did not imply the presence and absence theory to Emil von Dungern, it certainly did to an Englishman whose genetics came directly from William Bateson. Geoffrey Keynes, assistant in the surgical unit at St. Bartholomew's Hospital in London, who wrote a short textbook of blood transfusion in 1922, following the expansion of interest in it after the end of the war, explained von Dungern and Hirszfeld's theory of inheritance in superbly Batesonian manner:

> According to this theory [that is, Mendel's] each quality in an organism which can be isolated and investigated independently is termed a "unit-character," and the appearance of each such unit-character is determined by the presence of something called a "factor" in the sexual cells or "gametes." ... Further, these unit-characters are believed to occur in alternative pairs, and at first it was supposed that each alternative pair consisted of "dominant" and "recessive" characters, the second of which could only make its presence apparent if the dominant character were absent. Subsequently it was seen that the dominant and recessive character need not necessarily consist of two positive though opposite qualities, but might be better regarded as consisting of the presence of a character and its absence. ... To represent this idea more simply a conventional notation has been used according to which the large letters of the alphabet indicate the presence and the small letters the absence of each factor.
> In order to apply this theory to the case under consideration, it has been suggested that two pairs of factors are concerned:
> A the presence of the character producing Group II
> a the absence of the character producing Group II
> B the presence of the character producing Group III
> b the absence of the character producing Group III.[29]

Keynes does not, in fact, give any reference to Bateson, though perhaps he should have, since his explanation is almost a paraphrase.

The presence-and-absence hypothesis was not as "peculiar" as has been suggested – it was not, that is, peculiar to Bateson.[30] It was initially accepted by both of the American groups of geneticists, Thomas Morgan and his team at Columbia University and William Castle and Edward East at Harvard. But it made the terminology of the complex groups of mutations, such as those determining the eye colours of the fruit fly *Drosophila* in Morgan's laboratory, very difficult to handle.[31]

28 V. Dungern and Hirschfeld, "Vererbung gruppenspezifischer Strukturen II" (1910) (n. 13), 288.
29 Geoffrey Keynes, *Blood Transfusion*, in Series, Oxford Medical Publications (London: Frowde, 1922), 86–87.
30 Coleman, "Bateson" (1971) (n. 21), 506.
31 Thomas H. Morgan, "Factors and unit-characters in Mendelian heredity," *Am. Nat.*

The terminological problems of trying to fit the eye colours into a series of paired alleles were greatly reduced by an idea put forward by Alfred Sturtevant, a member of Morgan's group, towards the end of 1913.[32] He made the new suggestion that the three rabbit coat colours, wild, himalayan, and albino, might be controlled by three different alleles at the same locus. This multiple-allele concept made the descriptions much more straightforward and was soon generally adopted. It was not compatible with the presence-and-absence hypothesis.

The presence-and-absence hypothesis was a variant of the unit-character one, but its disappearance did not immediately affect the unit-character concept. The American William Castle continued to use the unit-character to describe the coat colours of his pure-line hooded rats. He looked on coat colour as Alfred Sturtevant did, as the product of a single genetic unit. If he could alter the outward appearance of the organism by selection within the pure line, that implied that the genetic unit was being altered. It was not until 1919 that he came round to admitting that the coat colours could have been due to selection for modifying genes and that the character might not have been determined by a single unit.[33]

Castle was among the last of the geneticists to cling to the unit-character. The geneticist-historian Elof Axel Carlson has discussed its history as a concept in *The Gene: a Critical History*. His opinion about it is summarized in his chapter headings: Chapter 4, "The Unit-character Fallacy," and Chapter 5, "The Demise of the Unit-character."

Carlson feels that the concept of the unit-character, the one-to-one identification of gene and visible character, did not survive the practical demonstration that more than one factor was involved in Castle's rat coat colours, and that a change in coat colour did not have to mean that the genetic unit itself had altered. "World war I had ended," says Carlson neatly, "but for Castle it was not armistice, it was unconditional surrender."[34]

Leslie C. Dunn, who also discusses this problem, felt that it had been shot down even earlier. The *coup de grâce*, he writes, was given it by the demonstration in 1909 that some characters, such as the red colour of wheat grains, depended on several paired alleles.[35] Dunn here quotes

47(1913):5–16. The *American Naturalist* for this year carries a large number of papers discussing the problem and the terminological difficulties.

32   Alfred H. Sturtevant, "The himalayan rabbit case with some observations on multiple allelomorphs," *Am. Nat.* 47(1913):234–239.

33   William E. Castle, "Piebald rats and the theory of genes," *Proc. Nat. Acad. Sci.* 5(1919):500–506.

34   Elof Axel Carlson, *The Gene: a Critical History* (Philadelphia, Pa.: Saunders, 1966), 38.

35   Dunn, *History of Genetics* (1965) (n. 14), 100. He is quoting Wilhelm Johannsen,

from the Danish geneticist Wilhelm Johannsen, writing in 1922; but Johannsen had already said the same thing in 1913:

> Today's research on inheritance . . . has freed itself from the morphological spirit. It has given up the concept of the marker or the unit-character [*Einfach-Eigenschaft*] as a unit [*Einheit*] or as conditioned by a unit, and seeks to define those factors in the constitution of gametes and zygotes whose combinations underlie the character or property with which morphology operates directly.[36]

But Dunn's *coup-de-grâce* did not take place in 1909, nor, as we have seen, was Johannsen historically correct in saying that by 1913 "today's research" had abandoned the concept. Bateson's second edition, which appeared in that same year, was still calling it "that concept of unit-characters which is destined to play so large a part in the development of genetics," as he had done in 1909.[37] But not even Carlson, who put its demise at the end of the war, had traced it to its end. It disappeared at this time from general genetics, from the vocabulary of the *Drosophila* workers and the plant breeders, but as a basic premise of the special field of blood group genetics, it persisted for far longer.

It is curious that neither Johannsen nor Bateson made any mention in 1913 of blood groups as a source of data on human inheritance. Although three years had passed since von Dungern and Hirszfeld's papers had appeared, Bateson was still able to say, as he had in 1909, that "of Mendelian inheritance of normal characteristics in man there is as yet little evidence. Only a single case has been established with any clearness, namely that of eye colour."[38]

But although von Dungern and Hirszfeld's demonstration of blood group inheritance did not create very much interest among geneticists, the position began to change when the Hirszfelds were able to connect blood groups with the subject of race.[39]

The opportunity for this arose in 1917 when Ludwik and Hanna Hirszfeld (who was a paediatrician in peacetime) were working as bacteriologists with the *Armée d'Orient*, the Serbian army in Salonika. The idea of investigating the blood group distribution in different races had occurred to von Dungern and Hirszfeld many years before in Heidel-

---

Biologi: Traek af de biologiske Videnskabens Udvikling i det nittende Aarhundrede (Copenhagen: Nordisk Vol, 1922), 191.

36 Wilhelm Johannsen, *Elemente der exakten Erblichkeitslehre mit Grundzügen der biologischen Variationsstatistik* (Jena: Fischer, 1st German ed., 1909; 2d ed., 1913), 413, 2d ed.

37 Bateson, *Mendel's Principles* (1909, 1913) (n. 22), 5, both editions.

38 Bateson, *Mendel's Principles* (1909, 1913) (n. 22), 205, both editions, citing Charles Chamberlain Hurst, "On the inheritance of eye colour in man," *Proc. Roy. Soc.* (London), *80*(1908):85–96.

39 William H. Schneider, "Chance and social setting in the application of the discovery of the blood groups," *Bull. Hist. Med.* *57*(1983):545–562.

berg, but it would have been a long and difficult project at that time.[40] In Salonika, however, they were in contact with troops of many "races" – English, French, Italian, Serbian, Bulgarian, Russian, Greek, Hindu, Annamite, Monegasque, and African – and they were able to collect up to 500 blood samples from each of sixteen separate nationalities. They found that type B, which was uncommon in Europe, was much more frequent in Asia and Africa, and they suggested that the A factor had originated somewhere in northern Europe and the B factor in India or Tibet.[41]

There had been very little interest in blood groups at all between von Dungern and Hirszfeld's papers of 1909–1910 and this of 1919. The second edition of the standard textbook of human genetics that appeared in 1923, "Bauer-Fischer-Lenz," as it was familiarly called, was indifferent to the new race-marker. It included elaborate discussions of inherited racial differences such as skull form, physiognomy, pigmentation, and hair colour, but made only one reference to blood, to von Dungern and Hirszfeld's early paper of 1910.[42] According to the bibliography collected by Michael Hesch in 1932, most of the small literature up to 1920 concentrated on blood groups as a transfusion problem.[43] But by 1925, the subject of blood groups, genetics and race had attracted more interest: it was possible to list eighty-five publications on race alone, and blood groups were becoming established as a means of investigating disputed paternity.[44] By 1932, genetics and race together accounted for about 18 percent of the blood group literature, for a total of about 550 publications. It was still not the major interest

40 Hirszfeld, *Historia* (1967) (n. 4), 62.
41 Hanka and Ludwik Hirschfeld, "Serological differences between the blood of different races: the result of researches on the Macedonian front," *Lancet* 180(1919):675–679; Hirszfeld and Hirszfeld, "Essai d'application des méthodes sérologiques au problème des races," *l'Anthropologie* 29(1919–20):505–537.
42 Erwin Baur, Eugen Fischer, and Fritz Lenz, *Grundriss der menschlichen Erblichkeitslehre und Rassenhygiene* (Munich: Lehmann, 2d ed., 1923), v. 1, *Menschliche Erblickeitslehre,* v. 2 *Menschliche Auslese und Rassenhygiene.*
43 Michael Hesch, "Das gesammte Schrifttum der Blutgruppenforschung in den drei ersten Jahrzehnten ihrer Entwicklung," in Paul Steffan, ed., *Handbuch der Blutgruppenkunde* (Munich: Lehmann, 1932), 539–647.
44 Leone Lattes, *L'Individualità del Sangue in Biologia in Clinica e in Medicina Legale* (Messina: 1923), translated by Fritz Schiff as *Die Individualität des Blutes in der Biologie in der Klinik und in der gerichtlichen Medizin, nach der umgearbeiteten italienischen Auflage bersetzt und ergänzt durch einen Anhang, die forensisch-medizinische Verwertbarkeit der Blutgruppen von Dr Fritz Schiff* (Berlin: Springer, 1925), 98–107; P. Introzzi, "La vita e le opere di Leone Lattes," in Instituto di Medicina Legale e delle Assicurazioni dell'Università Pavia, *Leone Lattes* (Pavia: Università di Pavia, 1954), 11–26. Lattes came to blood grouping through forensic pathology. He organized the first International Congress of Blood Transfusion in Rome in 1935. A Jew, he was able to spend the war years in Argentina and return to Italy in 1945. His book was the first textbook on blood groups.

of blood group serologists: most of the papers, 70 percent before 1920 and about 50 percent afterwards, continued to deal with the practical clinical problems of blood transfusion, and the rest with the equally practical problems of serological technique.

In 1926 the race interest acquired its own organization. The naval surgeon Paul Steffan and the anthropologist Otto Reche founded the Deutsche Gesellschaft für Blutgruppenforschung, and its organ, the *Zeitschrift für Rassenphysiologie*, with a view to organizing work on the blood groups as a race-marker and putting together a detailed map of blood group distribution in Europe. The map was to help define the homeland of the great group A race.[45] The political significance of this work is clear. It was felt that group A, more common in Europe, could be identified with the nordic or Aryan type, and that blood group studies would define the natural boundaries between the Aryans and the Slavs and Jews, tainted with the ideologically loaded group B that was infiltrating from the east.[46] The project was funded by the nationalist publisher Julius Friedrich Lehmann of Munich, well known as an enthusiastic promoter of the Nordic movement. The racist, rather than scientific, implications come across clearly in the literature: "It is Paul Steffan's great merit to have recognized the significance of the blood groups for race problems. He has drawn the attention of science to blood groups, where the Jews had tried to keep silent and to conceal their racial traits."[47]

Felix Bernstein, mathematician and director of the Institute for Mathematical Statistics at Göttingen, took up Hirszfeld's studies of race and inheritance from a different perspective. Bernstein had contributed to several areas of applied mathematics before becoming interested in blood group inheritance. He was born in 1878 in Halle, the son of Julius Bernstein, a well-known physiologist. He studied in several universities, including Göttingen, where he took his mathematics degree in 1901 under David Hilbert. His first job was at Halle, where as a nineteen-year-old-student in the seminar of Georg Cantor, he had proposed a solution to a problem in set theory that gained him general admiration. In 1907 he came back to Göttingen, where he presented a

45 Paul Steffan, "Die Arbeitsweise der Deutschen Gesellschaft für Blutgruppenforschung," *Z.f.Rassenphysiol.* 1(1928):8–11.

46 Pauline M. H. Mazumdar, "Blood and soil: the serology of the Aryan racial state," *Bull. Hist. Med.* 64(1990):187–219.

47 H. Gauch, "Beitrag zum Zusammenhang zwischen Blutgruppe und Rasse," *Z.f. Rassenphysiol.* 7(1937):116–122. This passage, together with some others of similar political and racial significance, is quoted by Hirszfeld, *Historia* (1967) (n. 3), 151. Hirszfeld himself was an assimilated Polish Jew. The last half of his autobiography describes his life in Warsaw during the Nazi occupation; it is impossible to read it without weeping.

Habilitationsschrift on actuarial science. In 1911 a special *ausserordentlich* professorship in mathematical statistics and actuarial science was created for him. In 1921 he was promoted to *Ordinarius*, a move that had nearly been blocked because his colleagues had some concerns about a savings-bond loan scheme that he had worked out with the mathematician Hjalmar Schacht. He had also helped resolve a problem involving the conduction of heat that was the starting point for the so-called Laplace transformation.[48] In short, Bernstein (Figure 14.2) was an exceptionally clever and active mathematician, both in an academic and practical sense.

Bernstein's interest in the inheritance of blood groups seems to have been an outgrowth of his interest in the Mendelian mathematics of the Stuttgart statistician Wilhelm Weinberg, of which he was highly critical. But Weinberg's methods were essential to mathematical Mendelism, and even Bernstein could not avoid using the most important of them.[49] He applied Weinberg's equation to demonstrate the unsatisfactory implications of the Swedish geneticist Herman Nilsson-Ehle's hypothesis on the inheritance of wheat colours. Nilsson-Ehle thought that they were determined by a series of paired Mendelian factors for the presence and absence of colour.[50] Bernstein's genetic hypothesis was that of Thomas Morgan, who had described his eye colour mutants in *Drosophila* as multiple alleles at a single locus, rather than a string of presence-and-absences, in accordance with Sturtevant's suggestion of 1913.[51]

Using Nilsson-Ehle's published figures, Bernstein compared the expected results on the two possible hypotheses, of two- or three-factorial inheritance, with the observed figures. Nilsson-Ehle provided data for three cases: the red grain colour in *Sammet* and *Buckel* wheats and the character *ligula* in oats. A fourth case, that of the capsule-form in *Bursa bursa-pastoris*, the shepherd's purse, came from a paper by the American geneticist George Shull. All four seemed to give better agreement be-

48 Henry Nathan, "Bernstein, Felix," in Charles C. Gillispie, ed., *Dictionary of Scientific Biography* (New York: Scribner's, 1970). v. 2, 58–59. Felix Bernstein, "Nekrolog," A. Peter, Dean, University of Göttingen, to Ministry of Sciences, Arts and Public Education, letter dated 15 July 1921; Becker, Ministry of Sciences, Arts and Public Education, to the Curators, University of Göttingen, letter dated 16 September 1921: Personalakt Felix Bernstein, Universitätsarchiv, Georg-August Universität, Göttingen.

49 Pauline M. H. Mazumdar, "Two models for human genetics: blood grouping and psychiatry in Germany between the wars," *Bull. Hist. Med. 69*(3) (1995).

50 Felix Bernstein, "Die Theorie der gleichsinnige Faktoren in der Mendel'schen Erblichkeitslehre vom Standpunkt der mathematischen Statistik," *Z. f. indukt. Abstamm.- u. Vererbl. 28*(1922):295–323. On Nilsson-Ehle, see Arne Müntzing, "Herman Nilsson-Ehle," in Charles C. Gillispie, ed., *Dictionary of Scientific Biography* (New York: Scribner's, 1974), 129–130.

51 Sturtevant, "Observations on multiple allelomorphs" (1913) (n. 32).

Figure 14.2. Felix Bernstein, Director of the Institute for Mathematical Statistics, deposed under Nazi law in 1933. Portrait photograph, n.d. (Courtesy of Niedersächsischen Staats- und Universitäts-bibliotek, Göttingen)

tween observed and expected figures on a hypothesis of three alleles at the same locus, rather than two presences with matching absences at two different loci.[52]

From the critique of two-factor inheritance in wheats it was a short step to the testing of Hirszfeld's hypothesis of two-factor inheritance in

52 These were the experiments, according to Johannsen, that gave the *coup-de-grâce* to the presence-and-absence theory in 1909. They are discussed by all the historians of genetics: Dunn, *History of Genetics* (1965) (n. 14), 100; Carlson, *The Gene* (1966) (n. 34), 27–28; Provine, *Population Genetics* (1971) (n. 14), 115 ff.

blood groups.[53] Hirszfeld himself had provided the data in his collection of figures from Salonika of 1919. Bernstein calculated expected gene frequencies using a modification of the Weinberg equilibrium and compared them with those observed.[54] He came to the conclusion that a better fit for the data was provided by an assumption of three alleles at a single locus, rather than two independent pairs.

According to Hirszfeld's hypothesis, the four phenotypes A, B, O, and AB distinguished by tests with anti-A and anti-B could have ten genotypes, which Bernstein transposed into gene frequencies from which to make his calculation (Table 14.1). However, if there were three alleles at one locus, as Bernstein proposed, there were only six possible genotypes (Table 14.2), where R represents the "recessive" or O component, regarded by Hirszfeld as "absence" at both loci. Bernstein saw the gene R determining group O as the original prototype gene, with two mutations A and B, which originated in different races that were now blending in Europe. Hirszfeld, too, thought of A and B as two separate races.

Bernstein converted each of the blood group phenotypes into its

53 Felix Bernstein, "Ergebnisse einer biostatistischen zusammenfassenden Betrachtung ber die Erbstrukturen der Menschen," *Klin. Wschr.* 3(1924):1495–1497; Bernstein, "Zusammenfassende Betrachtungen über erblichen Blutstrukturen des Menschen," *Z.f. indukt. Abstamm.- u. Vererbl.* 37(1925):237–270. These two papers are translated in F. R. Camp and F. R. Ellis, eds., in *Selected Contributions* (n. 4), v. 1, *Dunsford Memorial* (1966), *The ABO System* 83–90 (Bernstein, 1924); 91–138 (Bernstein, 1925).

54 The Hardy-Weinberg equilibrium was a deduction from Mendel's original binomial derived independently by the Cambridge mathematician Godfrey Harold Hardy in "Mendelian proportions in a mixed population," *Science*, n.s., 28(1908):49–50; and by Wilhelm Weinberg, a German statistician who founded the Stuttgart branch of the eugenics society, in "Ueber den Nachweis der Vererbung beim Menschen," *Jahresh. d. Ver. f. vaterl. Naturk.* (Württemberg) 64(1908):368–382. Mendel's own version of the binomial had supposed equal numbers of his two alleles:

$$Aa \times Aa = A^2 + 2Aa + a^2.$$

Weinberg incorporated the frequencies of the two alleles as variables in the equation and saw that a population in which the two alleles A and a occur in the frequencies p and q will consist, after one generation of random mating of the three genotypes AA, Aa and aa in equilibrium proportion:

$$pA + qa = 100 \text{ percent}$$
$$p^2AA + 2pqAa + q^2aa = 100 \text{ percent}.$$

For three alleles, this becomes:

$$p + q + r = 100 \text{ percent}$$
$$p^2AA + 2prRA = q^2BB + 2qrRB + 2pqAB + r^2RR = 100 \text{ percent}.$$

These papers are discussed by Provine, *Population Genetics* (1971) (n. 14), 133–136; and by Curt Stern, "The Hardy-Weinberg law," *Science* 97(1943):137–138; and in Stern, *Principles of Human Genetics* (San Francisco, Calif.: Freeman, 1st ed., 1949; 2d ed., 1960), 149–173 (p. 157); and by Mazumdar, "Two models" (1991) (n. 49).

Table 14.1. *Felix Bernstein's rendering of Ludwik Hirszfeld's two-locus theory of the inheritance of the ABO blood groups*

Phenotype	Genotype	Probable gene frequency	
Group O:	aa/bb	$(1-p)^2(1-q)^2$	$= p^2\,q^2$
Group A:	AA/bb or	$p^2(1-q)^2$	
	Aa/bb	$2p(1-p)(1-q)^2$	$= (1-p^2)q^2$
Group B:	aa/BB or	$(1-p)^2q^2$	
	aa/Bb	$2(1-p)^2\,q(1-q)$	$= p^2(1-q^2)$
Group AB:	AA/BB or	p^2q^2	
	Aa/Bb or	$2p(1-p)\,2q(1-q)$	
	AA/Bb or	$2p^2q(1-q)$	
	Aa/BB	$2p(1-p)q^2$	$= (1-p^2)(1-q^2)$

The four phenotypes are controlled by two loci, A and B, each with its "absence," a and b. Each phenotype has several possible genotypes: the probable frequency of the gene for A in the population is p, and that for a is $(1-p)$ or p, in Bernstein's notation, and similarly for B and b, represented by q and $(1-q)$, or q.

From the table,

$$
\begin{aligned}
\underline{A} + \underline{AB} &= (1-p^2)q^2 + (1-p^2)(1-q^2) &&= 1-p^2 \\
\underline{B} + \underline{AB} &= p^2(1-q^2) &&= 1-q^2 \\
\underline{AB} &= (1-p^2)(1-q^2) && \\
&\text{i.e., } (\underline{A} + \underline{AB}) \times (\underline{B} + \underline{AB}) &&= \underline{AB}
\end{aligned}
$$

equivalent in terms of genotypes and gene frequencies, according to the Weinberg formula. Using the frequencies (symbolized by the underline) of the blood groups in a population, he calculated that, if the two-locus hypothesis held,

$$(\underline{A} + \underline{AB}) \times (\underline{B} + \underline{AB}) = \underline{AB}$$

but that instead $(\underline{A}+\underline{AB}) \times (\underline{B}+\underline{AB})$ was always greater than $\underline{AB}$. The figures for Hirszfeld's different racial groups in no case matched the expectation (Table 14.1).

With three alleles at one locus, the sum of the probabilities for the frequencies of the three should be

$$p + q + r = 100 \text{ percent.}$$

With the same procedure (Table 14.2) and still using Hirszfeld's populations, Bernstein showed that his hypothesis accounted much better for the results: p + q + r added up to a figure between 98.9 and 102.2

Table 14.2. *Felix Bernstein's one-locus, three-allele theory of the inheritance of the ABO groups*

Phenotype	Genotype	Probable frequency		
Group O:	RR	r2		
Group A:	AR or	2pr		
	AA	p^2	=	$2pr + p^2$
Group B:	BR or	2qr		
	BB	q2	=	$2qr + p^2$
Group AB:	AB	2pq		

In this case there are three alternative alleles A, B and R at a single locus. Their gene frequencies in the population are represented by p, q and r, so that p + q + r = 100%

From this,

$$\underline{O} + \underline{A} \quad = \quad r^2 + 2pr\ p^2 \quad = \quad (r + p)^2$$
$$\underline{O} + \underline{B} \quad = \quad r2 + 2qr + q2 \quad = \quad (r + q)^2$$
$$p \quad = \quad 1 - \sqrt{\underline{O} + \underline{B}}$$
$$q \quad = \quad 1 - \sqrt{\underline{O} + \underline{A}}$$
$$r \quad = \quad \sqrt{\underline{O}}$$

percent for sixteen populations from which he had gathered data, well within the margins of error.

Following Bernstein's paper of 1925, a wave of Mendelian algebra broke over blood group serology, as the triple-allele hypothesis was debated in the world literature. Mathematically talented serologists and others, who like Bernstein himself were not serologists but enjoyed a mathematical problem, proposed their own variations. Tanemoto Furuhata of Japan claimed to have come upon the idea of three alleles independently. It was noted, however, that he used Bernstein's peculiar terminology of R for the gene that produced O.[55]

Some of the best-known serologists, such as Fritz Schiff of Berlin and the young Laurence Snyder of Boston, supported Bernstein at once. Karl Landsteiner, less inclined to rush to a conclusion, set his young co-worker Philip Levine to review and examine the problem and collect new data to test Bernstein's hypothesis. By the late twenties, he seems to have accepted the triple-allele solution, though he was still on the lookout for exceptions to it as late as 1931.[56]

55 Fritz Schiff to Felix Bernstein, letter d. July 4, 1926: Nachlass Felix Bernstein, Box 1a, Niedersächsischen Staats- und Universitäts-bibliotek, Göttingen.
56 Landsteiner (Levine) laboratory notes on tests of Hirszfeld vs Bernstein, dated February 1928; "Exceptions to Bernstein's theory," dated August 1931: Landsteiner Papers, 450 L239, Box 5, Folder 6, Rockefeller Archives.

Landsteiner wrote to Bernstein in respectful terms, pointing out that his findings of the presence of blood group B in non-human primates supported Bernstein's suggestion of a separate origin for the A and B races.[57] Schiff supported him both in published reviews and in a series of personal letters.[58] The Viennese engineer Siegmund Wellisch, on the other hand, only slowly came around, through many weighty papers, to accepting the triple-allele hypothesis, but without using Bernstein's terminology.[59] Wellisch's writing was part of the wave of papers appearing during the twenties in which people played mathematically with the published results of blood group studies, setting up and solving equations, comparing the results of different genetic schemes, and expressing them as equations, curves, and complex nomograms designed to demonstrate the congruence of one or other system with the available test results.

One of the crucial pieces of evidence was the existence of group O children of AB mothers. If A and B were at different loci, as in Hirszfeld's scheme, this was theoretically possible: the mother could have the genotype Aa/Bb, and so could have an aa/bb child. But if A and B were at the same locus, an AB mother must transmit either A or B to her child. There had been quite a few AB mothers with O children in the early literature, but as the serologists became aware of their significance as test cases, they became more self-conscious about publishing information on them, and fewer and fewer reports appeared. Landsteiner and Levine were among those who combed family material for examples of the critical O x AB mating. Tests were repeated, samples checked, and as time went on the positives began to be put down to technical errors, weak sera, labelling mistakes, or even non-maternity. The last case to appear in the literature was one of Hirszfeld's own of 1927. It was a last chance for him, and he studied it much more carefully than usual.[60]

The problem was reviewed over and over again: it was seen as immensely significant for human genetics. The argument about the inheritance of the blood groups generated a continually developing mathematical apparatus, as well as large series of test results involving

57 Landsteiner to Bernstein, letter dated 29 September 1925: Nachlass Felix Bernstein, 1a, Niedersächsischen Staats- und Universitäts-bibliotek, Göttingen.
58 Correspondence between Fritz Schiff and Bernstein, 1924–1930: Nachlass Felix Bernstein, Box 1a, Niedersächsischen Staats- und Universitäts-bibliotek, Göttingen.
59 Siegmund Wellisch, "Die Vererbung der gruppenbedingenden Eigenschaften des Blutes," in Steffan, *Handbuch* (1932) (n. 43), 112–230.
60 Hirszfeld, *Konstitionsserologie* (1928) (n. 4), 109. The serologist Fred H. Allen, in his introduction to the translation (1969), thinks that it might have been a case of an A antigen with weak B characteristics.

numbers never before seen in human material, numbers on the *Drosophila* scale. Where human genetics had had to rely on rare genetic diseases, the new marker was a normal one, so that any family could now provide material for Mendelian analysis. The relation between a blood group gene and its expression as a phenotype seemed to be perfectly clear: it was present at birth and unaltered by anything that happened later. The blood groups were the ideal human genetic marker, nature untouched by nurture.[61]

Two assumptions made it possible for Bernstein to calculate these gene frequencies from the results of the immunological tests. He is aware of one of these, but not the other.

His first assumption is that the relation of antigen to gene is direct and simple: "One can even go as far as to say that the agglutinins are . . . similar to secretions of the chromomeres."[62] There is, that is to say, a direct one-to-one relationship between the genetic "unit" and the blood group agglutinogen as "character." He is aware that this is not invariably and necessarily true, for he refers to "epigenetic factors such as Correns and Goldsmith have shown to occur."[63] But since the numerical ratios predicted are verified in practice, he accepts this assumption as valid.

His second assumption, of which he is unaware, is that the same one-to-one relationship exists between antibody and antigen. In the case of blood group tests, it is not the agglutinogen but the agglutination reaction that is the observable biological phenomenon, the "character" that represents the genetic "unit." When Bernstein writes:

$$\underline{A} = (AR + AA) \text{ or } (2pr + p^2)$$

he equates three stages. $\underline{A}$ is the frequency of positive results with a test serum anti-A, (AR + AA) represents the chemical structures on the cells, and $(2pr + p^2)$ the corresponding gene frequencies. It is his unquestioned acceptance of the receptor theory that makes the first stage possible. That he does in fact refer to Ehrlich for his immunology is shown by his citation of Ehrlich and Morgenroth's work on haemolysins, from which he adopts an explanation of why an individual whose cells carry A antigen does not have anti-A in his serum. Bernstein writes "We must assume, in accordance with the side-chain theory. . . ."[64]

61 Mazumdar, "Two models" (1991) (n. 46).
62 Bernstein, "Zusammenfassende Betrachtungen" (1925) (n. 53), 247 (p. 104 in translation).
63 Bernstein, "Zusammenfassenden Betrachtungen" (1925) (n. 53), 252 (p. 109 in translation).
64 Ehrlich and Morgenroth, "Ueber Haemolysine, III" (1900) (n. 11), 453–457, re-

From time to time throughout this long period, one finds the three assumptions – presence-and-absence, the unit-character, and the receptor – made explicit. Though the leading American geneticists had dropped the first two, all three persisted in the work of the blood groupers. Reuben Ottenberg, of the Mount Sinai Hospital in New York, for example, writing just before Bernstein, refers to the unit-character, though he points out that not-A and not-B are not *mere* absences:

> It would be equally possible and would lead to the same final result, to regard as unit-characters either the serum agglutinins or the susceptible substance on the red cells. . . . The quality A is dominant to Absent-A and B to Absent-B. The qualities Absent-A and Absent-B are not mere blanks but are definite qualities of red cells associated respectively with the development in the serum of and [that is, the agglutinins anti-A and anti-B].[65]

In the same year, 1921, Sanford Hooker and Lilian Anderson from the Evans Memorial Hospital in Boston presented a paper to the American Association of Immunologists in which they pointed out, as was often done, that the methods they used developed from those of Ehrlich and Morgenroth's haemolysin absorptions of 1899–1901. They began by citing "Ehrlich and Morgenroth, who showed (1899) that haemolysins were adsorbable; Bordet (1899) . . . and . . . Malkoff (1900), who demonstrated the same phenomenon," and they quoted other studies of 1920 and 1921 that used the differential absorption method. They then went on to use it themselves.[66]

Like Hooker and Anderson, the British serologist Charles Todd, working on animal blood groups at the National Institute for Medical Research in Hampstead, began a paper in 1930 by citing Bordet's species specificity studies, and then Ehrlich and Morgenroth's experiments with the absorption of multispecific sera with cells of different specificities. Todd had worked with R. G. White on the production of serum for the treatment of cattle plague in Egypt in 1910 and had had access to bovine sera that had been produced by isoimmunizing cattle with blood from other cattle. When one of these sera was "exhausted" with the red cells of an individual ox, it remained active against many others. When the cells of a different animal were used, its range of activity

ferred to by Bernstein, in "Zusammenfassenden Betrachtungen" (1925) (n. 53), 247–248 (p. 104 in translation).
65 Reuben Ottenberg, "Hereditary blood qualities: medicolegal application of human blood-grouping," *J. Immunol.* 6(1921):363–385 (p. 367–368). For Landsteiner's reaction to this paper, see Chapter 16 (n. 24).
66 Sanford B. Hooker and Lilian M. Anderson, "The specific antigenic groups of the four groups of human erythrocytes," *J. Immunol.* 6(1921):419–444.

against the various blood samples was different. Todd summarized his position in accordance with Ehrlich's side-chain theory:

This is entirely in accordance with the ingenious hypothesis put forward by Ehrlich, that every red blood-cell possesses a large number of haptophore groups, regarded as analogous to the side-chains of a complex organic molecule, each of which is able to combine in the animal body with fitting receptors. While the side-chain theory in its entirety may perhaps be regarded more as a schematic presentment of the collective phenomena of immunity than as a representation of the actual conditions, and must not be taken too literally, it is difficult to see how the serological behaviour of the red blood cells is to be explained without postulating a large number of separate affinities. . . .

Using Ehrlich's nomenclature, then, we may say that the red cell . . . can be shown to possess a large number of different receptors each of which is antigenic. . . . The receptors are different in the red cells of every individual . . . although many receptors are common to many individuals. As each receptor gives rise to a corresponding antibody . . . immunization with the red cells of different individuals will produce different antisera.[67]

Such an explicit statement of acceptance of the Ehrlich theory is unusual, however. More generally, the red cells are simply thought of as carrying "specific receptors," the specificity of the receptors implying that of the antibodies. In his massive review of 1933, Hirszfeld refers to "group specific receptors" and to the development of isoreceptors early in embryonal existence.[68] Oluf Thomsen, the Danish serologist, describes *das Thomsen'sche Phänomen* as the uncovering of a "latent receptor" that makes cells exposed to an infected medium polyagglutinable.[69]

The unit-character concept, though it had disappeared from general genetics, was thus retained in the special case of blood group genetics, and in company with the receptor theory it formed the basis of all blood group studies of populations until the 1950s. It is possible to say that in general both the immunology and the *Vererbungsmathematik*, the expanding apparatus of Mendelian mathematics applied to the blood groups, sat firmly on a base of Ehrlich's receptor theory, and on the differential absorption methods that he and Morgenroth had used in the studies on haemolysis of from 1899 to 1901.

Perhaps for this very reason Landsteiner had not been involved in

67 Charles Todd, "Cellular individuality in the higher animals, with special reference to the individuality of the red blood corpuscle," *Proc. Roy. Soc.* (London), B *106*(1930):22–44 (pp. 37–38).
68 Ludwik Hirszfeld, "Hauptprobleme der Blutgruppenforschung in den Jahren 1927–1933," in *Ergebn. d. Hyg. Bakt. Immunitätsf. 50*(1934):54–218 (pp. 54, 117).
69 Oluf Thomsen, "Ueber bakterielle Veränderung der Agglutinabilitätsverhältnisse der roten Blutkörperchen," *Acta Med. Scand.* (Stockholm) *70*(1929):436–448.

blood grouping until the mid-1920s. He had shown no interest in his own paper of 1901, which described the human blood groups, and apart from the work done by von Decastello and Sturli in the years immediately after its publication, he apparently never referred to it again until after his transfer to the Rockefeller Institute in New York. It seemed to lie outside his province: he could not work with such sharp specificities.

15

The Specificity of Cells and the Specificity of Proteins: Landsteiner's Temporary Compromise

Landsteiner had left Vienna to improve his prospects, but in Holland he found that the conditions at his new post were not good. He continued his efforts to find somewhere he could work: he sent some reprints, including a copy of his paper on antigens of known structure of 1918, to Simon Flexner, director of the Rockefeller Institute in New York. Flexner read Landsteiner's paper and was very impressed. He first offered financial help, which Landsteiner did not accept, saying guardedly that there were "obstacles in his present situation." Flexner then turned to a man he knew in Holland, Storm van Leeuwen (Figure 15.1), to find out what Landsteiner had meant by that.

Van Leeuwen wrote Flexner a long and graphic letter, telling him about Landsteiner's "present situation": he had a post as *Prosector*, or pathologist, in a small private Catholic hospital in The Hague, where he was responsible for all the routine clinical pathology of blood and urine, the bacteriology, the post-mortems, and the histological reports. It was work that he was, of course, well able to handle, since he had been doing a good part of it for many years at the Wilhelminenspital, and before that at Weichselbaum's institute in Vienna. But here the laboratory was one small room, where all the clinicians came to do their urine examinations and to chat over a cup of coffee. The only assistant was a friendly nun who was also the hospital church's organist, and who took time off for prayers. The position, as Landsteiner said, was not suited to carrying on experimental work. However, he was managing to do some work on two afternoons a week at the Pharmaco-Therapeutic Institute of the Rijks-Universiteit in Leiden, in Storm van Leeuwen's department. Van Leeuwen himself was working on the pathology of asthma and anaphylaxis.[1] After reading van Leeuwen's letter,

1 Karl Landsteiner to Simon Flexner, letter d. Den Haag, 6 December 1920: Storm van Leeuwen to Simon Flexner, letter dated Leiden, 7 May 1921: Landsteiner-Mackenzie Papers, B L23m, Box 1, American Philosophical Society; on Storm van Leeuwen's

Figure 15.1. Willem Storm van Leeuwen, director of the Pharmaco-Therapeutic Institute at the Rijks-Universiteit, Leiden, from 1920. Portrait photograph, n.d. Landsteiner worked in his laboratory two afternoons a week while he was at the R. K. Ziekenhuis, 1919–1921. (From the National Library of Medicine, Bethesda, Md.)

Flexner changed his offer of a grant to an invitation to spend a few weeks at the Rockefeller Institute, with a view to arranging a post there

work, see Hans Schadewaldt, *Geschichte der Allergie,* 4 v. (Munich-Deisenhofen: Dustri-Verlag, 1979), v. 1, 316–317.

Figure 15.2. Karl Landsteiner, at about the time he left Europe. (From the collection of George Mackenzie, American Philosophical Society, Philadelphia, Pa.)

for Landsteiner.[2] In April 1923, Karl Landsteiner, his wife Helene, and his son Ernst left The Hague for New York (Figures 15.2, 15.3). As a member of the Rockefeller Institute, head of his own laboratory, he took up one of its highest scientific positions.

The Rockefeller Institute (Figure 15.4) played an important part in

2 Simon Flexner to Storm van Leeuwen, letter dated 19 April 1921: Landsteiner-Mackenzie Papers, B L23m, Box 1, American Philosophical Society.

Figure 15.3. Helene Landsteiner, signed photograph, n.d. (From the collection of George Mackenzie, American Philosophical Society, Philadelphia, Pa.)

the history of American medicine. It was the flagship of the Rockefeller Foundation, one of the philanthropic foundations whose outpouring of funding for the new "scientific medicine" set the course of American medicine and its highly scientific style for the twentieth century.[3]

3 On the Rockefeller Institute, see Howard M. Berliner, *A System of Scientific Medicine:*

Figure 15.4. The Rockefeller Institute, New York, N.Y. (From a postcard in the collection of George Mackenzie, American Philosophical Society, Philadelphia, Pa.)

The research at the institute from its foundation in 1901 was intended to focus on the biological, chemical, and physical basis of physiology and pathology, especially on the new sciences of the early twentieth century, that is, on bacteriology and serology at first, then on immunology and immunochemistry.

Simon Flexner (Figure 15.5), director during most of Landsteiner's tenure, had been trained in bacteriology at Johns Hopkins under William H. Welch, the great importer of German scientific medicine into America. Flexner's early work at the institute was on the bacteriology of infantile diarrhoea, a study in the classic Koch manner, its roots in the clinic and its branches in the laboratory.[4] Beginning in 1900 with projects like Flexner's, the institute covered the whole spectrum of laboratory medicine: from clinical cases in the wards of the attached Rock-

Philanthropic Foundations in the Flexner Era (London: Tavistock, 1985), 100–138; George W. Corner, *A History of the Rockefeller Institute, 1901–1953: Origins and Growth* (New York: Rockefeller Institute Press, 1964), 237–248. Simon Flexner was Director from 1901 to 1935, and member of the Board of Trustees from 1910 to 1935, when he retired.

4   Simon Flexner and L. E. Holt, *Bacteriological and Clinical Studies of the Diarrheal Diseases of Infancy, with reference to the Bacillus dysenteriae (Shiga)* (New York: Rockefeller Institute, 1900).

Figure 15.5. Simon Flexner, director of the Rockefeller Institute for Medical Research, 1903–1935. Portrait photograph, n.d. (From the National Library of Medicine, Bethesda, Md.)

efeller Hospital, to bacteriology, immunology, and then the chemistry underlying immunological specificity. René Dubos, in his book on Oswald Avery and his research group at the institute from the mid-twenties, shows them working across just such a spectrum of linked projects.[5] Avery's projects were a more highly developed form of Flexner's dysentery studies of 1900, which had stopped at the level of serological identification and vaccine preparation. At the Rockefeller Institute, bacteriology, serology, immunology, and immunochemistry reached their apotheosis: they were the core studies of twentieth-century scientific medicine.

In many ways, a membership at the Rockefeller Institute was a uniquely privileged and protected position for a researcher. There was neither routine hospital work nor regular teaching to do: there was no need for Landsteiner to snatch a few hours away from the regular job of doing the hospital post-mortems to pursue his own projects, as he had always had to do in Vienna and in The Hague. He was protected, too, from the institutional power of groups such as Paltauf's at the serum institute in Vienna, and from the old controversies that still polarized thought in German-speaking immunology. He was protected from the growing anti-Semitism that, a decade later, was to silence so many of his contemporaries in Europe, including both Felix Bernstein and Ludwik Hirszfeld. His fellow members included a number of distinguished names in the biomedical sciences, and there was ample opportunity, and encouragement, to interact with them and to tap the expertise of other departments.

Landsteiner arrived at the Rockefeller with the most distinguished possible reputation, in a field that was accorded the highest possible status there. Flexner prefaced Landsteiner's first annual report to the Board of Scientific Directors with the following words:

Dr. Landsteiner, who came to the Rockefeller Institute by way of the Hague where he worked for a number of years, is one of the most eminent living immunologists. His work has been of fundamental character and he is responsible for important discoveries in immunology, not a few of which have found application to practical medicine. He is continuing in his chosen line of research at the Institute, and is exhibiting a wholesome co-operative spirit as the report, given in his own words, of his work which follows shows.[6]

He should have been happy. But Landsteiner was not by nature a happy man. He tended to feel isolated, persecuted, and depressed, and that did not change when he arrived at the Rockefeller. He did not fit kindly into its hierarchical structure, a structure dominated by Flexner,

5 René J. Dubos, *The Professor, the Institute and DNA* (New York: Rockefeller University Press, 1976), 35–46; 101–112.
6 Report of the Director of Laboratories, Division of Pathology and Bacteriology, 19 October 1923: Scientific Reports to the Corporation and the Board of Scientific Directors, v. XI (1922–1923), 47–50 (p. 47); Record Group 439, Box 4, Rockefeller Archives.

who guided both the research staff and the Board of Scientific Directors. Landsteiner was now never to be the director of his own institute, and the absence of the power and the responsibility that would have gone with it chafed him. He felt restricted by conditions in the institute, by having to prepare a budget, and by his inability to expand his research operation beyond what he was permitted.[7] He was irritated by having to write annual reports, and by being given a particular room: he felt that Flexner had it in for him, as Philip Miller, one of his first co-workers, remembered later.[8] Landsteiner did not like or respect Flexner and referred to him as "the Emperor." Authoritarian himself, he seemed to resent Flexner's authority in the institute hierarchy.

Flexner, on the other hand, greatly respected Landsteiner for his work but found him difficult to deal with. Flexner was keen to have the senior members cooperate with each other and interact socially with each other's groups, for example, by having lunch together. Landsteiner, however, was not a person to whom a co-operative spirit came easily. He soon slipped into the habit of lunching surrounded by his respectful juniors, whose conversation he dominated. He did not want to train a generation of American researchers, as Flexner had hoped: the co-workers in his laboratory were there to work on his projects under explicit direction, nothing more, though that was a problem that Flexner also had with several other distinguished heads of laboratories. No one called him by his first name, and he was seldom at ease in talking to others: Flexner himself said later that Landsteiner "did not have the warm qualities that made informal personal relations easy." He was "stiff and formal," and according to Flexner, "straightened up" whenever they met. But, Flexner said, one could not help admiring what he did.[9]

In spite of the difficult working conditions in Holland, and with only two afternoons a week at Storm van Leeuwen's department, Landsteiner had driven on with his research. He moved further in the same direction that he had established during his last days at the Wilhelminenspital. In a series of papers written in Dutch, he now distinguished conceptually between the antigen as a substance that produced an an-

7 Constantin Levaditi to George Mackenzie, letter n.d., in answer to one of 25 May 1946; Clara Nigg, conversation with Mackenzie, dated 11 April 1946; Clara Nigg, conversation with Mackenzie, dated 11 April 1946; Landsteiner-Mackenzie Papers, B L23m, Box 2, American Philosophical Society.
8 C. Philip Miller, conversation with George Mackenzie, dated ? May 1944; Landsteiner-Mackenzie Papers, B L23m, Box 1, American Philosophical Society.
9 Simon Flexner, conversation with George Mackenzie, dated 2 April 1944; Landsteiner-Mackenzie Papers, B L23m, Box 1, American Philosophical Society. The problem of training also arose with other members, such as Jacques Loeb, Phoebus Levene, and Alexis Carrel, according to Corner, *The Rockefeller Institute* (1964) (n. 3), 152–153.

tibody response when injected into an animal and the antigen as a substance that reacted with antibody.[10] He introduced the word "haptene" to describe the chemical grouping that, without itself being antigenic, would still react with antibody. The model for this was his azo-compound coupled to protein: the example in nature which it appeared to fit was the so-called heterogenetic antigen, described by the Swedish serologist John Forssman in 1911. Forssman had first found it in sheep cells, but had discovered later that something very close to the same antigen was present in several different species of animal.[11] Like Landsteiner's azo-protein, it consisted of two separable parts, which were only antigenic when they were able to work together.

The results of the Dutch experiments still conformed to the patterns of Landsteiner's earlier work. The boundaries between these model species, the azo-protein antigens, were smooth transitions: antigenic specificity was always a matter of more or less good fit. Up to his arrival in New York, the boundaries between species had always shown themselves to be smooth transitions between quantitative maxima, with no sharp-edged discontinuities. His experiments with his artificial azo-protein antigens, chemical models of the Forssman antigen, had all gone to support this point of view. But soon after his arrival at the Rockefeller, something occurred that brought this entire *parti pris* position into question. This was the "Serological examination of a species-hybrid," which he and his Rockefeller associate, chemist James van der Scheer, carried out early in 1924.[12]

It was the product of a contact with the physiologist Jacques Loeb, a member of the Rockefeller Institute and one of Landsteiner's new colleagues. Loeb had suggested that species characters might not be determined by Mendelian factors, but by the cytoplasm of the egg alone.[13] To investigate this notion, Landsteiner and van der Scheer took the

10 Karl Landsteiner, "Over heterogenetisch antigeen," *K. Akad. Wetenschappen te Amsterdam. Verslag van de gewone Vergaderingen der wisen natuurkundige Afdeeling* 29(1920–1921):1118–1121; Landsteiner, "Over het samenstellen van heterogenetisch antigeen uit hapteen en proteine," *K. Akad. Wetenschappen te Amsterdam, Versl. v.d. gew. Vergad. d. wis – en natuurk. Afd. 30*(1921):329–330.

11 John Forssman, "Die Herstellung spezifischer Schaf-hämolysine ohne Verwendung von Schafblut," *Biochem. Z. 37*(1911):78–115.

12 Karl Landsteiner and James van der Scheer, "Serological examination of a species-hybrid," *Proc. Soc. Exper. Biol. Med. 21*(1924):252, C. 191; Landsteiner and van der Scheer, "Serological examination of a species-hybrid, I. On the inheritance of species-specific qualities," *J. Immunol. 9*(1924):213–219, C. 194; Landsteiner and van der Scheer, "Serological examination of a species-hybrid, II. Tests with normal agglutinins," *J. Immunol. 9*(1924):221–226, C. 195.

13 Jacques Loeb, *The Organism as a Whole from a Physico-Chemical Point of View* (New York: Putnam, 1916), 8, 70; Loeb, "Is species-specificity a Mendelian character?" *Science 45*(1917):191–193.

most easily obtainable species hybrid, the mule, and its parents, a horse and donkey, and tested their red cells using a haemagglutination technique. With species so close together in the zoological scale, the smooth quantitative differences of the precipitin reaction using serum proteins would have made the investigation *possible*, if it had been strictly quantitative, but very difficult. Using the red cells and agglutination reactions, however, they found that it was disconcertingly easy: the specificities were sharply defined and could be separately absorbed from the antisera. There was no seamless blending here, even though these three "species" were so close together in the zoological scale. Landsteiner concluded that what was being distinguished here was not the species specificity, but specific inheritable isoagglutinogens, like those that "Mendelized" in the human families of von Dungern and Hirszfeld.

Sharp differences in isoantigens were not completely new to Landsteiner. He himself had discovered them in human blood, and he was, like everyone else, perfectly familiar with Ehrlich and Morgenroth's experiment of 1900 with the goat bloods. But this experience seems to have had a specially dislocating effect on his thinking, perhaps because it coincided with the dislocation he had experienced in moving to New York. Landsteiner's next project was highly significant for his personal development. It shows him prepared to re-examine and revise old antagonisms and habits of thought. He looked again at his own published objections to the technique of differential absorption as used by Ehrlich, and its extension in Malkoff's experiment of 1900.[14]

The conclusions that Ehrlich and Malkoff had drawn on the sharpness of receptor specificity and on the multiplicity of antibodies of separate specificity in an immune serum had been accepted by serologists all over the world. The result was twenty years of work on the specificity of haemolysins, just as Ehrlich himself had predicted in 1906.[15] These same conclusions on the sharpness of receptor specificity, passed on through von Dungern to Hirszfeld, had allowed this part of Ehrlich's theory to survive for many years after the Ehrlich chemical interpretation of receptor specificity had disappeared from current thought. As pointed out in Chapter 14, by 1924 a great deal of work was being done on red cell specificity on the basis of this concept, most of it referring

14 G. M. Malkoff, "Beitrag zur Frage der Agglutination der rothen Blutkörperchen," *Deutsch. med. Wschr.* 26(1900):229–231.
15 Paul Ehrlich, "A general review of the recent work in immunity," in Charles Bolduan, ed. and trans., *Studies in Immunity by Professor Paul Ehrlich and His Collaborators* (New York: Wiley, 1st ed., 1906; 2d. ed., 1910), 577–586 (p. 579).

directly to Ehrlich's goat experiments and his absorption technique, and sometimes referring to Malkoff's as well.

Landsteiner's long contest with Ehrlich on specific differences had resulted first in his criticism of Malkoff's experiment and the refutation of it that he had drawn from his own experiment with the elution of antibody from the absorbing cells, and then in the twenty years of work that had culminated in the successful demonstration that the specificity of his azo-protein antigens was not sharply defined but smoothly graded, a matter of *quantitative Abstufung*. It was this antithesis, this apparently irreconcilable difference in basic attitudes toward nature itself, that Landsteiner now had van der Scheer re-examine experimentally. Before the paper on the species hybrid was even written up, they had begun work on a direct comparison between the sharpness of the specificity of precipitins and of agglutinins. This comparison represents the point at which Landsteiner came to terms with the Ehrlich concept of receptor specificity and incorporated it into his own thinking, not as an antithesis now, but as a special case of his own concept.

Using what was basically the Malkoff subtraction-absorption technique, he and van der Scheer compared the sharpness of the specificity of antibodies against the azo-protein model antigen and against the natural antigens on the red cells of different species. The precipitin part of the comparison was done by adding to the antiserum a heterologous but related azo-protein antigen and comparing its titre before and after this procedure to the original homologous antigen against which it had been raised. They found, as they certainly expected, that the effect of the addition of the heterologous antigen was little different from that of dilution of the antiserum. The titre to the homologous antigen was lowered: the heterologous substance had absorbed a broad range of antibody specificities: there was no sharp edge, only a quantitative difference. The results, as they said, were those of the earlier series of experiments with azo-protein antigens that Landsteiner had done in his last years in Vienna.

The results with the cell antigens and the haemagglutinating antibody were quite different from those with the precipitins, however. The immune sera were raised by injecting rabbits with red cells of various other species. The antisera were then absorbed with the various cells and quantitatively tested for the specificities that remained, exactly as Malkoff had done in 1900, except that Malkoff's test had not been quantitative. And they found, just as Malkoff had done, that species specificity was sharp and distinct, that cells from closely related species such as mouse and rat, horse and donkey, and even monkey and human being, were easily and sharply distinguished, though conversely, some of the specificities were shared between species quite distant in the zoological scale.

Landsteiner's conclusions in the discussion of this experiment draw on the whole history of his and Ehrlich's contrary conceptions of the nature of specificity:

The conception, based on the findings of Ehrlich and Morgenroth, that the precipitins contain in general a multiplicity of antibodies, each acting on a single chemical group of the antigenic proteins, is not conclusively proven. . . . One has to account not only for the differing behaviour of a given immune serum against two antigens, but for the fact that precipitins in general show a maximum activity with the homologous protein, the reaction decreasing gradually in strength with the distance in the zoological scale. . . . The results of precipitin reactions with antigens containing binding groups of known chemical constitution leave no doubt as to the fact that a single precipitin will regularly react with other substances if their chemical structure is sufficiently near. . . .

In the hemagglutination experiments, a considerable part of the antibodies combined only with the homologous antigen, even when closely related species were examined. The question arises whether these antigens are simple species proteins, or whether other factors are concerned in their specificity. They may possibly consist of such proteins combined with other substances or groups whose structure is not in close correspondence with the zoological system. . . . Furthermore, the existence of similar groups in the blood corpuscles of distant species has already been demonstrated . . . by the method of splitting the combination of agglutinin and blood.

The peculiarities in specificity manifested by precipitinogens and agglutinogens suggest an essential difference in the chemical structures which determine the specificity of the two kinds of antigens.[16]

In 1902, when Landsteiner and Adriano Sturli originally did the experiment referred to here, in which antibody was eluted off the absorbing cells, this was not the interpretation they gave it. In those days, it seemed to be an antithesis to Malkoff's experiment, a demonstration of smooth transition and more or less good fit. Landsteiner had actually quoted the interpretation in terms of multiple distinct specificities as a kind of *reductio ad absurdum*: "You could suppose that on every single red cell there was a great number of different substances, and similarly thousands of distinct substances in the serum of every different kind of animal."[17]

This was the concept of multiple separate specificities on cells and matching separate antibodies in serum that Landsteiner had thought so "uneconomical" in 1902. In 1909, in his essay on the theories of

16   Karl Landsteiner and James van der Scheer, "On the specificity of agglutinins and precipitins," *J. Exper. Med. 40*(1924):91–107, C. 196 (p. 105–106).
17   Karl Landsteiner and Adriano Sturli, "Ueber Hämagglutinine normaler Sera," *Wiener klin. Wsch. 15*(1902):38–40, C. 19 (p. 39).

antibody formation, he had called Malkoff's interpretation misleadingly simple, badly thought out, and easy to refute.[18] Now, however, in 1924, he was able to accept it, and to incorporate the idea of the specific receptor into his own thinking by means of the carrier and haptene concept he had developed during his stay in Holland. The carrier protein molecule might represent species specificity with its smooth transitions, the haptene groups the separable, and perhaps non-protein, groups that might account for specific differences between individuals of the same species, and for similarities between animals far apart in the zoological scale. Landsteiner is now postulating two quite separate systems of specificity, the one depending on proteins, with smooth transitions representing his own, earlier thinking; the other depending on some other chemical structure and representing Ehrlich's sharp-edged cytotoxin specificity. It is remarkable how near, and how subtly far, Landsteiner's *haptene* is from Ehrlich's *haptophore*.

This conceptual acceptance of sharp *intra*species specificity that at the same time allowed Landsteiner to retain his earlier concept of smooth transitions in *inter*species relationships now opened up a new line of thinking for him. He was able to make use of the work that had grown up around Ehrlich's receptor theory, and that in the hands of von Dungern and the Hirszfelds had developed into the new field of blood grouping. It was as though he was able for the first time to accept his own discovery of the ABO blood groups of 1901.[19] He himself had almost completely ignored the discovery since that date, though it was often cited by the receptor-theorists, frequently, that is, after von Dungern and Hirszfeld remembered and resuscitated it in 1910.[20]

Now in the mid-twenties Landsteiner began, for the first time, to claim priority for his discovery and to respond very pointedly to what he saw as misconceptions and as infractions of his rights. The American serologist Reuben Ottenberg, of Mount Sinai Hospital in New York, put the following ambiguous words in a paper of 1921, in a description of the isoagglutination of red cells:

This phenomenon . . . was described by Greenbaum and by Shattuck in 1900, and mistakenly supposed to be the result of disease. Landsteiner reduced its

18 Karl Landsteiner, "Die Theorien der Antikörperbildung," *Wiener klin. Wschr.* 22(1909):1623–1631, C. 102 (p. 1625).
19 Karl Landsteiner, "Ueber Agglutinations-erscheinungen normaler menschlichen Blutes," *Wiener klin. Wschr. 14*(1901):1132–1134, C. 17. David R. Zimmerman, *Rh: the Intimate History of a Disease* (New York: Macmillan, 1973), puts this return to his scientific origins down to "a refugee's sense of alienation and deracination" (p. 13). It also, of course, coincided with the international wave of interest in blood group genetics that followed the appearance of Bernstein's triple-allele theory.
20 Emil v. Dungern and Ludwik Hirszfeld, "Ueber Vererbung gruppenspezifischer Strukturen des Blutes, II" *Z.f. Immunitätsf. 6*(1910):284–292.

318   *Absolute Specificity and Blood Group Genetics*

occurrence to a definite law. . . . DeCastello and Sturli the following year confirmed Landsteiner's findings. Among 155 persons examined by them were 4 who did not fit in any of Landsteiner's three groups. . . . This fourth group was only definitely recognised and named as such in 1907 by Jansky.[21]

Ottenberg's paper appeared in 1921, but it was not until 1924, with the work on the comparison of the specificity of agglutination and precipitation reactions, that Landsteiner began to respond to these slurs on his claim to priority. He uses Samuel Shattock's pseudo-agglutination as an illustration of one of the many misleading effects that are not true specific blood group agglutination, and he points out in a footnote that von Decastello and Sturli pursued their research with his express permission.[22]

These same shocking distortions of Landsteiner's contributions were put right in a massive review of all information to date on blood groups, prepared by the young Philip Levine (Figure 15.6).[23] Levine wrote it directly under Landsteiner's eye at the Rockefeller Institute in 1928, and Landsteiner translated it into German: "It is remarkable that some authors ascribe the discovery of normal isoagglutination and others even the discovery of the blood groups, to Shattock." Levine goes on to show that Shattock's agglutination was only so-called rouleaux formation, and not true immune agglutination at all; and elsewhere, that von Decastello and Sturli, who discovered the fourth group AB (not Jansky, as Ottenberg had thought) were actually co-workers of Landsteiner.[24]

The even more egregiously appalling remarks of Alexander S. Wiener, at that time a medical student at Brooklyn Jewish Hospital, appeared in print in 1929:

21 Albert S. F. Grünbaum, "The agglutination of red corpuscles," *Brit. Med. J.* i(1900): 1089, Report of a meeting of the Liverpool Medical Institution 19 April 1900; Samuel G. Shattock, "Chromocyte clumping in acute pneumonia and certain other diseases and the significance of the buffy coat in the shed blood," *J. Path. Bact.* 6(1900):303–314. On Shattock's claim, see Pauline M. H. Mazumdar, "The purpose of immunity: Landsteiner's interpretation of the human isoantibodies," *J. Hist. Biol.* 8(1975):115–133. Reuben Ottenberg, "Hereditary blood qualities: medicolegal application of human blood grouping," *J. Immunol.* 6(1921):363–385.
22 Karl Landsteiner and Dan H. Witt, "Observations on the human blood groups: irregular reactions, iso-agglutinins in sera of group IV, the factor A'," *J. Immunol.* 11(1926):221–247, 204, 211–212, n. 5.
23 For biographical material on Philip Levine, see Louis K. Diamond, "A tribute to Philip Levine" (letter), *Am. J. Clin. Path.* 74(1980):368–370; Anon., "Philip Levine: dedication," *Advances in Pathobiology* 7(1980):vii–xii; includes limited bibliography.
24 Philip Levine, "Menschliche Blutgruppen und individuelle Blutdifferenzen," *Ergebn. inn. Med.* 34(1928):111–153 (p. 121). Levine wrote this review at Landsteiner's request, and under his supervision, although Landsteiner's name does not appear on it (pers. commun., 1973).

Figure 15.6. Philip Levine, portrait photograph d. 1944, which appeared in the programme of the International Hematology and Rh Conference, 17–16 November 1946, Dallas, Tex. (From the National Library of Medicine, Bethesda, Md.)

Up to very recently Landsteiner's priority in discovering the phenomenon of isohemagglutination was unquestioned. In a paper written by Furuhata in 1927, however, it is stated that this reaction was in vogue in Japan and China in the thirteenth century. Furuhata even quotes medicolegal books of that time which discuss the use of this reaction to test blood relationship.[25]

This brought a telephone call from Levine, and an interview with Landsteiner that Wiener said he never forgot.[26] The next issue of the same journal contained Wiener and his colleagues' apologetic retraction, and a description of the ancient Chinese method of reading an oracle in the behaviour of drops of blood allowed to fall on a bone.[27]

Wiener's next article shows that he had now become one of Landsteiner's inner circle:

> The existence of individual differences of human blood was first recognized in 1900 by Landsteiner. . . . The following year he reduced this phenomenon to a definite law and demonstrated the existence of three human blood groups. With the consent of Landsteiner, von Decastello and Sturli (the latter a pupil of Landsteiner) continued the work and discovered the fourth and rarest group. . . .
>
> Many authors have incorrectly credited Shattock with the independent discovery of iso-hemagglutination, but one need only read this paper to recognise that he was studying rouleaux formation. . . . Furuhata's claim . . . has been disproved.[28]

From about 1920 or so, as discussed in Chapter 15, blood group serology had gained new impetus, both from the relation of blood group distribution to race, discovered by the Hirszfelds, and from the dispute between Ludwik Hirszfeld and Felix Bernstein on the nature of their inheritance. During the 1920s, serologists all over the world collected samples from different races and reworked Bernstein's arguments for the triple-allele hypothesis with their own material. Among

25 Alexander S. Wiener, Max Lederer, and Silik H. Polayes, "Studies in isohemagglutination, I. Theoretical considerations," *J. Immunol.* 17(1929):469–482.

26 Alexander S. Wiener, pers. commun. (1973). For biographical material on Alexander S. Wiener, see Richard E. Rosenfield, "*In memoriam*: Alexander S. Wiener, M.D.," *Haematologia* (Budapest), 11(1977):5–9; J. Moor-Jankowski, "Dr. Alexander S. Wiener 1907–1976," *Vox Sang.* 34(1978):189–190; Paul Speiser, "Nekrologia: *In memoriam* A. S. Wiener," *Blut* 35(1977):93–95; J. Hirschfeld, "Alexander Solomon Wiener (1907–1976)," *Int. Arch. Allergy Appl. Immunol.* 54(1977):191.

27 Alexander S. Wiener, Max Lederer, and Silik H. Polayes, "A note on the paper, 'Studies in isohemagglutination,'" *J. Immunol.* 17(1929):357–360.

28 Alexander S. Wiener, Max Lederer, and Silik H. Polayes, "Studies in isohemagglutination, III. On the heredity of the Landsteiner blood groups," *J. Immunol.* 18(1930): 201–221. Wiener is referring to a footnote in an earlier paper of Landsteiner's, "Zur Kenntnis der antifermentativen, lytischen und agglutinierenden Wirkung des Blutserums und der Lymphe," *Zbl. f. Bakt.* (Orig.), 27(1900):357–362, 361 n. 1.

them were Landsteiner and his young co-worker Philip Levine, who had probably accepted Bernstein's theory by 1928, though they were still on the watch for exceptions to it as late as 1931.[29] Landsteiner had written to congratulate Bernstein, and to draw his attention to a paper in which he himself had shown that chimpanzees had a B-like group, which he thought supported the idea of a two-race independent origin for A and B.[30] Landsteiner and Levine began collecting samples from New York hospital patients for what they called their "race paper," which was to investigate whether there was any difference in the blood group distribution of Blacks classified as very dark, dark, medium, and light skinned.[31]

The renewed interest that Landsteiner himself felt in his work may well have had drawn the attention to it that resulted in his being awarded the Nobel prize in 1930 for the paper of 1901, and by a curious extension, for all that had been built on it, in most of which he had played no part. Merrill Chase has often said, however, that Landsteiner would have much preferred to have been awarded it for his work on the azo-protein antigens.[32]

Landsteiner now had several separate lines of work in progress, each one with a different co-worker in the laboratory under him (Figure 15.7). The papers on protein specificity were done mainly with his first American collaborator at the Rockefeller, the chemist James van der Scheer, and those on red cell specificity with Philip Levine, who joined him in 1926 at the Rockefeller Institute itself, and later with Alexander Wiener, who collaborated from outside. There were others, too, who worked with him, some for many years. Clara Nigg, who was the only woman among his collaborators, did a series of projects connected with typhus and tissue culture, which ran from 1929 to 1937. John Jacobs worked on sensitization of guinea-pig skin with simple chemical compounds, along with Merrill Chase, from 1933. Jacob Fürth worked on the precipitins of the *Salmonella* group. According to Chase, the plan was to transform *Salmonella paratyphi* into *S. typhi*. One can hardly help thinking of Buchner's hay bacillus. But the experiments were not suc-

29 Landsteiner (Levine), laboratory notes on tests of Hirszfeld vs. Bernstein, dated February 1928. Exceptions to Bernstein's theory, dated August 1931: Landsteiner Papers, 450 L239, Box 5 , Folder 6, Rockefeller Archives.
30 Landsteiner to Bernstein, letter dated 1925: Nachlass Felix Bernstein, #1a, Göttingen, Niedersächsische Staats- und Universitäts-bibliotek.
31 Landsteiner (Levine), laboratory notes, lists of blood samples dated Haarlem Hospital, 22 November 1928; Landsteiner Papers, 450 L239-U, Paige Box 2: Rockefeller Archives.
32 Merrill W. Chase, personal communication 1974, and cited by Corner, *History of the Rockefeller Institute* (1964) (n. 3), 205.

Figure 15.7. Landsteiner's co-workers at the Rockefeller Institute: a photograph taken sometime after 1933, when Jacobs joined the laboratory, and 1939, when van der Scheer's post disappeared with Landsteiner's "retirement." Left to right: James van der Scheer, Clara Nigg, and John Jacobs. (From the collection of George Mackenzie, American Philosophical Society, Philadelphia, Pa.)

cessful. Fürth resented Landsteiner's dictation, and the collaboration came to an end after two years. Chase said that the transformation experiments were never published.[33]

33 Merrill W. Chase, "Notes about Dr Karl Landsteiner," typescript ms. dated 27 January 1944, p. 4. The papers that did appear were on the precipitable specific substances of the typhoid and proteus groups, 1927–1929: Jacob Fürth's name comes first on all of them. There is no indication that they were designed to do anything other than isolate the group-specific substances of the organisms, along the lines of Oswald Avery and Michael Heidelberger's isolation of the soluble specific substances of the *Pneumococci*, reported in 1925. There is no mention of transformation in Landsteiner's Rockefeller Reports for these years: Rockefeller Archives, Scientific Reports to the Corporation and Board of Scientific Divisions, Record Group 439 series 5 Reports of Directors of Laboratories, Path.-Bact. Div.: v. 15 (1926–1927), 129; v. XVI (1927–1928), 21 April 1928, 110; v. 17 (1928–1929), 20 April 1929, 86–90.

The new work on red cell specificity now proceeded along lines apparently identical to those of any born receptor-theorist. It was highly successful. Landsteiner and Levine described first the new blood group specificities M and N, then soon afterwards, another new system, the P group. The method they used was that of the immunization of rabbits with human red cells, and the absorption of immune sera with cells of known specificity: the remaining specificities in the serum then recognized previously unknown cellular antigens. Reagents sharply specific for any known group could, conversely, be produced by absorbing the immune serum with cells lacking the group. It was the Ehrlich–Malkoff differential subtraction method.

Landsteiner still hesitated, however. In some cases, the statement as to the sharp specificity of the antigens was slightly softened. The sera were not absolutely specific. They found, for example, that the titre of an anti-N serum might be lowered by repeated absorption with cells which supposedly were N-negative, according to an agglutination test. It is not certain, Landsteiner says, that each specificity corresponds to a special compound that could be isolated chemically.[34] He is working with the methods of the receptor theory, but he is aware that its assumptions may not all be completely dependable. The phenomena may not be as sharply separable as the terminology that he himself introduced would suggest.

The chapter that Landsteiner contributed to Edwin Jordan and Isidore Falk's *The Newer Knowledge of Bacteriology and Immunology* of 1928 carefully insists on the continuities between these apparently immunologically distinct substances. He cites work by others that shows quantitative differences in agglutinability of different group A red cells, and quantitative differences in titre in samples of Anti-A and anti-B from serum. The hypothesis of two pairs of agglutinating elements is "practically adequate but not strictly accurate": by absorbing O serum, which contains anti-A and anti-B, with cells of either type alone, part of the agglutinins for the other type is also removed.[35]

34 Karl Landsteiner and Philip Levine, "A new agglutinable factor differentiating human bloods," *Proc. Soc. Exper. Biol. and Med.* 24(1926–27):600–602, C. 220; Landsteiner and Levine, "Further observations on individual differences of human blood," *Proc. Soc. Exper. Biol. and Med.* 24(1926–1927):941–942, C. 225; Landsteiner and Levine, "On individual differences in human blood," *J. Exper. Med.* 47(1928):757–775, C. 236 (pp. 772–773). He makes the same point in Landsteiner, "The human blood groups," in Edwin O. Jordan and Isidore S. Falk, eds., *The Newer Knowledge of Bacteriology and Immunology* (Chicago, Ill.: University of Chicago Press, 1928), 892–908: "It is still doubtful however whether a special chemical substance corresponds to each of the serological factors which are found when several different antigens are tested with a number of serological reagents" (p. 904, C. 232).
35 Landsteiner, "Human blood groups" (1928) (n. 34), 896–897 (p. 894).

There is no such softening of the sharp edges in the work of Landsteiner's young friend A. S. Wiener.[36] Although Alexander Wiener was very close to Landsteiner – he said that he visited him every Wednesday evening for ten years and discussed papers and problems with him in great detail – there is at this time a distinct difference between their styles of thought.

Wiener's earliest stimulus to work on serology did not come from Landsteiner but from the published work of Felix Bernstein on the inheritance of the ABO groups, and from Bernstein's calculus of gene frequency. At this time, Wiener adopted whole-heartedly Bernstein's assumption of the one-to-one relationship of antibody to antigen to gene that was and remained typical of the thinking of the blood group mathematicians. Wiener's thinking had already taken this form before his telephone call from Levine and his meeting with Landsteiner. Unlike most of Landsteiner's collaborators, Wiener was already thinking independently before they met, even though he was only a student then, and he did not immediately adopt Landsteiner's style as his own.

Wiener's calculations are carried out in exactly the spirit of Bernstein's. He calculates gene frequencies from the data on test results with no hesitancy as to the specificity of the receptor and its genetic partner the unit-character, and even, perhaps, a touch of the presence-and-absence theory:

Let us consider the inheritance of a unit-character, such as a hypothetical dominant agglutinogen D. The inheritance of such a character depends on a single pair of allelomorphic genes, D and R, or D and not-D. . . .
Let dD be the frequency of gene D in a given population and rD be the frequency of gene R in that population, then,

$$dD + rD = 1$$

The frequencies of the phenotypes D+ and D− are:

PHENOTYPE:	GENOTYPE:	FREQUENCY:
D+	DD or DR	$(dD^2 + 2dDrD) = 1 - rD^2$
D−	RR	rD^2

36 Alexander S. Wiener, b. 1907, d. 1976; M.D. Long Island College of Medicine, N.Y., 1930; for biographical material, see n. 30. He was never a member of the Rockefeller Institute, but worked at the Jewish Hospital of Brooklyn in the Department of Pathology, and later at the Office of the New York City Medical Examiner. Wiener did not collaborate directly with Landsteiner on a project until after Levine had left the Rockefeller, in 1931. From this time Landsteiner did little work on red cell specificity until 1936, when he and Wiener did a paper on the agglutinogen M in the blood of the Rhesus monkey. From this there developed the work on the Rhesus blood groups, which was published from 1940 onwards. See Chapter 16.

Whence $rD = \sqrt{D-}$
$$D = 1 - rD = 1 - \sqrt{D-}. \ldots^{37}$$

It is Bernstein's method, and Bernstein's assumption. The test, the antigen, and the gene are in a direct one-to-one relationship, as in Bernstein's own papers. The receptor and the unit-character allow Wiener, like Bernstein and many others after him, to calculate gene frequency from the immunological discontinuity between D+ and D-, or D and not-D. That the receptor for Wiener is specific enough to make this calculation perfectly reliable can be seen from his analysis of the inheritance of chicken blood groups using the data of the British serologist Charles Todd.[38] Wiener writes:

> From an analysis of Todd's studies on immune isoagglutination reactions in three large families of chickens, it has been possible to prove that the almost complete serological individuality observed in the species can be explained on the basis of a relatively small number of different sharply defined agglutinogens. Thus Todd's polyvalent serum probably contains at the most 15 different specific agglutinins.[39]

Todd's serum, says Wiener, contains fifteen different sharply defined specific agglutinins. There is no *quantitative Abstufung* here.

Bernstein's and Wiener's earliest papers applied this method to the calculation of gene frequencies for the ABO blood groups. It is interesting to look at the way Bernstein deals with cases in which there is a systematic deviation from his relationship,

$$p + q + r = 100 \text{ percent}$$

37   Alexander S. Wiener, Max Lederer, and Silik H. Polayes, "Studies in isohemagglutination IV. On the chances of proving non-paternity with special reference to blood–groups," *J. Immunol.* 19(1930):259–282 (p. 260).

38   Charles Todd, "Cellular individuality in the higher animals with special reference to the individuality of the red blood corpuscle," *Proc. Roy. Soc.* (London), B 106(1930): 22–44. Todd's use of the receptor theory and his citation of Ehrlich are discussed in Chapter 14 (n. 67).

39   Alexander S. Wiener, "Individuality of the blood in higher animals II. Agglutination in red blood cells of fowls," *J. Genetics* 29(1934):1–8 (p. 5); Wiener, "Individuality of the blood in higher animals," *Z.f. indukt. Abstamm.- u. Vererbl.* 66(1933):31–48. He had used the same argument, in this paper for the *Zeitschrift für induktive Abstammungs – und Vererbungslehre*, the journal in which Bernstein generally published, and which carried a fair number of papers on blood group mathematics based on probability theory: this is the implication of the "inductive" in its title. This title was changed to *Zeitschrift für Vererbungslehre* in 1958, and changed again to *Molecular and General Genetics* in 1967.

that is, when the sum of the gene frequencies differ from unity by more than three times the standard deviation.

In the case of the sera being too weak, only a percentage of the true A and true B is detected. Bernstein calculates in this case, that

$$p + q + r > 100 \text{ percent.}$$

In the opposite circumstances, where the cells are too weak, the frequencies $\underline{O}$, $\underline{A}$, $\underline{B}$, and $\underline{AB}$ are represented by $\underline{O}'$, $\underline{A}'$, $\underline{B}'$, and $\underline{AB}'$, which also differ from the true values. Here, too, he shows that $p + q + r$ will be greater than 100 percent. He then goes on:

If we now look at those investigations in which this has occurred, those of Moss stand out as the extreme case, in which the figure differs from 1 by more than fifteen times the standard deviation in this direction. It is very probable that Moss, whose investigation is full of mistakes, was using too weak a serum. Moss's investigation is undoubtedly the worst that has been published in this area.[40]

Bernstein put these findings down to "technical errors." It does not occur to him, as it did to Landsteiner, that they might have an immunological explanation. For Bernstein, who was a mathematician, not a serologist, a "weak serum" or "weak cells" is a careless mistake, not an experimental finding. Other possibilities include inbreeding and inhomogeneity of population. These, too, may affect the sum of gene frequencies, says Bernstein, but they cannot be considered explanations unless technical reliability is beyond question.

The application of these methods to the ABO groups is not too difficult, but it is more problematic in the case of M and N groups where the results are not so clear-cut. Writing about the distribution of M and N in the population of New York, A. S. Wiener follows Bernstein's argument, though he does not refer directly to Bernstein's paper. Deviations from the distribution expected with a two-allele hypothesis may be due to inhomogeneity and non-random mating in the population – the population of New York, says Wiener, is anything but homogeneous and does not mate at all randomly. But deviations may also be due to the variability of the test sera, and in the strength of the N antigen, especially in the heterozygote MN. It is clear, however, that Wiener thought that with better techniques this difficulty would disappear. It was improvements in technique that did away with the fancied occur-

---

40   Felix Bernstein, "Fortgesetzte Untersuchungen aus der Theorie der Blutgruppen," *Z.f.indukt. Abstamm.- u. Vererbl.* 56(1930):233–273 (p. 252).

rence of O children of AB mothers, the argument upon which von Dungern and Hirszfeld's theory of inheritance had finally rested.

But in spite of Wiener's desire to use it as a unit of calculation, the N antigen did not appear in the agglutination tests as a single sharply defined receptor; this shows very plainly from his description of it:

By this technic, [Landsteiner's new one, in which cells and antiserum were centrifuged together in a tube, rather than being mixed on a slide] the distinction between M+ and M− bloods was always sharp. When testing for N however, difficulty was occasionally encountered. A diagnosis of N+ or N− could always be made even in doubtful cases however if the tests were repeated a sufficient number of times and with several different sera.[41]

Yet Wiener is still able to write,

If m and n represent the frequencies of genes M and N respectively, $m + n = 1$. The frequencies of the three phenotypes in terms of the frequencies of genes are then derived as follows:

PHENOTYPE	GENOTYPE	FREQUENCY
M+N−	MM	m^2
M−N+	NN	n^2
M+N+	MN	$2mn$[42]

For Wiener at this time, all is as black and white as von Dungern's axolotls. Landsteiner and Levine's treatment of the same problem is both more guarded and more subtle:

The results with a second property N, whose heredity was studied, varied depending on the particular immune serum used. Although the strongest agglutinations occurred with the same bloods there were differences in the reactions of minor strengths. . . .

In the following experiments two sera were selected which behaved identically and gave the smallest number of positive reactions. . . . Moreover, with the sera chosen there was a distinct break between positive and negative tests.[43]

The discontinuity, they are aware, is to some extent arbitrary. Nevertheless, calculation of the expected frequency of N+ from the more easily determined frequency of M+ on the supposition that they are determined by alleles gives quite good agreement. This is in fact the

41 Alexander S. Wiener and Maurice Vaisberg, "Heredity of the agglutinins M and N of Landsteiner and Levine," *J. Immunol.* 20(1931):371–388 (p. 373).

42 Wiener and Vaisberg, "Heredity of the agglutinins M and N" (1931) (n. 41), 388.

43 Karl Landsteiner and Philip Levine, "On the inheritance of agglutinogens of human blood demonstrable by immune agglutinins," *J. Exper. Med.* 48(1928):731–749, C. 242 (p. 734).

same calculation later used by Wiener, but Landsteiner is aware of its conceptual nature in a way that Wiener does not admit to being. It is evident that the properties are inherited, and separately they behave like Mendelian dominants, but,

in view of the limited number of families studied it would seem premature to attempt a final interpretation and to discuss further possibilities such as the existence of more than two allelomorphs. Also it has to be considered that the state of affairs may be complicated, e.g., by interacting or modifying effects of factors determining hitherto unknown agglutinable structures.[44]

During the twenties and thirties at the Rockefeller Institute, while Landsteiner was working with Levine on blood groups, he was also pursuing his earlier line of investigation on the specificity of azo-protein antigens, some of it using chemical groupings attached to cleaned red cell stroma, and perhaps designed as a model of a blood group.

The laboratory notes of Landsteiner's work with James van der Scheer show that though much of this work was not published, it never ceased to engage him. Among the dozens of yellowed packages in the cupboards of Landsteiner's old laboratory at the Rockefeller University, when I visited Merrill Chase there in 1973, there was one labelled "The one-or-two-antibody question." The experiments in it run from May 1930 to November 1939. The question is chewed over and over on sheets with this heading, in van der Scheer's neat pencil handwriting with an occasional comment from Landsteiner. There are protocols of experiments with results and readings, discussions of the implications of the experiments, lists of references, questions, interpretations.

On 23 September 1934, there is a three-page, 35 point

Program for one or two antibody question:

1. Whole I.S. dil. to compare with heterol. abs. I.S. as to specificity.
2. Compare same two solutions with regard to absorption with a new portion of heterologous antigen.
3. Compare specificity of: split from heterologous stroma then absorb the *het. abs.* fluid with homol. stroma and split again. Then compare both fluids.
4. Try to find paper where it says that albumin and globulin azo-proteins give antibodies of same specif. . . .
7. Make an *o*-aminobenzol sulphonic immune serum. This should not react with anthracil antigen. Then see whether a split from *o*-aminobenzol sulphonic stroma does react with anthracil antigen. This would prove that antibodies act differently in the serum when purified by splitting off (see exp. Feb. 18 1935).

---

44 Landsteiner and Levine, "Inheritance of agglutinogens" (1928) (n. 43), 748.

The experiments of February 1935 are listed on a sheet dated 12 February 1935: "Most essential experiments for two antibody question." There are six of these, and they are all cross-absorption, elution, and inhibition experiments, aimed at showing that although the reactions were strongest quantitatively with homologous antigen, the "splits" even from the heterologous azostroma antigen still reacted with the homologous antigen as well.[45]

This study seemed to show that even with these antigens with single determinant groups, a serum raised against one of them contained a multitude of cross-reacting antibodies. If the homologous antigen was

SO₃H

meta-aminobenzene-sulphonic-conjugate

NH₂

there were some sera that had a greater affinity for

As₃H₂

meta-amino-arsinic (Atoxyl)

NH₂

and some for

COOH

meta-amino-benzoic

NH₂

or for

SO₃H

ortho-amino-sulphonic

NH₂

45 These notes were kept for many years by Merrill W. Chase, one of Landsteiner's last collaborators, in Landsteiner's old laboratory at the Rockefeller University. He kindly allowed me to see and copy them several times and also helped me to find my way about in them. He has now deposited them in the Rockefeller Archives. See Landsteiner (van der Scheer), laboratory notes on "The one or two antibody question," file dated May 1930–November 1939: Landsteiner Papers, 450 L239-U, Box 8, Folder 1: Rockefeller Archives.

though all had maximal affinity for the original homologous conjugate. The cross-absorption experiments showed that the antibodies formed in response to this single structure "varied around a main pattern."[46]

Another series of experiments used what they called a "double-eagle" – a reminiscence of the Austrian flag – an azoprotein with two separate substituent groups. Absorption tests with either of these "partial antigens," as Landsteiner called them, removed from the serum the activity against the other head of the double-eagle. Van der Scheer's report of the experiment of 15 November 1937 is headed "Tests to show the different antibodies in azosera." Van der Scheer says: "It appears as if we are dealing with an antibody which has 2 affinities, one for the horse azo-group and one for the metanil group." Landsteiner has scrawled across it, "or one affinity for the eg. tyrosine-metanil," and the word "*significant.*" Little of this long series of experiments appeared in the published papers, but that was typical of Landsteiner's method of working over and over a problem.

The results of the work on the azo-protein model showed quite clearly that the supposition of a one-to-one "receptor" effect of antigen and antibody did not hold here at all. The single determinant raised antibody of more than one specificity, the "double-eagle" might produce a single loose-fitting specificity, or even conceivably, as van der Scheer had suggested, a single antibody with two linked specificities. There was no sign here of absolute specificity, or a one-to-one relationship of antigen and antibody.

It was in the late thirties that Landsteiner's old feeling for a physicochemical interpretation of specificity gained new support from his contact with the chemist Linus Pauling (Figure 15.8) of the California Institute of Technology at Pasadena.

Linus Pauling's work was both a culmination of the older physical or colloid chemistry and the starting point of a new chemical physics: it realized Wilhelm Ostwald's vision of the 1890s for an *allgemeine Chemie.* It showed the principles of chemistry to be logically dependent on those of physics and unified chemistry by creating a theoretical framework that linked all its branches, including structural organic chemistry. Ostwald had hoped to develop a general theory of chemical affinity: Pauling's quantum mechanics of the chemical bond provided a general explanation of why some chemical combinations are stable and some less so.[47]

46 Karl Landsteiner and James van der Scheer, "On the cross reactions of immune sera to azoproteins," *J. Exper. Med. 63*(1936):325–399, C. 303.
47 John W. Servos, *Physical Chemistry from Ostwald to Pauling: the Making of a Science in America* (Princeton, N.J.: Princeton University Press, 1990), 275–298 (pp. 293–294); Judith Goodstein, "Atoms, molecules and Linus Pauling," *Social Research 51*(1984): 691–708; James H. Sturdivant, "The scientific work of Linus Pauling," in Alexander

Figure 15.8. Linus Pauling, portrait photograph, n.d., but probably taken in the mid-1960s. (Courtesy of National Library of Medicine, Bethesda, Md.)

Pauling was thirty years younger than Landsteiner and full of spar-kling ideas. His important series of papers "The Nature of the Chem-

Rich and Norman Davidson, eds., *Structural Chemistry and Molecular Biology* (San Fran-cisco, Calif.: Freeman, 1968), 3–11.

ical Bond" had appeared between 1931 and 1933.[48] During the mid-thirties, he was working on the structure of haemoglobin and the nature of the bonding of the iron to the rest of the molecule, and he lectured on his findings at the Rockefeller Institute in 1936. According to Pauling's biographer Anthony Serafini, this was when Landsteiner met Pauling for the first time. Landsteiner invited him to come to his laboratory, and spent some time discussing problems of molecular structure with him. Pauling was already thinking about the role of weak short-range bonds, the so-called van der Waals and London forces, in the haemoglobin molecule and in the three-dimensional folding of proteins, which was to be the subject of his series of lectures at Cornell University the following year, in the autumn of 1937.[49] Landsteiner went to Cornell, which is in Ithaca in upstate New York, and talked to Pauling about serology.

Although Serafini thinks this talk led to nothing but a few references to Landsteiner in Pauling's lectures, it seems to me to have been a contact that was important to both of them. It turned Landsteiner's thoughts to the significance of bonding in the antigen–antibody reaction, and Pauling's to the problems of antigens and antibodies.[50] Pauling's ideas on short-range forces fitted Landsteiner's concept of the charge outline as the determinant of specificity.

Soon after his return to Pasadena from Cornell, Pauling wrote to Landsteiner for reprints, ready to start on a new set of problems.[51] Landsteiner sent him some, as well as a galley of a paper "in which the question of multiple antibody formation is discussed, and some remarks made on the peculiarity of antibody specificity, a subject which we touched upon when I was in Ithaca."[52] It was, most likely, a paper on the "one or two antibody question."

By 1940, Pauling had worked out a theory of the structure of antibodies and the mechanism of their production, in terms of secondary

48 Linus Pauling, "The nature of the chemical bond" (1931), reprinted in Rich and Davidson, eds., *Structural Chemistry* (1968), 849–884. The originals appeared in *J. Am. Chem. Soc.* 53(1931):1367–1400, 3225–3237; 54(1932):988–1003, 3570–3582; *J. Chem. Physics* 1(1933):362–374 (with G. W. Wheland), 606–617, 679–686 (with J. Sherman). For a complete bibliography of Pauling's work, see "Scientific publications of Linus Pauling," in Rich and Davidson, eds., *Structural Chemistry* (1968) (n. 47), 887–907.
49 Anthony Serafini, *Linus Pauling: A Man and His Science* (New York: Simon and Schuster, 1989), 73–74.
50 Pauling to Landsteiner, letter dated 12 March 1940: Landsteiner Papers, 450 L239, Box 4, Folder 15, Rockefeller Archives.
51 Pauling to Landsteiner, letter dated 16 March 1938: Landsteiner Papers, 450 L239, Box 4, Folder 15, Rockefeller Archives.
52 Landsteiner to Pauling, letter dated 28 March 1938: Landsteiner Papers, 450 L239, Box 4, Folder 15, Rockefeller Archives.

folding of the antibody onto the antigen outline, and was ready to come to New York and show it to Landsteiner. He presented it at the Rockefeller Institute in April 1940 and published it the following October. Pauling's idea was based on Landsteiner's charge outline concept of specificity. It was a development of Felix Haurowitz's template theory of antibody production and John Marrack's lattice theory of the antigen-antibody reaction, which had also been based on Landsteiner's concept of the charge outline.[53] In his published paper, Pauling acknowledges that his "interest in immunology was awakened by conversations with Dr. Karl Landsteiner."[54]

Landsteiner in turn was interested by Pauling's folding theory. He made contact with the physical chemists at the Rockefeller, and with the help of one of them, Alexandre Rothen, began to investigate the effect of unfolding the protein molecule on antigenic specificity. A preliminary report suggested, disappointingly, that the protein antigen was not affected by being spread out into a monomolecular film and losing its secondary folding.[55]

Landsteiner wrote carefully in his annual report to the Rockefeller Board for 1940:

In the current active discussion on protein structure, immunological phenomena have been called upon to support special conceptions. Thus it is assumed that an adequate basis for the specificity of proteins is not given by a linear arrangement of amino acids in straight chains, but that specificity requires arrangements due to foldings and secondary bonds of the peptide chain.[56]

53 Pauline M. H. Mazumdar, "The template theory of antibody formation and the chemical synthesis of the twenties," in Mazumdar, ed., *Immunology 1930–1980: Essays on the History of Immunology* (Toronto: Wall and Thompson, 1989), 13–32; Mazumdar, "Marrack, John Richardson, bacteriologist, immunologist," in F. Lawrence Holmes, ed., *Dictionary of Scientific Biography, Supplement II* (New York: Scribner's, 1991); Mazumdar, "Outline, lattice and template: the immunological theories of the thirties," presented at the Workshop on History of Immunology, VII International Congress of Immunology, Berlin 1989.

54 Pauling to Landsteiner (1940) (n. 50); Pauling, "On the structure and process of formation of antibodies," incomplete (spoken language), typescript version dated 6 March 1940 of his "A theory of the structure and process of formation of antibodies," *J. Am. Chem. Soc.* 62(1940):2643–2657; Landsteiner Papers, 450 L239-U, Paige Box 3, Rockefeller Archives. The published paper contains a note that "some of this material" was presented at the Rockefeller Institute on 17 April 1940.

55 Alexandre Rothen and Karl Landsteiner, "Adsorption of antibodies by egg albumin films," *Science* 90(1939):65–66, C. 323; Report of Dr. Rothen, in Reports of the Directors of Laboratories, Division of Chemistry, vol. 28 (1939–1940), 36–37, of Scientific Reports to the Corporation and Board of Scientific Directors: 439, Box 6, Rockefeller Archives.

56 Report of Dr. Landsteiner, in Reports of the Directors of Laboratories, Pathology and Bacteriology Division, vol. 28 (1939–1940), 49–53 (p. 52), of Scientific Reports

It appeared that antibodies, too, might still be active after being spread out into films. Ongoing discussions with Pauling, and Pauling's unwillingness to accept this result, seem to be behind the remarks in the last annual report that he was to submit. With his usual *sachliche Kühl*, in Erna Lesky's nice phrase,[57] which may have concealed excitement, hope, and disappointment, Landsteiner wrote:

Somewhat against expectation, films of pneumococcus antibodies proved to be reactive with the corresponding polysaccarides; however, the above hypothesis [Pauling's folding hypothesis] is not ruled out inasmuch as some folding may still exist in the flattened molecule (Pauling), and it is further to be considered that the results may be different with other antibodies.[58]

Besides asking Pauling to write a special chapter on bonding in relation to serology for the second edition of *The Specificity of Serological Reactions*, which was destined to be published posthumously in 1945, Landsteiner was even, it seems from his letters, considering moving to Pauling's institute at Pasadena when he reached retirement age at the Rockefeller:

It is very kind of you to consider the possibility of my working in Pasadena, an idea which certainly is attractive, especially since it would hold out the prospect of your cooperation or advice. I should be glad to have an opportunity to discuss the matter with you. It may offer some difficulties on account of the not inconsiderable requirements (concerning assistants) for my type of work, which probably are not proportional to its scientific value! As regards the R.I., I have to retire in June 1939, and afterwards I could still have some facilities, but restricted, on a year to year basis (for at least 2 years, I believe).[59]

It never happened: Landsteiner was to stay on at the Rockefeller until the end of his life. On the morning of 24 June 1943, he had a heart attack while working at the bench. Two days later, worrying about his work and struggling not to have to stay in bed, he died.[60] Helene Landsteiner, who already knew that she had cancer of the thyroid, survived him by six months.

to the Corporation and Board of Scientific Directors: 439, Box 6: Rockefeller Archives.

57 Erna Lesky, *Die Wiener medizinische Schule im XIX Jahrhundert* (Graz: Bohlaus, 1965), 600.

58 Alexandre Rothen and Karl Landsteiner, "Serological reactions of protein films on denatured proteins," *J. Exper. Med.* 76(1942):437–450, C. 341; Report of Dr. Landsteiner, in Reports of Division of Laboratories, Pathology and Bacteriology Division, vol. 31 (1942–1943), 52–55 (p. 54), of Scientific Reports to the Corporation and Board of Scientific Directors: 439, Box 7: Rockefeller Archives.

59 Pauling to Landsteiner (1938) (n. 51); Landsteiner to Pauling (1938) (n. 523).

60 Robert F. Watson, clinical notes on Landsteiner's last illness, dated 24–26 June 1943; Landsteiner-Mackenzie Papers, B L23 m, Box 3, American Philosophical Society.

During the twenties and thirties at the Rockefeller Institute Landsteiner was working in several distinct areas, each with its own coworker. The two most important of them, immunochemistry and blood grouping, operated on basic assumptions that seemed to be not just different but contradictory. In his book *The Specificity of Serological Reactions*, which came out first in German in 1933, these two fields – the specificity of proteins and the specificity of cellular antigens – are dealt with in two separate chapters. He makes the distinction that he first made in the paper on the serological examination of a species hybrid in 1924, between protein specificity, which reflects the smooth transitions of the zoological relationships between species, and the specificity of cellular antigens, which is part of a different system of relationships that is independent of species in the zoological sense:

It is a fact that, at least in the writer's experience the preparation of species specific fractions and the differentiation of nearly related species with precipitin sera is not so easy as with agglutinins and lysins. This effect seems to be explainable in that in the series of species the proteins vary smoothly, while the cell antigens often show variations without intermediate stages between.[61]

The cell antigens, he says, have a kind of mosaic structure; the elements that are detected by the serological reactions can be called "factors," a word that implies nothing about their chemical nature and is therefore not the same as the Ehrlich term "receptor," which contains the Ehrlich chemical hypothesis within it, and the idea of a firm chemical binding group. But it is no accident, he says, that this concept was applied mostly to cell antigens and agglutinin and lysin effects, although by definition it was supposed to apply to any immunological reaction.[62]

Landsteiner's work on the species hybrid that started off his career at the Rockefeller Institute opened the way to a partial, and temporary, compromise with the Ehrlich receptor theory. It was a compromise that made available to him all the work on blood group specificity that had grown up on this premise. During the early part of this period of acceptance, there is almost no difference between his work on red cell antigens and the work of those who had always accepted the sharp specificity of receptors. But this period of near-identity did not last long. He remained aware of the continuities between these seemingly

61 Karl Landsteiner *Die Spezifizität der serologischen Reaktionen* (Berlin: Springer, 1933); Landsteiner, *The Specificity of Serological Reactions, Revised Edition with a Chapter on Molecular Structure and Intermolecular Forces by Linus Pauling* (Cambridge, Mass.: Harvard University Press, 1946), 38.

62 Landsteiner, *Spezifizität* (1933) (n. 68), 53.

sharp specific differences, even while calculating gene frequencies in the usual Bernstein manner, in his work with Philip Levine during the twenties. But by the time he came to collaborate with A. S. Wiener, his earlier style of thought had reasserted itself, even in the case of cellular antigens. He had gone back to putting the emphasis of his discussions on the signs of continuity between the specificities rather than on the sharpness of their differences. The "one or two antibody" project may have been intended as a model system whose properties could have some bearing on the nature of blood group specificity. A. S. Wiener was certainly able to take it as such at a later date. Further, Pauling's bonding concepts seem to have taken the place that colloid chemistry had occupied for Landsteiner many years before. Weak, short-range forces contrasted with firm, covalent bonding offered a new interpretation of the charge outline that Landsteiner had contrasted with the firm chemical union associated with Ehrlich's side chains.

16

The Last Confrontation: The Controversy over the Rhesus System

The last of the series of conflicts to be discussed in this book began in about 1945 and lasted for the next twenty or thirty years. It took the form of a struggle between Alexander Wiener (Figure 16.1) in New York, who played the part of Landsteiner's student and upholder of unity and continuity, and the representatives of diversity and specificity, who in this case were the statistician Ronald A. Fisher and the serologist Robert R. Race in London and Cambridge. The controversy was not about the findings on the Rhesus blood groups, but about their interpretation. The two possible interpretations were contained in two rival systems of nomenclature, one of which implied the unitarian point of view, and the other the pluralistic.

As mentioned in Chapter 15, Wiener worked independently of Landsteiner, outside the Rockefeller Institute. He was never actually employed in Landsteiner's laboratory, though he had been in close touch with him from 1929 onwards: Wiener often said that he visited Landsteiner every Wednesday evening for ten years. In spite of this personal contact, Wiener's style of thought and his views, or rather, assumptions on specificity, started out from a position that was very unlike Landsteiner's.

The difference between the two styles of thought shows itself particularly plainly in the way each dealt with the findings on the MN blood group system. Landsteiner's emphasis on the continuous nature of the N specificity and his consciousness of the arbitrariness of the cut-off point between N+ and N− contrast with Wiener's feeling that such ambiguities would disappear as techniques improved.

In 1937 the first paper appeared in which Wiener acted as Landsteiner's co-worker. Wiener was only thirty years old, but he had already produced forty-seven papers.[1] The difference between this and earlier

1 Alexander S. Wiener, *Rh-Hr Blood Types: Applications in Clinical and Legal Medicine and Anthropology* (New York: Grune and Stratton, 1952); Wiener, *Advances in Blood-Grouping* (New York: Grune and Stratton, 1961); Wiener, *Advances in Blood-Grouping II, with a*

Figure 16.1. Alexander Solomon Wiener, portrait photograph, dated 1947.
(From the collection of the National Library of Medicine, Bethesda, Md.)

examples of his thought shows that at this point he was simply over-whelmed by a more powerful personality and way of thinking.[2]

According to Wiener, the problem was originally his. He had several samples of rabbit anti-human M sera, all of which gave strong results with human M+ cells; some of them also reacted with the red cells of *Macacus rhesus*, the Indian rhesus monkey. Landsteiner designed an experiment that tested the blood of fifteen primate species belonging to four types: man, chimpanzee, six Old World monkeys (*Cercopithedae*), and seven New World monkeys (*Platyrrhinae*), with six different anti-M sera. Though the original problem may have been Wiener's, the paper is clearly Landsteiner's, not only in style, but in that it repeats the de-sign that Landsteiner had used twelve years before in the group of papers on primate bloods that he wrote in collaboration with Philip Miller in 1925.[3] When this earlier series came out, Landsteiner's "dis-covery" of the receptor theory was still recent and was affecting him most deeply.

It is noteworthy that we have found from our studies in monkeys that whole genera or families appear to be characterized by the presence of certain sero-logical factors (so-called receptors) in their erythrocytes. A factor similar to B was found in all the bloods of the family *Platyrrhinae* examined. . . . This factor was not found in the blood of the *Cercopithedae*. . . . A peculiarity of the phe-nomenon unlike the phenomena of specificity in precipitin reactions on serum proteins is the lack of gradual transition from one family to another – its sharp discontinuity.[4]

In an accompanying figure, he showed the results in the form of an evolutionary tree with the groups written in beside the various branches

Section by Maurice Shapiro (New York: Grune and Stratton, 1965); Wiener, *Advances in Blood-Grouping III* (New York: Grune and Stratton, 1971). These four books are col-lections of Wiener's published work, arranged under subject headings and indexed. They do not include his earlier papers on the Bernstein hypothesis and the MN groups; the earliest one included is his No. 63, the first description of the Rh factor, in *Rh-Hr* (1954). It is as if Wiener felt that he had reached his intellectual maturity only with the work on the Rhesus system. A partial bibliography appears in Susan R. Hollán, *Current Topics in Immunohematology and Immunogenetics: Alexander S. Wiener Fest-schrift* (Budapest: Akadémiai Kiàdo, 1972), 17–30.

2 Karl Landsteiner and Alexander S. Wiener, "On the presence of M agglutinogens in the blood of monkeys," *J. Immunol.* 33(1937):19–25, C. 311.

3 Karl Landsteiner and C. Philip Miller, Jr., "Serological observations on the relation-ship of the bloods of man and the anthropoid apes," *Science* 61(1925):492–493, C. 202; Landsteiner and Miller, "Serological studies on the blood of the primates I. The differentiation of human and anthropoid bloods," *J. Exper. Med.* 42(1925):841–852, C. 207; Landsteiner and Miller, "Serological studies on the blood of the primates II. The blood groups in anthropoid apes," *J. Exper. Med.* 42(1925):853–862, C. 208; Land-steiner and Miller, "Serological studies on the blood of the primates III. Distribution of serological factors related to human isoagglutinins in the blood of lower monkeys," *J. Exper. Med.* 42(1925):863–877, C. 209.

4 Landsteiner and Miller, "Studies on the blood of primates, III." (1925) (n. 3), 872.

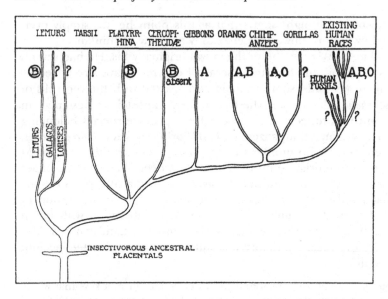

Figure 16.2 (*above and opposite*). Landsteiner's blood grouping of the primates. Above: In 1925, he shows his results in the form of an evolutionary tree with a simple dichotomy: the New World monkeys, the *Platyrrhinae*, have a B-like factor; the Old World *Cercopithae* do not. He sees no evidence of a gradual transition. From Landsteiner and Miller, "Studies on the blood of primates, III" (1925) (n. 3), fig. 1. Opposite: In 1937, he has gone back to *graduelle Abstufung*: he is able to arrange the results of typing with six different anti-M sera in the form of descending scores. At the top, the human blood gives strongly positive results with all the sera; then comes the chimpanzee with five, then the baboons and macaques from Africa and India with three, then the African green monkey with two, the South American spider monkey with only one, and the other *Platyrrhinae* of the New World with none. This series is no accidental effect: not only the monkeys but also the sera, "combined from several experiments," are arranged in descending order. From Landsteiner and Wiener, "M in monkeys" (1937) (n. 2).

(see Figure 16.2a). The New World monkeys, the *Platyrrhinae*, have a B-like factor; the *Cercopithedae*, the Old World monkeys, do not. There is, as he says, no evidence of a gradual transition: it was not, that is, evident to Landsteiner in 1925.

A comparison of this paper with the monkey-blood paper of 1937 shows how temporary was this surrender to the receptor theory. His earlier habits of thought had begun to reassert themselves by 1928, at the time of his work on the human MN groups.[5] By 1937, he had ceased to see the differences between species as discontinuous and was once more conscious only of smooth transitions between them. The layout

5 Karl Landsteiner and Philip Levine, "On individual differences in human blood," *J. Exper. Med.* 47(1928):757–775, C. 236, 549 ff.

Reactions of monkey blood with various anti-M sera*
(Combined from several experiments)

SOURCE OF BLOOD SPECIMENS†	ANTI-M IMMUNE SERUM‡					
	M5	M1	M21	M35	M2	M82
Human M..........................	+++	+++	++±	++±	++±	++±
Human N..........................	0	0	0	0	0	0
Chimpanzee........................	+++	+++	+++	++	+++	±
Sphinx baboon.....................	+++	++	++±	0	0	0
Drill baboon......................	+++	+++	++±	(±)	(+)	(+±)
Chacna baboon.....................	+++	+++	++±	0	tr.	0
M. rhesus.......................	+++	+++	++±	(+±)	±	tr.
Java macaque......................	+++	+++	+±	0	0	0
Sooty mangabey....................	+++	+++	++±	tr.	±	±
Green monkey......................	+++	+++	0	0	0	0
White spider monkey...............	++±	0	0	0	0	0
Black spider monkey...............	±	0	0	tr.		0
Wooly monkey......................	0	0	(±)	0	0	0
Brown ringtail (capuchin monkey).....	0	0	0	0	0	0
Moss monkey.......................	0	0	f.tr.	±		0
Lemur.............................	0	0	0	0	0	0
Average titer of testing fluids........	64	64	32	24	16	16

* The strength of the reactions is indicated by the number of plus signs, +++
representing the maximum reaction, that is, complete agglutination.

of the table of results of this later monkey-blood project shows how he
was able to arrange the differences between the various species as part
of a continuous spectrum, which he had not, in fact, tried to do in
1925. The human blood, which gives strongly positive results with all
six sera, appears at the top of the table; then comes the chimpanzee,
which reacts with the first five; then the baboons and macaques from
Africa and India, which react with the first three; then the African
green monkey with the first two; the South American spider monkey,
with the first one; and the rest of the *Platyrrhinae* from the New World
with none. The visual arrangement of the table, the grouping of the
ascending + scores, shows that Landsteiner saw the M antigen as approaching by *graduelle Abstufungen* through the lower primates to the
human type (Figure 16.2b). The discussion of these results is typical of
Landsteiner in its consciousness of the different possibilities, and the
many layers of personal experience that it displays, but also in its conclusion:

The formation of qualitatively different antibodies on immunizing rabbits with human M blood could be interpreted as a result of several discrete antigens in human M blood analogous to the assumption often held of several antigenic elements in human A blood – Forssman antigen and special human A antigen – . . . An alternative explanation would be the production of more than one antibody to a single antigen, possibly but not necessarily due to distinct group-ings in the antigen molecule. Against the assumption of several discrete sub-stances in human blood is the following . . . the supposed antigenic elements would show Mendelian segregation. . . .

The situation might be compared to the question . . . of the qualitative or quantitative nature of the difference between the human bloods A_1 and A_2. In favor of the qualitative difference in the present case is the observation that certain anti-M fluids after exhaustion with the blood of *M. rhesus* show only a very gradual diminution in the titer for human M erythrocytes.[6]

The subtlety and sensitivity of Landsteiner's thinking by this time is a reflection of the suggestions and reverberations of more than forty years of experience. In this case, it is also a reflection of his current work on the "one-or-two-antibody question" and the "double-eagle" project that he was working on with van der Scheer throughout the thirties, as their laboratory notes show.[7] The anti-M, though it is an agglutinin recognizing a cellular antigen, is being treated something like the protein precipitins of the azoprotein experiments, rather than in the receptor-theory manner that had been Wiener's up till this point, and that Landsteiner himself had come so very close to adopt-ing.

In 1924 a kind of receptor theory had seemed the only way to ex-plain the sharp differences between the bloods of the species hybrid and its parents. By 1937, it was the smooth transitions and more or less good fit of the specificities of the azoprotein models that provided the pattern for his interpretation of the monkey bloods. Writing many years later, Wiener said with complete conviction, "our findings proved that the agglutinogen M had a steplike evolution from mon-keys to ape to man."[8] It was this habit of thought that was now im-pressed upon Wiener, a habit that up till then had apparently been quite foreign to him.

According to Wiener, it was during the work on the monkey bloods – that is, in 1937 – that Wiener had the first indication of a rabbit anti-rhesus monkey serum, which, when the anti-M component was ab-sorbed out, still reacted with 85 percent of human red-cell samples. But

6 Landsteiner and Wiener, "M in monkeys" (1937) (n. 2), 23.
7 On the one-or-two-antibody question, see Chapter 15 (nn. 45, 46, 52).
8 Alexander S. Wiener, "Karl Landsteiner: his work and the Rhesus blood factor," Pas-sano Foundation Award Address, 1951; in his *Rh-Hr* (1954) (n. 1), 1–8 (p. 7).

it was weak and did not give very good reactions.[9] They did not publish this finding, Wiener says, until 1940, when Harold Peters of the University of Maryland in Baltimore consulted Wiener about two cases in which patients who had been transfused with blood of the correct ABO group had had unfavourable reactions to the transfusion.[10] Wiener and Landsteiner found that in these transfusion cases, and in some more collected by Wiener, the patients had in their serum an antibody whose reactions corresponded to those of the rabbit anti-rhesus serum, and gave positive agglutination with 85 percent of human red cell samples. Since it now seemed to have some clinical significance, they published a short report of the anti-Rhesus serum.[11]

A few months earlier Philip Levine and Rufus Stetson had published what the serologists called their historic paper.[12] It described the transfusion reaction that the mother of a stillborn foetus suffered when she was given ABO compatible blood from her husband. Levine and Stetson found that the mother's serum agglutinated 80 out of 104 samples from otherwise compatible donors, including one from her husband. They suggested that the mother had been immunized by the foetus against an antigen that it had inherited from the father.[13] The following

9   Alexander S. Wiener, "Karl Landsteiner MD History of the Rh-Hr blood group system," *N.Y. State J. Med.* *69*(1968):2915–2935 (pp. 2920–2921).
10 Alexander S. Wiener and Harold R. Peters, "Haemolytic reactions following transfusions of blood of the homologous group, with three cases in which the same agglutinogen was responsible," *Ann. Int. Med.* *13*(1940):2306–2322, and in *Rh-Hr* (1954) (n. 1), 25–41.
11 Karl Landsteiner and Alexander S. Wiener, "An agglutinable factor in human blood recognised by immune sera for rhesus blood," *Proc. Soc. Exper. Biol. Med.* *43* (1940): 223, and in *Rh-Hr* (1954) (n. 1), 24.
12 Philip Levine and Rufus E. Stetson, "An unusual case of intragroup agglutination," *J. Amer. Med. Assoc.* *113*(1939):126–127; reprinted with introduction by Johannes J. van Loghem, "Introduction to a historic paper," *Vox Sang.* *38*(1980):297–300.
13 Robert R. Race and Ruth Sanger, *Blood Groups in Man* (Oxford: Blackwell, 1st. ed., 1950; 6th ed., 1976). In each edition, the chapter on the Rh blood groups begins: "The discovery of the Rh groups by Landsteiner and Wiener in 1940, with which must be associated the work of Levine and Stetson in 1939, was undoubtedly the most important event in the blood group field since the discovery of the ABO system forty years before" (pp. different in each ed.). This tactful formulation conceals a priority dispute between Wiener and Levine that burned for many years. Wiener contended that Levine discovered the cause of *Erythroblastosis foetalis*, but not the Rhesus system. Levine proved that Landsteiner and Wiener's guinea-pig antibody specificity was not the same as that of human anti-Rh, and called it "anti-L W," to make his point clear. It was often said that they would have had a joint Nobel prize, but for this dispute. David R. Zimmerman, *Rh: the Intimate History of a Disease and Its Conquest* (New York: Macmillan, 1973), reviews the hearsay evidence for that rumour (p. 329). *Erythroblastosis foetalis,* a disease of the newborn, was also called *Hydrops foetalis, Icterus gravis neonatorum,* and congenital anaemia, or haemolytic disease of the newborn, the name depending on the presenting symptoms of the disease. In a

year Levine and his co-workers suggested that this was the mechanism behind *Erythroblastosis foetalis*, a disease of the newborn whose aetiology had been completely unknown.[14] In 1941, Landsteiner and Wiener drew all this together, described a way of producing a serum strong enough for practical use, and made a first attempt to work out the genetics of the new blood group system. The antiserum they used was produced by immunizing guinea-pigs with Rhesus monkey cells. This gave a more powerful serum than the original rabbit anti-Rhesus one, but in Landsteiner's usual way, they did not insist on the sharpness of the specificity. The position was rather like that with the anti-N sera: there was a continuous spectrum from weakly to strongly positive with human cells, but a usable serum could be made:

> The majority of animals [that is, the guinea-pigs producing the sera] were found to show a difference between the two sorts of [human] blood, Rh+ and Rh-, and in a group of ten animals usually one or more yielded sera suitable for practical diagnosis. While in the case of the immune rabbit sera the reagent was prepared by absorbing the diluted serum with negatively reacting human blood, it was found with several guinea-pig sera that absorption with human blood resulted merely in a nonspecific diminution of the agglutinin content no matter whether negative or positive blood was used.
>
> This led us to test the effect of simple dilution, and indeed it was found that a distinction between positive and negative blood could be made directly without absorbing the sera.[15]

"Positive" and "negative" appeared to differ only quantitatively, not qualitatively. It was possible to arrange the titre of the guinea-pig sera so that the results matched almost completely those obtained with a

slight or moderate case, the infant would be born jaundiced and anaemic, with many immature red cells or erythroblasts in the circulation, suggesting that the cause of the anaemia was blood destruction. The child might recover, or might die of heart failure due to the anaemia, or suffer cerebral damage due to the icterus. A more severe case might be stillborn, and show marked pitting oedema, with sloughing macerated skin. Typically, a mother had one or more normal children, then a succession of increasingly severely affected ones. See Margaret M. Pickles, *Hemolytic Disease of the New Born* (Springfield, Ill.: Thomas, 1949), 1–14, 71–116; includes illustrations of the characteristic appearance of the affected infants.

14 Philip Levine, Eugene M. Katzin, and Lyman Burnham, "Isoimmunization in pregnancy, its possible bearing on the etiology of *Erythroblastosis fetalis,*" *J. Am. Med. Assoc.* 116(1941):825–827; Levine, P. Vogel, Katzin, and Burnham, "Pathogenesis of *Erythroblastosis fetalis*: statistical evidence," *Science* 94(1941):371–372; Levine, Burnham, Katzin, and Vogel, "The role of isoimmunization in the pathogenesis of *Erythroblastosis fetalis,*" *Am. J. Obst. Gyne.* 42(1941):925–937.

15 Karl Landsteiner and Alexander S. Wiener, "Studies on an agglutinogen (Rh) in human blood reacting with anti-rhesus sera and with human isoantibodies," *J. Exper. Med.* 74(1941);309–320, C. 335, and in Wiener, *Rh-Hr* (1954) (n. 1), 311 (Original).

human serum, though the human sera were truly negative with the negatives.[16]

The frequency of negatives in Landsteiner and Wiener's population was 15.4 percent and the frequency of the "Rh-negative gene" was therefore 0.154 or 39.2 percent, and that of the positive gene was 60.8 percent. Positivity appeared to be inherited as a simple Mendelian dominant.

In 1941 Landsteiner and Wiener had already come across an anti-Rh serum that gave not 85 percent but 70 percent positive reactions. Levine had briefly mentioned one that seemed to react with all those that were negative with the 70 percent one. He thought it was "probably allelomorphic with an antigenic component of the Rh factor," and he called it Hr.[17]

After this there was no more new work until 1943. The English serologist Robert Race suggested that this was because those few people who were able to handle the new technology involved in detecting the Rhesus antibodies were swamped with requests for help from clinicians dealing with neonatal cases that might be *Erythroblastosis*, and with transfusion reactions that might be set down to anti-Rh.[18] The Rh system was turning out to be clinically highly significant.

In England, there had been very little interest in blood groups from either the race or the genetics perspective, even during the twenties when so much work was being done in this field in both America and in Germany. Interest in Britain seems to date from the discovery of the work of Felix Bernstein on the calculation of chromosome maps by the method of linkage to the blood groups. The discovery was made by Lancelot Hogben, professor of social biology at the London School of Economics. Hogben, a left-wing critic of the eugenics movement, thought that he could use linkage to a blood group to find out whether a seemingly hereditary disease was truly inherited in Mendelian fashion, or whether it was a result of environmental influences. The blood groups, nature untouched by nurture, would, if they were linked to a

---

16 It was this feature that led Levine in the course of the priority dispute to assert that Landsteiner and Wiener had not discovered the Rhesus system, and to refer to their antibody as anti-L W. The specificity that it recognized was later thought to be genetically independent of Rh, but perhaps to originate from a common precursor substance. See Robert R. Race, "Modern concepts of the blood group systems," *Ann. N.Y. Acad. Sci.* 127(1965):884–891.

17 Philip Levine, "The pathogenesis of fetal erythroblastosis," *N.Y. State J. Med.* 42(1942):1928–1934.

18 Race and Sanger, *Blood Groups* (1950) (n. 13), 96. On the changes in blood grouping technology needed to detect antibodies of the Rhesus system, see A. Derek Farr, "Blood group serology: the first four decades," *Med. Hist.* 23(1979):215–226.

disease, provide a guarantee of true inheritance that was unique in the confused and difficult field of human genetics and eugenics.[19]

The Galton Laboratory's Blood Group Research Unit at University College, London, was set up in 1935 to investigate the same problem, but not to use it as Hogben had done, to attack eugenics.[20] Ronald A. Fisher, an enthusiastic eugenist who had succeeded to the Galton Chair of Eugenics and Human Genetics at University College in 1933, was already well known for his contributions to statistics and population genetics.[21] Inspired by the critics, Fisher persuaded the Rockefeller Foundation that blood groups would be an important tool of human genetics, particularly for chromosome mapping. Within five years, the unit's mathematicians, led by Fisher, had taken Bernstein's method in several new directions. W. L. Stevens, Fisher's statistical assistant, applied Fisher's maximum likelihood method to the calculation of ABO gene frequencies from the data collected by Taylor.[22] A Fisher student, D. J. Finney, applied Fisher's *u*-statistics to linkage tests with blood groups, in a series of papers beginning in 1940, in which he extended Bernstein's method.[23]

In England, the new discoveries in blood group serology came in the

19 Pauline M. H. Mazumdar, *Eugenics, Human Genetics and Human Failings: The Eugenics Society, Its Sources and Its Critics in Britain* (London: Routledge, 1992), 146–195.

20 Mazumdar, *Eugenics, Human Genetics* (1992) (n. 19), 235–242.

21 Joan Fisher Box, *R. A. Fisher: the Life of a Scientist* (New York: Wiley, 1978); Frank Yates and Kenneth Mather, "Ronald Aylmer Fisher," *Biographical Memoirs of Fellows of the Royal Society* (London), 9(1963):91–120; Jerzy Neyman, "R. A. Fisher (1890–1962) an appreciation," *Science* 156(1967):1456–1460; William B. Provine, *The Origins of Theoretical Population Genetics* (Chicago, Ill.: Chicago University Press, 1971), esp. pp. 140–153; Norman T. Gridgeman, "Fisher, Ronald Aylmer," in Charles C. Gillispie, ed., *Dictionary of Scientific Biography* (New York: Scribners, 1972), v. 5, 7–11. Other memoirs on Fisher by his colleagues appear in *Biometrics*, 20(June 1964).

22 W. L. Stevens, "Estimation of blood group frequencies," *Ann. Eugen.* 8(1938):362–375; Stevens calculates the frequencies of A_1 A_2 B and O genes by Fisher's maximum likelihood method, from the data collected by George Taylor, then recalculates the frequencies of groups and subgroups from the gene frequencies. The *maximum likelihood* method appeared in Fisher, "Theory of statistical estimation," *Proc. Cambridge Phil. Soc.* 22(1925):700–725. In the words of Ian Hacking, "Fisher presents likelihood as if it were the predicate of an ordered pair, namely of a statistical hypothesis and an outcome; the likelihood is the chance of the outcome if the hypothesis is true." Ian Hacking, *Logic of Statistical Inference* (Cambridge: Cambridge University Press, 1965), 54–73 (p. 57); see also Donald A. Mackenzie, *Statistics in Britain 1865–1930: the Social Construction of Scientific Knowledge* (Edinburgh: Edinburgh University Press, 1981), 183–213 (esp. p. 204–209). George L. Taylor and Aileen M. Prior, "Blood groups in England I. Examination of family and unrelated material," *Ann. Eugen.* 8(1938):344–355; Taylor and Prior, "Blood groups in England II. Distribution in the population," *Ann.Eugen.* 8(1938):356–361.

23 D. J. Finney, "The detection of linkage," *Ann. Eugen.* 10(1940):171–214. A series of papers on linkage by Fisher appeared in 1935–1937, following up on the reception of the work of Felix Bernstein, and in dialogue with J. B. S. Haldane. These included Fisher, "The detection of linkage with dominant abnormalities," *Ann. Eugen.* 6(1935):187–201; Fisher, "The detection of linkage with recessive abnormalities,"

middle of the war, at a time when serologists were all impressed into the Emergency Blood Transfusion Service set up to deal with the large numbers of civilian injuries expected of the blitz. From 1939 to 1946 the Galton Laboratory Serum Unit, as it was called at this time, was ordered to evacuate from London to Cambridge and set to organizing the production of grouping sera for the service, as well as working out methods for 1 routine blood grouping of donors and methods of storing and preserving blood.

The unit was directed by the serologist George L. Taylor, who had moved to Cambridge with it from the Galton Laboratory.[24] The assistant director was Robert R. Race.[25] The unit's job was to provide a quality control service for hospital blood transfusion.[26] Samples of blood from

Ann. Eugen. 6(1935);339–351; Fisher, "Tests of significance applied to Haldane's data on partial sex linkage," *Ann. Eugen.* 7(1936):87–104.

24 Ronald A. Fisher, "G. L. Taylor M.D., Ph.D., F.R.C. P.," [Obituary] *Brit. Med. J.* i(1945):463–464; George L. Taylor was born in 1897 and took his medical degree in Manchester, where he was an outstanding student. In 1929 he gave up medical practice for research and took the Manchester M.D. in 1930 and the Cambridge Ph.D. in 1932. Taylor took charge of the Galton blood grouping laboratory under Fisher in 1935. When the war broke out in 1939 it was evacuated to Cambridge as the Galton Laboratory Serum Unit, where it concentrated on practical problems of grouping serum production and supply.

For an official account of the wartime activities of the Unit, see *History of the Second World War, UK Medical Series,* v. 2: Cuthbert L. Dunn, "The Emergency Medical Services," 334–355; F. H. K. Green and Major General Sir Gordon Covell, "Medical research," 97–110; bibliography of work on blood transfusion, etc., 132–137; Great Britain, Medical Research Council, *Medical Research in War: Report of the M.R.C. for the Years 1939–1945.*

25 Robert R. Race was born in 1907 and medically trained (L.R.C.P.; M.R.C.S.; Ph.D. Cambridge, 1947). He joined the Galton Laboratory Serum Unit in 1937, and from 1939 to 1946 he was assistant director. After the war ended, he was appointed director of the Medical Research Council Blood Group Research Laboratory, where he stayed until his retirement in 1974, mainly at the Lister Institute in London. See Sir Cyril Clarke, "Robert Russell Race, 28 November 1907–15 April 1984," *Biographical Memoirs of Fellows of the Royal Society* (London), 31(1985):453–492; Patricia Tippett, "In memoriam," *Vox Sang.* 47(1984):395–396.

Ruth Sanger was born in Australia in 1918 and took a London Ph.D. in 1948; she made her life at the Blood Group Reference Laboratory. In 1950 Robert Race and Ruth Sanger together produced the first edition of their book *Blood Groups in Man* (1950) (n. 13); the sixth and final edition came out in 1976. This famous book is one of the most beautifully and memorably written in the whole field of immunology, and it became better and better as the subject became more complicated and difficult to handle. Race and Sanger were husband and wife, and were indeed one flesh, or perhaps a chimera, as Sir Cyril Clarke genetically put it (*Biog. Mem.,* 482). They could even give a lecture together, as I saw myself. With their co-workers, they wrote some hundreds of articles in the field of blood group serology; their laboratory was a world reference centre, and their book the standard reference text for thirty years or more. Ruth Sanger succeeded Race as director of the Medical Research Council's Blood Group Unit in 1973. She was elected Fellow of the Royal Society in 1972. See *Who's Who of British Scientists 1971–1972* (Athens: Ohio University Press, 1972).

26 Box, *Fisher* (1978) (n. 21), 349–357.

all over Britain gave the Galton unit the chance to put together data from thousands of individuals and to calculate gene frequencies across the country. It was the first time in Britain that data had been available on that scale, although in Europe this kind of ethnological data was being collected since the early twenties.

The wartime "work of national importance," with its thousands of routine groupings, was dreary for the Cambridge serologists.[27] As Race said in 1946, they welcomed the chance to work on the new system:

There are now growing up very many young people who owe their lives to the researches of Dr Levine, the late Dr Landsteiner and Dr Wiener. I would like to express the thanks of another group of people, the English blood-transfusion workers, whose lives have not been saved, but have been made more interesting and exciting by the discovery of the Rh blood groups and their clinical associations. In England the news of the discovery came like sunlight in the drab endless routine of wartime transfusion work.[28]

According to Fisher, the British work on the Rhesus system dated from a letter of 1942 sent by Taylor and his colleague Patrick L. Mollison, professor of haematology at St Mary's Hospital in London, to the *British Medical Journal*.[29] In it, Taylor and Mollison appealed to the medical community for specimens of serum from mothers of children suffering from *Erythroblastosis foetalis*.[30] In 1943 Archibald J. McCall, pathologist from the North Staffordshire Hospital in Lancashire, sent them blood from a Mrs. St. who, although she was Rh positive, had erythroblastotic children. Taylor released the delighted Race to investigate the serum.[31] The St. serum was a powerful one that gave sharp results. It agglutinated all the Rhesus *negative* cell samples, and some of the positives, like Levine's anti-Hr, but to a total of 80 percent – which was higher than Levine's.[32] They interpreted this to be an antiserum that recognized the Rh-negative allele: it must be sorting out the heterozygotes. The next month, they had another paper in *Nature*, describing a second new specificity: samples of this one had come from two different sources, one from Kathleen Boorman and Barbara Dodd at the South

---

27 Fisher, "G.L. Taylor" (1945) (n. 24).
28 Robert R. Race, "The Rh genotypes and Fisher's theory," *Blood 3* (special issue 2)(1948):27–42, p. 27.
29 Fisher, "G. L. Taylor" (1945) (n. 24), 463.
30 George L. Taylor and Patrick L. Mollison, "Wanted: Anti-Rh sera," *Brit. Med. J.* i(1942):561–562.
31 Archibald J. McCall, Robert R. Race, and George L. Taylor, "Rhesus antibody in Rh positive mother," *Lancet* i(1944):214–215.
32 Robert R. Race and George L. Taylor, "A serum that discloses the genotype of some Rh positive people," *Nature* 152(1943):300.

London Blood Transfusion Centre, and one from the pathologist Charles V. Harrison of Liverpool University.[33] These two sera they called KJ. They agglutinated 30 percent of random cell samples; these were negative with the Rh negatives, and also with the St. negatives. Their specificity corresponded with one that Landsteiner and Wiener, and Levine, had recognized as distinguishing a subtype of Rh+, which they called Rh^2. Finally, in the autumn of 1943 Daniel F. Cappell, professor of pathology at Dundee, sent them a serum giving 70 percent of positive reactions, which was negative with *almost* all the Rh negatives and positive with most of the positives.[34]

Up to 1944, work on the Rh system went on in harmony on both sides of the Atlantic. The British workers at the Galton unit corroborated the American findings, and added three more sera to the series. At this point, twenty months after the letter to the journal, they had four antisera and could define seven alleles and eleven phenotypes. At the same time, A. S. Wiener published his latest results: he had three sera, which defined six alleles. The results of the two groups of workers corresponded perfectly, although they had been reached quite independently with different sera and a different population. Everybody was pleased.

Wiener's first interpretation of the heredity of the Rhesus system appeared in a short new chapter in the third edition of his textbook *Blood Groups and Blood Transfusion*, finished in January 1943, while Landsteiner was still alive.[35] He used as his model the pattern of the ABO blood groups, where there were four phenotypes and three alleles, which matched the existing information on the Rh system:

A   B   AB   O

Rh_1 Rh_2 Rh_1Rh_2 Rh neg (rh)

He even briefly suggested that the letters U, V, and W could be used as a terminology, like A, B, and O.[36] He had at this time three antisera to test with:

33   Robert R. Race, George L. Taylor, Kathleen E. Boorman, and Barbara Dodd, "Recognition of Rh genotypes in man," *Nature* 152(1943):563.
34   Robert R. Race, George L. Taylor, Daniel F. Cappell, and Marjorie N. McFarlane, "Recognition of a further common Rh genotype in man," *Nature* 153(1944):52–53.
35   Alexander S. Wiener, *Blood Groups and Blood Transfusion* (Springfield, Ill.: Thomas, 1st ed., 1935; 2d ed., 1939; 3d ed., 1943), 3d ed., 245–254.
36   Alexander S. Wiener, Eve B. Sonn, and R. B. Welkin, "Heredity and distribution of the Rh blood types," *Proc. Soc. Exper. Biol. Med.* 54(1943):238–240, and in *Rh-Hr* (1954) (n. 1), 367–368.

anti-Rh$_1$ giving 85 percent positives (the original guinea-pig anti-
                Rhesus monkey serum)
anti-Rh$_2$ giving 73 percent positives
anti-Rh$_{1,2}$ giving 87 percent positives

Wiener refers here to Levine's serum as "so-called Hr," which should correspond to an anti-O as it reacted with Rh negatives. The following year, when he had dropped the rest of the analogy with the ABO groups, he was still thinking of anti-Hr as a kind of anti-O, but neither Wiener nor Levine himself had really worked on it.

The next variant terminology appeared in Wiener and Landsteiner's paper of 1943 – it was Landsteiner's last completed work. In it they repudiated the earlier analogy with the ABO groups and changed the names of the antisera:

anti-Rh   giving 85 percent positives (the original guinea-pig anti-
             rhesus monkey)
anti-Rh$_1$ giving 73 percent positives
anti-Rh'   giving 87 percent positives[37]

The original Rh positive type was now divided into two subgroups by its reaction with the 73 percent anti-Rh$_1$ serum: the positives were Rh$_1$ and the negatives Rh$_2$. The 87 percent serum, instead of being seen as a summation of the other two, became a means of dividing the Rh$_1$ phenotypes again: the positives of this test formed a new sub-*sub*-class called Rh'.

The model was now group A and its subtypes, or group B and its subtypes, rather than the ABO system as a whole. A fourth serum, their third human one, reacted with some of the Rh$_2$ group. It was first called anti-Rh$_2$, but later, as its reactions seemed to parallel those of anti-Rh', its name was changed to anti-Rh", which suggested a three-level system of dichotomies. But that did not quite work, for both anti-Rh' and anti-Rh" reacted with some of the "so-called Rhesus negatives."[38]

It is not possible to illustrate this by a clear diagram of dichotomies. The mind struggles to find a pattern in these hints of symmetry, but every pattern is scribbled over by further, unpatterned findings.

In October 1943 Race and Taylor had a "personal communication" from Wiener, sending them a typescript of a paper in press that told them he had now demonstrated six alleles.[39] Race and Taylor wrote in

37 Alexander S. Wiener and Karl Landsteiner, "Heredity of variants of the Rh type," *Proc. Soc. Exper. Biol. Med. 53* 167–170 (1943); and in *Rh-Hr* (1954) (n. 1), 238–241 (p. 238). Rh' is read as Rh prime.
38 Alexander S. Wiener, "Distribution and heredity of the Rh types" *Science 98*(1943): 182–184; and in *Rh-Hr* (1954) (n. 1), 242–243. Rh" is read as Rh double prime.
39 Alexander S. Wiener, "Genetic theory of the Rh blood types," *Proc. Soc. Exper. Biol. Med. 54*(1943):316–319; and in *Rh-Hr* (1954) (n. 1), 244–247.

the "letter to *Nature*" in which they, with Boorman and Dodd, described their newest specificity, KJ, and its reactions: "We are at present calling our fourth and fifth allelomorphs Rhx and Rhy in the hope that a name and place is ready for them in Wiener's scheme."[40] In January 1944, along with the description of their fourth serum, they wrote: "We will use Wiener's names, which are Rh_1, Rh_2, rh, rh' . . . rh" . . . but for the gene Wiener calls Rh . . . we prefer Rh_o for Rh has for so long had a much wider meaning. The names we have used and are now abandoning [that is, the names of the serum donors, St. and KJ] are given in brackets." And further on they say, "We were stimulated by Wiener's letter to attempt to identify his types of sera with those we have used, and have found all of them represented in our collection."[41]

They published Wiener's table and their own, side by side. The two differed only in that Wiener had not had the St. serum, which allowed them to define one more allele and to distinguish heterozygotes within the Rh phenotype, since St. reacted with the rh or Rh-negative allele.

Fisher had not been in direct contact with the serum unit from 1939, when it was evacuated from London, up until 1943, when he, too, moved to Cambridge as Balfour Professor of Genetics, in succession to Reginald C. Punnett. Fisher was keen to work on the new system from the point of view of a genetic equilibrium that was under pressure from negative selection, but there was a heavy burden of routine work, and it was difficult for the unit to find time for it. At that period, Fisher was not in Cambridge, but he still kept in touch. As Race and Sanger loyally wrote, Fisher had "founded the laboratory, and had continuously inspired the workers in it, and taught them the habits of mind which were their most useful tool."[42]

Up till 1944, there was friendly co-operation between the Americans and the Galton unit. In spite of their late start, the British workers were contributing quite satisfactorily to the problems of blood group serology, and the work on the Rhesus system was making good progress between the two parties of workers. Race, a gentle and kindly man, allowed Wiener's nomenclature and systematization of the results to take precedence over his own, as we have seen. He was happy to find that results in the two laboratories duplicated each other exactly, except for the reactions of the 80 percent serum, St., which Wiener did not have, but which turned out to be an important clue to the genetics of the system. The results in 1944 are shown in Table 16.1.

When Fisher returned to Cambridge, contacts with the blood group-

40 Race et al., "Recognition of Rh genotypes" (1943) (n. 29), 564.
41 Race et al., "Recognition of a further Rh genotype" (1944) (n. 34), 52.
42 Race and Sanger, *Blood Groups* (1950) (n. 13), 99, 1st ed.

Table 16.1. *British and American anti-Rhesus sera (1944)*

Reference	Serum name	% Random samples +	Fisher's name	Wiener's name
McCall, Race & Taylor *Lancet i* 214 (1944)	anti-Rh	85%	anti-D	anti-Rh
Race & Taylor *Nature 152* 300 (1943)	St.	80% (all Rh− agglutinated)	anti-c	Anti-Hr (Levine)
Race, Taylor Boorman & Dodd *Nature 152* 563 (1943)	KJ	30% (very few Rh− are agglutinated)	Anti-E	anti-rh"
Race, Taylor Cappell & MacFarlane *Nature 153* 52 (1944)	A	70%	Anti-C	anti-rh'

With these 4 specificities Race and Taylor could define 11 phenotypes and 8 alleles. At this time Wiener was using 3 specificities and could define 6 alleles. Levine's anti-Hr had not been followed up.
See Figure XVI.3 for Fisher's interpretation of the inheritance of the system.

ing unit that had been his creation in earlier years were renewed. Race has described how the blood groupers used to meet Fisher almost every day for lunch at a pub called The Bun Shop, to discuss the latest complexities of the Rhesus system and its antisera. It was here that one day in January 1944 Fisher produced the synthesis of their findings that is illustrated in Figure 16.3. The beaten copper top of the pub table that it was written on was wet and uneven, and the writing was rather crooked. But Fisher's insight was clear: by arranging the results in this way he saw that the serum St. had a reciprocal relationship with the one called A. He suggested that the two sera recognized a pair of allelic antigens, which he called C and c. The other two sera, anti-Rh and KJ, had no such relationship: these antigens he called D and E, and he predicted that their alleles, d and e, would soon be discovered.[43]

Race and other British blood group workers at once adopted this

43 Robert R. Race, "Some notes on Fisher's contribution to human blood groups," *Biometrics 20*(1964):361–367, p. 362; Race to Mazumdar, pers. commun., August 1974.

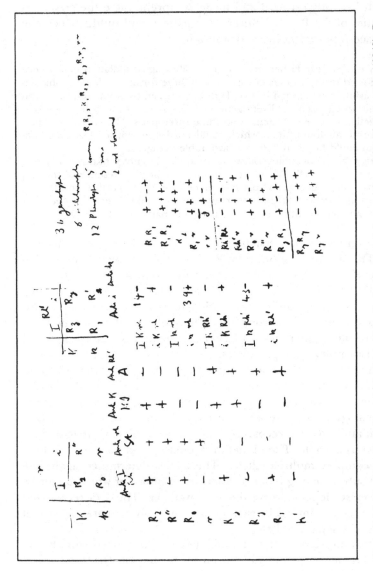

Figure 16.3. The Bun Shop Hypothesis: R. A. Fisher's elucidation of the genetics of the Rhesus system (1943). By arranging the results like this, Fisher saw that the serum A had a reciprocal relationship with serum St, suggesting to him a dominant and its recessive, or a presence and absence. He accordingly named them anti-C and anti-c. Since there was also an anti-D and an anti-E, he predicted that an anti-d and an anti-e would soon turn up. Anti-e soon did, but no anti-d has ever been confirmed. My thanks to Robert Race and Ruth Sanger for a copy of this scrap of paper.

scheme. It instantly clarified and made comprehensible the criss-cross complexities of the Rhesus antisera: the pattern had suddenly become visible. Race reported Fisher's idea in a letter to *Nature*:

In January of this year Fisher drew up the following formulation of the relationships found in the Rhesus factor. . . . The three forms of allelomorphic antigens are arbitrarily denoted by C c, D d, E e, chosen to avoid confusion with any symbols so far used. . . . Every gene of the system seems to be associated with a selection of three antigens from these three pairs. The system thus predicts an eighth allelomorph . . . which could not be recognised in a single individual but could be identified in a favourable pedigree.

It also suggests the possibility of two more antibodies not yet known, reacting with d and e, respectively. Wiener has supposed that the presence of the Rh_1 gene results in there being two "partial antigens" on the red cell (C and D . . .), and our recent work with St. and other sera seem to make three parts necessary to the total antigen resulting from the Rh_2 gene, namely, c, D, and E.[44]

It is not difficult to understand why this pattern had not occurred to Wiener. The strong St. serum that Race was using, which Fisher called anti-c, was Fisher's clue to the antithetical relationship of C and c, and Wiener did not have it. His sera, as Table 16.1 shows, were all "big-letter" sera – anti-D, anti-C, and anti-E, in Fisher's terminology – so there was not the same strong suggestion of paired reciprocities in Wiener's results as there was in Fisher's. Instead, his results suggested a series of complex multiple alleles.

This empiricist explanation is not enough to explain Wiener's interpretation, however, or the passion with which he maintained it. When he did acquire a sample of a strong St. serum, he still saw it as an "anti-O," by analogy with antisera to the multiple alleles of the $A_1 A_2 B O$ system, an antibody that recognized the recessive allele rh (cde in Fisher's terms) as a whole. The model in Wiener's mind was one of a long series of complex multiple alleles. The alleles determined agglutinogens with subtly varying blood factors, any number of which might be found to exist, depending on the sera available. The sera recognized "partial antigens," but each reaction did not imply a separate chemical structure, or a separate allele.

It was this impression of a modulating series of related complex alleles that Wiener's terminology was meant to suggest. His terms rh rh' rh" rh_y Rh_o Rh_1 Rh_2 Rh_z emphasize the wholeness and unity of each allele. However many specific sera are found to react, say, with Rh_1, it is still Rh_1.

For Fisher, on the other hand, there is not one locus with a series

44 Robert R. Race, "An 'incomplete' antibody in human serum," *Nature*, 153(1944): 771–772.

of modulating alleles, but a group of three loci, each with a single pair of alleles, as in the old Batesonian presence-and-absence pattern. Fisher's terms for the same series of Rh "chromosomes," to use his expression, are cde, Cde, cdE, CdE, cDe, CDe, cDE, and CDE. They emphasize the three separate antisera by which they are determined, the three separate antigens that compose them, and the three separate but linked genes that determine them. Each serological reaction implies an antigen, and each antigen a gene. The Rhesus factors are unit characters, not partial antigens.

Fisher's theory made three predictions. If there were three genetic loci, each with two alternatives, there must be eight different Rh "chromosomes." Seven had already been found: there would be one more, whose reactions could be predicted. Two more antisera would also be found, anti-d and anti-e. Strong confirmation for the theory came when Alfred Mourant at the North-East London Blood Supply Depot found the predicted anti-e, and the next when Louis Diamond of Children's Hospital, Boston, Massachusetts, found what seemed to be anti-d.[45] Fisher's predictions had worked out brilliantly.

A further implication was then seen in the theory. Fisher calculated the "chromosome" frequency by the maximum likelihood method, with Race's data.[46] He noticed that the frequencies could be arranged in three orders of magnitude: greater than 12 percent, around 1 percent, and very rare. He made what Race called "the fascinating suggestion" that the rarer chromosomes arose by crossing over from the three frequent forms, as shown in Figure 16.4.[47] Crossing over between the commonest heterozygotes would produce the second-order ones, and double crossing over, the rarest. The gene frequencies seemed to support this clever idea.

Fisher's inspired untangling of the Rhesus genes, antigens, and antibodies was welcomed with relief in Britain. The first edition of Robert Race and Ruth Sanger's book *Blood Groups in Man* came out in 1950. Its authoritative treatment of the whole subject of blood grouping gave a legitimacy to the Fisher–Race side of the controversy: the playing field

45 Arthur E. Mourant, "A new rhesus antibody," [anti-e] *Nature* 155(1945):542; Louis K. Diamond, "Physico-chemical and immunological characteristics of Rh antibodics," [anti-d] International Hematology and Rh Conference, 17–23 November 1946, Dallas, Tex. (unpublished data referred to by Haberman et al.). Several more reports of anti-d appeared soon after Fisher's prediction was made: for example, Sol Haberman, Joseph M. Hill, B. W. Everist, and J. W. Davenport, "The demonstration and characterization of the anti-d agglutinin and antigen predicted by Fisher and Race," *Blood* 3(1948):682–695. In the end, no true anti-d was ever confirmed.
46 Ronald A. Fisher and Robert R. Race, "Rh gene frequencies in Britain," *Nature* 157(1946):48–49.
47 Race, "Rh genotypes and Fisher's theory" (1948) (n. 28), 34.

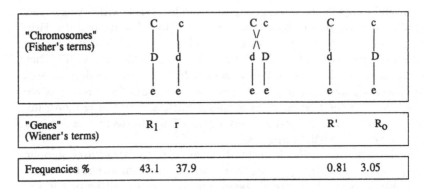

Figure 16.4. R. A. Fisher's suggestion for the origin of the less frequent Rh "chromosomes" by crossing over between the more frequent, supposed to occur on rare occasions. From Fisher and Race, "Rh gene frequencies in Britain" (1946) (n. 46).

was no longer even. The conflict was between an individual, on Wiener's side, and what was now an institution, Race and Sanger's book.

They wrote in a footnote to a chapter in which all of Fisher's suggestions were joyfully accepted:

The theory has now been completely confirmed in every detail and is accepted universally. Numerous attacks on the theory by Wiener have in fact been attacks concerning the highly academic and interesting point whether the three allelomorphic sites of Fisher are to be placed within or without the boundary of one gene. . . .

From these attacks the impression may unfortunately be received that Fisher's great contribution is wrong and has been disproved. The only real disagreement between the two schools . . . is over the academic point referred to and over the notation.[48]

Race and Sanger's footnote, however, plays down the depth of the disagreement. The argument about whether gene frequencies in different populations really did support Fisher's crossing-over idea or whether such crossing over and separation of the antigens had ever been found in a real family only touched the problem indirectly. The

48 Race and Sanger, *Blood Groups* (1950) (n. 13), 82. The section on the inheritance of the Rh groups contains the following sharp remark, directed at Wiener (p. 110): "This description of the inheritance does not call for any particular view concerning the mechanism of the inheritance. It serves equally well whether we passionately believe that the genes for the series of allelomorphic antigens are firmly attached to each other or separable. There is no need at this stage to become involved in what Professor Fisher once called the acrimony of a monophysite controversy." This was how Race saw the two conflicting views; and seen like this, the controversy is, as he felt, silly and trivial.

disagreement at its deepest level concerned the implications of Fisher's theory for the relationship of gene, antigen, and antibody. Fisher's view of this relationship is nicely expressed in the introductory paragraph of a paper that he proudly called "The Rhesus Factor: A Study in Scientific Method":

A number of genetic factors recognisable by serological tests are now known. Typically the red cells of some genotypes are agglutinated and those of others not agglutinated, by a test fluid, usually a serum containing a chemically active substance known as an antibody. . . . [I]t has been customary to designate the antibody by a Greek letter such as , corresponding with an antigen designated by the Latin letter A. Genotypes developing the antigen A may then be recognised by a test fluid containing . . . . is generally though perhaps not invariably true, that the genotypes which develop the antigen A are those which contain a particular hereditary particle of the germ plasm or gene . . . designated by the symbol G^A.[49]

For Fisher, as this passage shows, the relationship of antibody to antigen to gene is $1 : 1 : 1$. There is no difficulty in his mind about calculating the frequency of a "chromosome" as he and Race significantly called it, from the results of agglutination tests: the antibody recognizes the antigen A, which is determined by the gene G^A. Like the mathematician Bernstein, the mathematician Fisher accepts without question the only relationship between antibody antigen and gene that would make possible the use of the agglutination of red cells as a unit of calculation. If either of them had doubted this relationship in any way, they could not have built on it the towering structure of Mendelian population genetics that they did. No overlapping specificities, multiple antibodies, or more or less good fit would have supported such calculations: the biological phenomenon, the agglutination reaction, perfectly mirrors the gene. As Bernstein had written in 1924: "One can even go as far as to say that the agglutinogens are . . . similar to secretions of the chromomeres."[50]

That is to say, for both Bernstein and Fisher there is a direct one-to-one relationship between the genetic "unit" and the blood group "character." Bernstein had also written, "We must assume, in accordance with the side-chain theory," and he had cited Ehrlich and Mor-

49 Ronald A. Fisher, "The Rhesus factor: a study in scientific method," *Am. Sci.* 35(1947):95–102, 113.
50 Felix Bernstein, "Zusammenfassende Betrachtungen über erblichen Blutstrukturen des Menschen," *Z.f. indukt. Abstamms. u. Vererbl.* 37(1925):237–270 (p. 247). See the discussion of blood group genetics, the unit-character and the receptor in Chapter 15.

genroth's goat haemolysin experiment, and by implication, like so many blood group serologists after him, the Ehrlich receptor theory.

Fisher's variation on this idea appears at the end of his essay "The Rhesus Factor: A Study in Scientific Method," already quoted above. There may still be more Rh genotypes that have not yet been found. He writes: "Using reliable and unmixed testing fluids, each capable of identifying a single elementary antigen, this situation though complex, is by no means unmanageable."[51]

And, elsewhere:

A certain measure of order has been introduced into this complex system by the concept of elementary antigens, the presence of each of which is indicated unequivocally by a positive reaction with the corresponding antibody. Such antigens may well correspond with allelomorphic genes at three closely neighbouring loci.[52]

Fisher's elementary antigens, each of which is indicated unequivocally by a positive reaction, have a startlingly familiar sound. His "introduction" to the paper of 1946 repeats almost word for word the "conclusion" of Malkoff's paper of 1900: "The agglutinin has a specific binding affinity for the morphological element that it agglutinates. It is bound by this alone, and by no other."[53]

Fisher's mathematical essays on gene frequency in populations stand upon the same basis of absolute specificity as did Malkoff's differential absorption experiment of 1900. Malkoff's project was carried out at Koch's Institut für Infektionskrankheiten, under the supervision of August von Wassermann, close associate of both Koch and Ehrlich. It was designed to support the receptor theory of the specificity of goat blood haemolysins, of Ehrlich and Morgenroth. Whereas Bernstein, writing in 1924, had been close enough in time to refer directly to Ehrlich and Morgenroth's work of 1900, Fisher, writing in the 1940s, was now too distant. But the tradition persisted, even though the source had been forgotten.

As noted in Chapter 14, the receptor and the unit character continued in blood group genetics, with, but more often without, direct and overt references, long after they had disappeared from the fields of immunochemistry and general genetics. These are the two concepts

51 Fisher, "Study in scientific method" (1947) (n. 49), 113.
52 Ronald A. Fisher, "The fitting of gene frequencies to data on Rhesus reactions," *Ann. Eugen. 13*(1946):150–155 (p. 150).
53 G. M. Malkoff, "Beiträge zur Frage der Agglutination der rothen Blutkörperchen," *Deutsch. med. Wschr. 26*(1900):229–231 (p. 231). The significance of Malkoff's experiment and *das Malkoff'sche Phänomen* is discussed in many places: in Chapter 7 (nn. 19, 23); in Chapter 14 (nn. 7, 68); in Chapter 15 (n. 14).

that informed Fisher's understanding of the immunogenetics of the Rhesus system and that provided the intellectual basis for Wiener's rejection of it.

In June 1943 Landsteiner died suddenly. The paper he had done on the Rhesus system with Wiener in 1943 was his last completed work. There remained some incomplete projects that he had been doing with Merrill W. Chase, and the second edition of his book, *The Specificity of Serological Reactions*, which was complete textually, but not yet in press. Landsteiner's son Ernst, Chase, and Wiener pushed it through, together with Linus Pauling's chapter on bonding.[54]

A few weeks after Landsteiner's death, Wiener mentioned for the first time the Landsteiner-inspired interpretation of the serology and genetics of the Rhesus system that he was to struggle so passionately to establish. In a paper that came out on 30 August 1943, seven weeks after Landsteiner died, Wiener made his first reference to the idea of "partial antigens":

It was originally suggested that the standard guinea-pig anti-Rhesus (giving about 84% positive reactions) should be designated anti-Rh_1, and the anti-Rh serum giving 70% positive reactions as anti-Rh_2. This terminology has been abandoned however, because it was found that the hypothetical agglutinogens determined in this way do not Mendelize like A and B but are apparently "partial antigens" like the factors B_i B_{ii} B_{iii} of human group B, and the factors FA A_1 A_2 of group A blood.[55]

Again, a few months later,

As Wiener and Landsteiner have pointed out type Rh_1 and type Rh_2 bear serological and genetic relation to each other, similar to that of the subgroups A_1 and A_2. For example the reactions of type Rh_1 blood with the agglutinins Anti-Rh and Anti-Rh_1 are not due to the two genetically distinct antigens Rh and Rh_1 in such blood but to an agglutinogen Rh_1 containing two partial antigens inherited as a unit by means of a corresponding gene.[56]

54 Ernst Landsteiner, "Foreword," to Karl Landsteiner, *The Specificity of Serological Reactions*, 2d ed. (Cambridge: Harvard University Press, 1945).
55 Wiener, "Distribution and heredity of the Rh types" (1943) (n. 38), 242. The terminology B_i B_{ii} B_{iii} for the subgroups of B comes from a paper by V. Friedenreich and S. With, "Ueber B-Antigen und B-Antikörper bei Menschen und Tieren," *Z. f. Immunitätsf.* 78(1933):152–172; see also Landsteiner, *Serological Reactions* (1945) (n. 54), 116. Landsteiner discusses the idea of "partial antigens and antigenic mosaics," as an explanation of the difference between human and animal B antigens. It is a passage to which Wiener often refers; see n. 87 for other Landsteiner passages, and the references to them in Wiener's papers.
56 Wiener, "Genetic theory of the Rh blood types" (1943) (n. 39), 244. Wiener also saw the reactions of Levine's anti-Hr serum as representing an anti-O. See Wiener, "Analogy between Hr and O," *Science* 100(1944):595–597, and in *Rh-Hr* (1954) (n. 1), 251–252.

These two early references to "partial antigens" are portents of what was to come. They are echoes of Landsteiner's last work on specificity, the "one-or-two-antibody question," and the "double-eagle" project that had occupied him through the thirties. Although these two early papers make no reference to Landsteiner as the source of the idea, Landsteiner himself uses the phrase in the paper in which he discusses the "double-eagle" antigens:

Aromatic amino compounds containing two different acid groupings were synthesized and used for preparing conjugated antigens. The immune sera obtained by injection of these azoantigens were examined for the presence of multiple antibodies by means of "partial" antigens made from compounds containing only one of the acid groups.[57]

Further on in the same paper, he makes the connection between the azo-protein model antigen and its counterpart in nature: "Of some significance too for the problem of the relation of immune bodies to natural antigens are the two instances reported, in which an appreciable amount of antibody was produced for one only of the two groups present, though each had formerly been found to serve as a serological determinant."[58]

Wiener's references to Landsteiner's work as a model for the explanation of the Rhesus system thus started immediately after Landsteiner's death. It was upon this idea that Wiener built his interpretation of the immunogenetics of the Rhesus system and his terminology.

During 1944 Wiener published several papers on the technical problems of detecting the antibodies, problems that he himself dealt with every day in his work on medicolegal blood grouping for the New York Medical Examiner's Office. His "Review of Knowledge up to 1945" discusses the genetics of the system, but in more practical terms, and does not mention Landsteiner.[59]

The idea of actually designing a terminology to suggest partial antigens does not seem to have occurred to him until 1946, when the New York Academy of Sciences met to discuss the problem of a nomenclature for the new system. Wiener spoke, and there was some discussion by Philip Levine, Silik Polayes (Wiener's old collaborator), and the Harvard immunologist William C. Boyd, who had been a fairly close friend

---

57 Karl Landsteiner and James van der Scheer, "On cross reactions of immune sera to azoproteins II. Antigens with azocomponents containing two determinant groups," *J. Exper. Med.* 67(1938):709–723 (p. 710), C. 316.

58 Landsteiner and v. d. Scheer, "On cross-reactions, II." (1938) (n. 57), 722.

59 Alexander S. Wiener, "The Rh blood types and some of their applications," *Am. J. Clin. Path.* 15(1945):106–121, and in *Rh-Hr* (1954) (n. 1), 63–78.

of Landsteiner's.[60] A transcript of the proceedings was sent to Race in England for his views. Race replied amiably that the Americans were being more cautious than the Fisher school of thought, and they might well be right.[61]

The further step, which really is only the more explicit statement of the earlier ones, was to oppose directly a statement by Landsteiner to Fisher's genetic hypothesis. Landsteiner stated that simple chemical compounds could give rise to several distinct but specific antibodies, and that the number of antibodies was not directly correlated with the number of separate chemical structures, while Fisher's hypothesis and Fisher's terminology implied that each antibody recognized a separate distinct structure. Echoing Landsteiner and opposing Fisher, Wiener writes: "a single letter should be used to designate agglutinogens behaving like units, and separate letters should be used only for agglutinogens that separate genetically."[62]

His earliest recognition of Fisher's hypothesis as a possible danger to his own appears in a paper of November 1945, which starts off: "The hypothesis proposed by Fisher, that there are three different varieties of antisera capable of agglutinating Rh-negative blood, corresponding to the three varieties of anti-Rh sera, appears to be gaining ground."[63]

By 1948, Wiener's objections to Fisher's genetic hypothesis and Fisher's terminology had more or less acquired their definitive form, already systematized by four years of discussion. A "Letter to the Editor" of the *British Medical Journal* sets out sixteen theses against Fisher, beginning with No. 1, that his own terminology has priority. But No. 11 (as in the case of Marx's theses against Feuerbach) is the most important one, and the most germane to this interpretation of the conflict: "11. They [that is, Wiener's own symbols] do not involve any incorrect assumption of one-to-one correspondence between genes and partial antigens."[64] It was this idea, here merely one of sixteen objections to

60 Alexander S. Wiener and Eve B. Sonn, "The Rh series of genes with special reference to nomenclature," *Ann. N.Y. Acad. Sci.* 46(1946):969–992; discussion by Philip Levine, Silik H. Polayes, and William C. Boyd. Wiener's first thorough-going attack on the Fisher–Race nomenclature seems to have been Alexander S. Wiener, "Theory and nomenclature of the Rh types, subtypes and genotypes," *Brit. Med. J.* i(1946): 982–984.

61 Robert R. Race, contribution to the discussion on Rh nomenclature, *New York Acad. Sci.* (1946) (n. 60), 988.

62 Alexander S. Wiener, "The Rh system in the chimpanzee," *Science* 104(1946):578–579. He quotes directly from Landsteiner's *Serological Reactions* (1945) (n. 54), 114–116, here as in many other places; see n. 87 for other references by Wiener.

63 Alexander S. Wiener, "Theory and nomenclature of the Hr blood factors," *Science* 102(1945):479–482.

64 Alexander S. Wiener, "Anti-Rh serum nomenclature," *Brit. Med. J.* i(1948):805–822, and in *Rh-Hr* (1954) (n. 1), 319–322.

the Fisher–Race CDE terminology, that Wiener gradually developed, and through which he linked himself to Landsteiner. In 1951, in an account of the history of the Rhesus blood types, and under the subheading LANDSTEINER FOUNDER OF THE SCIENCE OF IMMUNOCHEMISTRY, Wiener wrote:

By studying the cross-reactions of antibodies against relatively simple chemical structures, Landsteiner showed that antigens elicited not merely a single corresponding antibody, but a whole spectrum of antibodies of different specificities. Similarly, by studying the cross-reactions of different human anti-A and anti-B blood on blood of lower animals, it has been possible to demonstrate that agglutinogens A and B behave serologically as if they had a mosaic structure. This "mosaic structure" has proved to be a universal property of agglutinogens, not only A and B, but also M and N, Rh-Hr, etc., and represents, at least in part an artifact due to the ability of a single antigen to elicit and react with a multiplicity of antibodies, as demonstrated by Landsteiner and his coworkers in their studies on antigens of known chemical composition.[65]

In the same year, Wiener had written in a speech at the acceptance of an award that he was given in 1951, "I trust that our future work will justify this token of your confidence and the tradition laid down by the great pathfinder Karl Landsteiner, whose many contributions I have attempted to describe to you from the viewpoint of a pupil, co-worker, friend and admirer."[66]

Wiener seems to be alone among all of those who worked with Landsteiner in seeing himself in this light. Those who worked in Landsteiner's laboratory seldom thought of themselves as his friends, and their respect for him was often mixed with resentment. Only Wiener regarded himself as Landsteiner's pupil, and as carrying on a kind of thought transmitted to him by Landsteiner. Since this is also the only one of this series of teacher–student relationships of which we have any first-hand knowledge, it is interesting to look a little more closely at it, and at Landsteiner's relationships with some of his co-workers at the Rockefeller Institute.

Philip Levine worked directly under Landsteiner as his assistant for about seven years, beginning in 1925. Their projects extended over the late twenties and were all connected with blood grouping: their papers contain first descriptions of the P groups and the MN groups. It was

65 Alexander S. Wiener, "History of the Rhesus blood types," *J. Hist. Med.* 7(1952):369–383; and in *Rh-Hr* (1954) (n. 1), 9–23 (p. 12). In the references (p. 23), Wiener writes, under his n. 44, "For a basic understanding of the Rh-Hr types, one must read Landsteiner's concept of the mosaic structure of agglutinogens," and he refers to Landsteiner, *Serological Reactions* (1945) (n. 54), 113–116. See n. 87 for Wiener's references to Landsteiner.
66 Wiener, "Landsteiner his work" (1951) (n. 8), 8.

Levine who was set to telephone Wiener about his *gaffe* concerning Furuhata's Chinese forensic blood grouping method, and, working under Landsteiner's close supervision, to review the literature on Bernstein's triple-allele hypothesis. The laboratory assistants were not allowed to introduce problems of their own, but, says Levine, Landsteiner was very receptive to any ideas concerning the problem in hand.[67] It was a pattern of work that left not only leadership but all decisions strictly up to Landsteiner. In the words of Alexandre Rothen, the physical chemist who collaborated with Landsteiner during his Pauling period, within his own group Landsteiner was a dictator.[68] There was no room here for preparing young minds for independent thought and investigation. Levine's dependency on Landsteiner was suddenly ruptured in 1930, when he left the Rockefeller Institute to join the University of Wisconsin. Landsteiner, he said, made him swear not to work on blood groups – a custom that is not unusual in commercial laboratories, but uncommon outside. Levine obediently turned at first to the specificity of bacteriophage and antibodies to *Salmonella*, though he later went back to blood group serology.[69] The rupture was very hard on him: he entered psychoanalysis, fantasizing that Landsteiner was his father, and in the course of the analysis, came to hate Landsteiner and to spit on his picture.[70] In May 1943, however, Levine was to name his son Karl, after Landsteiner: the ambivalent relationship was not quite over.[71]

Merrill Chase joined the laboratory in 1933, as assistant, coming after James van der Scheer in status, whose title was associate. In June 1939, when Landsteiner officially retired from the Rockefeller Institute, van der Scheer's post was abolished, but Chase stayed on, still as assistant.

Chase's memory of life at the laboratory, written in 1950, is filled with the sense of powerlessness and resentment of a long-time coworker:

Our laboratory centered as completely about the central figure as did the choice of the evening victuals served at Dr Landsteiner's home.
  Indeed, I am of the opinion that the greatest criticism of Dr Landsteiner may be directed at the one point of his failure to train, deliberately, rather than

67 Philip Levine, interview with George Mackenzie, d. 1944; Landsteiner–Mackenzie Papers, B L23m, Box 1: American Philosophical Society.
68 Alexandre Rothen, interview with Mackenzie, dated 7 February 1948.
69 Philip Levine, "Dedication," and "Response," *Advances in Pathobiology* 7(1980):vii–xii (p. xii); Levine to Mazumdar, pers. comm., 1978.
70 Levine, interview with Mackenzie, n.d.; Landsteiner–Mackenzie Papers, B L23m, Box 6: American Philosophical Society.
71 Levine, interview with Mackenzie (1944) (n. 70.)

incidentally, or to encourage independent scientific productivity in his long train of associates.[72]

Wiener, the Benjamin among the collaborators, never worked under Landsteiner in his own laboratory. Their relationship, unlike the others, was that of master and pupil, not master and servant, and Wiener proudly described it many times, as he did in the speech already quoted, which he used as an historical introduction to his book *Rh-Hr Blood Types*.[73]

Each of Wiener's books was prefaced with some tribute to Landsteiner, as "the father of immunohaematology,"[74] or, as in *Rh-Hr Blood Types*, by an historical introduction explaining Landsteiner's important contributions, and associating Wiener with him. Each of Wiener's books had a picture of Landsteiner at the laboratory bench as a frontispiece, a copy of which he gave me (Figure 16.5). The only exception is *Blood Groups and Blood Transfusion*, the third and last edition of which came out in early 1943, during Landsteiner's lifetime.[75]

This one was written, according to Wiener, with Landsteiner's help and advice. Landsteiner would have found photographs and adulatory remarks extremely embarrassing: he would most certainly not have permitted any to appear in the book:

About this time [1930] I decided to undertake the preparation of a book on blood groups, and here again I received unexpected and generous help from Dr. Karl Landsteiner. One evening a week was set aside for a regular visit to his home, and this continued for more than a decade. At these meetings, I had not only the advantage of his invaluable constructive criticisms of my manuscript materials, but also the opportunity to learn at first hand from this great scientist his latest work and ideas, thus gaining an insight into the subject that stood me in good stead in later years when studying the serology and genetics of the Rh-Hr types. Those pleasant meetings, the interludes during which we chatted of lighter matters, such as the scientist who had twins, baptised one and kept the other as a control, the visits to Dr. Landsteiner's summer homes in Nantucket and Newfane where we pored over galley proofs together, are fond memories which will remain with me for the rest of my days.[76]

It is not clear how Landsteiner responded to this admiration. Levine, jealous of his sibling Wiener and of Wiener's relationship with Land-

72 Merrill W. Chase to Mackenzie, letter dated 16 December 1950; Landsteiner–Mackenzie Papers, B L23m, Box 3, American Philosophical Society; Chase to Mazumdar, pers. commun., 1973.
73 Wiener, "Landsteiner his work" (1951) (n. 8).
74 Alexander S. Wiener and Irving B. Wexler, *Heredity of the Blood Groups* (New York: Grune and Stratton, 1958), x.
75 Wiener, *Blood Groups and Blood Transfusion* (1943) (n. 35).
76 Wiener, "Landsteiner his work" (1951) (n. 8), 5.

Figure 16.5. Karl Landsteiner, a photograph probably taken at the Institute for Pathological Anatomy, where he worked 1897–1907. Landsteiner gave a signed copy of this photograph to A. S. Wiener, who used it as a frontispiece in each of his books and made copies of it to give his friends. (My thanks to A. S. Wiener.)

steiner, tells a story of Landsteiner at a party at Wiener's house in Brooklyn, stiff and uncomfortable, and as Landsteiner told Levine, miserable. Levine said snidely that Wiener was trying to show him off. Levine, too, would work on papers at Landsteiner's house, and Landsteiner, said Levine, would ask him not to tell Wiener that he had been there.[77] Wiener and Levine were in jealous competition for the favour of a demanding, highly introverted personality, who found favour, or

---

77 Levine, interview with Mackenzie, n.d. (n. 70); Mackenzie dates this story to somewhere around 1941.

even warmth, very difficult. It is likely that both Wiener and Levine, and not Levine alone, fantasized their relationship with him to some extent.

It is not hard to see where this love and admiration shows itself in Wiener's work, for Wiener himself makes it as clear as he can. In accordance with Landsteiner's partial antigen idea, Wiener interpreted the numerous different serological reactions of the Rhesus agglutinogens as effects of the partial antigens of a complex, the complex being inherited as a unit, determined by a single gene. The gene R_1 determined a complex Rh_1, which gave serological reactions with several different antiserums: these for the present included anti-Rh_o, anti-rh', and anti-rh". Any number of others might be found later. These serological specificities he called "blood factors," which was Landsteiner's term, and like Landsteiner, he refused to see them as separate entities as they were not inherited separately.

The key passage in Landsteiner's writing, to which Wiener often refers in his explanations of this interpretation, appears in *Specificity of Serological Reactions* of 1945.[78] Here Landsteiner points out that "blood factor" does not mean the same as Ehrlich's "receptor," and that a plurality of separate antigenic specificities need not indicate separate structures, especially when they fail to Mendelize separately. Echoing Landsteiner, Wiener writes:

A single antigen molecule can react with several antibodies of different specificities, and the unknown attributes of the antigen molecule which determine these serological reactions may be termed "serological factors" of the antigen. Since even a single simple antigen may have many such "serological factors," antigens behave in general as if they have mosaic structures.[79]

Later on, in a paper that appeared in 1961, Wiener developed the idea that the blood factor is a phenomenon only and not a *Ding-an-sich*. Its recognition by antibody is all we know of it, and with different antibodies the blood factors of a given agglutinogen take on a different

78 Landsteiner, *Serological Reactions* (1945) (n. 54), 113–114. This is the same reference that Wiener gives in "Landsteiner his work" (1951) (n. 8); he advises his reader to go to Landsteiner's book if he wants to understand more of the problem he is discussing (see n. 87 for other references to the same). Landsteiner used the terms "mosaic" and "blood factor" in the 1933 German edition of his book, and not only in the second one. Landsteiner, *Die Spezifizität der serologische Reaktionen* (Berlin: Springer 1933), 53.

79 Alexander S. Wiener and Irving B. Wexler, "The mosaic structure of red blood cell agglutinogens," *Bact. Revs. 16*(1952):69–87, and in *Rh-Hr* (1954) (n. 1), 618–636 (p. 619). Landsteiner actually uses the word "mosaic" to mean a group of separable entities; Wiener uses it to mean entities that may only *seem* to be separable. Landsteiner, *Serologische Reaktionen* (1933) (n. 78), 53.

appearance. If there were no antibodies, there would be no phenomenon.[80] In 1964, he elaborated on the same idea:

The existence of extrinsic attributes depends entirely on the availability of agents by means of which such attributes can be detected. . . . Colors, tastes and odors exist only by virtue of persons capable of experiencing those sensations, and there would be no serological specificities (blood factors) were there no animals capable of producing specific antisera with which to test them.[81]

In this same paper, he draws an exact parallel between his interpretation of the Rhesus antigens and the "one-or-two-antibody question" that Landsteiner had worked on during the thirties. He rearranges the table from Landsteiner's paper to match the reactions of the antimetanilic azo-horse serum with the different azo-proteins to a generalized scheme of the reaction of an antigen with three "blood factors." These three blood factors may represent three different substances, three combining groups on a single substance, or three serological specificities that are only aspects of a single substance, just as the antiserum for metanilic acid-azo-horse serum consisted of a family of antibody specificities all directed preferentially to different aspects of what was known to be a single chemical entity.[82] Landsteiner had concluded:

Specificity is determined by the structure of the metanilic molecule as a whole. Clearly there are not several specific groupings in the immunizing antigen, identical with the groupings contained in the positively reacting heterologous antigens, to which the antibodies could be specifically related. . . .
The results under discussion have some bearing on absorption experiments with natural antigens. Thus one must expect that a separation of antibody fractions will be possible even when an inciting natural antigen does not contain several determinant structures.[83]

Wiener's conclusion is the same: "It is clear that the expression 'cross-reacting antibody,' which is so often used, is redundant, because all antibodies can cross-react to a greater or less degree, provided that suitable antigens are available to test them."[84]

80 Alexander S. Wiener, "Principles of blood group serology and nomenclature: a critical review," *Transfusion* 1(1961):308–320, and in *Adv. II* (1965) (n. 1), 16–28.
81 Alexander S. Wiener, "Fundamentals of immunogenetics, with special reference to the human blood groups I. Heredity of the blood groups; agglutinogens and blood factors; and nomenclature," *Med. Proc.* 10(1964):559–573, and in *Adv. II* (1965) (n. 1), 1–15.
82 Karl Landsteiner and James van der Scheer, "On cross-reactions of immune sera to azo-proteins," *J. Exper. Med.* 63(1936):325–339, C. 303. The table Wiener rearranges is Table V, p. 331.
83 Landsteiner and v. d. Scheer, "On cross-reactions" (1936) (n. 82), 337.
84 Wiener, "Fundamentals of immunogenetics" (1964/1965) (n. 81), 8.

The controversy over the Rhesus system has the same form as the other controversies discussed in this book. Fisher and Race stand for absolute specificity, in the Koch–Ehrlich tradition, and Wiener for unity and the blurring of sharp distinctions, as he was taught by Landsteiner. The two rival nomenclatures carry within them not only different interpretations of the findings on the Rhesus blood group system, but the whole history of the dialogue on the nature of species and specificity. Within the arguments of the 1950s and 1960s, the layers of past arguments can still be discerned. The Ehrlich receptor theory persists in form although reference to Ehrlich himself and to his chemical explanation of absolute specificity is no longer made, just as the unit-character of Batesonian genetics is still there in spirit, though unaccompanied by Bateson himself.[85] Fisher's early experience of blood group immunology was linked to his experience of a Batesonian genetics mathematized by Felix Bernstein. It has been a connection of the greatest importance for British blood group serology and genetics, and for human population genetics during the period of its *Blütezeit*, from 1935 to about 1960.[86]

Wiener's connection with the other, the unitarian, side of the controversy is equally plain. His loyalty to Landsteiner and to Landsteiner's teaching is clearly the source of his interpretation. It is striking that although there were many later studies on the chemistry of specificity, and on the nature and size of the antibody combining site, Wiener never cited any of them. His references are always to Landsteiner alone and the same three citations nearly always serve as examples for the point he wishes to make.[87] Wiener's attachment is to Landsteiner himself, as much as to his intellectual standpoint. It was Landsteiner who taught him to reject sharply specific distinctions and to see instead *graduelle Abstufung*, series of modulating alleles, rather than the well-

85 Race and Sanger do in fact refer to Bateson in each edition of their book – though only to his statement of 1909 that there was little evidence of Mendelian inheritance in man (*Blood Groups* [n. 13], p. 1 all editions); they had a copy of Bateson's first (1909) edition on a shelf in their laboratory, as they showed me in July 1976.

86 Mazumdar, *Eugenics, Human Genetics* (1992) (n. 19), 238–242; Mazumdar, "Two models for human genetics: blood grouping and psychiatry in Germany between the wars," *Bull. Hist. Med.* 69(3) (1995).

87 Wiener's three citations from Landsteiner are Landsteiner, *Serological Reactions*, 2d ed. (1945) (n. 54), 113–117; Landsteiner and v. d. Scheer, "On cross reactions of immune sera to azoproteins," *J. Immunol.* 63(1936):325–339, C. 303 (n. 82); and Landsteiner and v. d. Scheer, "On cross reactions of immune sera to azo-proteins, II. Antigens with azocomponents containing two different groups," *J. Immunol.* 67(1938):709–723, C. 316 (n. 57). The first is Landsteiner's discussion of the mosaic structure of cellular antigens, and the second two are the publications on the "one-or-two-antibody question" and the "double-eagle" antigens. See Wiener's references in nn. 8, 55, 62, 65, 78, 79, and elsewhere.

defined elementary antigens of the Fisher–Race interpretation. It was the same intellectual preference that Gruber had taught Landsteiner and Nägeli had taught Gruber, and probably, though Nägeli did not like to admit it, that Schleiden had taught Nägeli. Wiener's loyalty to his teacher is part of that tradition of teacher–student relationships, a tradition seemingly much less powerful in America or Britain, than in the German-speaking world.[88]

In spite of this intellectual, or ideological, content, the controversy over the Rhesus terminology was carried on mainly on three quite different fronts. Wiener tried to argue on theoretical grounds, but for the other people involved in the dispute, the controversy was seen in terms of a priority dispute first; second, a dispute about which terminology was easier to teach; and third, a dispute over which terminology should be used for the labelling of bottles of antiserum. The last two are problems of great practical importance in hospital laboratories, where the intellectual problem hardly matters, and in fact was barely comprehensible.

The first organized meeting on the subject seems to have been the one that was held at the New York Academy of Sciences, on 18 May 1946. Wiener sent Race a copy of his paper, so that Race was able to join in the discussion by correspondence. His reply was published along with the discussion of those who were present in person: his attitude to Wiener's claim then was that Wiener's was the more conservative approach, and it might well prevail.[89] However, interest in the Rh factor was not confined to Race, Wiener, and Levine, the first principals in the arguments about it. It had been quickly accepted as a new discovery of central importance in blood transfusion and clinical haematology. In November 1946 a meeting entitled "International Hematology and Rh Conference" was arranged, to take place partly in Dallas, Texas, and partly in Mexico City. It was organized by practical men, a blood banker and a clinical haematologist: Sol Haberman and Joseph Hill, from the William Buchanan Blood Center, Baylor Hospital, and the Department of Clinical Pathology, Southwestern Medical College, in Dallas, Texas. The meetings were to be "little get-togethers to discuss the Rh factor," but they expanded into a large International Congress, out of which developed the International Society of Hematology. The most important papers were published as a special supplement to the

88 Karl R. Rothschuh, *Geschichte der Physiologie* (Berlin: Springer, 1953), translated by Guenther B. Risse as *History of Physiology* (Huntington: Krieger, 1973), 310–311: "Table of Johannes Müller's disciples," and "Carl Ludwig and his disciples."
89 Wiener and Sonn, "The Rh series and nomenclature" (1946) (n. 62); Race, reply (1946) (n. 63).

new journal *Blood*, which had begun in 1946 and was edited by the clinical haematologist William Dameshek from Boston. Dameshek's introduction to the volume gives a feeling of the enthusiasm and hopefulness of this first year of peace following six years of war. He begins by acknowledging Landsteiner, Wiener, and Levine as the "pioneers" and then writes:

The great interest in this subject was well brought out in the Dallas–Mexico City Congress of November 1946. Here were brought together many of the recognised leaders in this and related fields, and the discussions which took place and which are reported in this special issue of *Blood* were highly stimulating. This was particularly true in regard to such controversial matters as those relating to nomenclature; the Fisher–Race theory of three gene loci *vs* the Wiener theory of multiple alleles. . . .

It cannot be said of the various individuals working in this field that they lack enthusiasm or desire to spread their various theses across the world's literature. Although some of their discussions have on occasion seemed to descend to acrimonious levels one cannot deny that they have often stimulated the interested investigators to renewed endeavors which have at times led to valuable discoveries.[90]

It is already possible to perceive that Wiener's side was losing ground to the Fisher–Race theory. The two first papers in this collection are by Levine and Race. Levine uses both the terminologies, but he refers to the "brilliance of Fisher's contribution" and to its support by the finding of the predicted anti-e serum.[91] The frequency of 96 percent positive reactions with this serum fitted almost perfectly the predicted gene frequency of 97.3 percent for e. Race's paper has already been referred to: naturally enough it too stresses the brilliance of Fisher's theory, and the fulfilment of its predictions.[92]

Haberman and Hill, the organizers of the conference, used the CDE nomenclature in one of their papers.[93] In the other, they referred to "the fascinating story of the discovery and use of sera of different specificities to unravel the intricate relationships of the different Rh antigens," as told by Dr. Race.[94] The following year, Haberman and Hill

90 William Dameschek, "Preface: the Rh factor," in Joseph M. Hill and William Dameschek, eds., *The Rh Factor in the Clinic and the Laboratory*, Special Issue No.2, *Blood: the Journal of Hematology* (New York: Grune and Stratton, 1948).

91 Philip Levine, "A survey of the significance of the Rh factor," in *The Rh Factor* (1948) (n. 90), 3–26 (p. 10).

92 Race, "The Rh genotypes and Fisher's theory," in *The Rh Factor* (1948) (nn. 90, 28).

93 Ernest E. Muirhead, Arvel E. Haley, Sol Haberman, and Joseph M. Hill, "Acute renal insufficiency due to incompatible transfusion and other causes, with particular emphasis on management," in *The Rh Factor* (1948) (n. 90), 101–138.

94 Joseph M. Hill, Sol Haberman, and F. Jones, "Hemolytic Rh globulins: evidence for a possible third order of antibodies incapable of agglutination or blocking," in *The Rh Factor* (n. 90), 80–100 (p. 80).

submitted to *Blood* a paper called "The Demonstration and Character-ization of the anti-d Agglutinin and Antigen Predicted by Fisher and Race."[95] Hill also wrote an editorial for the *American Journal of Clinical Pathology* in which he supported the CDE terminology.[96] The practical men of the blood bank and the hospital laboratory had gone over to the Fisher–Race side.

Wiener was not present at the Dallas–Mexico meeting, although he was on the programme. Philip Levine suggested meanly that he "must be having one of his unfortunately frequent breakdowns, or possibly he may have been engaged in a quarrel with one or more of his emi-nent colleagues."[97] Race, who then felt, according to Levine, that Wie-ner had the best mind of all those whom he had met in the United States, called in on him in New York on his way back to England.[98] Wiener did come to the next meeting to consider the problem of the Rh terminology, which was held in October 1947 in Washington, D.C., at the request of the surgeon-general of the U.S. Public Health Service. An official advisory review board of the Division of Biologic Standards was set up. Its members were two clinical haematologists, William B. Castle and Maxwell M. Wintrobe, and a blood group serologist, Laur-ence H. Snyder, who had been among Bernstein's first supporters in America. Their mandate was to decide on a standardized label for the bottles of Rh antiserum that were now being produced and that were subject to legislation on the control of biological substances.

The standardization board produced a report that discussed both terminologies, summarized the points in favour and against each of them, and concluded:

for the present a compromise must be made and both systems must be used on the container labels. . . . Because the Wiener system has priority and is un-derstood by everyone in the United States concerned with the production and use of anti-Rh serums, the Board recommends that the Wiener terminology appear first on the label, followed by the Fisher–Race terminology in paren-theses.[99]

The reviewers were not oblivious to the theoretical reasoning behind the two systems of nomenclature. The report states:

95 Haberman, Hill, Everist, and Davenport, "Anti-d as predicted by Fisher and Race" (1948) (n. 45).
96 Joseph M. Hill "Editorial: The complexities of the Rh problem some suggestions for clarification," *Am. J. Clin. Path.* 17(1946):494–501.
97 Clarke, "Robert Russell Race" (1985) (n. 25), citing letter from Bruce Chown, 465.
98 Clarke, "Robert Russell Race" (1985) (n. 25), citing letter from Levine, 478.
99 William B. Castle, Maxwell M. Wintrobe, and Laurence H. Snyder (Chairman), "On the nomenclature of the Anti-Rh typing serums: Report of the Advisory Review Board," *Science* 107(1948):27–31 (p. 30).

Another difference between the two systems of nomenclature involves the basic problem of the one-to-one correspondence between gene and antigen. On the Fisher system the one-to-one correspondence is upheld, a view which is rapidly being incorporated into genetic theory and into the beliefs of immunologists. On the Wiener system certain genes result in the production of a single antigen, but others produce two or more.[100]

Fisher's view, as they said, was widely accepted in England, and was gaining ground in the States. Their compromise solution implied that though Wiener's terminology had priority, Fisher's was expected gradually to replace it.

The American report was the subject of editorials by both the *British Medical Journal* and the *Lancet* on 28 February 1948. Both of them saw the compromise solution of the report as a stage on the way to complete acceptance of the Fisher–Race terminology. The *British Medical Journal* wrote:

While the immediate effect of the report in the United States will be to ensure the temporary confinement within brackets of the Fisher–Race descriptions, the Board's comments must none the less have gone a long way to encouraging its wider use by the United States and other workers, and thus rendering the present compromise transitional.[101]

The *Lancet* wrote: "If, as seems possible, the Fisher–Race terminology is finally adopted for international use this will not imply an lack of recognition of the magnificent contribution which Wiener has made to the development of the subject."[102]

The report came as a nasty surprise to Wiener.[103] By his own account, it had been sent to him in a draft form and at that stage it had recommended acceptance of his terminology alone. He sent a protest to the journal *Science*, but it was refused. When the *British Medical Journal*'s editorial appeared, he sent his protest to them, to the *Lancet*, and to the *Australian Medical Journal*, all of whom published it.[104]

Wiener had lost in his effort to have his terminology accepted as the international one. The International Society of Hematology, although it was a largely American organization, inclined towards the British

100 Castle, Wintrobe, and Snyder, "Report of Advisory Review Board" (1948) (n. 99), 29.
101 Editorial, "Anti-Rh serum nomenclature," *Brit. Med. J.* i(1948):400.
102 Editorial, "Terminology of the Rh factor," *The Lancet* i(1948):329.
103 Alexander S. Wiener, "History of blood group nomenclature with a questionnaire on Rh-Hr nomenclature," *J. Forensic Med.* 14(1967):3–12 (p. 5).
104 Wiener, "Anti-Rh serum nomenclature" (1948) (n. 64); Wiener, "Nomenclature of Rh factors," *Lancet* i(1948):343; Wiener, "Conference on nomenclature of the Rh factors," *Med. J. Australia* 35(1948):530–531.

terminology. At the society's next meeting – in Buffalo, New York, in August 1948 – a committee on nomenclature, which did not include Wiener, proposed that the CDE terminology should be adopted as the international one.[105] The recommendation was made shortly after Wiener had left the room and would have been adopted except that someone rushed out to tell him and he returned quickly enough to catch the eye of the chairman, Sol Haberman, and prevent it.[106]

Wiener was losing in the Division of Biologic Standards, and he was losing in the International Society. The society was influenced no doubt by Sol Haberman, its president, and Joseph Hill, its secretary, who had organized the Dallas meeting, and who were both inclined to the CDE terminology. But Wiener had one other organizational resource left, the American Medical Association. Its Committee on Medicolegal Problems became involved in the controversy because of the use of blood grouping tests in cases of disputed paternity. The members of this committee were Israel Davidsohn, Philip Levine, and Alexander Wiener: in their first report in 1952, they use both terminologies; with the Fisher–Race one in parentheses following Wiener's, as the Division of Biologic Standards had recommended.[107] But in 1956 a new committee produced a new "tentative supplementary report" advocating Wiener's terminology alone.

It was clearly written by Wiener. The first page contains several of Wiener's diagnostic *Sätze*, such as: "It is most important to distinguish clearly between agglutinogens and their serologic attributes, the blood factors" and "The C-D-E notations for the Rh-Hr types make no allowance for the difference between a blood factor and an agglutinogen, and incorporate the tacit, incorrect, assumption that every agglutinogen has but a single corresponding antibody."[108]

A footnote to this report makes the following odd statement:

A preliminary draft has been sent to 48 leading workers in the field. . . . An effort has been made to take all the comments received into account.

In arriving at its conclusions, this fact-finding committee has necessarily *ex-*

---

105 This meeting was announced, and the programme detailed, in Blood 3 720–723 (1948) and briefly reported in *Blood 3*(1948):1313, but no proceedings appeared. This account is taken from Wiener, "History of blood group nomenclature" (1967) (n. 103).
106 Wiener, "History of blood group nomenclature" (1967) (n. 103), 5, fn.
107 Israel Davidsohn, Philip Levine and Alexander S. Wiener, "Medicolegal application of blood grouping tests," *J. Am. Med. Ass. 149*(152):699–706.
108 Alexander S. Wiener, Richard D. Owen, Clyde Stormont and Irving B. Wexler, "Medico-legal application of blood grouping tests," *J. Am. Med. Ass. 161*(1956):233–239 (p. 233).

cluded all comments that in its opinion were incorrect and unsound, and has included only those that the committee believes to have a firm scientific basis [Emphasis added].

The report, as they said, was controversial.[109] It was circulated, and in 1957 a second, "revised and final version," was published. The committee for this was the same, and so were the conclusions. But its members had tried to produce a semblance of fairness by including a section on dissent. Of the 49 letters received in reply to their circulation of the earlier version, they said, 43 expressed classifiable opinions, of which 23 favoured CDE, 17 Rh-Hr, and 5 said they used both. Some of those who objected to Wiener's terminology did so not because of the theory it embodied, but because it was extremely difficult to teach. However, the report still concluded that it was desirable to have a single terminology, and that the majority of the votes had been for Wiener's.[110]

The next year *Science* published a manifesto signed by thirty American immunohaematologists and one Italian, declaring that they rejected this report, that it was highly biased, and that they did not recognize the competence of the committee. They stated that they would go on using the CDE nomenclature.[111]

The next effort to reach agreement was a two-day meeting at Princeton in 1960, convened by the National Research Council's Committee on Blood and Related Problems. The only agreement reached was to meet again in two years' time, but the second meeting was never held. The results of the two days of arguments at Princeton was instead the production of a third terminology, one in which the reactions were designated by numbers. Its senior author was Richard E. Rosenfeld, who had proposed such a coding system at Princeton, and who now – with his colleagues Shaul Kochwa from the Mount Sinai Hospital in New York, Fred H. Allen from the Blood Grouping Laboratory in Boston, and Scott Swisher from the University of Rochester – worked out the kind of coding system he had suggested then. It was presented first at a meeting of the American Association of Blood Banks in Chicago in October 1961 and subsequently published in their new official journal, *Transfusion*, then in its second year. All the authors except Kochwa, who was a protein biochemist, were blood group serologists working – like Wiener, Levine, and Race and Sanger – on the Rh system at its most advanced level, and all of them had signed the letter to *Science* of

109   Alexander S. Wiener, Richard D. Owen, Clyde Stormont and Irving B. Wexler, "Medico-legal applications of blood grouping tests," *J. Am. Med. Ass. 164*(1957): 2306–2043 (p. 2036).
110   Wiener, Owen, Stormont and Wexler, "Medicolegal applications" (1957) (n. 118).
111   Fred H. Allen, Jr., and thirty others, "Blood group nomenclature," *Science 127*(1958):1255–1256.

1950 protesting Wiener's attempt to make his own terminology official.[112]

Wiener had been defeated at every turn in his efforts to get his terminology adopted exclusively. It became more and more difficult for him to arrange meetings or to find an organization to sponsor him, or even to get a journal to publish his papers. But as it became more difficult for him, he became more and more passionately determined to try to do it. He still hoped to be able to organize the ultimate meeting at which his colleagues would at last agree with him. He still hoped, by circulating a questionnaire, to be able to prove to them that the majority did agree with him, and that the others were too confused to know that they should.

The final episode was the meeting he organized under the auspices of the section of Biological and Medical Sciences of the New York Academy of Sciences to report on his last questionnaire. It was chaired by John Miale, a recent convert, who was professor of pathology at the University of Miami.[113] Following this meeting, Wiener tried to arrange one in London, but the proposal was rejected. He wrote angrily:

The CDE protagonists behave as though they fear that their only chance to survive is by keeping the lay and scientific public in ignorance of the facts and by allowing the labels of Rh-Hr antisera to speak for them. It is significant that similarly, chiropractors seldom argue their case "scientifically," they merely place a sign in front of their office.

Further on in the same paper, which appeared the year after this disappointment, he wrote:

Many sponsors of the CDE and 123 notations have positions of influence as editors of scientific journals and of referees of scientific articles. Such referees take advantage of their anonymity to place obstacles in the way of the publication in journals of wide circulation of any scientifically critical articles dealing with Rh-Hr nomenclature. The present author has during the past 20 years experienced difficulty with the publication of such articles, which because of

---

112 This meeting and its discouraging results are described in Wiener, "The Rh-Hr blood types: The anatomy of a controversy," *J. Forensic Med.* 15(1968):22–40, and in Richard E. Rosenfield, Fred H. Allen, Jr., Scott N. Swisher, and Shaul Kochwa, "A review of Rh serology and a presentation of a new terminology," presented in part at the Annual Meeting of the American Association of Blood Banks, Chicago, October 1961: *Transfusion* 2(1962):287–312.

113 Alexander S. Wiener, "Nomenclature of blood groups with special reference to the Rh-Hr blood types," *Trans. NY Acad. Sci.*, II 29(1967):875–886; John B. Miale, "A pathologist's critique of the controversy over blood group serology and nomenclature," *Trans. NY Acad. Sci.*, II 29(1967):887–891; Wiener, "Final results of Rh-Hr nomenclature questionnaire," *Trans. NY Acad. Sci.*, II 29(1967):982–987.

such censorship have often had to be printed in journals other than those for which they were intended.[114]

It has a sad and bitter sound. Even this last paper was printed in a small South African journal, edited by his friend Maurice Shapiro, which is not carried by U.S. university libraries. Wiener's campaign continued, but it was reduced to the level of angry letters to editors and to Race and Sanger, none of which were published: their rising tone of personal abuse, and personal distress, made them private documents.

Dear Rob and Ruth,
    Is it true that you are coming out with a sixth edition? If so, when? and why? Your book is now notorious as the "Sourcebook of Blood Group Mythology." Still, it is used as a bible by those who do not realize that blood grouping is a science & not a religion.
    For your own sake & the sake of your readers, I hope the 6th edition will not merely repeat all the false assertions (Fisher's "amazing" predictions; LW; little d; triple inheritance; Stetson-Levine false citation; P^k, etc.) of previous editions. . . . To repeat the same baloney would be unconscionable unless you are simply very obtuse & not so bright. However, that remains to be seen – The new reviews of your book can now be as blunt as this letter.
    Sincerely
    Al
    I forgot to mention the fraud of Fisher's (sic) shorthand[115]

For Race and Sanger, these letters were a constant persecution. They destroyed a large file of them when the Medical Research Council's unit moved out of the Lister Institute in 1975.[116] Shapiro and Wexler alone remained faithful to Wiener.

It was the new specificities, which were reported every year, that finally made the Fisher hypothesis untenable in its original form. It had survived the demonstration that there was no anti-d, and that neither separate inheritance nor crossing over between its three loci ever occurred. It survived the demonstration that there were many more than two alternatives at each of the three loci, as long as these could be shown to be separate from each other. But it could not survive the demonstration of specificities depending on position effects, cis-trans effects, compound antigens, "missing" antigens, and other genetically determined subtleties. By the time of Race and Sanger's fifth edition of 1968, the original five antisera and eight "chromosomes" of 1945

114 Wiener, "Anatomy of a controversy" (1968) (n. 114), 37.
115 Alexander S. Wiener to Robert Race and Ruth Sanger, letter dated September 27 [1973]; letter in my possession. My thanks to Race and Sanger for permission to quote.
116 Clarke, "Robert Russell Race" (1985) (n. 25), 465.

had become thirty antisera and more than forty known gene complexes. Now, eighteen years after their original edition of 1950, Race and Sanger themselves were not so enthusiastic:

> The question arises whether the CDE scheme has anything more to offer, whether it represents only so small a part of the pattern that thinking about some of the latest observations is handicapped if attempts are made to fit them into the pattern. Avant garde opinion is that the time of usefulness of such committed symbols as C and c and rh' and hr', which denote alternatives, has passed, as far as pioneer work is concerned. It is felt that for further advance the notation must be freed of all interpretative meaning and record only reactions, plus, minus and weak, between cells and numbered antisera. . . .
> We think that a full account can still be communicated in terms of CDE, and what is more, we think that this is the only notation in which a detailed account can be *communicated*: on the other hand, it can be stored more efficiently by the numerical notation, but it has to be thawed out and reconstituted in familiar terms, at any rate for the present generation of workers.[117]

The neatness of the CDE system has again been scribbled over by new complexities. The receptor does not seem to mate with only one specific antibody. As the traces of the presence-and-absence theory have faded, so has its partner the unit-character: the relation of genes and antigens is no longer as simple as it once was. As Race and Sanger themselves wrote in 1968, the closest they could possibly be was separated by one enzyme, and there are several blood group antigens now known to depend on a series of enzymes. But there has still been no chemical explanation of the Rhesus antigens. Landsteiner's remark of 1945 still stands: "The interpretation of cross-reactions and multiple serological characters must remain hypothetical until chemical confirmation is available."[118] It occurs in one of Wiener's favorite passages.

In the course of this controversy, Wiener wrote innumerable papers explaining his intellectual point of view and calling on Landsteiner's authority to witness, as well as claiming priority. But his colleagues became steadily less sympathetic to his claims, and the constant repetition of the same arguments over twenty-five years made it increasingly difficult for him to get space in important refereed journals. Many of his papers and letters were refused, as he himself admitted by the late sixties.[119] This was already starting to happen in the fifties, as Wiener lost his hold on the audience he was trying to convince.

The journals were not Wiener's only resource. He also had access to organizations such as the International Society for Hematology, the

---

117 Race and Sanger, *Blood Groups* (1950) (n. 13), 227.
118 Landsteiner, *Serological Reactions* (1945) (n. 54), 116.
119 Wiener, "Landsteiner: history of Rh-Hr" (1968) (n. 9).

Division of Biologic Standards, the National Research Council's Committee on Blood, and the American Medical Association: through their committees and reports he tried to gain the support of the official community. Many of his arguments took place orally and at meetings, where he confronted increasingly sceptical audiences in person, until they became afraid of his anger and desperation, and shut him out.

Half a century earlier, Max von Gruber, in a similar situation, facing the unbeatable strength of the Koch–Ehrlich axis, of Ehrlich's immunological theories, and of their institutional support, resigned his chair in Vienna and moved to Munich. There, after one final savage and witty attack on Ehrlich, he gave up immunology altogether and turned to hygiene and eugenics. Landsteiner, denied professional success in Vienna, perhaps for the same reason, left for Holland and later for America. But Wiener persisted, although his efforts became less and less credible to his peers. He made repeated attempts to organize the international community and to persuade it to accept his arguments and recognize his terminology, but no final agreement was ever reached in his lifetime.

The problem was never really resolved, although it has ceased to generate the interest it once did. Wiener died in 1976, Race in 1984, and Levine soon after. The main protagonists are now gone, and blood group serology itself is no longer at the forefront of human genetics or immunology. Most of the laboratories – that is, those known to the author – went on using both terminologies, a modified form of Wiener's for speaking and the Race and Fisher letters that were not very clear in speech, for writing the results of tests on laboratory protocols. The numerical terminology has been adopted for the new specificities for which there were no terms in either of the older systems, but in practice it has only been used for punched-card or computerized data collections.

The determining factor was probably not the immunogenetical implications of a nomenclature, but its practical usefulness. That was what mattered to those who were not interested in immunogenetics *per se*, but in talking about clinical problems and in labelling test sera and bottles of blood for transfusion.

Conclusion: Fragment of an Agon

Now that I have completed this work, my hope is that its structure is clear enough to make its message plain. Buried, perhaps, in masses of detail – detail that God and this historian love for its own sake as well as for its value as evidence – there is a simple, solid armature, a pattern and meaning in the history of these events. The pattern is that of a dialogue in which the same roles are taken up and played out in five successive generations.

As Kant said, the students of nature can be divided into two groups, which I have called pluralists and unitarians. The pluralists are the analytical thinkers, the dissectors, and dichotomisers, for whom understanding consists in the uncovering of unsuspected diversity, and who try to define and separate species. The unitarians, on the other hand, are those for whom understanding means the union of the seemingly diverse into continuous wholes, covered by a common law, and joined by stepwise transitions. The differences they perceive are relative rather than absolute; they prefer the idea of series to that of species.

There are no revolutionary dislocations, no radical changes, in this perception of the flow of scientific thought. The most striking thing about it is its continuity. It is, to use Nägeli's phrase, a monument to *unendlich Theilbarkeit*. The form of the problem, and the form of the arguments about it, remains the same. Only the content that flows through them changes and metamorphoses from generation to generation. There would seem at first sight to be little in common between the discussions of the blood group geneticists of the 1950s and the botanists of a hundred years earlier. But there is no break between them, only the *graduelle Abstufung* of a century of dialogue between these two groups of thinkers, the pluralists and the unitarians.

How is it that this form has remained constant, as it has done for generation after generation? The answer seems to be that science, like other human cultural activities, is passed on from teacher to student. The perception of the problems to be faced, and the adoption of patterns of thought about them, seems to be less a matter of spontaneous choice than a part of the training by example in scientific thinking that is given to each student along with the current technology. It is the equivalent of the acculturation to social and intellectual values and table manners that the family and the school give the student in the earlier years of formation. It is also true that each of the students had more than one teacher, but each seems to have picked out from them a single individual to honour with lifelong admiration.

It is my feeling that in the culture that I have discussed here, which is mainly that of nineteenth- and early twentieth-century German science, it is almost impossible to exaggerate the determining effect of this mixture of technology and intellectual patterning that is passed from teacher to student. It is a patterning that outweighs in many cases the desire to have a successful career: in the social climate of the time, it was the pluralists, the Koch–Ehrlich team, who were guaranteed a good position. Unitarians, from Nägeli to Gruber and Buchner, to Landsteiner, to Wiener, all suffered the fate of outsiders, yet they persisted in their opposition. The young people were given a sketch, as it were, of their life's work. They were taught to look at the empirical data of their science in a particular way, to expect a certain structure in nature, and to feel that they had made a successful achievement when they had found such a structure.

Ludwig Fleck, in his *Genesis and Development of a Scientific Fact*, first published in 1935 but only recently gaining recognition, presents just such a scene of pedagogical moulding in his discussion of the introductory lecture given by the serologist Julius Citron to his students in 1910. Citron tells the students,

We shall presently have an opportunity to discuss the nature of specificity in detail, and thus get to know its limitations. *For the time being, however, I would ask you to commit firmly to your memory the law that every true antibody is specific, and that all non-specific substances are not antibodies. The law of specificity is the precondition of serodiagnostics.*[1]

Fleck himself, as befits an advanced thinker writing in 1935, is conscious of this as the induction of the students into a particular pattern of thinking, a *Denkstil*, as he calls it. He himself rejects the treatment of "toxins, amboceptors and complements as chemical entities . . . this primitive scheme . . . is being progressively discarded in accordance with current physico-chemical and colloidal theories in other fields."[2]

1   Ludwig Fleck, *Genesis and Development of a Scientific Fact* (1935), Thaddeus J. Trenn and Robert K. Merton, trans. Frederick Bradley and Th. J. Trenn; foreword by Thomas S. Kuhn (Chicago, Ill.: University of Chicago Press, 1979), 58.
2   Fleck, *Genesis and Development of a Scientific Fact* (1935/1979), 63.

It was probably his awareness of the difference between these two conceptions of specificity, the pluralistic approach of Citron and his own physico-chemical or colloid, and so unitarian style of thought, that led Fleck to his insight into the socially conditioned nature of scientific facts. The Citron students were being drawn into a system of beliefs about serology that Fleck did not share.

For the historian, the understanding of science from this point of view depends, as perhaps it did for Fleck, on the ability to reconstruct the dialogue. The meaning of a statement to its author or its hearers, cannot be understood in isolation: it is not simply a response made to nature itself. It is part of a dialogue about nature in which the scientist is playing a part. Statements are responses, not just to nature, but also to those playing opposite in the drama. It is impossible for the historian to understand these statements without knowing to whom they were addressed, who were the other characters in the play, the opponents and the supporters. Their arguments take place within this human context of loyalty to one group and opposition to the other. Their experiments are as much concerned to wring an admission of defeat from their antagonists as an answer from nature. They turn to nature as a witness that truth is on their side.

This interpretation of the history of science has a highly metaphorical sound; but it is not a metaphor. The dialogue is a lived dialogue.

The biologists in this drama inhabited a world of arguments, many of which took place at the meetings of the various societies to which they belonged, at which the figurative dialogue became actual, and the drama was staged before a real audience. It was a scene of which I became very conscious while I was writing this book: the space in my head was full of shouting voices, arguing and struggling. Many of the most important and revealing incidents took place in public: I am thinking of Nägeli's speech at the Naturforscherversammlung of 1877; the conflict between Gruber and Pfeiffer at Wiesbaden in 1896; the paper, rather than the speech, in which Gruber wanted to attack Ehrlich in 1901 – a piece of lived dialogue that Naunyn, and Ehrlich himself, were afraid to hear performed; the festival lecture of 1905 at which Pauli endorsed Landsteiner as the successor to Ehrlich; Landsteiner's own lecture to the International Medical Congress of 1909 in Budapest, on the theories of antibody production; and, of course, the five meetings at which Wiener failed to persuade an audience of his colleagues that his interpretation of the data was the right one.

It was never enough, to convince this audience, simply to announce an ingenious idea at a meeting. The ideas had to be fitted into a social power structure in order to acquire authority. The power structure built up within the Koch–Ehrlich group may be one of the most effective ever formed in science. Membership of this mutually supporting group ensured professional promotion and the self-confidence that went with it to its students. It had the opposite effect for those who were not members, wherever its power and influence reached, that is, mainly within the German *Sprachgebiet.* We have seen the career problems and the personal distress of the workers whose loyalties and convictions kept them outside this protective group.

Chairs and directorships were not the only manifestations of this institutional power. There were also journals. For the historian, the opinions and affiliations of the editors of journals are among the most valuable clues to the structure of the dialogue, especially at their beginnings. For the protagonists in the dialogue themselves, the journal acts as a megaphone through which their voices come amplified by the editor's endorsement, as long as they belong to the editor's group. If they do not, it is an impenetrable fortress, defended by editors and referees, which locks them out. This, more than anything, was what defeated Wiener.

It is hard to say how general to make this perception of the history of science. I hesitate to claim that such a pattern could be found in another era. It belongs, I think, specifically to an age of institutes, directors of institutes, and journals as centres of power. At an earlier stage in history, it is likely that the individual thinker was more isolated, that the "scientific community" was less powerful, because less powerfully organized.

It may be, however, that some features will be found useful in the understanding of other times and places. I am thinking particularly of the dialogue, the statement about nature that is a stylized response to other statements and that can be understood by the historian only in this context. It is a heuristic maxim, perhaps, for the historian of science.

Bibliography

This bibliography is arranged in a rather unusual way. The first two sections cover the bibliographic sources, and the general historical writing on the subject. The third section is arranged under the names of authors of the primary sources, in each case listing secondary and bibliographic material relating to the person and their work, followed by a list of their own papers. I have done this so that the historians who succeed me may have an introit into this very under-researched subject.

1. Bibliographical Sources

Allgemeine deutsche Biographie herausgeben von der historische Commission bei der Adademie der Wissenschaften. Leipzig: Duncker, 1875–1912, 1st ed.; Duncker and Humbolt, 1930.

Association of Research Libraries. *Catalog of Books Represented by Library of Congress Printed Cards.* Ann Arbor, Mich.: Edwards, 1942, vv. 1–167; *Supplement,* vv. 1–42, 1942–1947.

British Library. *General Catalogue of Printed Books.* London: British Library, 1965, vv. 1–263.

Bugge, Günther. *Das Buch der grossen Chemiker.* 2 v. Berlin: Verlag Chemie, 1930. Contains "Biographische Bibliographie," v. 2, 463–516, on chemists included in both vols.

Deutsches Bucherverzeichnis: eine Zusammenstellung der im deutschen Buchhandel erschienenen Bücher, Zeitschriften und Landkarten. Leipzig: Verlag des Borsenvereins der Deutschen Buchhändler, continuing.

Edwards, Paul, ed. *Encyclopedia of Philosophy.* London: Macmillan, 1969.

Farber, Eduard. *Nobel Prize Winners in Chemistry 1900–1950.* New York: Schumann, 1953.

Fischer, Isidor. *Biographisches Lexikon der hervorragenden Aerzte der letzten fünfzig Jahre.* Vienna: Urban, 1933.

Gillispie, Charles C., ed. *Dictionary of Scientific Biography.* New York: Scribner's, 1970–present. Continuing.

National Union Catalog Pre-1956 Imprints. London: Mansell, 1968, vv. 1–424.

Neue deutsche Biographie, herausgeben von der historischen Kommission bei der Bayerischen Akademie der Wissenschaften. Berlin: Duncker and Humbolt, 1953–present. Successor to *Allgemeine deutsche Biographie.*

383

Poggendorf, Johann Christian, *Biographisch-literarisches Handworterbuch zur Geschichte der exacten Wissenschaften-enthaltend Nachweisungen über Lebensverhaltnisse und Leistungen von Mathematikern, Astronomen. Physikern, Chemikern, Mineralogen, Geologen usw. aller Volker und Zeiten.* Leipzig: Barth, 1863–. Continued by Zaunick and Salié, below.

Royal Society of London. *Catalogue of Scientific Papers (1800–1863).* London: Eyre and Spottiswoode, 1867, vv. 1–6. *Catalogue of Scientific Papers (1864–1873),* vv. 7–8; *Catalogue of Scientific Papers (1874–1883),* vv. 9–11; *Supplement* (1800–1883), v. 12; *Catalogue of Scientific Papers (1884–1900),* vv. 13–19.

Titus, Edna Brown. *Union List of Serials in Libraries of the United States and Canada.* New York: Wilson, 1965, 3d ed.

Who's Who of British Scientists. Athens: Ohio University Press, 1971–present.

Zaunick, Rudolph, and J. C. Salié. *Poggendorfs biographisch-literarisches Handwörterbuch der exakten Naturwissenschaften, unter Mitwirkung der Akademie der wissenschaften zu Berlin, Göttingen, Heidelberg, München und Wien, herausgeben von der sächsischen Akademie der Wissenschaften zu Leipzig, Berichtsjahre 1932–1953.* Berlin: Akad. Verlag, 1956–1971, v. 7a. In 5 parts and Supplement.

2. General and Historical Works

Abderhalden, Emil, ed. *Handbuch der biologischen Arbeitsmethoden.* Berlin: 1920.

Ansbacher, Heinz Ludwig, and Rowena R. Ansbacher. *The Individual Psychology of Alfred Adler: A Systematic Presentation in Selections from his Writings.* New York: Basic Books, 1956.

Baker, John R. "The cell theory a restatement, history and critique." *Q. J. Microsc. Sci.* 89(1948):103–125; 90(1949):87–108.

Bäumler, Ernst. *Ein Jahrhundert Chemie.* Dusseldorf: Econ, 1975.

Beller, Steven. *Vienna and the Jews 1867–1938: A Cultural History.* Cambridge: Cambridge University Press, 1989.

Berry, Arthur J. *From Classical to Modern Chemistry: Some Sketches of its Historical Development.* Cambridge: Cambridge University Press, 1954.

Bowler, Peter J. *The Eclipse of Darwinism: anti-Darwinian Evolution Theories in the Decades around 1900.* Baltimore, Md.: Johns Hopkins University Press, 1983.

Bowler, Peter, J. *The Mendelian Revolution: The Emergence of Hereditarian Concepts in Modern Science and Society.* Baltimore, Md.: Johns Hopkins University Press, 1989.

Bulloch, William. *The History of Bacteriology.* London: London University Press, 1938.

Camp, Lt Col. Frank R., and Col. Frank R. Ellis. *Selected Contributions to the Literature of Blood Groups and Immunology.* Blood Transfusion Division, U.S. Army Medical Research Laboratory, Fort Knox, Ky. Springfield, Va.: National Technical Information Service, 1971.

Carlsen, Elof Axel. *The Gene: a Critical History.* Philadelphia, Pa: Saunders, 1966.

Corner, George W. *A History of the Rockefeller Institute 1901–1953, Origins and Growth.* New York: Rockefeller Institute Press, 1964.

Crombie, Alistair C., ed. *Scientific Change, a Symposium on the History of Science, Oxford 1961.* New York: Basic Books, 1963.

Crosland, Maurice P. *Historical Studies in the Language of Chemistry.* London: Heinemann, 1962.

Cullen, Michael J. *The Statistical Movement in Early Victorian Britain: the Foundations of Empirical Social Research.* Hassocks, Sussex: Harvester, 1975.

DeJager, Timothy F. "G. R. Treviranus and the Biology of a World in Transition." Ph.D. diss., University of Toronto, 1990.

Duchesneau, François. *Genèse de la Théorie Cellulaire.* Montreal: Bellarmin, 1987.

Dunn, Cuthbert L. *The Emergency Medical Services: v. 1 England and Wales.* London: HMSO, 1952. In Series, History of the Second World War, United Kingdom Medical Series, Sir A. S. MacNalty, general editor.

Dunn, Leslie C. *A Short History of Genetics: The Development of Some of the Main Lines of Thought 1864–1939.* New York: McGraw-Hill, 1965.

Faber, Knud. *Nosology in Modern Internal Medicine, with an Introductory Note by Rufus Cole.* New York: Hoeber, 1923.

Farley, John. "The spontaneous generation controversy (1700–1860) I. The origin of parasitic worms." *J. Hist. Biol. 5* 95–125 (1972).

Farley, John. "The spontaneous generation controversy (1859–1880) II. British and German reactions to the problem of abiogenesis." *J. Hist. Biol.* 5(1972):285–319.

Farr, A. Derek. "Blood group serology: the first four decades (1900–1939)." *Med. Hist. 23* 215–226 (1979).

Flinn, Michael W. Introduction to *Report on the Sanitary Condition of the Labouring Population of Great Britain by Edwin Chadwick.* Edinburgh University Press, 1965, 1—73.

Foster, William D. *A History of Medical Bacteriology and Immunology.* London: Heinemann, 1970.

Fruton, Joseph S. *Contrasts in Scientific Style: Research Groups in the Chemical and Biomedical Sciences.* Philadelphia, Pa: American Philosophical Society, 1990.

Geison, Gerald L. *Michael Foster and the Cambridge School of Physiology: the Scientific Enterprise in Late Victorian Society.* Princeton, N.J.: Princeton University Press, 1978.

Giere, Ronald N., and Richard S. Westfall, eds. *Foundations of Scientific Method.* Bloomington: Indiana University Press, 1973.

Graebe, Carl. *Geschichte der organischen Chemie.* 2 v. Berlin: Springer, 1920), v. 1.

Grainger, Thomas H. *A Guide to the History of Bacteriology.* New York: Ronald, 1958.

Green, Francis Henry Knethel, and Major-General Sir Gordon Covell. *Medical Research.* London: HMSO, 1953. United Kingdom medical series, in History of the Second World War, Sir A. S. MacNalty, general editor.

Hacking, Ian. *Logic of Statistical Inference.* Cambridge: University Press, 1965.

Ihde, Aaron J. *The Development of Modern Chemistry.* New York: Harper, 1964.

Jacobs, Natasha X. "From unit to unity: protozoology, cell theory and the new concept of life." *J. Hist. Biol.* 22(1989):215–242.

Jahn, Ilse. *Grundzüge der Biologiegeschichte.* Jena: Fischer, 1990.

Janik, Alan, and Steven Toulmin. *Wittgenstein's Vienna.* New York: Simon, 1973.

Langer, William L. *Political and Social Upheaval 1832–1852.* In Series, Rise of Modern Europe. New York: Harper, 1969.

Leicester, Henry Marshall. *Source Book in Chemistry 1900–1950.* Cambridge, Mass.: Harvard University Press, 1968.

Leicester, Henry Marshall, and Herbert S. Klickstein. *A Source Book in Chemistry 1400–1900.* New York: McGraw-Hill, 1952.

Lenoir, Timothy. *The Strategy of Life: Vital Materialism in XIXth Century German Physiology.* Dordrecht: Reidel, 1983.

Lesky, Erna. *Die Wiener medizinische Schule im XIX Jahrhundert.* Graz, Böhlaus, 1965.

Lewin, Kurt. "The conflict between Aristotelian and Galilean modes of thought." *J. Gen. Psychol.* 5(1931):141–177. Also in Kurt Lewin, *A Dynamic Theory of Personality. Selected Papers*, translated by D. K. Adams and K. E. Zener. New York: McGraw-Hill, 1935.

Mazumdar, Pauline M. H. *Eugenics, Human Genetics and Human Failings: The Eugenics Society, Its Sources and Its Critics.* London: Routledge, 1992.

Mazumdar, Pauline M. H. "Immunology, a history." In Kenneth F. Kiple, ed., *Cambridge History and Geography of Human Disease.* Cambridge: Cambridge University Press, 1993.

Mazumdar, Pauline M. H. "Marrack, John Richardson. " In F. Larry Holmes, ed., *Dictionary of Scientific Biography* (1991) Supplement 2.

Mazumdar, Pauline M. H. "The template theory of antibody formation and the chemical synthesis of the twenties." In Mazumdar, ed., *Immunology 1930–1980: Essays on the History of Immunology.* Toronto: Wall, 1989.

Mazumdar, Pauline M. H. "The antigen-antibody reaction and the physics and chemistry of life." *Bull. Hist. Med.* 48(1974):1–21.

Mazumdar, Pauline M. H. "Immunity in 1890." *J. Hist. Med.* 27(1972):312–324.

Moulin, Anne-Marie. *Le dernier Langage de la Médecine: Histoire de l'Immunologie de Pasteur au SIDA.* Paris: Presses Universitaires de France, 1991.

Olby, Robert C. *The Origins of Mendelism.* Introduction by C. D. Darlington. New York: Schocken, 1966.

Partington, James R. *History of Chemistry.* 4v. London: Macmillan, 1964.

Provine, William B. *The Origins of Theoretical Population Genetics.* Chicago: University of Chicago Press, 1971.

Ramsay, O. Bertrand. *Stereochemistry.* London: Heyden, 1981.

Rothschuh, Karl E. *Geschichte der Physiologie.* Berlin: Springer, 1953. Translated by Gunther B. Risse as *History of Physiology.* Huntington: Krieger, 1973.

von Sachs, Julius. *Geschichte der Botanik von 1530–1860.* Munich: Oldenbourg, 1875. V. 15 in series Geschichte der Wissenschaften in Deutschland neuere Zeit, herausgeben durch die historische Commission bei der königliche Academie der Wissenschaften in Bayern. Translated by H. E. F. Garnsey, revised by I. B. Balfour, as *History of Botany 1530–1860.* Oxford: Clarendon, 1906.

Schneider, William H. "Chance and social setting in the application of the discovery of the blood groups." *Bull. Hist. Med.* 57(1983):545–562.

Servos, John W. *Physical Chemistry from Ostwald to Pauling: the Making of a Science in America.* Princeton, N.J.: Princeton University Press, 1990.

Silverstein, Arthur M. *A History of Immunology.* New York: Academic Press, 1989.

Sloan, Philip R. "John Locke, John Ray and the natural system." *J. Hist. Biol.* 5(1972):1–53.

Sloan, Philip R. "Buffon, German biology and the historical interpretation of species." *Brit. J. Hist. Sci.* 12(1979):109–153.

Sloan, Philip R. "Darwin, vital matter and the transformism of species." *J. Hist. Biol.* 19(1986):369–445.

Stafleu, Frans Antonie. *Linnaeus and the Linnaeans: The Spreading of their Ideas in Systematic Botany.* Utrecht: Oasthoek, 1971.

Stern, Curt. "The Hardy-Weinberg law." *Science* 97(1943):137–138.
Stern, Curt. *Principles of Human Genetics.* San Francisco, Calif.: Freeman, 1st ed. 1949, 2d ed. 1960.
Stern, Curt, and E. R. Sherwood. *The Origins of Genetics: A Mendel Sourcebook.* San Francisco, Calif.: Freeman, 1966.
Swinburne, R. C. "The presence and absence theory." *Ann. Sci.* 18(1962):131–145.
Topley, William Whiteman Carlton, and Sir Graham S. Wilson. *The Principles of Bacteriology and Immunity.* Baltimore, Md.: Wood, 1st ed. 1929, 2d ed. 1936.
Vandervliet, William G. *Microbiology and the Spontaneous Generation Debate during the 1870s.* Lawrence, Kans.: Coronado Press, 1971.
Wohl, Anthony S. *Endangered Lives: Public Health in Victorian Britain.* London: Methuen, 1983.
Woodward, William Ray, and Robert S. Cohen. *World Views and Scientific Discipline Formation.* Dordrecht: Kluwer, 1991.
Yost, W. "The first 45 years of physical chemistry in Germany." *Ann. Rev. Phys. Chem.* 17(1966):1–14.
Zimmermann, David R. *Rh, the Intimate History of a Disease.* New York: Macmillan, 1973.

3. General Bibliography

Bio-bibliographical and secondary sources where used are listed first, followed by primary sources.

Allen, F. Hal. See Richard E. Rosenfield, F. Hal Allen Jr., Scott N. Swisher, and Shaul Kochwa (1962) (Bibl. sec. 3).
Allen, F. Hal, Jr., and thirty others: "Blood group nomen-clature." *Science* 127(1958):1255–1256.
Apelt, Ernst Friedrich:
    Glasmacher, Thomas. *Fries-Apelt-Schleiden: Verzeichnis der Primär- und Secundärliteratur 1798–1988.* Cologne: Dinter, 1989.
    Laudan, L. L. (Larry). "Apelt, Ernst Friedrich." In Charles C. Gillispie, ed., *Dictionary of Scientific Biography.* New York: Scribner's, 1970, v. 1.
Apelt, Ernst Friedrich. *Theorie der Induktion.* Leipzig: Engelmann, 1854.
Apolant, Hugo. "Ueber die Empfindlichkeit von Krebs-mäusen gegen introperitoneale Injektionen." *Z. f. Immunitätsf.* 3(1909):108–114.
Apolant, Hugo. "Ueber die Beziehungen der Milz zur aktiven Geschwulstimmunität." *Z. F. Immunitätsf.* (1913).
Apolant, Hugo. See also under Paul Ehrlich: Hugo Apolant et al., *Ehrlich-Festschrift* (1914) (Bibl. sec. 3).
Aquinas, St. Thomas. "De ente et essentia." In Robert P. Goodwin, ed., *Selected Writings of St. Thomas Aquinas.* New York: Bobbs-Merrill, 1965.
Arrhenius, Svante:
    Bernstein, Robert S. "Svante Arrhenius and ionic dissociation: a revaluation." In G. Dubpernell and J. H. Westbrook, *Electrochemistry* (1977) (Bibliog. sec. 2).
    Fruton, Joseph S. *Contrasts in Scientific Style* (1990), 241–262 (Bibliog. sec. 2).
    Mazumdar, Pauline M. H. "The antigen-antibody reaction and the physics and chemistry of life." *Bull. Hist. Med.* 48(1974):1–21.

Neisser, Max. "Kritische Bemerkungen zur Arrhenius'schen Agglutinin Ver-teilungs-formel." *Zbl. f. Bakt. 36*(1904):671–676.

Nernst, H. Walther. "Ueber die Anwendbarkeit der Gesetze des chemischen Gleichgewichts auf Gemische von Toxin und Antitoxin." *Z. f. Elektrochemie 10*(1904):377–380.

Ostwald, Wilhelm. "Svante August Arrhenius." *Z. f. phys. Chem. 69*(1909):v–xx.

Palmaer, W. "Arrhenius (1859–1927)." In G. Bugge, *Das Buch der grossen Chemiker*. Berlin: Verlag Chemie, 1930, v. 2, 443–462 (Bibl. sec. 2).

Riesenfeld, E. H. "Svante Arrhenius." *Ber. d. Deutsch. chem. Ges. 63*(1930):1–40.

Rubin, Lewis P. "Styles in scientific explanation: Paul Ehrlich and Svante Arrhenius on immunochemistry." *J. Hist. Med. 35*(1980):397–425.

Snelders, H. A. M. "Arrhenius, Svante August." In Charles C. Gillispie, ed., *Dictionary of Scientific Biography*. New York: Scribner's, 1970, v. 1, 296–302.

Wells, Harry Gideon. [Review of] Svante Arrhenius. *Immunochemistry: the Application of the Principles of Physical Chemistry to the study of the Biological Antibodies*. New York: Macmillan, 1907. in *J. Am. Chem. Soc. 30* 650–652 (1908).

Arrhenius, Svante. "Ueber die Dissociation der in Wasser gelösten Stoffe." *Z. f. physik. Chem. 1*(1887):631–648.

Arrhenius, Svante. *Textbook of Electrochemistry*. London: Longmans, Green 1902. The English translation was by J. McCrae from the German translation, which in turn was a revised version of lectures given in Swedish in 1897.

Arrhenius, Svante. "Zur physikalische Chemie der Agglutinine." *Z. f. physik. Chem. 46*(1903):415–462.

Arrhenius, Svante. *Immunochemistry: the Application of the Principles of Physical Chemistry to the Study of the Biological Antibodies*. New York: Macmillan, 1907. German ed., Leipzig: Acad. Verlag, 1907.

Arrhenius, Svante, and Thorvald Madsen. "Physical chemistry applied to toxins and antitoxins." In Carl Julius Salomonsen, ed., *Festskrift ved indvielsen of Statens Seruminstitut 1902: Contributions from the University Laboratory for Medical Bacteriology to Celebrate the Inauguration of the State Serum Institute*. Copenhagen, Olsen, 1902, 1–111.

Bamberger, Eugen von:

Blangey, Louis. "Eugen Bamberger 1857–1932." *Helv. chem. Acta 16*(1923): 644–676. Includes bibliography, pp. 676–685.

Gienapp, R. A. "Bamberger, Eugen." In Charles C. Gillispie, ed., *Dictionary of Scientific Biography* (1970), v. 1, 426–427 (Bibl. sec. 1).

Partington, James R. (1964), v. 4, 840–847 (Bibl. sec. 3).

von Bamberger, Eugen. "Weiters über Diazo- und Iso-diazo-Verbindungen VII Mittheilung über Diazoverbindungen." *Ber. d. Deutsch. chem. Ges. 27*(1894):914–917.

von Bamberger, Eugen, and Karl Landsteiner. "Das Verhalten des Diazobenzols gegen Kaliumpermanganat." *Ber. d. Deutsch. chem. Ges. 26*(1893):482–495, C. 3.

von Basch, Samuel:

Lesky, Erna. "Kompensationslehre und denkökonomisches Prinzip." *Gesnerus 23*(1966):97–108.

von Basch, Samuel. "Die Compensationslehre von erkenntnisstheoretisch

Standpunkte." *Verhandlng d. XIII Congress f. innere Medicin.* Wiesbaden, 1895, pp. 433–447.

Bateson, William:

Bateson, Beatrice. *William Bateson, F.R.S., Naturalist: His Essays and Addresses, together with a Short Account of His Life.* Cambridge: Cambridge University Press, 1928.

Coleman, William. "Bateson, William." In Charles C. Gillispie, ed., *Dictionary of Scientific Biography* (1971), v. 1, 505–506 (Bibl. sec. 2).

Bateson, William. *Mendel's Principles of Heredity.* Cambridge: Cambridge University Press, 1st ed. 1909, 2d ed. 1913.

Baur, Erwin, Eugen Fischer, and Fritz Lenz. *Grundriss der menschlichen Erblichkeitslehre und Rassenhygiene.* Munich: Lehmann, 2d ed., 1923. V. 1, "Menschliche Erblichkeitslehre"; v. 2, "Menschliche Auslese und Rassenhygiene."

Bechhold, Heinrich:

Zaunick, Rudolph, and J. C. Salié, eds. "Bechhold, Heinrich (1866–1937)." In *Poggendorfs Handworterbuch* (1956), v. 7a, pt. 1, 113–114 (Bibl. sec. 1). Includes bibliography, list of biographical sources.

Bechhold, Heinrich. "Die Bakterienagglutination ein physikalisch-chemisches Phänomen: Vortrag, in Gemeinschaft mit Prof. Dr Max Neisser und Dr [Ulrich] Friedemann." *Verh. d. Ges. Deutsch. Naturf. u. Aerzte,* 75 Versammlung, Cassel, 1903. Leipzig: Vogel, 1904, pt. 1, 487–488.

Bechhold, Heinrich. "Die Ausflockung von Suspensionen bzw. Kolloiden und die Bakterienagglutination." *Z. f. physik. Chem.* 48(1904):385–423.

Bechhold, Heinrich. [Review of] Pauli, Wolfgang, "Wandlungen in der Pathologie durch die Forschritte der allgemeinin Chemie." Festival Address at the 3d Annual Meeting of the k. k. Gesellschaft der Aerzte in Vienna, 24 March 1905. In *Wiener klin. Wschr.* 18(1905):550–551.

Bechhold, Heinrich. "Ungelöste Fragen Ueber den Anteil der Kolloid-chemie an der Immunitätsforschung." *Wiener klin. Wschr.* 18(1905):666–668.

Bechhold, Heinrich. "Strukturbildung in Gallerten." *Z. f. physik. Chem.* 52(1905):185–199.

Bechhold, Heinrich. "Kolloidstudien mit der Filtrationsmethod." *Z. f. physik. Chem.* 60(1907):257–318.

Bechhold, Heinrich. "Die elektrische Ladung von Toxin und Antitoxin." *Münch. med. Wschr.* 54(1907):1921–1922.

Benedikt, Rudolf, and Max Bamberger. "Ueber die Einwirkung von Jodwasserstoffsäure auf schwefelhaltige Substanzen." *Monatsh. f. Chem.* (Vienna) 12(1891):1–4.

Berczeller, L. "Soll die Wassermann'sche Reaktionen mit aktiven oder inaktivierten Patientenserum ausgeführt werden?" *Z. f. Immunitätsf.* 27(1918): 305–325.

Bernstein, Felix:

Mazumdar, Pauline M. H. "Two models for human genetics: blood grouping and psychiatry in Germany between the wars." *Bull. Hist. Med.* 69(1995).

Nathan, Henry. "Bernstein, Felix." In Charles C. Gillispie, ed., *Dictionary of Scientific Biography* (1970), v. 2, 58–59 (Bibl. sec. 1).

Bernstein, Felix. "Die Theorie der gleichsinnige Faktoren in der Mendelschen Erblichkeitslehre vom Standpunkt der mathematischen Statistik." *Z. f. indukt. Abstamm. u. Vererbl.* 28(1922):295–323.

Bernstein, Felix. "Ergebnisse einer biostatistischen zusammenfassenden Betrachtung über die Erbstrukturen der Menschen." *Klin. Wschr.* *3*(1924): 1495–1497. Translated in *Selected Contributions* (1966), F. R. Camp and F. R. Ellis, eds., v. 1, 83–90 (Bibl. sec. 2).

Bernstein, Felix. "Zusammenfassende Betrachtungen über erbliche Blutstrukturen des Menschen." *Z. f. indukt. Abstamm. u. Verebs.* *37*(1925):237–270. Translated in *Selected Contributions* (1966), F. R. Camp and F. R. Ellis, eds., v. 1, 91–138 (Bibl. sec. 2).

Bernstein, Felix. "Fortgesetzte Untersuchungen aus der Theorie der Blutgruppen." *Z. f. indukt. Abstamm. u. Vererbs.* *56*(1930):233–273.

Bertheim, Alfred. "Chemie der Arsenverbindungen." In Apolant, Hugo et al., *Ehrlich-Festschrift* (1914) 447–476 (Bibl. sec. 3).

Billitzer, Jean. "Theorie der Kolloide, II." *Z. f. physik. Chem.* *51*(1905):129–166.

Biltz, Wilhelm Eugen:
   Zaunick, Rudolph, and J. C. Salié, eds. "Biltz, Wilhelm Eugen (1877–1943)." In *Poggendorfs Handworterbuch* (1956), v. 7a, pt. 1, 185–187. Includes bibliography, list of biographical sources (Bibl. sec. 1).

Biltz, Wilhelm. "Ein Versuch zur Deutung der Agglutinierungsvorgänge." *Z. f. phys. Chemie* *48*(1904):615–623.

Bordet, Jules:
   Beumer, J. "Jules Bordet 1870–1961." *J. Gen. Microbiol.* *29*(1962):1–13.
   Oakley, C. L. "Jules Jean Baptiste Vincent Bordet, 1870–1961." *Biographical Memoirs of the Fellows of Royal Society* (London) *8*(1962):19–25. Includes "select" bibliography.
   Vieuchange, J. "Bordet, Jules." In Charles C. Gillispie, ed., *Dictionary of Scientific Biography* (1970) 300–301 (Bibl. sec. 1).
   Volume Jubilaire de Jules Bordet. *Ann. de l'Inst. Pasteur 79*(5) (1950):479–520; E. Renaux, "Discours," 479–491; M. Spaelant, 492–506; P. Bordet, "L'Institut Pasteur de Bruxelles," 507–520.

Bordet, Jules. "Sur le mode d'action des sérums préventifs." *Ann. de l'Inst. Pasteur 10*(1896):193–219. Translated in *Bordet-Gay*, 81–103.

Bordet, Jules. "Agglutination et dissolution des globules rouges par le sérum: deuxième mémoire." *Ann. de l'Inst. Pasteur 13*(1899):273–297.

Bordet, Jules. "Les sérums hémolytiques, leurs anticorps et les théories des sérums cytolytiques." *Ann. d. l'Inst. Pasteur 14*(1900):257–296. Translated in *Bordet-Gay*, 186–216.

Bordet, Jules. "Sur le mode d'action des antitoxines sur les toxines." *Ann. d. l'Inst. Pasteur 17*(1903):161–186. Translated in *Bordet-Gay*, 259–279.

Bordet, Jules, and Frederick P. Gay. "L'absorption de l'alexine et le pouvoir antagonist des sérums normaux." *Ann. de l'Inst. Pasteur 22*(1908):625–643. Translated in *Bordet-Gay*, 398–413.

Bordet, Jules. "A general résumé of immunity" (1909). In *Bordet-Gay*, 496–530.

Bordet, Jules, and Oswald Streng. "Les phénomènes d'absorption et la conglutinine du sérum de boeuf." *Z. f. Bakt.* (1 Abt. Orig.) *49*(1909):260–276. Translated in *Bordet-Gay*, 440–461.

[Bordet, Jules] Gay, Frederick P., ed. and trans. *Studies in Immunity by Professor Jules Bordet and His Collaborators.* New York: Wiley, 1909.

Bordet, Jules. *Traité de l'Immunité dans les Maladies Infectieuses.* Paris: Masson, 1st ed. 1920, 2d ed. 1939.

Bredig, Georg. *Anorganische Fermente.* Leipzig: Engelmann, 1901.

Buchner, Hans:
von Gruber, Max, "Hans Buchners Anteil an der Entwicklung der Bakteriologie." *Münchner med. Wschr. 50*(1903):564–568.
Kohler, Robert E. "The reception of Eduard Buchner's discovery of cell-free fermentation." *J. Hist. Biol. 5*(1972):327–353.
Buchner, Hans. See also Max von Gruber, [Replies to Richard Pfeiffer] (1896) (Bibl. sec. 3).
Buchner, Hans. *Ueber die experimentelle Erzeugung des Milzbrandcontagiums aus den Heupilzen nebst Versuchen über die Entstehung des Milzbrandes durch Einathmung.* Die medicinschen Facultät der kgl. Ludwig-Maximilians-Universität München, Pro venia legendi, vorgelegt von Dr Hans Buchner k.b. Assistentarzt, *Sitzungsberichte der math.-phys. Classe der kgl. bayerische Akademie der Wissenschaften,* 367–413; 413–423 (1800).
Buchner, Hans. "Ueber die experimentell Erzeugung des Milzbrandcontagiums, II. Mittheilung." *Sitzungsberichte der math-phys. Classe der kgl. bayerische Akademie der Wissenschaften,* 147–169 (1882).
Buchner, Hans. "Zur Nomenclatur der Spaltpilze." *Sitzungsberichte der Gesellschaft f. Morphologie und Physiologie* (Munich) *1*(1885):121–129.
Buchner, Hans. "Zur neueren Literatur über die Frage vom genetischen Zusammenhang der Milzbrand und Heubacterien." *Sitzungsberichte der Gesellschaft f. Morphologie und Physiologie* (Munich) *1*(1885):27–30.
Buchner, Hans. "Zur Kenntniss der Alexine, sowie der specifisch-bactericiden und specifisch hämolytischen Wirkung." *Münch. med. Wschr. 47*(1900):277–283.
du Bois-Reymond, Emil. "Gedächtnissrede auf Johannes Müller, gehalten in der Akademie der Wissenschaften am 8 Juli 1858." In du Bois-Reymond, *Reden von Emil duBois-Reymond.* 2v. Leipzig: Veit, 1912, v. 2, 143–334.
du Bois-Reymond, Emil. *"Ueber die Grenzen des Naturerkennens." und 'Die sieben Welträthsel." zwei Vorträge.* Leipzig: Veit, 1898.
Castle, William B., Maxwell M. Wintrobe, and Laurence H. Snyder (Chairman). "On the nomenclature of the anti-Rh typing serums: Report of the Advisory Review Board." *Science 107*(1948):27–31.
Castle, William E. "Piebald rats and the theory of genes." *Proc. Nat. Acad. Sci. 5*(1919):500–506.
Coca, Arthur F.:
Fischer, Isidor. "Coca, Arthur Fernandez." *Biographisches Lexikon* (1933) v. 2, 256 (Bibl. sec. 1).
Coca, Arthur F. *Essentials of Immunology for Medical Students.* Baltimore, Ms.: Williams, 1925.
Cohn, Ferdinand J.:
Cohn, Pauline. *Ferdinand Cohn: Blätter der Erinnerung Zusammengestellt von seiner Gattin Pauline Cohn mit Beitragen von Professor F. Rosen.* Breslau: Kerns, 1901.
Geison, Gerald L. "Cohn, Ferdinand Julius." In Charles C. Gillispie, ed., *Dictionary of Scientific Biography* (1971), v. 3, 336–341 (Bibl. sec. 1).
Cohn, Ferdinand J. "Untersuchung über die Entwicklungsgeschichte der microscopischen Algen und Pilze." *Acad. Caes. Leop. Nova Acta 24*(1854):101–256.
Cohn, Ferdinand J. "Ueber Pilze als Ursache von Thierkrankheiten." *Jahres Berichte der schlesischen Gesellschaft f. vaterl. Cult. 32*(1854):43–48.
Cohn, Ferdinand J. "Ueber den Ursprung der Schlesischen Flora." *Jahres Berichte der schlesischen Gesellschaft f. vaterl. Cult. 38*(1860):110–126.

Cohn, Ferdinand J. "Ueber die Algen des Carlsbader Sprudels, und deren Antheil an der Bildung des Sprudelsinters." *Jahres Berichte der. schlesischen Gesellschaft f. vaterl. Cutl. 40*(1862):65–67.

Cohn, Ferdinand J. "Beiträge zur Physiologie der Phycochromaceen und Florideen." *Arch. f. Microsc. Anat. 3*(1867):1–60.

Cohn, Ferdinand J. [In discussion following] Schneider, "Ueber Hallier's Cholerapilze und dessen Entwicklung." *Jahres Berichte der schlesisch. Gesellschaft f. vaterl. Cult. 45*(1867):114–125.

Cohn, Ferdinand J. "Conspectus familiarum cryptogamarum secundum methodum naturalem dispositarum auctore Ferdinand Cohn." *Hedwigia 11*(1872):17–20.

Cohn, Ferdinand J. "Untersuchung über Bacterien." *Beitr. z. Biol. d. Pflanzen 1*(1875):127–222.

Cohn, Ferdinand J. "Untersuchung über Bacterien." *Beitr. z. Biol. d. Pflanzen 1*(1872):127–222 (Heft 2). Published 1875.

Cohn, Ferdinand J. *Lebensfragen: Rede gehalten am 22 Sept. 1886 in der 2 allgemeinen Sitzung der 59 Versammlung deutscher Naturforscher und Aerzte zu Berlin.* Berlin: Hirschwald, 1887.

Cohnheim, Julius. *Gesammelte Abhandlung von Julius Cohnheim.* Edited by E. Wagner. Berlin: Hirschwald, 1885.

Correns, Carl. "G. Mendels Regel über das Verhalten der Nachkommenschaft der Rassenbastarde." *Berichte d. Deutsch. bot. Ges. 18*(1900):158–161.

Correns, Carl. "Ueber die dominierenden Merkmale der Bastarde." *Berichte d. Deutsch. bot. Ges. 21*(1903):133–147.

Cuénot, Lucien. "L'hérédité de la pigmentation chez les souris." *Arch. de Zool. Expér. et Gén. 1* (4th ser.)(1903):33–41.

Dameshek, William. "Preface: the Rh factor." In Joseph M. Hill and William Dameshek, eds., *The Rh Factor in the Clinic and the Laboratory.* Selection of papers presented at the International Hematology and Rh Conference, Dallas, TX and Mexico City, November 1946. *Blood* Special Issue no. 2 (1948).

Danysz, Jean. "Contribution a l'étude de l'immunité. Propriétés des mélanges des toxines avec leurs antitoxines. Constitution des toxines." *Ann. de l'Inst. Pasteur 13*(1899):581–595.

Danysz, Jean. "Contribution a l'étude des propriétés et de la nature des toxines avec leurs antitoxines." *Ann. de l'Inst. Pasteur 16*(1902):331–345.

Diamond, Louis K. "A history of blood transfusion." In Maxwell M. Wintrobe, ed., *Blood Pure and Eloquent: a Story of Discovery, of People and of Ideas.* New York: 1980, 659–690.

Diamond, Louis K. "The story of our blood groups." In Wintrobe, *Blood Pure and Eloquent* (1980), 691–718.

von Dungern, Emil, Freiherr:
Hirszfeld, Ludwik. *Historia* (1957) (Bibl. sec. 3).

von Dungern, Emil. "Spezifität der Antikörperbildung." *Koch-Festschrift* (1903):1–16.

von Dungern, Emil. *Die Antikörper.* Jena: Fischer, 1903.

von Dungern, Emil, and Ludwik Hirszfeld. "Ueber Nachweis und Vererbung biochemischer Strukturen." *Z. f. Immunitätsf. 4*(1910):531–546.

von Dungern, Emil. "Ueber Nachweis und Vererbung biochemischer Strukturen und ihre forensische Bedeutung." *Münch. med. Wschr.* 57(1910):293–295.

von Dungern, Emil. "Rezeptorenspezifität." In Apolant, Hugo *et al.*, *Ehrlich-Festschrift* (1914) 162–165.

von Dungern, Emil. "Beiträge zur Immunitätslehre." *Münch. med Wschr.* 47(1900):I, 677–680; II, 962–965.

von Dungern, Emil, and Ludwik Hirszfeld. "Ueber eine Methode das Blut verschiedener Menschen serologisch zu unterschieden." *Münch. med. Wschr.* 57(1910):741–742.

von Dungern, Emil, and Ludwik Hirszfeld. "Ueber Vererbung gruppenspezifischer Strukturen des Blutes, II." *Z. f. Immunitätsf.* 6(1910):284–292.

von Dungern, Emil, and Ludwick Hirszfeld. "Ueber gruppen-spezifische Strukturen des Blutes, III." *Z. f. Immunitätsf.* 8(1911):526–562.

Durham, Herbert E. "Immunitas gegen Cholera und Typhus: demonstration zu dem Vortrag Grubers." *Verhandlungen d. Congr. f. innere Med. (XIII Congr.)* Wiesbaden, 1896. Wiesbaden: Bergmann, 1896, 228–230.

Durham, Herbert E. "On a special action of the serum of highly immunized animals." *J. Path. Bact.* 4(1897):13–43.

Editorial. "Anti-Rh serum nomenclature." *Brit. med. J.* 1(1948):400.

Editorial. "Terminology of the Rh factor." *Lancet* i(1948):329.

Ehrenberg, Christian Gottfried:

Jahn, Ilse. "Ehrenberg, Christian Gottfried." In Charles C. Gillispie, ed., *Dictionary of Scientific Biography.* New York: Scribner's, 1971, v. 4, 288–292.

Winsor, Mary P. *Starfish, Jellyfish and the Order of Life: Issues in XIX Century Science.* New Haven, Conn.: Yale University Press, 1976, 28–43.

Churchill, Frederick B. "The guts of the matter: infusoria from Ehrenberg to Bütschli." *J. Hist. Biol.* 22(1989):189–213.

Ehrenberg, Christian Gottfried. *Die Infusionsthierchen als volkommene Organismen: ein Blick in das tiefere organische Leben der Natur.* Leipzig: Voss, 1838.

Ehrlich, Paul:

Apolant, Hugo, et al. *Paul Ehrlich: eine Darstellung seines wissenschaftlichen Wirkens: Festschrift zum 60 Geburtstages des Forschers.* Jena: Fischer, 1914. Contains a bibliography by Hans Sachs (pp. 625–657) that also includes work done under Ehrlich's direction, up to February 1914.

Aschoff, Ludwig. *Ehrlich's Seitenkettentheorie und ihre Anwendung auf die knstlichen Immunisierungsprozesse: zusammenfassende Darstellung.* Jena: Fischer, 1902. Reprint from *Z. f. allgem. Physiol.* 1 (Sammelreferate)(1902):69–249.

Bäumler, Ernst. *Paul Ehrlich: Scientist for Life.* New York: Holmes, 1984.

Bolduan, Charles, ed. and trans. *Studies in Immunity by Professor Paul Ehrlich and his Collaborators.* New York: Wiley, 1st ed. 1906, 2d ed. 1910.

Himmelweit, Felix, Martha Marquardt, and Sir Henry Dale, eds. *The Collected Papers of Paul Ehrlich in Four Volumes including a Complete Bibliography.* London: Pergamon, 1956. The fourth volume with the bibliography has never appeared.

Koch, Richard. "Paul Ehrlich (1854–1915)." In Günther Bugge, *Das Buch der grossen Chemiker* (1930) v. 2, 421–442 (Bibl. sec. 1).

Liebenau, Jonathan. "Paul Ehrlich as a commercial scientist and research administrator." *Medical Hist.* 34(1990):65–78.

Marquardt, Martha. *Paul Ehrlich.* New York: Schumann, 1915.

Michaelis, Leonor. "Zur Erinnerung an Paul Ehrlich: seine wiedergefundene Doktor Dissertation." *Naturwissenschaften* 7(1919):165–168.

Parascandola, John, and Roland Jasensky. "Origins of the receptor theory of drug action." *Bull. Hist. Med.* 48(1974):199–220.

Salomonsen, Carl Julius. "Lebenserinnerungen aus dem Breslauer Sommer-semester 1877." *Berl. klin. Wschr.* 51(1914):485–490.

Ehrlich, Paul. "Beiträge zur Theorie und Praxis der histologischen Färbung: Inaugural Dissertation, Universität Leipzig 1878." In Himmelweit, Marquardt, and Dale, eds., *Collected Papers* 29–64 (Ger.) 65–94 (Engl.)

Ehrlich, Paul. "Ueber die specifische Granulationen des Blutes: Vortrag, gehalten den 16 mai, in Verhandlungen der Berliner physiologische Gesellschaft." *Arch. f. u. Physiol.* (Physiol. Abt.) (1878–1879):571–579. Also in Ehrlich, *Farbenanalytische Untersuchungen,* 5–16. Also in Himmelweit, Marquardt, and Dale, eds., *Collected Papers,* v. 1, 117–123 (not translated).

Ehrlich, Paul. "Methodologische Beiträge zur Physiologie und Pathologie der verschiedenen Formen der Leukocyten." *Z. f. klin. Med.* 1(1880):553–560. Also in Ehrlich, *Farbenanalytische Untersuchungen,* 42–50. Also in Himmelweit, Marquardt, and Dale, eds., *Collected Papers,* v. 1, 124–129 (Ger.).

Ehrlich, Paul. "Ueber das Methylenblau und seine klinische-bakterioskopische Verwertung." *Z. f. klin. Med.* 2(1881):710–713. This has not been reprinted either in Ehrlich, *Farbenanalalytische Untersuchungen* or in Himmelweit, Marquardt, and Dale, eds., *Collected Papers.*

Ehrlich, Paul. "Ueber die Färbung der Tuberkelbacillen." In Verein f. inn. Med. 1 mai 1882. *Deutsche med. Wschr.* 8(1882):269–270. Also in Himmelweit, Marquardt, and Dale, eds., *Collected Papers,* v. 2, 311–313, as "Modification der von Koch angebenen Methode der Färbung von Tuberkelbacillen" (Ger.).

Ehrlich, Paul. "Ueber eine neue Harnprobe." *Z. f. klin. Med.* 5(1881):285–288. Also in Himmelweit, Marquardt, and Dale, eds., *Collected Papers,* v. 1, 619–629 (Ger.).

Ehrlich, Paul. *Farbenanalytische Untersuchung zur Histologie und Klinik des Blutes: gesammelte Mittheilungen.* Berlin: Hirschwald, 1891.

Ehrlich, Paul. "Die Wertbemessung des Diphtherieheilserums, und deren theoretische Grundlagen." *Klin. Jahrb.* 6(1897–1898):299–326. Also in Himmelweit, Marquardt, and Dale, eds., *Collected Papers,* v. 2, 86–106 (Ger.) 107–125 (Engl.).

Ehrlich, Paul. "Ueber die Beziehungen von chemische Constitution, Vertheilung und pharmakologischer Wirkung: Vortrag, gehalten im *Verein f. innere Med.,* Berlin, December 12 1898. Reported in *Münch. med. Wschr.* 45(1898):1654–1655. First published in full in *Internationale Beiträge zur inneren Medicin: Ernst von Leyden zur Feier seines 70-jährigen Geburtstages, am 20 April 1902, gewidmet von seinen Freunden und seinen Schülern.* Berlin: Hirschwald, 1902, v. 1, 645–679. Also in Himmelweit, Marquardt, and Dale, eds., *Collected Papers,* v. 1, 570–595 (Ger.); and in *Ehrlich-Bolduan* 404–442.

Ehrlich, Paul. "Die Schutzstoffe des Blutes." *Verhandlungen d. Ges. Deutsch. Naturf. u. Aerzte* 73 Versammlung zu Hamburg. Leipzig: Vogel, 1901, pt. I, 250–275. Also Himmelweit, Marquardt, and Dale, eds., *Collected Papers,* v. 2, 298–315 (Ger.).

Ehrlich, Paul. "Ueber die Beziehungen von chemischer Constitution, vertheilung und pharmakologischer Wirking." *Internationale Beiträge zur inneren*

Medicin: Ernst von Leyden zur Feier seines 70 Jährigen Geburtstages am 20 April 1902, gewidmet von seinen Freunden und seinen Schülern. Berlin: Hirschwald, 1902, v. 1, 645–679.

Ehrlich, Paul. "A general review of the recent work in immunity" (1906). In *Ehrlich-Bolduan*, 577–586.

Ehrlich, Paul. "Robert Koch." *Z. f. Immunitätsf.* 6(1910) (unpaginated, preceding title page).

Ehrlich, Paul, and Sahashiro Hata. *Die experimentelle Chemotherapie der Spirillosen.* Berlin: Springer, 1910. Translated by A. Newbold and Rev. R. W. Felkin as *The Experimental Chemotherapy of Spirilloses.* London: Rebman, 1911.

Ehrlich, Paul, and Julius Morgenroth. "Zur Theorie der Lysinwirkung." *Berl. klin. Wschr.* 36(1899):6–9. (1899). Also in Himmelweit, Marquardt, and Dale, eds., *Collected Papers,* v. 2, 143–149 (Ger.), 150–155 (Engl.).

Ehrlich, Paul, and Julius Morgenroth. "Ueber Hämolysine: zweite Mittheilung." *Berl. klin. Wschr.* 36(1899):481–486. Also in Himmelweit, Marquardt, and Dale, eds., *Collected Papers,* v. 2, 156–164 (Ger.), 165–172 (Engl.).

Ehrlich, Paul, and Julius Morgenroth. "Ueber Hämolysine: dritte Mittheilung." *Berl. klin. Wschr.* 37(1900):453–458. Also in Himmelweit, Marquardt, and Dale, eds., *Collected Papers,* v. 2, 196–204 (Ger.), 205–212 (Engl.).

Ehrlich, Paul, and Julius Morgenroth. "Ueber Hämolysine: vierte Mittheilung." *Berl. klin. Wschr.* 37(1900):681–687. Also in Himmelweit, Marquardt, and Dale, eds., *Collected Papers,* v. 2, 213–223 (Ger.), 224–233 (Engl.).

Ehrlich, Paul, and Julius Morgenroth. "Ueber Hämolysine: fünfte Mittheilung." *Berl. klin. Wschr.* 38(1901):251–257. Also in Himmelweit, Marquardt, and Dale, eds., *Collected Papers,* v. 2, 234–245 (Ger.), 246–255 (Engl.).

Ehrlich, Paul, and Julius Morgenroth. "Ueber Hämolysine: sechste Mittheilung." *Berl. klin. Wschr.* 38(1901):569–574, 598–604. Also in Himmelweit, Marquardt, and Dale, eds., *Collected Papers,* v. 2, 256–277 (Ger.), 278–297 (Engl.).

Eisenberg, Philipp. See also Volk, R., "Ueber quantitiven Grundlagen der Bindungsverhältnisse" (1927) (Bibl. sec. 3).

Eisenberg, Philipp, and Richard Volk. "Untersuchungen über die Agglutination: vorläufige Mittheilungen." *Wiener klin. Wschr.* 14(1901):1221–1223.

Eisenberg, Philipp, and Richard Volk. "Untersuchung über Agglutination." *Z. f. Hyg. u. Infectionskr.* 40(1902):155–195. For comments on this material, see Svante Arrhenius, 1903; Jules Bordet, 1903; Wilhelm Biltz, 1904; Jean Billitzer, 1905; Karl Landsteiner, 1905; Clemens von Pirquet (manuscript).

Endlicher, Stephan. *Genera plantarum secundum ordines naturales disposita, auctore Stephano Endlicher.* Vienna: Beck, 1836–1840.

Endlicher, Stephan [Istvan Laszlo] and Franz Unger. *Grundzüge der Botanik, entworfen von Stephan Endlicher und Franz Unger.* Vienna: Gerold, 1843.

Exner, Siegmund. *Entwurf zu einer physiologischen Erklarung der pychischen Erseheinungen.* Leipzig: Deuticke, 1894.

Finney, D. J. "The detection of linkage." *Ann. Eugen.* 10(1940):171–214.

Fischer, Emil H.:

Bergmann, Max. "Emil Fischer (1852–1919)." In Günther Bugge, *Das Buch der grossen Chemiker* (1930), 408–420 (Bibl. sec. 1).

Farber, Eduard. "Fischer, Emil Hermann." In Charles C. Gillispie; ed., *Dictionary of Scientific Biography* (1972), v. 5, 1–5 (Bibl. sec. 1).

396   *Bibliography*

Fruton, Joseph S. *Contrasts in Scientific Style: Research Groups in the Chemical and Biochemical Sciences.* Philadelphia, Pa.: American Philosophical Society, 1990. 163–229; 375–403.

Hoesch, Kurt. *Emil Fischer sein Leben und sein Werk im Auftrage der Deutschen chemischen Gesellschaftdargestellt.* Berlin: Verlag Chemie, 1921.

Fischer, Emil. "Synthesen in der Zuckergruppe, I." *Berichte der Deutsch. chem. Ges.* 23(1890):2114–2141. Also Fischer, *Kohlenhydrate und Fermente* (1909), 1–29.

Fischer, Emil. "Die Chemie der Kohlenhydrate und ihre Bedeutung für Physiologie: Rede, gehalten zu Feier des Stiffungstages der militärztlichen Bildungsanstalten am 2 August 1894." In Fischer, *Kohlenhydrate und Fermente* (1909) 96–115.

Fischer, Emil. "Synthesen in der Zuckergruppe, II." *Berichte der Deutsch. chem. Ges.* 27(1894):3189–3232. Also Fischer, *Kohlenhydrate und Fermente* (1909), 30–75.

Fischer, Emil. "Einfluss der Konfiguration auf die Wirkung der Enzyme, I." *Berichte der Deutsch. chem. Ges.* 27(1894):2985–2993. Also Fischer, *Kohlenhydrate und Fermente* (1909), 836–844.

Fischer, Emil. "Bedeutung der Stereochemie für die Physiologie." *Z. f. physiol. Chem.* (Hoppe-Seylers) 26(1898):60–87. Also Fischer, *Kohlenhydrate und Fermente* (1909), 116–137.

Fischer, Emil. *Untersuchungen über Kohlenhydraten und Fermente 1884–1908.* Berlin: Springer, 1909.

Fischer, Emil, and Karl Landsteiner. "Ueber den Glykolaldehyde." *Berichte der Deutsch. chem. Ges.* 25(1892):2549–2554, C. 2.

Fischer, Emil, and Hand Thierfelder. "Verhalten der verschiedenen Zucker gegen reine Hefen." *Berichte der Deutsch. chem. Ges.* 27(1894):2031–2037. Also Fischer, *Kohlenhydrate und Fermente* (1909), 829–835.

Fisher, Ronald Aylmer:

Box, Joan Fisher. *R. A. Fisher: the Life of a Scientist.* New York: Wiley, 1978.

Gridgman, Norman T. "Fisher, Ronald Aylmer." In Gillispie, C. C., ed., *Dictionary of Scientific Biography* (1975), v. 5, 7–11 (Bibl. sec. 1).

Neyman, Jerzy, "R. A. Fisher (1890–1962) an appreciation." *Science* 156(1967):1456–1460.

Race, Robert R. "Some notes on Fisher's contribution to human blood groups." *Biometrics* 20(1964):361–367.

Ronald Aylmer Fisher. Special Issue of *Biometrics 20* June issue (1964).

Yates, Frank, and Kenneth Mather. "Ronald Aylmer Fisher." *Obituary Notices of Fellows of Roy. Soc.* (London) 9(1963):91–120.

Fisher, Ronald A. See also under Race, R. R., "Fisher's theory" (1944); Race, "Fisher's theory" (1948) (Bibl. sec. 3); Provine, W. B., *Population Genetics* (1971) (Bibl. sec. 2); Hacking, I., *Statistical Inference* (1965) (Bibl. sec. 2).

Fisher, Ronald A. "Theory of statistical estimation." *Proc. Cambridge Phil. Soc.* 22(1925):700–725.

Fisher, Ronald A. "The detection of linkage with 'dominant' abnormalities." *Ann. Eugen.* 6(1935):187–201.

Fisher, Ronald A. "The detection of linkage with 'recessive' abnormalities." *Ann. Eugen.* 6(1935):339–351.

Fisher, Ronald A. "Tests of significance applied to Haldane's data on partial sex linkage." *Ann. Eugen.* 7(1936):87–104.

Fisher, Ronald A. "The fitting of gene frequencies to data on *Rhesus* reactions." *Ann. Eugen. 13*(1946):150–155.

Fisher, Ronald A., and Robert R. Race. "Rh gene frequencies in Britain." *Nature 157*(1946):48–49.

Fisher, Ronald A. "The Rhesus factor a study in scientific method." *Am. Sci. 35*(1947):95–102, 113.

Fleck, Ludwik. "Sérologie constitutionelle." In Hirszfeldowa, Kelus, and Milgrom (1956) 146–149.

Forssman, John. "Die Herstellung spezifischer Schaf-hämolysine ohne Verwendung von Schafblut." *Biochem. Z. 37*(1911):78–115.

Forster, J., Friedrich Hofmann, and Max von Pettenkofer. "Vorwort." *Arch. f. Hyg. 1*(1883):1–3.

Friedberger, Ernst:
  Fischer, Isidor. "Friedberger, Ernst." In *Biographisches Lexikon* (1933), v. 1, 449 (Bibl. sec. 1).

Friedberger, Ernst, and S. Girgolaff. "Ueber Anaphylaxie XVII. Mitteilung: die Bedeutung sessiler Rezeptoren für die Anaphylaxie." *Z. f. Immunitätsf. 9*(1911):575–582.

Friedemann, Ulrich:
  Fischer, Isidor. "Friedemann, Ulrich." In Fischer, *Biographisches Lexikon* (1933), v. 1, 450 (Bibl. sec. 2).
  Zaunick, Rudolph, and J. C. Salié, eds. "Friedemann, Ulrich." In *Poggendorfs Handwörterbuch* (1956–71), v. 7a, pt. 2, 121–122. Includes bibliography and list of biographical sources.

Friedemann, Ulrich. See also Hirszfeld, Ludwik (1957); Neisser, Max, and Friedemann, Ulrich(1904) (Bibl. sec. 3).

Friedemann, Ulrich. "Ueber die Fallung von Eiweiss durch andere Colloide und ihre Beziehungen zu den Serum-körperreaktionen." *Arch. f. Hyg. 55*(1906):361–389. Summary and critique by Karl Landsteiner in *Zbl. f. Physiol. 20*(1906):171.

Friedemann, Ulrich. "Infektion und Immunität." In Max Rubner, Max von Gruber, and Marin Ficker, eds., *Handbuch der Hygiene*. Leipzig: Hirzel, 1913, v. 3, pt. 1, 661–810.

Friedemann, Ulrich. "Die bedeutung der Lehre Kochs für die Klinik der Infektionskrankheiten." *Deutsche med. Wschr. 58*(1932):495–497.

Friedemann, Ulrich, and Hans Friedenthal. "Beziehung der Kernstoffe zu den Immunkörpern." *Zbl. f. Physiol. 20*(1906):585–587.

Friedenreich, V., and S. With. "Ueber B-Antigen and B-Antikörper bei Menschen und Tieren." *Z. f. Immunitätsf. 78*(1933):152–172.

Friedenthal, Hans. "Weitere Versuche über die Reaktion auf Blutverwandtschaft." In Verhandlungen der Berl. physiolog. Ges. *Arch. f. Physiol.* (Engelmanns) (20 vol. nos.) (1904):387–388.

Friedenthal, Hans. See also Friedemann, U., and Friedenthal, H. (1906) (Bibl. sec. 3).

Fries, Jacob Friedrich:
  Glasmacher, Thomas. *Fries-Apelt-Schleiden: Verzeichnis der Primär- und Secundärliteratur 1798–1988*. Cologne: Dinter, 1989.
  Gregory, Frederick, "Die Kritik von J. F. Fries an Schellings Naturphilosophie." *Sudhoffs Archiv 67*(1983):145–157.
  Jahn, Ilse. "Matthias Jacob Schleiden an der Universität Jena." *Naturwissen-*

schaft, Tradition, Fortschritt supplement to *NTM: Z. f. Geschichte der Naturwissenschaften* (1963):63–72.

Jahn, Ilse. "The influence of Jacob Friedrich Fries on Matthias Schleiden." In Woodward, Wm Ray, and Cohen, Richard S., eds., *World Views and Scientific Discipline Formation.* Dordrecht: Kluwer, 1991.

Mourelatos, Alexander P. "Fries, Jacob Friedrich." In Paul Edwards, ed., *Encyclopedia of Philosophy* (1969), v. 3 (Bibl. sec. 1).

Nobis, H. M. "Fries, Jacob Friedrich." In Charles C. Gillispie, ed., *Dictionary of Scientific Biography* (1972), v. 5 (Bibl. sec. 1).

Zaunik, Rudolph, and J. C. Salié. "Fries, Jacob Friedrich (1773–1843)." In *Poggendorfs Handwörterbuch* (1971), v. 7a, Supplement, 224–226. Includes bibliography and critical and secondary sources (Bibl. sec. 1).

Fries, Jacob F. *System der Philosophie als evidente Wissenschaft aufgestellt.* Leipzig: Hinricks, 1804.

Fries, Jacob F. In Ulrich Charpa, ed. and introduced, *Matthias Jacob Schleiden, Wissenschaftsphilosophische Schriften mit kommentierenden Texten von Jacob Friedrich Fries, Christian Nees von Esenbeck und Gerd Buchdahl.* Cologne: Dinter, 1989.

Gaffky, Georg:

Evans, Richard J. *Death in Hamburg: Society and Politics in the Cholera Years 1830–1910.* Oxford: Clarendon Press, 1987, 265–266, 497–498, 508.

Gaffky, Georg. See under Koch, R. (Bibl. sec. 3): Möllers, B. (1950), 368–370 (biog.); Gaffky, G., Pfuhl, W., and Schwalbe, J., eds. (1912).

Gaffky, Georg. "Experimentell erzeugte Septicämie mit Rücksicht auf progressive Virulenz und accomodative Züchtung." *Mitth. aus dem kaiserl. Gesundheitsamte 1* 1–54 (Article no. 3).

Gauch, H. "Beitrag zum Zusammenhang zwischen Blutgruppe und Rasse." *Z. f. Rassenphysiol.* 7(1937):116–122.

Gelmo, P., and Wilhelm Suida. "Studien über die Vorgänge beim Färben animalischer Textilfasern: vorgelegt in der Sitzung am 25 Oktober 1906 des k. Akad. der Wissenschaften in Wien." *Monatsh. f. Chem.* (Wien) 27(1906):1193–1198.

Gengou, Octave. "Récherches sur l'agglutination des globules rouges par les précipités chimiques et sur la suspension de ces précipités dans les mediaux colloidaux." *Ann. de l'Inst. Pasteur* 18(1904):678–700. Translated in *Bordet-Gay*, 312–332.

Gengou, Octave. "Contribution a l'étude de l'adhésion moléculaire et de son intervention dans les phénomènes biologiques." *Arch. Internat. de Physiol.* 7(1908):1–210. Résumé 178–210; translation of résumé in *Bordet-Gay*, 414–439.

Gibbs, Josiah Willard. "On the equilibrium of heterogeneous substances." *Trans. Conn. Acad.* 3(1875–1876):108–248. Also Bumstead, H. A. and van Name, R. G. eds., *The Scientific Papers of J. Willard Gibbs.* London: Longmans, 1906. Reprinted with additions, New York: Dover, 1961. Translated by Wilhelm Ostwald, *Thermodynamische Studien von J. Willard Gibbs unter Mitwirkung des Verfassers aus dem englischen bersetzt.* Leipzig: Engelmann, 1892.

von Goethe, Johann Wolfgang. *Die Wahlverwandschaften* (1808). Munich: D.T.V. -Gesamtausgabe, 1972, v. 19.

Graham, Thomas H. "On the diffusion of liquids: Bakerian Lecture to the Royal

Society of London, December 1849." *Phil. Trans. Roy. Soc.* (London) *140*(1850):1–46.

Graham, Thomas H. "Liquid diffusion applied to analysis." *Phil. Trans. Roy. Soc.* (London) *151*(1861):183–199.

Grassberger, Roland:
Fischer, Isidor. "Grassberger, Roland." In *Biographisches Lexikon* (1933), v. 1, 529 (Bibl. sec. 1).
Lesky, Erna. *Die Wiener medizinische Schule* (1965), 602 (Bibl. sec. 2).

Grassberger, Roland, and Arthur Schattenfroh. *Ueber die Beziehungen von Toxin und Antitoxin.* Vienna: Deuticke, 1904.

von Gruber, Max:
"Festschrift für Max von Gruber." *Arch. f. Hyg. 93*(1923). Articles reflect his interests, but are not about his work.
Grassberger, Roland. "Max von Gruber." *Wiener klin. Wschr. 40*(1927):1304–1306.
Gruber, G. B. "Lebensbild: Max von Gruber, 6:7:1853–16:9:1927." *Münch. med. Wschr. 95*(1953):806–807.
von Gruber, Max. "Kleine Mitteilungen: der 70 Geburtstag Max von Gruber." *Münch. med. Wschr. 70*(1923):1038–1039. Autobiographical essay.
Lesky, Erna. *Die Wiener medizinschen Schule* (1965), 595 ff. (Bibl. sec. 2).
Süpfle, K. "Max von Gruber zum Gedächtnis." *Deutsche med. Wschr. 53*(1927): 1869.

von Gruber, Max. "Ueber den augenblicklichen Stand der Bakteriologie der Cholera." *VIII Congrès Internationale d'Hygiène et Démographie: Comptes Rendus et Mémoires,* Budapest 1894. Budapest: Pesti Konyvnyomda-Reszvenytarasag, 1896, v. 2, pt. 1, 266–278.

von Gruber, Max. "Pasteurs Lebenswerk im Zusammenhang mit der gesammten Entwicklung der Microbiologie." *Wiener klin. Wschr. 8*(1895):823–828, 844–848, 836–866.

von Gruber, Max. "Erwiderung auf R. Pfeiffers Kritik meines Vortrages 'Pasteurs Lebenswerk im Zusammenhang mit der gesammten Entwicklung der Microbiologie,' " *Deutsche med. Wsch. 22*(1896):94–95.

von Gruber, Max. "Theorie der activen und passiven Immunität gegen Cholera, Typhus und verwandte Krankheitsprocesse." *Münch. med. Wschr. 43*(1896): 206–207.

von Gruber, Max. "Ueber active und passive Immunität gegen Cholera und Typhus." *Verhandlungen der Congress f. innere Medicin, XIII Congr.,* Wiesbaden 1896. Wiesbaden: Bergmann, 1896, 207–217.

von Gruber, Max. "Zur Theorie der Antikörper, I. Ueber die Antitoxin-Immunität; II. Ueber Bakteriolyse und Hämolyse: Vortrag gehalten in der k. k. Gesellschaft der Aerzte in Wien 25 oktober 1901." *Münch. med. Wschr. 48*(1901):1827–1830, 1294–1297, 1965–1968. Complete text of lecture. Summary in *Wiener klin. Wschr. 14*(1901):1093–1094, 1142–1143. Report of meeting, with discussions by Rudolf Kraus, 1190–1192; Friedrich Wechsberg, 1192–1195; R. Kretz 1195–1196; Max von Gruber 1214–1215; Richard Paltauf 1215–1218 (1901).

von Gruber, Max. [Two letters, Bernhard Naunyn to v. Gruber] *Münch. med. Wschr. 48*(1901):1933, 1933–1944.

von Gruber, Max. "Neue Früchte der Ehrlich'schen Toxinlehre." *Wiener klin. Wschr. 16*(1903):791–793.

von Gruber, Max. "Hans Buchners Anteil an der Entwicklung der Bakteriologie." *Münch. med. Wschr. 50*(1903):564–568.

von Gruber, Max. "Geschichte der Entdeckung der spezifischen Agglutination." In Kraus, Rudolf, and Levaditi, Constantin, *Handbuch der Immunitätsforschung und experimentelle Therapie*, etc. Jena: Fischer, 1914, 2d ed., 150–154. And as "Agglutination." In *Wiener med. Wschr. 77*(1927):742–743. "Wiener mikrobiologische Forschung." Special Number, Rudolf Kraus, ed.

von Gruber, Max. See also Durham, H. E. "On a special action of the serum" (1896); Grünbaum, A. S. F., "Agglutination of red cells" (1900) (Bibl. sec. 3).

von Gruber, Max. [Letter to Hermann von Schlesinger, 15 December 1908]. *Wiener med. Wschr. 81*(1931):309.

von Gruber, Max. "Der 70 Geburtstag Max von Grubers." *Münch. med. Wschr. 70*(1923):1038–1039.

von Gruber, Max. "Lord Lister und Deutschland." *Münch. med. Wschr. 74*(1927):592–593.

von Gruber, Max, and Hans Buchner. [Replies to Pfeiffer] *Deutsche med. Wschr. 22*(1896):128.

von Gruber, Max, and Herbert E. Durham. "Eine neue Method zur raschen Erkennung des Choleravibrio und des Typhusbacillus." *Münch. med. Wschr. 43*(1896):285–286.

von Gruber, Max, and Clemens Freiherr von Pirquet. "Toxin und Antitoxin." *Münch. med. Wschr. 50*(1903):1193–1263.

Grünbaum, Albert S. F. "Blood and the identification of bacterial species." *Science Progress* n.s., *1*(5)(1897).

Grünbaum, Albert S. F. "The agglutination of red corpuscles: report of a meeting of the Liverpool Medical Institution, 19 April 1900." *Brit. med. J.* i(1900):1089.

Haberman, Sol. See also Muirhead, E. E., Haley, A. E., Haberman, S. and Hill, J. M., "Anti-d." (1948) (Bibl. sec. 3).

Haberman, Sol, Joseph M. Hill, B. W. Everist, and J. W. Davenport. "The demonstration and characterization of the anti-d agglutinin and antigen predicted by Fisher and Race." *Blood 3*(1948):682–695.

Haeckel, Ernst. *Die Welträthsel: gemeinverständliche Studien über monistische Philosophie, mit einem Nachworte: das Glaubensbekenntniss der reinen Vernunft.* Bonn: Strauss, 1903.

Haecker, Valentin. *Wandtafeln zur allgemeine Biologie.* Leipzig: Quelle und Meyer, 1907.

Haecker, Valentin. "Ueber Axolotlkreuzung." *Verhandlungen der Deutsch. zool. Ges.* 18 Jahresversammlung, Stuttgart 9–11 June 1908, 194–205.

Haecker, Valentin. *Allgemeine Vererbungslehre.* Braunschweig: Vieweg, 1911.

Hantzsch, Arthur:

Costa, Albert B. "Hantzsch, Arthur Rudolf." In Charles C. Gillispie, ed., *Dictionary of Scientific Biography* (1972), v. 6, 102–109 (Bibl. sec. 1).

Hein, Friedrich. "A. Hantzsch." *Z. f. Elekrochem. 42*(1936):1–48.

Partington, James R. *History of Chemistry* (1964), v. 4, 842–847 (Bibl. sec. 2).

Ramsay, O. Bertrand. *Stereochemistry* (1981) (Bibl. sec. 2).

Hantzsch, Arthur. *Grundriss der Stereochemie.* Breslau: Trewendt, 1st ed., 1893; Leipzig: Barth, 2d ed., 1904.

Hantzsch, Arthur. "Ueber Stereoisomerie bei Diazoverbindungen und die Natur der 'iso-diazokörper,' " *Berichte der Deutsch. chem. Ges.* 27(1894):1702–1731.

Hantzsch, Arthur. "Die Diazoverbindungen." In F. B. Ahrens, ed., *Sammlung chemischer und chemisch-technischer Vorträge.* Stuttgart: Enke, 1903, 1–82.

Hardy, William B. "Mendelian proportions in a mixed population." *Science* n.s. 28(1908):49–50.

Heidelberger, Michael and Oswald T. Avery. "The soluble specific substance of *Pneumonococcus.*" *J. Exper. Med.* 40(1923):301–306.

Henle, Jacob. *Pathologische Untersuchungen.* Berlin: Hirschwald, 1840.

Hertz, Heinrich Rudolph. *Die Principien der Mechanik in neuen Zusammenhange,* dargestellt mit einem Vorwort von Hermann von Helmholtz. In series "Gesammelte Werke H. R. Hertz." V. 3. Leipzig: Barth, 1894. Translated by Daniel E. Jones and John Thomas Wally as *Principles of Mechanics Presented in a New Form;* preface by H. v. Helmholtz (1899); introduction by Robert S. Cohen. New York: Dover, 1956.

Hertz, Paul and Moritz Schlick, eds. and comm. *Hermann von Helmholtz Schriften zur Erkenntnistheorie.* Berlin: Springer, 1921.

Hertzig, J., and Karl Landsteiner. "Ueber die Methylierung von Eiweisstoffen." *Biochem. Z.* 61(1914):458–463, C. 148.

Hill, Joseph M. "Editorial: the complexities of the Rh problem, and some suggestions for clarification." *Am. J. Clin. Path.* 17(1946):494–501.

Hill, Joseph M., and William E. Dameshek, eds. "The Rh factor in the clinic and the laboratory." A selection of papers presented at the International Hematology and Rh Conference, Dallas, Texas and Mexico City, November 1946. In *Blood* (1948), Special Issue no. 2.

Hill, Joseph M., Sol Haberman, and F. Jones. "Hemolytic Rh globulins: evidence for a third order of antibodies incapable of agglutination or blocking." Paper presented at the International Hematology and Rh Conference, Dallas Texas and Mexico City, November 1946. In Joseph M. Hill and William Dameshek, "The Rh factor in the clinic and the laboratory." In *Blood* (1948), Special Issue no. 2, 80–100.

Hill, Joseph M. See also Haberman, S., Hill, J. M., Everist, B. W. and Davenport, J. W., "Ant–d agglutinin." (1948); Muirhead, E. E, Haley, A. E., Haberman, S. and Hill, J. M., "Renal insufficiency due to incompatible transfusion" (1948) (Bibl. sec. 3).

Hirschfeld, Ludwig. See Hirszfeld, Ludwik (Bibl. sec. 3).

Hirszfeld, Ludwik (also spelled Hirschfeld, Ludwig).

Hirszfeld, Ludwik. *Historia Jednego Zycia.* Edited by Hanna Hirszfeld. Warsaw: Instytut Wydawniczy Pax, 1st ed. 1957, 2d ed., 1967.

Hirszfeld, Ludwik [Hirschfeld, Ludwig]. "Untersuchung über die Hämagglutination und ihre physikalische Grundlagen." *Arch. f. Hyg.* 63(1907):237–286. And as "Untersuchung über die Hämagglutination und ihre physikalischen Grundlagen." Inaugural Dissertation, Berlin, 1907. Munich, 1908.

Hirszfeld, Ludwik [Hirschfeld]. See also v. Dungern, E. and Hirschfeld, L., "Nachweis und Vererbung biochemischer Strukturen" (1910); "Eine Methode das Blut serologisch zu unterschieden" (1910); "Vererbung gruppenspezifischer Strukturen, II" (1910); "Vererbung gruppen-spezifischer Strukturen, III" (1911) (Bibl. sec. 3).

Hirszfeld, Ludwik. *Konstitutionsserologie und Blutgruppenforschung.* Berlin: Springer, 1928. Translated by F. C. Farnham Co., Philadelphia, Pa., as *Constitutional Serology and Blood Group Research.* In series "Selected Contributions." F. R. Camp and F. R. Ellis, eds. (Bibl. sec. 1).

Hirszfeld, Ludwik. "Hauptprobleme der Blutgruppenforschung in den Jahren 1927–1933." *Ergebnisse der. Hygiene Bakteriologie u. Immunitätsf. 50*(1934):54–218.

Hirszfeld, Ludwik. Translated by Hanna Hirszfeldowa. *Les Groupes Sanguines leur Application à la Biologie, à la Medecine et au Droit.* Paris: Masson, 1938.

Hirszfeld, Hanna and Ludwik. "Serological differences between the blood of different races: the result of researches on the Macedonian front." *Lancet 180*(1919):675–679.

Hirszfeld, Hanna and Ludwik. "Essai d'application des méthodes serologiques au problème des races." *l'Anthropologie 29*(1919–1920):505–537.

Hirszfeldowa, Hanna, Andrzey Kelus, and Fellix Milgrom. *Ludwik Hirszfeld.* Warsaw: Prace Wroclawskiego Towarzstwa Naukowego, 1956. Includes bibliography, some articles in French.

van't Hoff, Jacobus Henricus:
Cohen, Ernst Julius. *Jacobus Henricus van't Hoff sein Leben und Wirken.* Leipzig: Akadem. Verlagsgesellschaft, 1912.

Cohen, Ernst Julius. "van't Hoff (1852–1911)." In Günther Bugge, ed., *Das Buch der grossen Chemiker* (1930), v. 2, 391–407 (Bibl. sec. 1).

Farber, Eduard. "Jacobus Henricus van't Hoff." In his *Nobel Prize Winners* (1953), 3–6 (Bibl. sec. 1).

Ramsay, O. Bertrand. "Molecular models in the early development of stereochemistry, I. The van't Hoff model; II. The Kekulé models and the Baeyer strain theory." In O. B. Ramsay, O. B., *van't Hoff-LeBel Centenniel: Symposium arranged by the Division of the History of Chemistry of the American Chemical Society,* 11–12 September 1974. Washington, D.C.: American Chemical Society, 1975, 74–96.

Snelders, H. A. M. "J. H. van't Hoff's research school in Amsterdam, 1877–1895." *Janus 71*(1984):1–30.

van't Hoff, Jacobus H. "L'équilibre chimique dans les systèmes gazaux on disson à l'état dilué." *Arch. neelandaises des Sciences Exactes et naturelles 20*(1866):239–302.

van't Hoff, Jacobus H. "Ueber feste Lösungen und Moleculargewichtsbestimmung an festen Körpern." *Z. f. phys. Chem. 5*(1890) 322–339.

van't Hoff, Jacobus H. *La Chimie dans l'Espace.* Rotterdam: Bazendijk, 1875. Translated by F. Herrmann as *Die Lagerung der Atome im Raume, mit einem Vorwart von Dr Johannes Wislicenus.* Braunschweig: Vieweg, 1877, 2d ed., 1894.

von Hoffmann, August Wilhelm:
Lepsius, B. "August Wilhelm von Hoffmann (1818–1892)." In Bugge, Günther, *Das Buch der grossen Chemiker* (1930), v. 2, 136–153 (Bibl. sec. 2).

von Hoffmann, August Wilhelm. Report of session of 28 Jan. 1890. *Ber. d. Deutsch. chem. Ges. 23*(1890):97–99.

Hooker, Sanford B., and Lillian M. Anderson. "The specific antigenic groups of the four groups of human erythrocytes." *J. Immunol. 6*(1921):419–444.

Hurst, C. C. "On the inheritance of eye colour." *Proc. Roy. Soc.* (London) B *80*(1908):85–96.

Huxley, Thomas Henry:
Foster, Sir Michael. "Thomas Henry Huxley." *Obituary Notices of Fellows of the Royal Society* (London) *59*(1895–96):xlvi–lxvi.
Foster, Sir Michael, and E. Ray Lankester, eds., *Scientific Memoirs of T. H. Huxley.* 4 v. London: Macmillan, 1898–1902.
Huxley, Leonard. *Life and Letters of Thomas Henry Huxley.* 2 v. New York: Appleton, 1900.
Mitchell, P. Chalmers. *Thomas Henry Huxley a Sketch of his Life and Work.* London: Putnam's, 1900. Contains bibliography, xv–xvii.
Rehbock, P. Fritz. "Huxley, Haeckel and the oceanographers: the tragicomedy of *Bathybius haeckelii.*" *Isis* *66*(1975):503–533.
Huxley, Thomas H. "On some organisms living at great depth in the north Atlantic Ocean." *Q. J. Microsc. Sci.* *8*(1868):203–212.
Huxley, Thomas H. "On the relations of *Penicillium, Torula,* and *Bacterium*: special report of an address to the Biological Section of the British Association for the Advancement of Science, September 13 1870." *Q. J. Microsc. Sci.* *10*(1870):355–362.
Huxley, Thomas H. "Biogenesis and abiogenesis." Presidential address to the British Association for the Advancement of Science for 1870. In Huxley, *Discourses Biological and Geological Essays.* New York: Appleton, 1877, 229–271.
Johannsen, Wihelm. *Elemente der exakten Erblichkeitslehre mit Grundzugen der biologischen Variationsstatistik.* Jena: Fischer, 1st Ger. ed. 1909, 2d ed. 1913.
Kant, Immanuel. *Kritik der reinen Vernunft.* Riga: Hartknoch, 1st ed. 1781, 2d ed. 1787. Translated by Norman Kemp Smith 1929. Toronto: Macmillan, 1965.
Kant, Immanuel. *Metaphysische Anfangsgründe der Naturwissenschaft.* Leipzig: Neueste Auflag, 1794.
von Kekulé, August:
Ramsay, O. Bertrand. "Molecular models in the early development of stereochemistry. I. The van't Hoff model. II. The Kekulé models and the Baeyer strain theory." In O. B. Russell, ed., *Van't Hoff-LeBel Centenniel: Symposium arranged by the Division of the History of Chemistry of the American Chemical Society,* 11–12 September 1974. Washington, D.C.: Americal Chemical Society, 1975, 74–96.
Schultz, G. "Berichte über den Feier der Deutschen chemischen Gesellschaft zu Ehren August Kekulés." *Berichte der Deutsch. chem. Ges.* *23*(1890):1265–1312.
Keynes, Geoffrey Langdon. *Blood Transfusion.* London: Frowde, 1922.
Kiliani, Heinrich. "Ueber das Cyanhydrin der Laevulose I Mittheilung." *Berichte der Deutsch. chem. Ges.* *18*(1885):3066–3072.
Kiliani, Heinrich. "Ueber die Constitution der Dextrose-carbonsäure." *Ber. d. Deutsche. chem. Ges.* *19*(1886):1128–1130.
Kiliani, Heinrich. "Ueber die Einwirkung von Blausäure auf Dextrose." *Ber. d. Deutsch. chem. Ges.* *19*(1886):767–772.
Kiliani, Heinrich. "Ueber das Cyanhydrin der Laevulose II Mittheilung." *Ber. d. Deutsch. chem. Ges.* *19*(1886):221–227.
Klebs, Edwin. "Ueber die Umgestaltung der Medizinischen Anschauungen in den letzten drei Jahrzehnten." *Amtlicher Ber. der 50. Versamml. Deutscher Naturf. u. Aerzte* Munich (September 1877):41–45.

Koch, Robert:
Biewend, R. "Aus der Familienchronik von Robert Koch." *Deutsche Revue* 15(1891):179–186.
Brock, Thomas D. *Robert Koch: a Life in Medicine and Bacteriology.* Madison, Wisc.: Science Tech, 1988.
Carter, F. Codell. "Koch's postulates in relation to the work of Henle and Klebs." *Med. Hist.* 29(1985):353–374.
Cohn, Ferdinand. "Ein Brief über Koch." *Deutsche Revue* 15(1891):30–31.
Coleman, William B. "Koch's comma bacillus: the first year." *Bull. Hist. Med.* 61(1987):315–342.
Ehrlich, Paul. "Robert Koch +." *Z. f. Immunitätsf.* 6 (unpaginated) (1910).
Evans, Richard J. *Death in Hamburg: Society and Politics in the Cholera Years, 1830–1910.* Oxford: Clarendon Press, 1987.
Festschrift [special number for 60th birthday of Robert Koch]. *Deutsche med. Wschr.* 29(50) (1903).
Festschrift zum sechsigsten Geburtstage von Robert Koch herausgeben von seinen dankbaren Schülern. Jena: Fischer, 1903.
Friedemann, Ulrich. "Die Bedeutung der Lehre Robert Kochs für die Klinik der Infektionskrankheiten." *Deutsche med. Wschr.* 58(1932):495–497, 50th anniversary of Koch's discovery of the tubercle bacillus, special number.
Gaffky, Georg, W. Pfuhl, and J. Schwalbe, eds. *Gesammelte Werke von Robert Koch.* 2 v. in 3. Leipzig: Thieme, 1912.
Goldscheider, A. "Der Einfluss der Lehre Robert Kochs auf den klinischen Unterricht." *Deutsche med. Wschr.* 581932:476–478, 50th anniversary of Koch's discovery of the tubercle bacillus, special number.
Hamel, C. "Das Lebenswerk von Robert Koch in seinen Auswirkung auf die öffentliche Gesundheitspflege." *Deutsche med. Wschr.* 58(1932):487–488, 50th anniversary of Koch's discovery of the tubercle bacillus, special number.
Heymann, Bruno. *Robert Koch I Teil 1843–1882.* Leipzig: Akademische Verlagsgesellschaft, 1932. V. 12 in Series, "Grosse Männer: Studien zur Biologie des Genies," Wilhelm Ostwald ed. Part 2 never published.
King, Lester S. "Dr Koch's postulates." *J. Hist. Med.* 7(1952):350–361.
Kirchner, M. *Robert Koch.* Berlin: Springer, 1924, v. 5. In series "Meister der Heilkunde," Max Neuburger ed.
Kolle, Kurt. "Die Bedeutung von Robert Koch für die experimentelle Therapie und Prophylaxie der Infektionskrankheiten." *Deutsche med. Wschr.* 58(1932):493–495, 50th anniversary of Koch's discovery of the tubercle bacillus, special number.
Kolle, Kurt, ed. *Robert Koch: Briefe an Wilhelm Kolle.* Stuttgart: Thieme, 1959.
Kühne, Willy. "Zur Erinnerung an Julius Cohnheim." In Wagner, E. ed., *Gesammelte Abhandlung von Julius Cohnheim.* Berlin: Hirschwald, 1885.
Loeffler, Friedrich. "Robert Koch zum 60 Geburtstage." *Deutsche med. Wschr.* 29(1903):937–943 (Festschrift Number).
Lubarsch, Otto. "Der Einfluss Kochs auf die pathologische Morphologie und allgemeine Pathologie." *Deutsche med. Wschr.* 58(1932):478–481, 50th anniversary of Koch's discovery of the tubercle bacillus, special number.
Möllers, Bernhard. *Robert Koch Personlichkeit und Lebenswerk 1843–1910.* Hannover: Schmore, 1950.
Pfeiffer, Richard. "Der Einfluss Robert Kochs auf die Immunitätslehre." *Deut-*

sche med. Wschr. *58*(1932):490–493, 50th anniversary of Koch's discovery of the tubercle bacillus, special number.

Pfuhl, E. "Privatbriefe von Robert Koch: Vortrag, gehalten April 4 1911, in der Gesellschaft für Natur-und Heilkunde in Berlin." *Deutsche med. Wschr.* *37*(1911):1399–1400, 1443–1444, 1483–1485, 1524–1526.

Das Reichsgesundheitsamt 1876–1929: Festschrift herausgeben vom Reichsgesundheitsamte aus Anlass seines fünfzigjährigen Bestehens. Berlin: Springer, 1926.

Schopohl. "Robert Koch und die preussische Medizinalverwaltung." *Deutsche med. Wschr.* *58*(1932):489–490, 50th anniversary of Koch's discovery of the tubercle bacillus, special number.

Schwalbe, J. "Feuilleton: Brief aus vergangenen Tagen" [2 letters from Robert Koch, 3 from Emil von Behring, 2 from Paul Ehrlich]. *Deutsche med. Wschr.* *55*(1929):1772–1773.

Unger, H. *Robert Koch Roman eines grossen Lebens.* Berlin: Deutsche Buch-Gemeinschaft, 1961.

Koch, Robert. *Untersuchungen über die Atiologie der Wund infektionskrankheiten.* Leipzig: Vogel, 1875). Also Gaffky, G., et al., *Gesammelte Werke* v. 1, 61–108.

Koch, Robert. "Verfahren zur Untersuchung, zum Konservieren und Photographieren der Bakterien." *Beitr. z. Biol. d. Pflanzen* *2*(1877):399–434. Also Gaffky, G., et al., *Gesammelte Werke,* v. 1, 27–60.

Koch, Robert. [Review of] von Nägeli, Carl, "Die niederen Pilze in ihren Beziehungen zu den Infectionskrankheiten. München 1877) und Dr Hans Buchner, 'Die Naegeli'sche Theorie der Infectionskrankheiten in ihren Beziehungen zur medicinischen Erfahrung,' (Leipzig: 1871) Besprochen von Kreisphysicus Dr Koch in Wallstein, *[sic.]* " *Deutsche med. Wschr.* *4*(1878):7–8.

Koch, Robert. "Zur Untersuchung von pathogenen Organismen." *Mittheilung aus dem kaiserl. Gesundheitsamte 1* Art 1 (1881). Also Gaffky, G., et al., *Gesammelte Werke,* v. 1, 112–163.

Koch, Robert. "Ueber die Ätiologie der Tuberkulose." In *Verhandlungen der Kongress f. inneren Med..* Wiesbaden: Bergmann, 1882, 56–68, with discussion following 68–78. Also Gaffky, G., et al., *Gesammelte Werke,* v. 1, 446–453.

Koch, Robert, and Carl Flügge. "Zur Einführung." *Z. f. Hyg. u. Infektionskr.* *1*(1885):1–2.

Koch, Robert. "Antrittsrede in der Akademie der Wissenschaften am 1 Juli 1901." Gaffky, G., et al., *Gesammelte Werke,* v. 1, 1–4.

Koestler, Arthur. *The Case of the Midwife Toad.* London: Hutchinson, 1971.

Kolle, Wilhelm:

Fischer, Isidor. "Kolle, Wilhelm." In *Biographisches Lexikon* (1933), v. 1, 798 (Bibl. sec. 1).

Laubenheimer, K. "Wilhelm Kolle zum Gedächtnis." *Münch. med. Wschr.* *82*(1935):919–920.

Kolle, Wilhelm. See also under Koch, R. (Bibl. sec. 3), Kolle, W., "Bedeutung von Robert Koch." (1932); Kolle, K., *Kochs Briefe an Kolle* (1959).

Kolle, Wilhelm, and August von Wassermann. *Handbuch der Pathogenen Mikroorganismen unter Mitwirkung von R. Abel.* 4 v. in 5, 2 supplements. Jena: Fischer, 1902–1904; 2d ed., 1912–1913 (8 v.). *Handbuch der pathogenen Mikroorganismen, begründet von Wilhelm Kolle und August von Wassermann, 3 erweiterte Auflage mit Einschluss der Immunitätslehre und Epidemiologie sowie der mikrobiologischen Diagnostik und Technik von Fachgelehrten neu bearbeitet und herausgeben von Wilhelm Kolle, Rudolf Kraus, und Paul Uhlenhuth.* Jena: Fischer, 3d ed., 1928–1931.

von Korányi, Sandor (Alexander) Baron, and Paul Friedrich Richter, eds., *Physikalische Chemie und Medizin: ein Handbuch.* 2 v. in 1. Leipzig: Thieme, 1907.

Kraus, Rudolf:
  Eisler, M. "Rudolf Kraus." *Wiener klin. Wschr. 45*(1932):1072–1073.
  Fischer, Isidor. "Kraus, Rudolf." In *Biographisches Lexikon* (1933), v. 1, 816 (Bibl. sec. 1).
  Lowenstein, E. "Professor Dr Rudolf Kraus." *Wiener med. Wschr. 82*(1932): 1016–1017.

Kraus, Rudolf. "Ueber specifische Reactionen in Keimfreien Filtraten aus Cholera, Typhus und Pestbouillon, erzeugt durch homologes serum." *Wiener klin. Wschr. 10*(1897):736–738.

Kraus, Rudolf, ed., "Die Wiener mikrobiologische Forschung und ihre Ergebnisse." Joint meeting of the Wiener mikrobiologische Gesellschaft and the Deutsche Vereinigung für Mikrobiologie, 7–9 June 1927. Special Number of *Wiener med. Wschr. 77*(23)(1927). See under Rudolf Kraus (1927); Max von Gruber (1927/1914); Karl Landsteiner (1927) C. 226; Clemens von Pirquet (1927); Richard Volk (1927) (Bibl. sec. 3).

Kraus, Rudolf. "30 Jahre Präzipitinlehre." *Wiener med. Wschr. 77*(1927):743–744; "Wiener mikrobiologische Forschung." Special Number, Rudolf Kraus ed.

Kraus, Rudolf. See also Kolle, W., Kraus, R., and Uhlenhuth, P., *Handbuch der pathogenen Organismen* (1928–31) (Bibl. sec. 3).

Kraus, Rudolf, and J. M. de la Barrera. "Studien über Flecktyphus in Sdamerika." *Z. f. Immunitätsf. 34*(1922):1–35.

Kraus, Rudolf, and P. Clairmont. "Ueber Hämolysine und Antihämolysine." *Wiener klin. Wschr. 13*(1900):49–56.

Kraus, Rudolf, and Constantin Levaditi. *Handbuch der Technik und Methodik der Immunitätsforschung.* Jena: Fischer, 1908.

Kraus, Rudolf, and Constantin Levaditi. *Handbuch der Immunitätsforschung und experimentelle Therapie, mit besondere Bercksichtigung der Technik und Methodik Bearbeitet,* von Emil Abderhalden, B. Aschner u.a., herausgeben von Prof. Dr Rudolf Kraus und Dr Constantin Levaditi, neu bearbeitete und erweiterte 2 Auflag des *Handbuchs der Technik und Methodik der Immunitätsforschung.* Jena: Fischer, 1914.

Kraus, Rudolf, and L. Löw. "Ueber Fadenbildung." *Wiener klin. Wschr. 12*(1899):761–764.

Kraus, Rudolf, and L. Löw. "Ueber Agglutination." *Wiener klin. Wschr. 12*(1899): 95–98.

Kraus, Rudolf, and Paul Uhlenhuth. *Handbuch der mikrobiologischen Technik, unter Mitarbeit hervorragender Fachgelehrten.* Berlin: Urban, 1923–1924.

Kühne, Willy, and Carl von Voit. "An unsere Leser." *Z. f. Biol. 1*(or 19)(1883): 1–4.

de Lamarck, Jean-Baptiste A. *Philosophie Zoölogique, ou Exposition des Considérations Relatives à l'Histoire Naturelle des Animaux; la Diversité de leur Organization et des Facultés qu'ils en obtiennent; aux Causes Physiques qui maintiennent en eux la Vie, et donnent lieu aux Mouvements qu'ils exécutent, enfin, celles qui Produisent, à les unes le Sentiment et les autres l'Intelligence de ceux qui en sont doués.* Paris:

Dentu, 1809. Translated and introduced by H. Elliott as *Zoological Philosophy, an Exposition with Regard to the Natural History of Animals, the Diversity of their Organization and the Faculties Which They Derive from It; The Physical Causes Which Maintain Life within Them and Give Rise to Their Various Movements; Lastly, Those Which Produce Feeling and Intelligence in Some among Them.* New York: Hafner, 1963.

Landsteiner, Karl:

Anonymous [Chase, Merrill W.]. "Karl Landsteiner 1868–1943." *J. Immunol.* *48*(1944):1–16. Includes bibliography pp. 5–16; (C. + number) following Landsteiner papers refer to the numbering in Chase's bibliography.

Arzt, L. "Karl Landsteiner, geb. 14 Juni 1868 gest. 26 Juni 1943." *Wiener klin. Wschr.* *60*(1948):557–558.

Groedel, F. "Session in memory of the late honorary member Karl Landsteiner." *Proc. R. Virchow Med. Soc.* *3*(1945):98–99.

von Gruber, Max [Letter to Prof. Dr Hermann von Schlesinger]. *Wiener med. Wschr.* *81*(1931):309.

Holzer, Franz Joseph. "Karl Landsteiner." *Wiener med. Wschr.* *98*(1948):378–379.

Holzer, Franz Joseph. "Persönliche Erinnerungen an Karl Landsteiner." *Beitr. z. gerichtl. Med.* *26*(1969):143–147.

Keating, Peter. "The problem of the natural antibodies." *J. Hist. Biol.* *24*(1991):245–263.

Landsteiner, Karl. Manuscripts and laboratory notes, Rockefeller Archives, Tarrytown, Record Group 450, Series L239; Reports of Director of Laboratories, 1922–1943: Record Group 439.

Levine, Philip. "Dr Karl Landsteiner – an appreciation." *Proc. R. Virchow Med. Soc.* *3*(1945):99–102.

Levine, Philip. "Landsteiner's concept of the individuality of human blood." *Proc. R. Virchow Med. Soc.* *3*(1945):102–110.

Mackenzie, George. Incomplete biographical ms, with notes letters and photographs relating to Karl Landsteiner, B L23m, Boxes 1–6, American Philosophical Society's Archives, Philadelphia, Pa.

Mazumdar, Pauline M. H. "The purpose of immunity: Landsteiner's explanation of the human isoantibodies." *J. Hist. Biol.* *8*(1975):115–134.

Mazumdar, Pauline M. H. "Karl Landsteiner in America." *Proc. Congr. int. XXIV Hist. Artis Medicinae* 25–31 August 1974, Budapest. Budapest: Museum, Bibliotheka et Archivum Historiae Artis Medicinae de Ph. Semmelweis Nominata, 1976.

Prokop, Otto. "Karl Landsteiner zum Gedächtnis." *Beitr. z. gerichtl. Med.* *26*(1969):138–142.

Rous, Francis Peyton. "Karl Landsteiner." *Obituary Notices of Fellows of Roy. Soc.* (London) *5*(1947):295–312. Includes bibliography by Merrill W. Chase; (C. + number) following Landsteiner papers refer to the numbering in Chase's bibliography.

Speiser, Paul and Ferdinand G. Smekal. *Karl Landsteiner Entdecker der Blutgruppen und Pioneer der Immunologie: Biographie eines Nobelpreisträgers aus der Wiener Medizinischen Schule.* Vienna: Hollinek, 1st ed., 1961, 2d ed. 1975. Translated by Richard Rickett as *Karl Landsteiner: The Discoverer of the Blood Groups and a Pioneer in the Field of Immunology: Biography of a Nobel Prize Winner of the Vienna Medical School.* Vienna: Hollinek, 1975. Contains biographical material on Landsteiner's colleagues and co-workers of both his Vienna and his New York periods, and includes the complete bibliography by Merrill W. Chase; (C. + number) following Landsteiner papers refer to the numbering in Chase's bibliography.

Speiser, Paul. "Karl Landsteiner." In Charles C. Gillispie, ed., *Dictionary of Scientific Biography* (1973), v. 7, 622–625 (Bibl. sec. 1).

Speiser, Paul. "Zum 100 Geburtstag Karl Lansteiners." *Wiener klin. Wschr.* 80(1968):37–40. Includes list of secondary articles, memoires, etc. on Landsteiner.

Wiener, Alexander S. "Karl Landsteiner MD – History of the Rh-Hr Blood Group System." *New York J. Med.* 69 (22)(1969):2915–2935.

Wiener, Alexander S. "Karl Landsteiner: his work and the Rhesus blood factor." Passano Foundation Award Address, 1951. In his *Rh-Hr* (1954) 1–8.

Landsteiner, Karl. See also Fischer, E., and Landsteiner, K., "Glycolaldehyde" (1892) (C. 2) (Bibl. sec. 3).

Landsteiner, Karl. See also von Bamberger, E., and Landsteiner, K., "Verhalten des Diazobenzol gegen Kaliumpermanganat" (1893) (C. 3) (Bibl. sec. 3).

Lansteiner, Karl. See also Scholl, R., and Landsteiner, K., "Reduction der Pseudonitrol" (1896) (C. 7) (Bibl. sec. 3).

Landsteiner, Karl. "Ueber die Folgen der Einverleibung sterilisirter Bakterienculturen." *Wiener klin. Wschr.* 10(1897):439–444, . 8.

Landsteiner, Karl. "Ueber die Wirkung des Choleraserums ausserhalb des Thierkörpers." *Zbl. f. Bakt.* (Orig.) 23 (1898):847–852, C. 11.

Landsteiner, Karl. "Zur Kenntnis der spezifisch auf Blutkörperchen wirkenden Sera." *Zbl. f. Bakt.* 25(1899):546–549, C. 12.

Landsteiner, Karl. "Zur Kenntniss der antifermentativen, lytischen und agglutinierenden Wirkung des Blutserums und der Lymphe." *Zbl. f. Bakt.* (Orig.) 27(1900):357–362, C. 13.

Landsteiner, Karl. "Ueber Agglutinationserscheinungen normalen menschlichen Blutes." *Wiener klin. Wschr.* 14(1901):1132–1134, C. 17.

Landsteiner, Karl. "Beobachtungen über Hämagglutination." *Wiener klin. Rundschau* 16(1902):774, C. 25.

Landsteiner, Karl. "Ueber Serumagglutinine." *Münch. med. Wschr.* 49(1902): 1905–1908, C. 26.

Landsteiner, Karl. "Bemerkung zur Mittheilung von Jean Billitzer, 'Theorie der Kolloide, II.'." *Z. f. phys. Chemie 51*(1905):741–742, C. 45.

Landsteiner, Karl. "Bemerkung zu der Mitteilung von U. Friedemann und H. Friedenthal, 'Beziehungen der Kernstoffe zu den Immunkörpern'." *Zbl. f. Physiol.* 20(1907):657–658, C. 67.

Landsteiner, Karl. "Zu der Erwiderung von Friedemann und Friedenthal." *Zbl. f. Physiol.* 20(1907):806, C. 69.

Landsteiner, Karl. "Die Theorien der Antikörperbildung, nach einem Referat für den XVI Internationalen medizinischen Kongress in Budapest." *Wiener klin. Wschr.* 22(1909):1623–1631. Also *Ergebn. wiss. Med.* 1(1909–1910):185–207, C. 102.

Landsteiner, Karl. "Bemerkungen zu der Abhandlung von Traube: 'Die Resonanztheorie, eine physikalische Theorie der Immunitätserscheinungen.'" *Z. Immunitätsf.* (Orig.) 9(1911):779–786, C. 122.

Landsteiner, Karl. "Over de serologische Specifiteit van het haëmoglobine van verschillende diersoorten." *K. Akad. Wetenschappen te Amsterdam, Verslag van de Gewone Vergaderingen der Wis- en Natuurkundige Afdeeling* 29(1920–1921): 1029–1034, C. 175.

Landsteiner, Karl. "Over heterogenetisch antigeen." *K. Akad. Wetenschappen te Amsterdam, Verslag van de Gewone Vergaderingen der Wis- en Natuurkundige Afdeeling* 29(1920–1921):1118–1121, C. 176.

Landsteiner, Karl. "Over het samenstellen van heterogenetisch antigeen mit hapteen en proteine." *K. Akad. Wetenschappen te Amsterdam, Verslag van de Gewone Vergaderingen der Wis- en Natuurkundige Afdeeling* 30(1921):329–330, C. 180.

Landsteiner, Karl. "Onderzoekingen over anaphylaxie door azoproteinen." *K. Akad. Welenschappen te Amsterdam, Verslag van de Gewone Vergaderingen der Wis- en Natuurkundige Afdeeling* 31(1922):54–55, C. 181.

Landsteiner, Karl. "Serologische Individualdifferenzen und die menschliche Blutgruppen." *Wiener klin. Wschr.* 77(1927):744–745. In "Wiener mikrobiologische Forschung." Special Number, Rudolf Kraus ed.

Landsteiner, Karl. "The human blood groups." In Edwin O. Jordan and Isidore S. Falk, eds., *The Newer Knowledge of Bacteriology and Immunology.* Chicago: University of Chicago Press, 1928, 829–908, C. 232.

Landsteiner, Karl. "Cell antigens and individual specificity." Presidential address to the American Association of Immunologists. *J. Immunol.* 15(1928): 589–600, C. 244.

Landsteiner, Karl. *Die Spezifizität der serologischen Reaktionen.* Berlin: Springer, 1st ed. 1933, C. 289. Translated as *The Specificity of Serological Reactions.* Springfield, Ill.: Thomas, 1st ed. 1936, C. 309. 2d ed. titled *The Specificity of Serological Reactions, revised edition, with a Chapter on Molecular Structure and Intermolecular forces by Linus Pauling.* Cambridge, Mass.: Harvard University Press, 1946, C. 346. Title is not exactly as cited in C. 346 by M. W. Chase.

Landsteiner, Karl. "Ueber die Antigeneigenschaften von methyliertem Eiweiss: VII. Mitteilung über Antigene." *Z. f. Immunitätsf.* 26(1917):122–133, C. 162.

Landsteiner, Karl. See Rothen, A., and Landsteiner, K., "Adsorption of antibodies by egg albumin films." (1939) (C. 323); Rothen, A., and Landsteiner, K., "Serological reactions of protein films and denatured proteins" (1942) (C. 341) (Bibl. sec. 3).

Landsteiner, Karl. See also under Wiener, A. S.: Wiener, Alexander S. and Landsteiner, K., "Heredity of Rh." (1943) (C. 343) (Bibl. sec. 3).

Landsteiner, Karl, J. Dee Herzig, and K. Landsteiner. "Methylierung von Eiweisstoffen" (1914) (C. 148) (Bibl. sec. 3).

Landsteiner, Karl, and Nikolaus von Jagić. "Ueber die Verbindungen und die Entstehung von Immunkörpern." *Münch. med. Wschr.* 50(1903):764–768, C. 31.

Landsteiner, Karl, and Nikolaus von Jagić. "Ueber Analogien der Wirkung kolloidaler Kieselsäure mit der Reaktionen der Immunkörper und verwandte Stoffe." *Wiener klin. Wschr.* 17(1904):63–64, C. 34.

Landsteiner, Karl, and Nikolaus von Jagić. "Ueber Reactionen anorganischer Kolloide und Immunkörperreaktionen." *Münch. med. Wschr.* 51(1904): 1185–1189, C. 39.

Landsteiner, Karl, and B. Jablons. "Ueber die Bildung von Antikörpern gegen verändertes arteigenes Serumeiweiss, V. Mitteilung über Antigene." *Z. f. Immunitätsf.* 20(1914):618–621, C. 145.

Landsteiner, Karl, and B. Jablons. "Ueber die Antigeneigenschaften acetyliertem Eiweiss, VI. Mitteilung über Antigene." *Z. f. Immunitätsf.* 21(1914): 193–201, C. 146.

Landsteiner, Karl, and Hans Lampl. "Ueber Antigene mit verschiedenartigen Acylgruppen: X. Mitteilung über Antigene." *Z. f. Immunitätsf.* *26*(1917): 256–176, C. 166.

Landsteiner, Karl, and Hans Lampl. "Ueber die Antigeneigenschaften von Azoproteinen. XI. Mitteilung über Antigene." *Z. f. Immunitätsf.* *26*(1917):293–304, C. 167.

Landsteiner, Karl, and Hans Lampl. "Ueber die Abhängigkeit der serologischen Spezifizität von der chemischen Struktur. (Darstellung von Antigenen mit bekannter chemischer Konstitution der specifischen Gruppen) XII. Mitteilung über Antigene." *Biochem. Z.* *86*(1918):343–394, C. 169.

Landsteiner, Karl, and Philip Levine. "A new agglutinable factor differentiating human bloods." *Proc. Soc. Exper. Biol. Med.* *24*(1926–1927):600–602, C. 220.

Landsteiner, Karl, and Philip Levine. "Further observations on individual differences of human blood." *Proc. Soc. Exper. Biol. Med.* *24*(1926–1927):941–942, C. 225.

Landsteiner, Karl, and Philip Levine. "On individual differences in human blood." *J. Exper. Med.* *47*(1928):757–775, C. 236.

Landsteiner, Karl, and Philip Levine. "On the inheritance of agglutinogens of human blood demonstrable by immune agglutinins." *J. Exper. Med.* *48*(1928):731–749, C. 242.

Landsteiner, Karl, and Charles Philip Miller, Jr. "Serological observations on the relationships of the bloods of man and the anthropoid apes." *Science* *61*(1925):492–493, C. 202.

Landsteiner, Karl, and Charles Philip Miller, Jr. "Serological studies on the blood of the primates, I. The differentiation of human and anthropoid bloods." *J. Exper. Med.* *42*(1925):841–852, C. 207.

Landsteiner, Karl, and Charles Philip Miller, Jr. "Serological studies on the blood of the primates II. The blood groups of anthropoid apes." *J. Exper. Med.* *42*(1925):853–862, C. 208.

Landsteiner, Karl, and Charles Philip Miller, Jr. "Serological studies on the blood of the primates, III. Distribution of serological factors related to human isoagglutinins in the blood of lower monkeys." *J. Exper. Med.* *42*(1925):863–877, C. 209.

Landsteiner, Karl, and Wolfgang Pauli. "Elektrische Wanderung der Immunstoffe." *Verhandlungen der Kongr. f. inneren Medizin* XXV Kongress (Vienna 1908), 571–575, C. 85.

Landsteiner, Karl, and Emil Prášek. "Ueber die Aufhebung der Artspezifizität von Serumeiweiss, IV. Mitteilung über Antigene." *Z. f. Immunitätsf.* *20*(1913):211–237, C. 142.

Landsteiner, Karl, and M. Reich. "Ueber die Verbindungen der Immunkörper." *Centr. Bakt.* (Orig.) *39*(1905):83–93, C. 48.

Landsteiner, Karl, and James van der Scheer. "Serological examination of a species-hybrid." *Proc. Soc. Exper. Biol. Med.* *21*(1924):252, C. 191.

Landsteiner, Karl, and James van der Scheer. "Serological examination of a species-hybrid, I. On the inheritance of species-specific qualities." *J. Immunol.* *9*(1924):213–219, C. 194.

Landsteiner, Karl, and James van der Scheer. "Serological examination of a species-hybrid, II. Tests with normal agglutinins." *J. Immunol.* *9*(1924):221–226, C. 195.

Landsteiner, Karl, and James van der Scheer. "On the specificity of agglutinins and precipitins." *J. Exper. Med.* *40*(1924):91–107, C. 196.

Landsteiner, Karl, and James van der Scheer. "On the cross-reactions of immune sera to azoproteins." *J. Exper. Med.* *63*(1936):325–399, C. 303.

Landsteiner, Karl, and James van der Scheer. "On cross-reactions of immune sera to azoproteins, II. Antigens with azocomponents containing two determinant groups." *J. Exper. Med.* *67*(1939):709–723, C. 316.

Landsteiner, Karl, and R. Stanković. "Ueber die Adsorption von Eiweisskörpern und über Agglutininverbindungen." *Zentralbl. f. Bakt.* (Orig.) *41*(1906): 108–117, C. 55.

Landsteiner, Karl, and Adriano Sturli. "Ueber Hämagglutinine normaler Sera." *Wiener klin. Wschr.* *15*(1902):38–40, C. 19.

Landsteiner, Karl, and Alexander Wiener. "On the presence of M agglutinogens in the blood of monkeys." *J. Immunol.* *33*(1937):19–25, C. 311.

Landsteiner, Karl, and Alexander S. Wiener. "An agglutinable factor in human blood recognised by immune sera for Rhesus blood." *Proc. Soc. Exper. Biol. Med.* *43*(1940):223, C. 325. Also Wiener, A. S., *Rh-Hr* (1954) 24 (Bibl. sec. 3).

Landsteiner, Karl, and Alexander S. Wiener. "Studies on an aglutinogen (Rh) in human blood reacting with anti-Rhesus sera and with human isoantibodies." *J. Exper. Med.* *74*(1941):309–320, C. 335. Also Wiener, A. S., *Rh-Hr* (1954) 42–53 (Bibl. sec. 3).

Landsteiner, Karl, and Dan H. Witt. "Observations on the human blood groups: irregular reactions; iso-agglutinins in sera of group IV; the factor A^1." *J. Immunol.* *11*(1926):221–247, C. 211.

Lattes, Leone. *L'Individualità del Sangue in Biologia in Clinica e in Medicina Legale.* Messina, 1923. Translated by Fritz Schiff as *Die Individualität des Blutes in der Biologie in der Klinik und in der gerichtlichen Medizin, nach der umgearbeiteten italienischen Auflage bersetzt und ergänzt durch einen Auhang, Die forensisch-medizinische Verwertbarkeit der Blutgruppen, von Dr Fritz Schiff.* Berlin: Springer, 1925.

Lattes, Leone:
   Introzzi, P. "La vita e le opere di Leone Lattes." In Istituto di Medicina Legale e delle Assicurazioni dell' Universitá Pavia, *Leone Lattes.* Pavia: University of Pavia, 1954.

LeBel, Joseph-Achille:
   Ramsay, O. Bertrand, ed. *Van't Hoff-LeBel Centenniel: Symposium arranged by the Division of the History of Chemistry of the American Chemical Society, September 11–12 1974.* Washington, D.C.: American Chemical Society, 1975, 74–96.

LeBel, Joseph-Achille. "Sur les relations qui existent entre les formules atomiques des corps atomiques et le pouvoir rotatoire de leurs dissolutions." *Bull. de la Soc. Chim.* (Paris) *22*(n.s.)(1874):337–347.

Levine, Philip:
   Anonymous. "Philip Levine: Dedication." *Advances in Pathobiology* 7(1980): vii–xii.

   Diamond, Louis K. "A tribute to Philip Levine." *Am. J. Clin. Path.* 74(1980): 368–370.

   Rosenfield, Richard E. "The William Allan Memorial Award presented to

Philip Levine and Alexander S. Wiener at the Annual Meeting of the American Society of Human Genetics, Baltimore Maryland, October 10 1975." *Am. J. Hum. Genet. 28*(1976):101–106. Biographical sketches of Levine and Wiener.

Levine, Philip. "Menschliche Blutgruppen und individuelle Blutdifferenzen." *Ergebn. inn. Med. 34* 111–153 (1928).

Levine, Philip. "An unusual case of intragroup agglutination." *J. Amer. Med. Assoc. 113* 126–127 (1939).

Levine, Philip. "The pathogenesis of fetal erythroblastosis." *NY State J. Med. 42*(1942):1928–1934.

Levine, Philip. "A survey of the significance of the Rh factor." Paper presented at the International hematology and Rh Conference, Dallas Texas and Mexico City, November 1946. In Joseph M. Hill and William Dameshek, eds., "The Rh factor in the clinic and the laboratory." *Blood* (1948). Special Issue no. 2, 3–26.

Levine, Philip, Lyman Burnham, Eugene M. Katzin, and P. Vogel. "The role of isoimmunization in the pathogenesis of *Erythroblastosis fetalis.*" *Am. J. Obst. Gynec. 42*(1941):925–937.

Levine, Philip, Eugene M. Katzin, and Lyman Burnham. "Isoimmunization in pregnancy, its possible bearing on the etiology of *Erythroblastosis fetalis.*" *J. Am. Med. Assoc. 116*(1941):825–827.

Levine, Philip, P. Vogel, Eugene M. Katzin, and Lyman Burnham. "Pathogenesis of *Erythroblastosis fetalis*: statistical evidence." *Science 94*(1941):371–372.

Loeb, Jacques:

Loeb, Jacques. 20 letters to Ernst Mach, 1885–1905; in collection of Ernst Mach Institut der Fraunhofer-Gesellschaft, Freiburg-im-Breisgau.

Osterhout, Winthrop John Vanleuven. "Jacques Loeb." *J. Gen. Physiol.* (Loeb Memorial Volume) *8*(1928):ix–lix.

Pauly, Philip J. *Controlling Life: Jacques Loeb and the Engineering Ideal in Biology.* Berkeley: University of California Press, 1987.

Robertson, T. Brailsford, "The life and work of a mechanistic philosopher: Jacques Loeb." *Science Progress in XX Century 21*(1926):114–129.

Rothberg, M. *The Physiologist Jacques Loeb (1859–1924) and His Research Activities.* Inaugural Dissertation, Zurich. Winterthur: Schellenbeg, 1965.

Loeb, Jacques. *Der Heliotropismus der Thiere und seine Uebereinstimmung mit dem Heliotropismus der Pflanzen.* Würzburg: Herz, 1890.

Loeb, Jacques. "Zur Theorie der physiologischen Licht- und Schwerkraftwirkungen." *Arch. f. die ges. Physiol. 66*(1897):439–459.

Loeb, Jacques. *The Mechanistic Conception of Life: Biological Essays.* Chicago: University of Chicago Press, 1912. Reprinted with introduction by Donald Fleming. Cambridge, Mass.: Harvard University Press, 1964.

Loeb, Jacques. *The Organism as a Whole from a Physico-chemical Point of View.* New York: Putnam, 1916.

Loeb, Jacques. "Is species-specificity a Mendelian charater?" *Science 45*(1917): 191–193.

Loeb, Jacques. *Proteins and the Theory of Colloidal Behavior.* New York: McGraw-Hill, 1922.

Loeffler, Friedrich:

"Festschrift gewidmet zum 60 Geburtstage des Herrn Geheimrats Dr Friedrich Loeffler von Schülern und Mitarbeitern, sowie von der Redaktion und dem Verlag des Zentralblatts für Bakteriologie." *Zbl. f. Bakt. 64* (Orig.) 1912. Papers reflect his interests but are not about him.

Fischer, Isidor. "Loeffler, Friedrich." *Biographisches Lexikon* (1933), v. 1, 929 (Bibl. sec. 1)

Möllers, Bernhard. *Robert Koch* (1950) 370–372 (see Koch, Bibl. sec. 3).

Loeffler, Friedrich, *Vorlesungen über die geschichtliche Entwicklung der Lehre von den Bacterien: Erster Theil: bis zum Jahre 1878.* Leipzig: Vogel, 1887. No further parts appeared.

Löffler, Friedrich (Loeffler). "Zur Immunitätsfrage." *Mitth. aus dem Kaiserl. Gesundheitsamte 1* (Art. 2)(1881):1–54.

Loos, Adolf:

Banham, Reyner. *Theory and Design in the First Machine Age.* London: Architectural Press, 1960, 88–97.

Münz, Ludwig, and Gustav Knstler. *Adolf Loos Pioneer of Modern Architecture* (1964), with an introduction by Nikolaus Pevsner and an appreciation by Oskar Kokoschka. New York: Praeger, 1966. Contains chronological catalogue of projects, and translation of Loos's essay "Ornament und Verbrechen."

Schmalenbach, Fritz. "Der Name, 'neue Sachlichkeit,' " in his *Kunshistorische Studien.* Basel: Schudel, 1941, 22–32.

Schmalenbach, Fritz, *Die Malerei der "neuen Sachlichkeit."* Berlin: Mann, 1973.

Loos, Adolf. *Adolf Loos sämtliche Schriften in zwei Bänden*, Franz Glück, ed. Vienna: 1962.

McCall, Archibald J., Robert R. Race, and George L. Taylor. "Rhesus antibody in Rh positive mother." *Lancet 1*(1944):214–215.

Mach, Ernst:

Bahr, Hermann. "Mach." In his *Bilderbuch.* Vienna: Wila, 1921, 35–41.

Blackmore, John T. *Ernst Mach His Life Work and Influence.* Berkeley: University of California Press, 1972. Contains bibliography, including material on as well as by Mach to date.

Capek, M. "Ernst Mach's biological theory of knowledge." *Synthèse 18*(1968): 171–191.

Frank, Philipp. *Between Physics and Philosophy.* Cambridge, Mass.: Harvard University Press, 1941.

Haller, Rudolf, and Friedrich Stadler. *Ernst Mach – Werk und Wirkung.* Vienna: Hälder-Pichler-Tempsky, 1988.

Heller, Karl Daniel. *Ernst Mach Wegbereiter der modernen Physik, mit ausgewählten Kapiteln aus seinem Werk.* Vienna: Springer, 1964.

Henning, Hans. *Ernst Mach als Philosoph Physiker und Psycholog.* Leipzig: Barth, 1915.

Janik, Alan, and Steven Toulmin. *Wittgenstein's Vienna.* New York: Simon, 1973.

Kreidl, Alois. "Ernst Mach." *Wiener klin. Wschr. 29*(1916):394–396.

Lesky, Erna. "Kompensationslehre und denkökonomisches Prinzip." *Gesnerus 23*(1966):97–108.

[Letters to Ernst Mach]. Archives of the Ernst-Mach-Instituts der Fraunhofer Gesellschaft, Freiburg-im-Breisgau.

Musil, Robert. *Beitrag zur Beurteilung der Lehren Machs: Inaugural-Dissertation der Doktorwrde genehmigt von der philosophischen Facultät der Friedrich Wilhelms Universität zu Berlin.* Berlin: Dissertations-Verlag Karl Arnold, 1908. Translated by Kevin Mulligan as *On Mach's Theories.* Washington, D.C.: Catholic University Press, 1982.

Thiele, Joachim. "Ernst Mach: Bibliographie." *Centaurus* 8(1963):189–237.

Thiele, Joachim. "Zur Wirkungsgeschichte der Methodenlehre Ernst Machs." In *Symposium aus Anlass des 50 Todestages von Ernst Mach, veranstaltet am 11–12 März 1966 vom Ernst Mach Institut Freiburg i Br.*. Freiburg-im-Breisgau: Ernst Mach Institut, 1966.

Mach, Ernst. *Die Mechanik in ihrer Entwicklung, historisch-enkritisch dargestellt.* 1st ed. 1883, 4th ed 1901. Translated by Thomas J. McCormack as *The Science of Mechanics: a Critical and Historical Account of its Development.* Chicago: Open Court, 1st ed. 1893, 3d ed. 1907.

Mach, Ernst. *Beiträge zur Analyse der Empfindungen.* Jena: 1st ed. 1886. Translated by C. M. Williams as *Contributions to the Analysis of the Sensations.* Chicago: Open Court, 1897. Later editions were enlarged and extended, and the title changed to *Die Analyse der Empfindungen und das Verhältnis des Physischen zum Psychischen.* Translated by C. M. Williams and revised, Waterlow, Sydney, from 5th ed. 1906 as *The Analysis of Sensations and the Relation of the Physical to the Psychical.* Chicago: Open Court, 1914. This is the basis for the new reprint, 1959.

Mach, Ernst. *Populär-wissenschaftliche Vorlesungen.* Leipzig: Barth, 1st ed. 1896, 5th ed. 1923. Translated by Thomas J. McCormack as *Popular Scientific Lectures.* Chicago: Open Court, 1895.

Mach, Ernst. *Erkenntnis und Irrtum: Skizzen zur Psychologie der Forschung.* Leipzig: Barth, 1st ed. 1905, 5th ed. 1926. Translated by Thomas J. McCormack (chaps. XXI and XXII) and Paul Foulkes, with intro. by Erwin N. Hiebert, from 5th ed., as *Knowledge and Error: Sketches on the Psychology of Enquiry.* Dordrecht: Reidel, 1976. In Series Vienna Circle Collection, v. 3.

Mach, Ernst. *"Die Leitgedanken meiner naturwissenschaftlichen Erkenntnislehre und ihre Aufnahme durch die Zeitgenossen." und "Sinnliche Elemente und naturwissenschaftliche Begriffe." zwei Aufsätze.* Leipzig: Barth, 1919.

Madsen, Thorvald:

Cockburn, W. Charles. "The international contribution to the standardization of biological substances, I. Biological Standards and the League of Nations, 1921–1946." *Biologicals* 19(1991):161–169.

Fischer, Isidor, ed. "Madsen, Thorvald." In *Biographisches Lexikon* (1933), v. 2 970 (Bibl. sec. 1)

Schelde-Møller, E. *Thorvald Madsen i Videnskabens og Menneskehedens Tjeneste.* Copenhagen: Nyt Nordisk Forlag, 1970.

Madsen, Thorvald. "Ueber Tetanolysin." *Z. f. Hyg. u. Infektionskr.* 32(1899):214–237.

Madsen, Thorvald. "Experimentelle Undersøgelser over Difteriegiften." Dissertation, Copenhagen, 1896. Published as "Ueber Messung der Stärke des anti-diphtherischen Serums." *Z. f. Hyg. u. Infectionskr.* 24(1897):425–442.

Madsen, Thorvald. "Ueber Heilversuche im Reagenzglas." *Z. f. Hyg. u. Infektionskr.* 32(1899):239–245.

Madsen, Thorvald. See also Arrhenius, Sv. , and Madsen, Th., "Physical chemistry applied to toxins and antitoxins" (1902) (Bibl. sec. 3).

Malkoff, G. M. "Beitrag zur Frage der Agglutination der rothen Blutkörperchen." *Deutsche.med. Wschr.* 26(1900):229–231.

Mansfeld, G. "Eine physiologische Erklarung der Agglutination." *Z. f. Immunitätsf.* 27(1918):197–212.

Marx, Karl. "Luther als Schiedsrichter zwischen Straus und Feuerbach." In Series *Karl Marx-Friedrich Engels Werke.* Berlin: Dietz Verlag, 1964, v. 1, 26–27.

McCall, Archibald J., Robert R. Race, and George L. Taylor. "Rhesus antibody in Rh positive mother." *Lancet 1*(1944):214–215.

Medical Research Council (Great Britain). *Medical Research in War: Report of the Medical Research Council for the Years 1939–45.* London: HMSO, 1947, Cmd 7335.

Meyer, Victor:
   Fruton, Joseph S. *Contrasts in Scientific Style: Research Groups in the Chemical and Biochemical Sciences.* Philadelphia, Pa.: American Philosophical Society, 1990, 122–124, 145–157, 232.

Meyer, Richard E. *Victor Meyer: Leben und Wirken eines deutschen Chemikers und Naturforschers 1848–1897.* In Series, "Grosse Männer: Studien zur Biologie des Genies," Wilhelm Ostwald, ed. Leipzig: Akad. Verlag, 1917, v. 4, 250–251.

Meyer, Victor, E. Demole, and W. Michler. "Ueber die Nitroverbindungen der Fettreihe, II. Abhandlung." *Ann. d. Chem.* (Leibigs) *175*(1875):88–164.

Meyer, Victor, J. Tscherniak, J. Locher, and M. Lecco. "Untersuchung über die Verschiedenheiten der primären, secundären und tertiären Nitroverbindungen." *Ann. d. Chem.* (Leibigs) *180*(1876):111–206.

Meyer, Victor. "Einführung Stickstoffhaltiger Radicale in Fettkörper." *Ber. d. Deutsch. chem. Ges.* 10(1877):2075–2078.

Meyer, Victor. "Chemische Probleme der Gegenwart." Vortrag, gehalten September 1889, *Deutsche Naturforscher und Aerzte, Tageblatt der 62 Versammlung,* Heidelberg. Heidelberg: Universitätsbuchdruckerei Hörning, 1890, 126–134.

Meyer, Victor. "Ergebnisse und Ziele der stereochemischen Forschung." *Ber. d. Deutsch. chem. Ges.* 23(1890):567–619.

Meyer, Victor, and L. Oelkers. "Ueber die negative Natur organischer Radicale: Untersuchung des Desoxybenzoins." *Ber. d. Deutsch. chem. Ges.* 21(1888): 1295–1306.

Miale, John B. "A pathologist's critique of the controversy over blood groups and serology and nomenclature." *Trans. N. Y. Acad. Sci. II 29*(1967):887–891.

Michaelis, Leonor:
   Fruton, Joseph S. *Contrasts in Scientific Style: Research Groups in the Chemical and Biochemical Sciences.* Philadelphia, Pa.: American Philosophical Society, 1990, 249, 252–256, 260.

Michaelis, Leonor. "Elektrische Uberführung von Fermenten, I. Das Invertin." *Biochem. Z. 16*(1908):81–86.

Michaelis, Leonor. "Physikalische Chemie der Kolloide." In Alexander von Korányi, [Sandor] Baron, and Paul Friedrich Richter, eds., *Physikalische Chemie und Medizin: ein Handbuch.* 2 v. in 1. Leipzig: Thieme, 1908, v. 2, 341–453.

Michaelis, Leonor. See also under Ehrlich, P,: Michaelis, L., "Ehrlichs wiedergefundene Doktor-dissertation" (1919) (Bibl. sec. 3).

Morgan, Thomas Hunt. "Factors and unit-characters in Mendelian heredity." *Am. Nat. 47*(1913):5–16.

Morgenroth, Julius:
   Fischer, Isidor. "Morgenroth, Julius." In *Biographisches Lexikon* (1933), v. 2, 1068 (Bibl. sec. 1).

Morgenroth, Julius. See also Ehrlich, P., and Morgenroth, J. "Mittheilungen über Agglutination, I–VI" (1899, 1900, 1901) (Bibl. sec. 3).

Morgenroth, Julius. "Ueber die Wiedergewinnung von Toxin aus seiner Antitoxinverbindungen." *Berl. klin. Wschr.* 42(1905):1550–1554.

Morgenroth, Julius. "Chemotherapeutische Studien." In *Ehrlich-Festschrift* (1914), 541–582.

Morgenroth, Julius. "Zur Kenntnis der Beziehungen zwischen chemischer Konstitution und chemotherapeutischer Wirkung." *Berl. klin. Wschr.* 59(1917): 55–63.

Morgenroth, Julius, and Hans Sachs. "Ueber die quantitativen Beziehungen von Ambozeptor, Komplement und Antikomplement." *Berl. klin. Wschr.* 39(1902):817–822. Translated in *Ehrlich-Bolduan* (1910), 250–266 (Bibl. sec. 2.).

Mourant, Arthur E. "A new Rhesus antibody" [anti-e]. *Nature* 155(1945):542.

Müller, Johannes Peter:

du Bois-Reymond, Emil Heinrich. "Gedächtnissrede auf Johannes Müller, gehalten in der Akademie der Wissenschaften am 8 Juli 1858" (1858). In Emil du Bois-Reymond, *Reden von Emil du Bois-Reymond.* 2v. Leipzig: Veit, 1912, v. 2, 143–334.

Haberling, Wilhelm. *Johannes Müller, das Leben des rheinischen Naturforschers auf Grund neuer Quellen und seiner Briefe.* Leipzig: Akad. Verlagsgesellschaft, 1924. In series, "Grosse Männer, Studien zur Biologie des Genies." Wilhelm Ostwald, ed., v. 9.

Koller, Gottfried. *Das Leben des Biologen Johannes Müller 1801–1858.* Stuttgart: Wissenschaftliche Verlagsgesellschaft, 1958. Contains bibliography of Müller's own work and of secondary work on Müller to 1958.

Mazumdar, Pauline M. H. "Johannes Müller on the blood, the lymph and the chyle." *Isis* 66(1975):242–253.

Steudel, J. "Müller, Johannes Peter." In Gillispie, C. C. ed., *Dictionary of Scientific Biography* (1974), v. 9, 567–574 (Bibl. sec. 1).

Muirhead, Ernest E., Arvel E. Haley, Sol Haberman, and Joseph M. Hill. "Acute renal insufficiency due to incompatible transfusion and other causes, with particular emphasis on management." Paper presented at the International Hematology and Rh Conference, Dallas Texas and Mexico City, November 1946. In Joseph M. Hill and William Dameshek, eds., *The Rh Factor in the Clinic and the Laboratory.* Special Issue no. 2. *Blood* (1948): 101–138.

Müller, Johannes. *Handbuch der Physiologie des Menschen für Vorlesung.* Coblenz: Hölscher, lst ed. 1834–1840, 3d ed., 1850.

Musil, Robert:

von Allesch, Johannes. "Robert Musil in der geistigen Bewegung seiner Zeit." In R. Dinklage, ed., *Musil* (1960), 133–142.

Dinklage, Karl, ed. *Robert Musil: Leben Werk, Wirkung.* Vienna, Amalthea-Verlag, 1960.

Roseberry, Robert L. *Robert Musil, ein Forschungsbericht.* Frankfurt-am-Main: Athenäum-Fischer Taschenbuch Verlag, 1974. Contains bibliography of Musil and of scholarly work on him to date.

Musil, Robert. *Beitrag zur Beurteilung der Lehren Machs: Inaugural-Dissertation zur Erlangung der Doktorwrde, genehmigt von der philosophischen Fakultät der Friedrich Wilhelms-Universität zu Berlin.* Berlin: Dissertations-Verlag Arnold, 1908.

Translated by Kevin Mulligan as *On Mach's Theories*. Introduction by Georg Henrick von Wright. Washington, D.C.: Catholic University Press, 1982.

Musil, Robert. *Der Mann ohne Eigenschaften*. 3 v. Berlin: Rowohlt, 1931. Translated by Eithne Wilkins and Ernst Kaiser as *The Man without Qualities*. London: 1953. Reprinted with Foreword by E. Wilkins and E. Kaiser. 1 v. New York: Capricorn, 1965.

von Nägeli, Carl:

Cramer, C. "Prof. C. von Nägeli." *Actes de la Soc. Helv. des Sciences Naturelles* (Freibourg). *Comptes Rendus*, 74th Session 184–188 (1890–1891).

Kölliker, Rudolf Albert. *Erinnerungen aus meinem Leben*. Leipzig: Engelmann, 1899.

Olby, Robert. "Nägeli, Carl Wilhelm." In Charles C. Gillispie, ed., *Dictionary of Scientific Biography* (1971), v. 9, 600–602 (Bibl. sec. 2).

Schwendener, Simon. "Carl Wilhelm von Nägeli." In *Ber. d. Deutsch. bot. Ges.* *9*(1891):26–42.

Wilkie, John S. "Nägeli's work on the fine structure of living matter." *Ann. Sci.* *16*(1960):11–14, 171–202, 209–239; *17*(1961):27–62.

von Nägeli, Carl. "Ueber die gegenwartige Aufgabe der Naturgeschichte, insbesondere der Botanik." *Z. f. wiss. Bot.* *1*(1)(1844):1–33; *1*(2)(1844):1–45.

von Nägeli, Carl. "Die neueren Algensysteme und Versuch zur Begründung eines Eigenen Systems der Algen und Florideen." *Neue Denkschr. der Schweizer. naturforschende Gesellsch.* *9*(1847).

von Nägeli, Carl. *Gattungen einzelliger Algen, physiologisch und systematisch bearbeitet*. Zurich: Schulthess, 1848.

von Nägeli, Carl. "Die Stärkekörner." In C. Nägeli and C. Cramer, *Pflanzenphysiologische Untersuchungen*. Zurich: Schulthess, 1855–1858, v. 2(1858), 1–623.

von Nägeli, Carl. "Ueber die aus Protein-substanzen bestehenden Crystalloide in der Paranuss: vorgetragen am 11 Juli 1862." *Bot. Mitth.* *1*(1863):217. Cited by Wilkie, "Nägeli's work" (1961)

von Nägeli, Carl. "Die Individualität in der Natur, mit vorzglicher Berucksichtigung des Pflanzenreiches." *Monatsschr. d. wissensch. Vereins* (Zurich) *2*(1856):171–212.

von Nägeli, Carl. *Entstehung und Begriff der naturhistorischen Art*. Munich: Verlag der kgl. Akademie, 1865.

von Nägeli, Carl. *Die niederen Pilze in ihren Beziehungen zu den Infectionskrankheiten und der Gesundheitspflege*. Munich: Oldenbourg, 1877.

von Nägeli, Carl. *Theorie der Gärung: ein Beitrag zur Molecularphysiologie*. Munich: Oldenbourg, 1879.

von Nägeli, Carl. "Die Schranken der naturwissenschaftlichen Erkenntniss." Lecture to the 50th Versammlung Deutscher Naturforscher und Aerzte: Munich, 1877. In his *Mechanisch-physiologische Theorie der Abstammungslehre, mit einem Anhang: 1. Die Schranken der naturwissenschaftlichen Erkenntniss. 2. Kräfte und Gestaltungen im molecularen Gebiet*. Munich: Oldenbourg, 1884.

Naunyn, Bernhard. "Die Entwicklung der inneren Medicin mit Hygiene und Bakteriologie im XIX Jahrhundert." *Verhandlungen d. Ges. Deutsch. Naturf. u. Aerzte*, 72 Versammlung, Aachen. Leipzig: Vogel, 1900, pt. I, 59–70.

Naunyn, Bernhard. *Erinnerungen Gedanken und Meinungen*. Munich: Bergmann, 1925.

Neisser, Max. See Bechhold, H., "Bakterienagglutination" (1903) (Bibl. sec. 3).
Neisser, Max. "Kritische Bemerkungen zur Arrhenischen Agglutininverteilungs-
formel." *Zbl. f. Bakt.* *36*(1904):671–676.
Neisser, Max, and Friedemann Ulrich. "Studien über Ausflockungserscheinun-
gen, I." *Münch. med. Wschr.* *51*(1904):446–469. "Studien über Ausflock-
ungserscheinungen, II. Beziehungen zur Bakterien-agglutination."
51(1904):827–831.
Neisser, Max, and Hans Sachs. "Ein Verfahren zum forensischen Nachweis der
Herkunft des Blutes." *Berl. klin. Wschr.* *42*(1905):1388–1389.
Nernst, H. Walther:
Barkan, Diana L. Kormos. "Walther Nernst and the Transition to Modern
Physical Chemistry." Thesis, Harvard 1990. Ann Arbor, Mich.: University
Microfilms, 91-13131. In copyright.
Farber, Eduard. "Walther Nernst." In his *Nobel Prize Winners* (1953), 77–81
(Bibl. sec. 1).
Fruton, Joseph S. *Contrasts in Scientific Style* (1990), 247–248 (Bibl. sec. 2).
Hiebert, Erwin N. "Nernst, Hermann Walther." In Charles C. Gillespie, ed.,
Dictionary of Scientific Biography (1978), v. 15, 432–453 (Bibl. sec. 1).
Zaunik, Rudolph, and J. C. Salié, eds. "Nernst, Hermann Walther C. (1864–
1941)." In *Poggendorfs Handwörterbuch* (1958), v. 7a, pt. 3, 405. Includes
bibliography, list of biographical sources to date. (Bibl. sec. 1).
Nernst, H. Walther. *Theoretische Chemie vom Standpunkt der Avogadro'schen Regel
und der Thermodynamik.* Stuttgart: 1st ed. 1893, 3d ed. 1900, many later eds.
Nernst, H. Walther. "Ueber die Anwendbarkeit der Gesetze des chemischen
Gleichgewichts auf Gemisch von Toxin und Antitoxin." [Review of Arrhe-
nius, Sv. , "Physikalische Chemie der Agglutinine." (1903)]. *Z. f. Elektro-
chemie 10*(1904):377–380.
Nernst, H. Walther, "Das Institut für physikalsche Chemie und besonders Elektro-
chemie an der Universität Göttingen." *Z. f. Elektrochemie 2*(1896):629–636.
Obermayer, Friedrich. "Nucleoalbumin-ausscheidung im Harn." *Wiener klin.
Wschr.* *4*(1891):966–967.
Obermayer, Friedrich. "Ueber Xanthoprotein. Vorläufige Mittheilung." *Zbl. f.
Physiol.* *6*(1892):300–301.
Obermayer, Friedrich. "Färben thierischer Fasern und Gewebe unter Erzeu-
gung von Azoderivaten ihrer Eiweissartigen Bestandtheilie." *Ber. d. Deutsch.
chem. Ges.* *27*(4) (Referate) (1894):354–355.
Obermayer, Friedrich, and Ernst Peter Pick. "Biologisch-chemische Studie über
das Eiklar: ein Beitrag zur Immunitätslehre." *Wiener klin. Rundschau
16*(1902):277–279.
Obermayer, Friedrich, and Ernst Peter Pick. "Ueber den Einfluss physikalischer
und chemischer Zustandsänderungen präcipitogener Substanzen auf die
Bildung von Immunpräcipitinen, Vortrag, gehalten den 22 mai 1903, k. k.
Ges. der Aerzte in Wien." *Wiener klin. Wschr.* *16*(1903):659–660.
Obermayer, Friedrich, and Ernst Peter Pick. "Beiträge zur Kenntnis der Prä-
zipitin-bildung." *Wiener klin. Wschr.* *17*(1904):265–267.
Obermayer, Friedrich, and Ernst Peter Pick. "Ueber die chemischen Grundlagen der Arteigenschaften der Eiweisskörper. Bildung von Immunpräzipitin
durch chemisch veränderte Eiweisskörper." *Wiener klin. Wschr.* *19*(1906):
327–333.

Oppenheimer, Carl. "Toxine und Schutzstoffe." *Biol. Zbl.19*(1899): 799–814.

Ostwald, Wilhelm:
Farber, Eduard. "Wilhelm Ostwald." In his *Nobel Prize Winners* (1953), 37–41 (Bibl. sec. 1).
Fruton, Joseph S. *Contrasts in Scientific Style* (1990) 241–262 (Bibl. sec. 2).
Körber, Hans-Günther. "Ostwald, Friedrich Wilhelm." In Charles C. Gillispie, ed., *Dictionary of Scientific Biography.* New York: Scribner's, 1978, v. 15/ Supplement 1, 455–469.
Nernst, H. Walther. "Das Institut für physikalsche Chemie und besonders Elektrochemie an der Universität Göttingen." *Z. f. Elektrochemie* 2(1896): 629–636.
Servos, John W. *Physical Chemistry from Ostwald to Pauling* (1990), 3–87 (Bibl. sec. 2).
Zaunick, Rudolph, and J. C. Salié, eds. "Ostwald, Wilhelm Friedrich (1853–1932)." In *Poggendorfs Handwörterbuch* (1971), v. 7a, Supplement, 476–482. Includes bibliography and list of biographical sources to date.

Ostwald, Wilhelm. *Lebenslinien: eine Selbstbiographie.* 3 v. in 1. Berlin: Klasing, Volksausgabe, 1933.
Ostwald, Wilhelm. *Lehrbuch der allgemeinen Chemie.* 2 v. Leipzig: Engelmann 1885–1887.
Ostwald, Wilhelm. *Abhandlungen und Vorträge allgemeinen Inhaltes 1887–1903.* Leipzig: Veit, 1904.
Ostwald, Wilhelm. *Grundriss der allgemeinen Chemie.* Leipzig: Engelmann, 1899. Translated by J. Walker as *Outlines of General Chemistry.* London: Macmillan, 1890.
Ostwald, Wilhelm. *Der Werdegang einer Wissenschaft: sieben gemeinverständliche Vorträge aus der Geschichte der Chemie.* Leipzig: Akad. Verlag, 2d ed., 1908. First published as *Leitlinien der Chemie,* a series of lectures held in autumn 1905 at Massachussets Institute of Technology.

Ostwald, Wolfgang:
Körber, Hans-Günther. "Ostwald, Carl Wilhelm Wolfgang." In Charles C. Gillispie, ed., *Dictionary of Scientific Biography.* New York: Scribner's, 1974, v. 10, 251–252 (Bibl. sec. 1) lists obituary notices.
Lottermooser, A. "Wolfgang Ostwald 60 Jahre alt." *Kolloidzeitschr. 103*(1943): 89–94.
Servos, John W. *Physical Chemistry from Ostwald to Pauling* (1990), 300–308.

Ostwald, Wolfgang. *Grundriss der Kolloidchemie.* Dresden: Steinkopf, 1909.
Ostwald, Wolfgang. *Die neuere Entwicklung der Kolloidchemie: Vortrag, geh. auf der 84 Versamml. Deutscher Naturforscher und Aerzte zu Münster-i-W. 1912.* Dresden: Steinkopf, 1912.
Ottenberg, Reuben. "Hereditary blood qualities: medicolegal application of human blood-grouping." *J. Immunol. 6*(1921):363–385.

Paltauf, Richard:
Eiselsberg. "Nachruf des Präsidenten der Gesellschaft: Richard Paltauf, gestorben 21 April 1924. Gedenkfeier der Gesellschaft der Aerzte in Wien am 2 Mai 1924." *Wiener klin. Wschr. 37*(1924):487–488.
Fischer, Isidor. "Paltauf, Richard." In *Biographisches Lexikon* (1933), v. 2, 1167 (Bibl. sec. 1).
Kraus, Rudolf. "Richard Paltauf und das Serotherapeutische Institut." *Wiener*

med. Wschr. 77(1927):739–741. In "Wiener mikrobiologische Forschung." Special Number, Rudolf Kraus, ed.

Lesky, Erna. *Wiener medizinische Schule* (1965), 577–578 (Bibl. sec. 2).

Maresch, R. "Richard Paltauf: Gedenkfeier der Gesellschaft der Ärzte in Wien, Mai 1924." *Wiener klin. Wschr.* 37(1924):488–491.

Paltauf, Rudolf. "Ueber Agglutination und Präcipitation." *Deutsche med. Wschr.* 29(1903):946–950.

Pasteur, Louis:

Carter, K. Codell. "The development of Pasteur's concept of disease causation and the emergence of specific causes in nineteenth century medicine." *Bull. Hist. Med.* 65 528–548 (1991).

Carter, K. Codell. "The Koch-Pasteur dispute on establishing the cause of Anthrax." *Bull. Hist. Med.* 62(1988):42–57.

Dagognet, Franois. *Méthodes et Doctrines dans l'Oeuvre de Pasteur.* Paris: Presses Universitaires, Editions Galien, 1967.

Dubos, René J. *Louis Pasteur, Freelance of Science.* Boston: Little, Brown, 1950.

Duclaux, Emile. *Pasteur, Histoire d'un Esprit.* Sceaux: Charaire, 1896. Translated by E. F. Smith and F. Hedges as *Pasteur History of a Mind* (1920). Reprinted by Library of N.Y. Academy of Medicine. Metuchen: Scarecrow Reprints 1973.

Duclaux, Emile. *Traité de Microbiologie.* 4 v. Paris: Masson, 1898.

Farley, John, and Gerald L. Geison. "Science, politics and spontaneous generation in nineteenth century France: the Pasteur-Pouchet debate." *Bull. Hist. Med.* 48(1974):161–198.

Gaffky, Georg. "Experimentelle erzuegte Septicämie" (1881) (Bibl. sec. 3).

Geison, Gerald J. "Pasteur, Louis." In Charles C. Gillespie, ed., *Dictionary of Scientific Biography* (1974), v. 10, 350–416 (Bibl. sec. 1).

Geison, Gerald J. *The Private Science of Louis Pasteur* (in preparation).

von Gruber, Max. "Pasteurs Lebenswerk" (1895), "Erwiderung" (1896) (Bibl. sec. 3).

Koch, Robert, ed. *Mittheilungen aus dem k. Gesundheitsamte* (1881) (Bibl. sec. 3).

Mollaret, H. H. "Contribution à la connaissance des relations entre Koch et Pasteur." *NTM: Schriftenreihe f. Naturwissenschaft, Technik u. Med.* 20(1983): 57–65.

Pfeiffer, Richard. "Kritische Bemerkungen" (1896) (Bibl. sec. 3).

Valléry-Radot, Pasteur. *La Vie de Pasteur.* Paris: Hachette, 1900.

Valléry-Radot, Pasteur, ed. *Oeuvres de Pasteur réunies par Valléry-Radot.* 8 v. in 9. Paris: Masson, 1933.

Pasteur, Louis. "Sur les maladies virulentes et en particulier sur la maladie appelée vulgairement choléra des poules." *C. R. de l'Acad. des Sci.* 90(1880): 239–248. Also Valléry-Radot, ed., *Oeuvres de Pasteur,* v. 6, 291–313.

Pasteur, Louis. "De l'atténuation du virus du choléra des poules." *C. R. de l'Acad. des Sci.* 91(1880):673–680. Also Valléry-Radot, ed., *Oeuvres de Pasteur,* v. 6, 323–331.

Pasteur, Louis. "De l'extension de la théorie des germes à l'étiologie de quelques maladies communes." *C. R. de l'Acad. des Sci.* 90(1880):1033–1044. Also Valléry-Radot, ed., *Oeuvres de Pasteur,* v. 6, 147–158.

Pasteur, Louis. "De l'atténuation des virus et de leur retour à la virulence." *C. R. de l'Acad. des Sci.* 92(1881):429–435. Also Valléry-Radot, ed., *Oeuvres de Pasteur,* v. 6, 332–338.

Pasteur, Louis. "Sur la virulence du microbe du choléra des poules." Also Val-léry-Radot, ed., *Oeuvres de Pasteur*, v. 7, 52–54.

Pasteur, Louis. "Expériences faites avec la salive d'un enfant mort de la rage." Receuil de Médecine Vétérinaire, *Bull. de la Soc. Centrale de Méd. Vét.* *58*(1881):150–155. Also Valléry-Radot, ed., *Oeuvres de Pasteur*, v. 6, 553–558.

Pasteur, Louis. "De l'atténuation des virus." *IV Congrès Internationale de l'Hygiene et de Démographie* Geneva 7-9 September 1882: Dunant, P. L. ed., *Comptes Rendus et Mémoires*. Geneva: Georg, 1883, v. 1, 127–145. Also Valléry-Radot, ed., *Oeuvres de Pasteur*, v. 6, 391–411.

Pasteur, Louis. "Microbes pathogènes et vaccins." *VII Congrès Periodique Internationale des Sci. Méd.*, Copenhagen, 8 August 1884. Copenhagen, 1886, v. 1 19–28. Also Valléry-Radot, ed., *Oeuvres de Pasteur*, v. 6 part 2, 590–602.

Pauli, Wolfgang:

Bechhold, Heinrich. "[Review of] Pauli, 'Wandlungen in der Pathologie.'" *Wiener klin. Wschr.* *18*(1905):550–551.

Blackmore, John T. *Ernst Mach.* (1972):73–83, 315 (Bibl. sec. 3).

Fischer, Isidor. "Pauli, Wolfgang." In *Biographisches Lexikon* (1933), v. 3, 1182 (Bibl. sec. 1).

Fruton, Joseph S. *Contrasts in Scientific Style* (1991), 221–222, 260 (Bibl. sec. 3).

Mazumdar, Pauline M. H. "The antigen-antibody reaction and the physics and chemistry of life." *Bull. Hist. Med.* *48*(1974):1–21.

Smutny, František. "Ernst Mach and Wolfgang Pauli's ancestors in Prague." *Gesnerus* *46*(1989):183–194.

Zaunik, Rudolph, and J. C. Salié, eds. "Pauli, Wolfgang Josef (1869–1955)." In *Poggendorfs Handwörterbuch* (1958), v. 7a, pt. 3, 517. Lists biographical sources, but no bibliography (Bibl. sec. 1).

Pauli, Wolfgang. *Ueber physikalisch-chemische Methoden und Probleme in der Medizin: Vortrag, gehalten den 10 November 1899 in der Gesellschaft der Aerzte in Wien.* Vienna: Perles, 1900. Translation in his *Physical Chemistry* (1907), 1–23.

Pauli, Wolfgang. "Allgemeine physiko-chemie der Zellen und Gewebe." *Ergebn. d. Physiol.* *1.* 1 Abt. Biochemie) (1902):1–14. Translation in his *Physical Chemistry* (1907), 23–44.

Pauli, Wolfgang. "Der kolloide Zustand und die Vorgänge der lebende Substanz: Vortrag, gehalten den 13 Mai 1902, in Morphologischphysiologische Gesellschaft in Wien." *Naturwiss. Rundschau* *17*(1902):312. Translation in his *Physical Chemistry* (1907), 44–71.

Pauli, Wolfgang. "Wandlungen in der Pathologie durch die Fortschritte der allgemeinen Chemie: Festival address at the 3d Annual Meeting of the k. k. Ges. d. Aerzte in Vienna, 24 March 1905. Translation in his *Physical Chemistry* (1907), 101–137.

Pauli, Wolfgang. "Ueber den Anteil der Kolloidchemie an der Immunitätsforschung." *Wiener klin. Wschr.* *18*(1905):665–666.

Pauli, Wolfgang. "Untersuchung über physikalische Zustandsänderungen der Kolloide: V. Mitteilung: Die elektrische Ladung von Eiweiss." *Beitr. z. chem. Physiol. u. Path.* (Hofmeisters) 7(1906):531–547.

Pauli, Wolfgang. *Physical Chemistry in the Service of Medicine: Seven Addresses.* Translated by Martin H. Fischer. New York: Wiley, 1907.

Pauli, Wolfgang. *Kolloidchemie der Eiweisskörper.* Dresden: Steinkopf, 1920. Trans-

lated by P. C. L. Thorne as *Colloid Chemistry of Proteins.* Philadelphia, Pa:
Blackiston's, 1922.
Pauling, Linus:
Serafini, Anthony. *Linus Pauling: a Man and his Science.* New York: Simon And
Schuster, 1989.
Servos, John W. *Physical Chemistry form Ostwald to Pauling: the Making of a Science
in America.* Princeton, N.J.: Princeton University Press, 1990.
Sturdivant, James H. "The scientific work of Linus Pauling." In Alexander
Rich and Norman Davidson, *Structural Chemistry and Molecular Biology.* San
Francisco, Calif.: Freeman, 1968, 3–11. Rich and Davidson's volume con-
tains a bibliography of Pauling's work to date.
Pauling, Linus. "The nature of the chemical bond." *J. Am. Chem. Soc. 53*(1931):
1367–1400, 3225–3237; *54*(1932):988–1003, 3570–3582. *J. Chem. Phys.* 362–
374; with G. W. Wheland, 606–617; with J. Sherman, 679–686. 1933).
Reprinted in A. Rich and N. Davidson, *Structural Chemistry and Molecular
Biology* (1968), 849–884.
Pauling, Linus. "A theory of the structure and process of formation of antibod-
ies." *J. Am. Chem. Soc. 62*(1940):2643–2657.
Pauling, Linus. "Molecular structure and intermolecular forces." In K. Land-
steiner, *Specificity of Serological Reactions* (1946) (Bibl. sec. 3).
Pauly, Hermann. "Ueber die Konstitution des Histidins, I. Mitteilung." *Z. f.
physiolog. Chem.* (Hoppe–Seylers) *42*(1904):508–518.
von Pettenkofer, Max Joseph:
Dolman, Claude E. "Pettenkofer, Max Josef von." In Charles C. Gillispie, ed.,
Dictionary of Scientific Biography. New York: Scribner's, 1974 v. 10, 556–563
(Bibl. sec. 1).
Emmerich, Rudolf. *Max Pettenkofer's Bodenlehre der Cholera indica. Experimentell
begründet und weiter ausgebaut von Rudolf Emmerich, mit Beiträgen von Ernst
Angerer et al.* Munich: Lehmann, 1910.
Evans, Richard J. *Death in Hamburg: Society and Politics in the Cholera Years 1830–
1910.* Oxford, Clarendon Press, 1987, 237–273, 475–507.
Eyler, John M. *Victorian Social Medicine: the Ideas and Methods of William Farr.*
Baltimore, Md.: Johns Hopkins University Press, 1979, 97–122.
Flügge, Carl. *Micro-organisms with Special Reference to the Etiology of Infectious
Disease.* Translated by William Watson Cheyne, from 2d ed. of his *Fermente
und Mikroparasiten.* London: The Sydenham Society, 1890, 415–470. Com-
pares Pettenkofer's and Koch's theories of cholera transmission.
von Gruber, Max. "Max von Pettenkofer." *Berichte der Deutsch. chem. Ges.*
36(4)(1903):4512–4572.
Hume, Edgar E. *Max von Pettenkofer: his Theory of the Etiology of Cholera, Typhoid
Fever and Other Intestinal Diseases: Review of His Arguments and Evidence.* New
York: Hoeber, 1927. Includes bibliography of his work on public health.
Pelling, Margaret. *Cholera, Fever and English Medicine.* Oxford: Oxford University Press, 1978. Chap. 4 on Pettenkofer's sources.
von Pettenkofer, Max. *Ueber den Werth der Gesundheit für eine Stadt: zwei Vorlesungen gehalten den 26 und 29 März 1873 im Verein für Volksbildung in München.*
Braunschweig: Vieweg, 1873. Translated by Henry E. Sigerist as "On the
value of health to a city." *Bull. Hist. Med. 10*(1941):473–486, 487–503, 594–
613.
von Pettenkofer, Max. "Ueber Cholera, mit Berücksichtigung der jüngsten Chol-

era Epidemie in Hamburg." *Münch. med. Wschr. 39*(1892):807–817. See 808 ff. for an account of swallowing cholera culture; also discussed by Evans, *Death in Hamburg* (1987).

von Pettenkofer, Max. See Forster, J., and Hofmann, Fr., "Vorwort" (1883) (Bibl. sec. 3).

Pfeiffer, Hermann. "Des biologische Blutnachweis." In Emil Abderhalden, ed., Handbuch der biologischen Arbeitsmethoden (Berlin: 1920–), sec. 4. *Angewandte chenische und physikalische Methoden* (1923), v. 12, p. 1, 105–176.

Pfeiffer, Hermann. "Beiträge zur Lösung des biologisch-forensischen Problems der Unterscheidung von Spermaeiweiss gegenber dem anderen Eiweissarten derselben Species durch die Präzipitinmethode." *Wiener klin. Wschr. 18*(1905):637–641.

Pfeiffer, Richard:
Fischer, Isidor. "Pfeiffer, Richard." In *Biographisches Lexikon.* (1933), v. 3, 1206–1207 (Bibl. sec. 1).

Möllers, Bernhard. *Robert Koch* (1950), 380–383 (see under Koch, Bibl. sec. 3).

Pfeiffer, Richard, "Kritische Bemerkungen zu dem Aufsatze Max Grubers 'Pasteurs Lebenswerk im Zusammenhang mit der gesammten Entwicklung der Mikrobiologie,'" *Deutsche med. Wschr. 22* 15–16. 1896.

Pfeiffer, Richard. "Kritische Bemerkungen zu Grubers Theorie der activen und passiven Immunität gegen Cholera, Typhus, und verwandte Krankheitsprocesse." *Deutsche med. Wschr. 22*(1896):232–234.

Pfeiffer, Richard. "Bemerkung zu vorstehender Erwiderung." *Deutsche med. Wschr. 22*(1896):95.

Pfeiffer, Richard. "Ein neues Grundgesetz der Immunität." *Deutsche med. Wschr. 22*(1896): 97–99; 119–122.

Pfeiffer, Richard, and Ernst Friedberger. "Ueber die im normalen Ziegenserum erhaltenen bacteriolytischen Stoffe (Ambiceptoren Ehrlichs)." *Deutsche med. Wschr. 27*(1901):834–836.

Pfeiffer, Richard, and Isaeeff. "Ueber die spezifische Bedeutung der Choleraimmunität." *Z. f. Hyg. u. Infektionskr. 17*(1894):355–400.

Pfeiffer, Richard, and Wilhelm Kolle. "Ueber die specifische Immunitätsreactionen der Typhusbacillen." *Z. f. Hyg. u. Infektionskr. 21*(1896):203–246.

Pfeiffer, Richard, and Wilhelm Kolle. "Zur Differentialdiagnose der Typhusbacillen vermittels Serums der gegen Typhus immunisirten Thiere." *Deutsche med. Wschr. 22*(1896):185–186.

Pick, Ernst Peter:
Brücke, F., A. Lindner, and W. Weis. "Prof. Dr Ernst Peter Pick zum Gedächtnis." *Wiener klin. Wschr. 72*(1960):109–110.

Molitor, H. "Ernst Peter Pick." *Arch. Int. Pharm. Therap. 132*(1961):205–221.

Pick, Ernst Peter. "Darstellung der Antigene mit chemischen und physikalischen Methoden." In Rudolf Kraus and Constantin Levaditi, *Handbuch der Technik und Methodik der Immunitätsforschung.* Jena: Fischer, 1908, v. 1, 531–586.

Pick, Ernst Peter. "Biochemie der Antigene, mit besonderer Bercksichtigung der chemischen Grundlagen der Antigenspezifizität." In Wilhelm Kolle and August von Wassermann, eds., *Handbuch der pathogenen Mikroorganismen.* Jena: Fischer, 2d ed. 1912, v. 1, 685–868.

Pick, Ernst Peter. See also Obermayer, Friedrich, "Nucleoalbumin Ausscheidung im Harn" (1891); Obermayer, F., "Färben von Azoderivaten thierischer Fasern," 1894; Obermayer, F., and Pick, E. P. "Biologisch-chemisch Studie über das Eiklar" (1902, 1903); "Zur Kenntnis der Präzipitinbildung" (1904); "Chemische Grundlagen der Arteigenschaft" (1906) (Bibl. sec. 3).

Pickles, Margaret M. *Hemolytic Disease of the Newborn.* Springfield, Ill.: Thomas, 1949.

Pirquet von Cesenatico, Clemens, Freiherr:
Pirquet von Cesenatico, Clemens Freiherr. Papers 1903–1921 in National Library of Medicine, Bethesda, Md., ms C 141. Includes a collection of reprints of published work.

Wagner, Richard. *Clemens von Pirquet His Life and Work.* Baltimore, Md.: Johns Hopkins University Press, 1968.

von Pirquet, Clemens. "Zur Geschichte der Allergie." *Wiener klin. Wschr.* 77(1927):745–748. "Wiener mikrobiologische Forschung." Special Number, Rudolf Kraus, ed.

Race, Robert R.:
Clarke, Sir Cyril. "Robert Russell Race, 28 November 1907—15 April 1984." *Biographical Memoirs of Fellows of the Royal Society* (London) *31*(1985):453–492.

"Race, Robert R." In *Who's Who of British Scientists 1971–1972.* Athens: Ohio University Press, 1972.

Tippett, Patricia. "In Memoriam." *Vox Sang.* 47(1984):395–396.

Race, Robert R. "An 'incomplete' antibody in human serum." *Nature* *153*(1944):771–772.

Race, Robert R. "The Rh genotypes and Fisher's theory." Paper presented at International Hematology and Rh Conference, Dallas, Texas and Mexico City, November 1948. In Joseph M. Hill and William Dameshek, eds., "The Rh factor in the Clinic and the Laboratory." *Blood* (1948), Special Issue no. 2, 27–42.

Race, Robert R. "Some notes on Fisher's contribution to human blood groups." *Biometrics 20*(1964):361–367. Special Issue on Ronald Aylmer Fisher.

Race, Robert R. "Modern concepts of the blood group systems." *Ann. N. Y. Acad. Sci. 127*(1965):884–891.

Race, Robert R. See also McCall, Archibald J., Race, R. R. and Taylor, G. L., "Rh antibody in Rh + mother." 1944); Fisher, R. A., and Race, R. R., "Rh gene frequencies" (1946) (Bibl. sec. 3).

Race, Robert R. and Ruth Sanger. *Blood Groups in Man.* Oxford: Blackwell, 1st ed. 1950, 6th ed. 1976.

Race, Robert R., and George L. Taylor. "A serum that discloses the genotype of some Rh positive people." *Nature 152*(1943):300.

Race, Robert R., George L. Taylor, Kathleen E. Boorman, and Barbara E. Dodd. "Recognition of the Rh genotypes in man." *Nature 152*(1943):563.

Race, Robert R., George Taylor, Daniel F. Cappell, and Marjorie N. McFarlane. "Recognition of a further common Rh genotype in man." *Nature 153*(1944):52–53.

Rosenfield, Richard E. "The William Allan Memorial Award presented to Philip Levine and Alexander S. Wiener at the annual meeting of the American Society of Human Genetics, Baltimore, Maryland October 10, 1975." *Am. J. Hum. Genet.* 28(1976):101–106.

Rosenfield, Richard E., Fred Hal Allen, Jr., Scott N. Swisher, and Shaul Kochwa. "A review of Rh serology and a presentation of a new terminology." Presented in part at the Annual Meeting of the American Association of Blood Banks, Chicago, October 1961. *Transfusion* 2(1962):287–312.

Rothen, Alexander, and Karl Landsteiner. "Adsorption of antibodies by egg albumin films." *Science* 90(1939):65–66, C. 323.

Rothen, Alexander, and Karl Landsteiner. "Serological reactions of protein films and denatured proteins." *J. Exper. Med.* 76(1942):437–450, C. 341.

Rubner, Max, Max von Gruber, and Martin Ficker, eds. *Handbuch der Hygiene.* Leipzig: Hirzel, 1913.

Sachs, Hans:
Fischer, Isidor. "Sachs, Hans." In *Biographisches Lexikon* (1933), v. 2, 1349–1350 (Bibl. sec. 1)

Sachs, Hans, and G. Bolkowska. "Beiträge zur Kenntnis der komplexen Konstitution der Komplemente." *Z. f. Immunitätsf.* 7(1910):778–786.

Sachs, Hans, and L. Omorokow. "Ueber die Wirkung des Kobragiftes auf die Komplemente, II. Mitteilung." *Z. f. Immunitätsf.* 11(1911):770–774.

Salomonsen, Carl Julius, ed. *Festskrift ved indvielsen af Statens Seruminstitut 1902: Contributions from the University Laboratory for Medical Bacteriology to Celebrate the Inauguration of the State Serum Institute.* Copenhagen: Olsen, 1902.

Salomonsen, Carl Julius. "Lebenserinnerungen aus dem Breslauer Sommersemester 1877." *Berl. klin. Wschr.* 51(1914):485–490.

Sanger, Ruth. See Race, R. R., and Sanger, R., *Blood Groups in Man* (1950–1976).

Schattenfroh, Arthur:
Fischer, Isidor. "Schattenfroh, Arthur." In *Biographisches Lexikon* (1933), v. 2, 1375 (Bibl. sec. 1).
Lesky, Erna. *Die Wiener medizinische Schule* (1965), 602 (Bibl. sec. 2).

Schattenfroh, Arthur. See under Grassberger, R., and Schattenfroh, A., *Toxin und Antitoxin* (1904) (Bibl. sec. 3).

Schleiden, Matthias Jacob:
Buchdahl, Gerd. "Leading principles and induction: the methodology of Matthias J. Schleiden." In Ronald A. Giere and Richard S. Westfall, eds., *Foundations of Scientific Method.* Bloomington: Indiana University Press, 1973, 23–52.
Charpa, Ulrich. Introduction to Matthias Jacob Schleiden, *Wissenschaftsphilosophische Schriften mit kommentierenden Texten von Jacob Friedrich Fries, Christian Nees von Esenbeck und Gerd Buchdahl.* Cologne: Dinter, 1989.
Glasmacher, Thomas. *Fries-Apelt-Schleiden: Verzeichnis der Primär- und Sekondär Literatur 1798–1988.* Cologne: Dinter, 1989.
Jahn, Ilse. "Matthias Jacob Schleiden an der Universität Jena." *Naturwissenschaft, Tradition, Fortschritt:* supplement to *NTM: Z. f. Geschichte der Wissenschaften* (1963):63–72.
Jahn, Ilse. "The influence of Jacob Friedrich Fries on Matthias Schleiden." In William Ray Woodward and Richard S. Cohen, eds., *World Views and Scientific Discipline Formation.* Dordrecht: Kluwer, 1991, 357–365.
Wunschmann, E. "Schleiden, Matthias J." In *Allgemeine deutsche Biographie* (1930), v. 31, 417–421 (Bibl. sec. 1).

Schleiden, Matthias J. "Beiträge zur Phytogenese." *Arch. f. Anat. u. Physiol.* 2(1838):137–176. Translated by H. Smith. In *Theodor Schwann: Microscopical*

Researches into the Accordance in the Structure and Growth of Plants (1839). London: Sydenham Society, 1847, 231–263.

Schleiden, Matthias J. *Grundzüge der wissenschaftliche Botanik, nebst einer methodologischen Einleitung, als Einleitung zum Studium der Pflanze.* Leipzig: Engelmann, 1st ed. 1842–43, 2d ed. 1845, 3d ed. 1849–50 (illustrated). Yranslated by Edwin Ray Lankester as *Principles of Scientific Botany, or, Botany as an Inductive Science.* London: Longmans, 1849.

Schoenberg, Arnold:
Adorno, Theodor. "Schoenberg and progress" (1941). In his *Philosophy of Modern Music,* translated by A. G. Mitchell and W. V. Blomster. New York: Seabury Press, 1973.

Schoenberg, Arnold. *Die glückliche Hand* Scenario-libretto-opera, Op. 18 (1910–1913). Vienna: Universal Editions, 1926. Recorded by R. Craft, conducting the Columbia Symphony Orchestra, with R. Oliver. In Series, Music of Arnold Schoenberg, v. 1. Columbia M 2S 679, 1963.

Schoenberg, Arnold. *Musikalisches Taschenbuch.* Vienna: 1911. Cited by Adorno, "Scheonberg and progress" (1941/1973) (Bibl. sec. 3).

Schoenberg, Arnold. *Gurrelieder von Jens Peter Jacobsen, Deutsch von Robert Franz Arnold, für Soli, Chor und Orchester.* Vienna: Universal Editions, 1920. Recorded by M. Napier, J. Thomas, Yvonne Minton, et al., BBC Symphony Orchestra and Chorus. Conducted by Pierre Boulez (Columbia M 2 33303, 1975).

Scholl, Roland. "Umwandlungen von Ketoximen in Pseudonitrole." *Ber. d. Deutsch. chem. Ges.* 21(1888):506–510.

Scholl, Roland, and Karl Landsteiner. "Reduction der Pseudonitrole zu Ketoximen." *Ber. d. deutsch. chem. Ges.* 29(1896):87–90, C. 7.

Schultz, G. "Bericht über den Feier der Deutschen chemischen Gesellschaft zu Ehren August Kekulés." *Ber. d. Deutsch. chem. Ges.* 23(1890):1265–1312.

Shattock, Samuel G.:
Keating, Peter. "The problem of the natural antibodies." *J. Hist. Biol.* 24(1991):245–263.

Mazumdar, Pauline M. H. "The purpose of immunity: Landsteiner's explanation of the human isoantibodies." *J. Hist. Biol.* 8(1975):115–134.

Shattock, Samuel G. "Chromocyte clumping in acute pneumonia and certain other diseases, and the significance of the buffy coat in the shed blood." *J. Path. Bact.* 6(1900):303–314.

Steffan, Paul:
Mazumdar, Pauline M. H. "Blood and soil: the serology of the Aryan racial state." *Bull. Hist. Med.* 64(1990):187–219, on the Deutsche Gesellschaft f. Blutgruppenforschung.

Steffan, Paul. "Die Arbeitsweise der Deutschen Gesellschaft für Blutgruppenforschung." *Z. f. Rassenphysiol.* 1(1928):8–11.

Stevens, W. L. "Estimation of blood group frequencies." *Ann. Eugen.* (London) 8(1938):362–375.

Sturtevant, Alfred H. "The Himalyan rabbit case with some observations on multiple allelomorphs." *Am. Nat.* 47(1913):23–239.

Suida, Wilhelm. "Ueber das Verhalten von Teerfarbstoffen gegenu:ber Stärke, Kieselsäure und Silikaten, vorgelegt in der Sitzung am 16 Juni 1904 des k. Akad. der Wissenschaften in Wien, Math.-Naturw. Klasse." *Monatsh. f. Chem.* (Vienna) 25(1904):1107–1143.

Suida, Wilhelm. "Ueber den Einfluss der aktiven Atomgruppen in den Textil-fasern auf das Zustandgekommen von Färbungen: vorgelegt in der Sitzung am 12 Jänner 1905 des k. Akad. der Wissenschaften in Wien, Math.-Na-turw. Klasse." *Monatsh. f. Chem.* (Vienna) *26*(1905):413–427.

Suida, Wilhelm. See also Gelmo, P. and Suida, W., "Färben animalischer Tex-tilfasern." (1906) (Bibl. sec. 3).

Taylor, George L.:
Fischer, Ronald A. "G. L. Taylor, M.D., Ph.D., F.R.C.P." [obituary] *Brit. Med. J. i*(1945):463–464.

Taylor, George, L. See also McCall, A. J., Race, R. R. and Taylor, G. L., "Rh antibody in Rh + mother" (1944); Race, R. R. and Taylor, G. L., "Serum that discloses the genotype of some Rh+s" (1943); Race, R. R. Taylor, G. L., Boorman, K. E. and Dodd, B. E., "Recognition of the Rh genotypes" (1943); Race, R. R., Taylor, G. L., Cappell, D. F. and McFarlane, M. N., "Recognition of a further common Rh genotype" (1944) (Bibl. sec. 3).

Taylor, George L., and Patrick L. Mollison. "Wanted: anti-Rh sera." *Brit. Med. J. i*(1942):561-562.

Taylor, George L., and Aileen M. Prior. "Blood groups in England, I. Exami-nation of family and unrelated material." *Ann. Eugen.* *8*(1938):344–355.

Taylor, George L., and Aileen M. Prior. "Blood groups in England, II. Distri-bution in the population." *Ann. Eugen.* *8*(1938):356–361.

Thomsen, Oluf. "Ueber bakterielle Veränderung der Agglutinabilitätsverhält-nisse der roten Blutkörperchen." *Acat med. Scand.* (Stockholm) *70*(1929): 436–448.

Tiselius, Arne:
Farber, Eduard. "Arne Tiselius." In his *Nobel Prize Winners* (1953), 194–198 (Bibl. sec. 1).

Kay, L. E. "Laboratory technology and biological knowledge: the Tiselius electrophoresis apparatus." *Hist. Phil. Life Sci.* *10*(1988):51–72.

Kekwick, R. A., and Kai O. Pedersen. "Arne Tiselius." *Biographical Memoirs of Fellows of the Royal Society* (London) *20*(1974):401–428.

Tiselius, Arne. "The moving-boundary method of studying the electrophoresis of proteins." Inaugural Dissertation. Uppsala, 1930.

Tiselius, Arne. "A new apparatus for electrophoretic analysis of colloidal mixtures." *Trans. Faraday Soc.* *33*(1937):524–531.

Todd, Charles. "Cellular individuality in the higher animals, with special ref-erence to the individuality of the red blood corpuscle." *Proc. Royal Society* (London) B *106*(1930):22–44.

Traube, Jsidor. "Die Resonanztheorie, eine physikalische Theorie der Immu-nitätserscheinungen." *Z. f. Immunitätsf.* *9*(1911):246–272.

Traube, Jsidor. See Landsteiner, K., "Bemerkungen zu der Abhandlung von Traube" (1911) (C. 122) (Bibl. sec. 3).

Uhlenhuth, Paul:
Fischer, Isidor. "Uhlenhuth, Paul." In *Biographisches Lexikon* (1933), v. 2, 1594 (Bibl. sec. 1).

Uhlenhuth, Paul. "Zur Lehre von der Unterscheidung verschiedener Eiweis-sarten mit Hilfe specifischer Sera." In *Koch-Festschrift* (1903), 49–74 (Bibl. sec. 3).

Uhlenhuth, Paul. "Ein Verfahren zur biologischen Unterscheidung von Blut verwandter Tier." *Deutsche med. Wschr.* *31*(1905):1673–1678.

Uhlenhuth, Paul. See also Kolle, W., Kraus, R., and Uhlenhuth, P., *Handbuch der pathogenen Organismen* (1928–1931); Kraus, R. and Uhlenhuth, P., *Handbuch der mikrobiologischen Technik* (1923–1924) (Bibl. sec. 3).

Verworn, Max. *Allgemeine Physiologie eine Grundriss der Lehre vom Leben.* Jena: Fischer, 1st ed. 1895, 5th ed. 1909.

Volk, Richard. "Ueber die quantitativen Grundlagen der Bindungsverhältnisse zwischen Agglutinin und Bakterien (Bindungsgesetz von Eisenberg und Volk)." *Wiener med. Wschr.* 77(1927):748–749, "Wiener mikrobiologische Forschung." Special Number, R. Kraus, ed.

de Vries, Hugo. "Sur la loi de disjonction des hybrides." *C. R. de l'Acad. des Sci.* 130(1900):845–847.

von Wassermann, August. See Kolle, W., and, *Handbuch der pathogenen Mikroorganismen* (1902, 1913, 1929) (Bibl. sec. 3).

Weichselbaum, Anton:
Ghon, Anton. "Anton Weichselbaum." *Wiener med. Wschr.* 77(1927):738–739, "Wiener mikrobiologische Forschung." Special Number, R. Kraus, ed.

Weinberg, Wilhelm. "Ueber den Nachweis der Vererbung beim Menschen." *Jahresh. d. Ver. f. vaterl. Naturk.* (Wrttemberg) 64(1908):368–382.

Wells, Harry Gideon:
Fisher, Isidor. "Wells, Harry Gideon." In *Biographisches Lexikon* (1933), v. 2, 1665 (Bibl. sec. 1).

Wells, H. Gideon. "The chemical basis of immunological specificity." *J. Immunol.* 9(1924):291–308.

Wells, H. Gideon. *The Chemical Aspects of Immunity.* New York: Chemical Catalog Co., 1924, 2d ed. 1929. In series American Chemical Society Monographs, W. A. Noyes, ed.

Wells, H. Gideon. "Studies on the chemistry of anaphylaxis." *J. Infect. Dis.* 5(1908):449–483.

Wells, H. Gideon. [Review of] Arrhenius, Sv., *Immunochemistry* (1907). In *J. Am. Chem. Soc.* 30(1908):650–652.

Wernich, Albrecht L. A.:
Fischer, Isidor. "Wernich, Albrecht." In *Biographisches Lexikon* (1933), v. 2, 1668–1669 (Bibl. sec. 1).

Wernich, Albrecht L. A. *Der Abdominaltyphus: Untersuchungen über sein Wesen, seine Tödlichkeit und seine Bekämpfung.* Berlin: Hirschwald, 1882.

Wernich, Albrecht L. A. *Die Entwicklung der organisirten Krankheitsgifte, nebst einem offenen Briefe an Herrn Professor Klebs in Prag.* Berlin: Riemer, 1880.

Wernich, Albrecht L. A. *Die Medicin der Gegenwart in ihrer Stellung zu den Naturwissenschaften und zur Logik: ein Beitrag zu den Zeitfragen unserer Wissenschaft.* Berlin: Riemer, 1881.

Wiener, Alexander Solomon:
Davidsohn, Israel, Philip Levine, and Alexander S. Wiener. "Medico-legal application of blood-grouping tests." *J. Am. Med. Assoc.* 149(1952):699–706.

Hirschfeld, J. "Alexander Solomon Wiener, 1907–1976." *Int. Arch. Allergy Appl. Immunol.* 54(1977):191.

Hollán, Susan R. *Current Topics in Immunohematology and Immunogenetics: Alexander S. Wiener Festschrift.* Budapest: Akadémiai Kiádo, 1972. Contains bibliography to date.

Moor-Jankowski, J. "Dr Alexander S. Wiener, 1907–1976." *Vox Sang 34*(1978): 189–190.

Rosenfeld, Richard E. "William Allan Memorial Award." (1976) (Bibl. sec. 3).

Rosenfield, Richard E. "*In memoriam:* Dr Alexander S. Wiener, MD." *Haematologia* (Budapest) *11*(1977):5–9.

Speiser, Paul. "Nekrologia: *in memoriam* A. S. Wiener." *Blut 35*(1977):93–95.

Wiener, Alexander S. "Individuality of the blood in higher animals." *Z. f. indukt. Abstamm. u. Vererbs. 66*(1933):31–48.

Wiener, Alexander S. "Individuality of the blood in higher animals II. Agglutination in red blood cells of fowls." *J. Genetics 29*(1934):1–8.

Wiener, Alexander S. *Blood Groups and Blood Transfusion.* Springfield, Ill.: Thomas, 1st ed. 1935, 2d ed. 1939, 3d ed. 1945.

Wiener, Alexander S. "Distribution and heredity of the Rh types." *Science 98*(1943):182–184. In his *Rh-Hr* (1952), 242–243.

Wiener, Alexander S. "Genetic theory of the Rh blood types." *Proc. Soc. Exper. Biol. Med. 54*(1943):316–319. In his *Rh-Hr* (1952), 244–247.

Wiener, Alexander S. "Analogy between Hr and O." *Science 100*(1944):595–597. In his *Rh-Hr* (1952), 251–252.

Wiener, Alexander S. "The Rh blood types and some of their applications." *Am. J. Clin. Path. 15*(1945):106–121. In his *Rh-Hr* (1952), 63–78.

Wiener, Alexander S. "Theory and nomenclature of the blood factors." *Science 102*(1945):479–482.

Wiener, Alexander S. "Theory and nomenclature of the Rh types, sub-types and genotypes." *Brit. Med. J. 1*(1946):982–984.

Wiener, Alexander S. "The Rh system in the chimpanzee." *Science 104*(1946):578–579.

Wiener, Alexander S. "Anti-Rh serum nomenclature." *Brit. Med. J. 1*(1948):805–822. In his *Rh-Hr* (1952), 319–322.

Wiener, Alexander S. "Conference on nomenclature of the Rh factors." *Med. J. Australia 35*(1948):530–531.

Wiener, Alexander S. "Nomenclature of Rh factors." *Lancet 1*(1948):343.

Wiener, Alexander S. "Karl Landsteiner his work and the Rhesus blood factor: Remarks made on the occasion of the Passano Award, Atlantic City, New Jersey June 13, 1951." *Current Med. Digest* (August 1951):29–32 (September 1951):49–54. In his *Rh-Hr* (1952), 1–8.

Wiener, Alexander S. "History of the Rhesus blood types." *J. Hist. Med. 7*(1952):369–383. In his *Rh-Hr* (1952), 9–23.

Wiener, Alexander S. *Rh-Hr Blood Types: Applications in Clinical and Legal Medicine and Anthropology. Selected Articles in Immunohematology.* New York: Grune, 1952. Contains bibliography up to 1952.

Wiener, Alexander S. *Advances in Blood Grouping.* New York: Grune, 1961. Contains bibliography 1953–1960.

Wiener, Alexander S. "Principles of blood group serology and nomenclature: a critical review." *Transfusion 1*(1961):308–320. Also his *Advances in Blood Grouping II* (1965) 16–28.

Wiener, Alexander S. "Fundamentals of immogenetics, with special reference to the human blood groups, I. Heredity of the blood groups; agglutinogens and blood factors; and nomenclature." *Med. Proc. 10*(1964):559–573. Also his *Advances in Blood Grouping II* (1965) 1–15.

Wiener, Alexander S. *Advances in Blood Grouping II,* with a section by Maurice Shapiro. New York: Grune, 1965. Contains bibliography 1960–1965.

Wiener, Alexander S. "History of blood group nomenclature with a question-naire on Rh-Hr nomenclature." *J. Forensic Med. 14*(1967):3–12.

Wiener, Alexander S. "Nomenclature of blood groups with special reference to the Rh-Hr blood types." *Trans. N. Y. Acad. Sci. II 29*(1967):875–886.

Wiener, Alexander S. "Final results of Rh-Hr nomenclature questionnaire." *Trans. N. Y. Acad. Sci. II 29*(1967):892–897.

Wiener, Alexander S. "The Rh-Hr blood types: the anatomy of a controversy." *J. Forensic Med. 15*(1968):22–40.

Wiener, Alexander S. "Karl Landsteiner, M.D. History of the Rh-Hr blood group system." *N. Y. State J. Med. 69*(1968):2915–2935.

Wiener, Alexander S. *Advances in Blood Grouping III.* New York: Grune, 1971. Contains bibliography 1965–1971.

Wiener, Alexander S. See also under Landsteiner, K.: Landsteiner, K. and Wiener, A. S., "Presence of M agglutinogens in monkeys" (1937) (C.311); Landsteiner, L. and Wiener, A. S., "Agglutinable factor in human blood recognised by immune sera for *Rhesus* blood" (1940) (C. 325); Landsteiner, K. and Wiener, A. S., "Studies on an agglutinogen (Rh)" (1941) (C. 335) (Bibl. sec. 3).

Wiener, Alexander S., and K. Landsteiner. "Heredity of variants of the Rh type." *Proc. Soc. Exper. Biol. Med. 53*(1943):167–170, C.343. In his *Rh-Hr* (1952), 238-241.

Wiener, Alexander S., Max Lederer, and Silik H. Polayes. "Studies in iso-hemagglutination I. Theoretical considerations." *J. Immunol. 17*(1929): 469–482.

Wiener, Alexander S., Max Lederer, and Silik H. Polayes. "A note on the paper, 'Studies in isohemagglutination,' " *J. Immuno.l 17*(1929):357–360.

Wiener, Alexander S., Max Lederer, and Silik H. Polayes. "Studies on iso-hemagglutination III. On the heredity of the Landsteiner blood groups." *J. Immunol. 18*(1930):201–221.

Wiener, Alexander S., Max Lederer, and Silik H. Polayes. "Studies in iso-hemagglutination IV. On the chances of proving non-paternity with special reference to blood groups." *J. Immunol. 19*(1930):259–282.

Wiener, Alexander S., Richard D. Owen, Clyde Stormont, and Irving B. Wexler. "Medico-legal application of blood-grouping tests." *J. Am. Med. Assoc. 161*(1956):233–239.

Wiener, Alexander S., Richard D. Owen, Clyde Stormont, and Irving B. Wexler. "Medico-legal applications of blood-grouping tests." *J. Am. Med. Assoc. 164*(1957):2036–2043.

Wiener, Alexander S., and Harold R. Peters. "Hemolytic reactions following transfusions of blood of the homologous group, with three cases in which the same agglutinogen was responsible." *Ann. Int. Med. 13*(1940):2306–2322. In his *Rh–Hr* (1952), 25–41.

Wiener, Alexander S., and Eve B. Sonn. "The Rh series of genes with special reference to nomenclature." *Ann. N. Y. Acad. Sci. 46*(1946):969–992.

Wiener, Alexander S., Eve B. Sonn, and R. B. Welkin. "Heredity and distribu-tion of the Rh blood types." *Proc. Soc. Exper. Biol. Med. 54*(1943):238–240. In his *Rh-Hr* (1952,) 367-368.

Wiener, Alexander S., and Maurice Vaisberg. "Heredity of the agglutinins M and N of Landsteiner and Levine." *J. Immunol. 20*(1931):371–388.

Wiener, Alexander S., and Irving B. Wexler. "The mosaic structure of red blood cell agglutinogens." *Bact. Rev. 16*(1952):69–87. In his *Rh-Hr* (1952), 618–639.

Wiener, Alexander S., and Irving B. Wexler. *Heredity of the Blood Groups.* New York: Grune, 1958.

Wintrobe, Maxwell M. *Blood Pure and Eloquent: A Story of Discovery, of People and of Ideas.* New York: McGraw-Hill, 1980.

Zinsser, Hans:
    Fischer, Isidor. "Zinsser, Hans." In *Biographisches Lexikon* (1933), v. 2, 1726 (Bibl. sec. 1).

Zinsser, Hans. *Infection and Resistance: An Exposition of the Biological Phenomena underlying the Occurrence of Infection and the Recovery of the Animal Body from Infectious Disease . . . with a Chapter on Colloids and Colloidal Reactions by Prof. Stewart W. Young.* New York: Macmillan, 1st ed. 1914, 2d ed. 1918, 3d ed. 1923.

Zinsser, Hans. *As I Remember Him; the Biography of R. S..* Boston, Mass.: Little, Brown, 1940. Autobiography of Hans Zinsser himself.

Index

Printed in the United States
By Bookmasters